Bio-Based Epoxy Polymers, Blends, and Composites

Bio-Based Epoxy Polymers, Blends, and Composites

Synthesis, Properties, Characterization, and Applications

Edited by

Jyotishkumar Parameswaranpillai
Sanjay Mavinkere Rangappa
Suchart Siengchin
Seno Jose

WILEY-VCH

Editors

Dr. Jyotishkumar Parameswaranpillai
King Mongkut's University of Technology North Bangkok
Center of Innovation in Design and Engineering for Manufacturing
1518 Pracharaj 1
Wongsawang Road, Bangsue
10800 Bangkok
Thailand

Dr. Sanjay Mavinkere Rangappa
King Mongkut's University of Technology North Bangkok
Department of Mechanical and Process Engineering
1518 Pracharaj 1
Wongsawang Road, Bangsue
10800 Bangkok
Thailand

Prof. Dr.-Ing.habil. Suchart Siengchin
King Mongkut's University of Technology North Bangkok
Department of Mechanical and Process Engineering
1518 Pracharaj 1
Wongsawang Road, Bangsue
10800 Bangkok
Thailand

Dr. Seno Jose
Government College Kottayam
Department of Chemistry
Nattakom P O
Kottayam
686013 Kerala
India

Cover
Cover Image: © 985 Bilder/Pixabay

Library of Congress Card No.:
applied for

British Library Cataloguing-in-Publication Data
A catalogue record for this book is available from the British Library.

Bibliographic information published by the Deutsche Nationalbibliothek
The Deutsche Nationalbibliothek lists this publication in the Deutsche Nationalbibliografie; detailed bibliographic data are available on the Internet at <http://dnb.d-nb.de>.

Print ISBN: 978-3-527-34648-6
ePDF ISBN: 978-3-527-82359-8
ePub ISBN: 978-3-527-82361-1
oBook ISBN: 978-3-527-82360-4

Typesetting SPi Global, Chennai, India
Printing and Binding
CPI Group (UK) Ltd, Croydon CR0 4YY
Printed on acid-free paper

10 9 8 7 6 5 4 3 2 1

Dedication

Editors are honored to dedicate this book to their family members and friends. Editors would like to thank King Mongkut's University of Technology North Bangkok (KMUTNB), Thailand for the support through Grant No. KMUTNB-BasicR-64-16.

Contents

Preface

Epoxy polymers are thermosetting polymers widely used in construction and building, automobile, aerospace, and marine industries. However, traditional epoxy systems are nonbiodegradable and cannot be recycled. The plastic waste is generating at an alarming rate and only 9% is recycled, and the remaining 91% is either landfilled or dumped in the natural environment. Moreover, plastic waste accumulation in the seawater, especially in the Pacific Ocean, is increasing at an alarming rate. Therefore, researchers and scientists are concentrating to develop biodegradable polymers. One of the inventions in this area is the production of bioepoxy resins from different plant sources. A surge in the production of bioepoxy resins was observed in the past decade. The polymer manufacturers (Sicomin, Gougeon Brothers, Wessex Resins, Bitrez, etc.) are now producing bioepoxy resins. Generally, bioepoxy resins are brittle, which limits their application in advanced composite industries. Studies have shown that the incorporation of fillers/fibers can significantly enhance the thermomechanical, electrical, and other physical properties of the bioepoxy composites. The high acceptance level of bioepoxy resins in automobile, aerospace, construction, and marine industries all across the world is expected to increase the production of bioepoxy resins in the coming years.

The research in the field of bioepoxy resins and their blends and composites are flourishing. This leads to an upsurge in the number of publications. However, no books have been published in the area of bioepoxy resins. Therefore, we believe that it is important to edit a book on "Bio-Based Epoxy Polymers, Blends, and Composites: Synthesis, Properties, Characterization, and Applications." We hope that the present book will benefit scientists, engineers, academic staff, and students working in the area of bio-based epoxy polymers, blends, and composites.

This book consists of 12 chapters that describe "Bio-Based Epoxy Polymers, Blends, and Composites: Synthesis, Properties, Characterization, and Applications." The chapter "Synthesis of Bioepoxy Resins" summarizes the synthesis of fully bio-based epoxy resins, their properties, and potential uses. The chapter "Natural/Synthetic Fiber-Reinforced Bioepoxy Composites" focuses on the different bioepoxy resins and natural/synthetic fiber-reinforced bioepoxy composites, their mechanical properties, and applications. The chapter "Polymer Blends Based on Bioepoxy Polymers" reviews the preparation of bio-based epoxy blends for various applications. The chapter "Cure Kinetics of

Bioepoxy Polymers, Their Blends, and Composites" emphasizes the importance of kinetics studies of the bioepoxy/hardener reaction. The chapter "Rheology of Bioepoxy Polymers, Their Blends, and Composites" discusses the role of rheological studies on the processing of the bioepoxy composites. The chapter "Dynamical Mechanical Thermal Analysis of Bioepoxy Polymers, Their Blends, and Composites" describes the viscoelastic properties of bioepoxy blends and composites. The chapter "Mechanical Properties of Bioepoxy Polymers, Their Blends, and Composites" discusses the factors influencing the mechanical aspects of bioepoxy composites. The chapter "Bioepoxy Polymer, Blends, and Composites Derived Utilitarian Electrical, Magnetic, and Optical Properties" reviews the electrical, electronic, magnetic, and optical properties of bioepoxy systems. The chapter "Spectroscopy and Other Miscellanies Techniques for the Characterization of Bioepoxy Polymers, Their Blends, and Composites" gives an overview of the characterization of bioepoxy systems using various techniques such as Fourier transform infrared spectroscopy, nuclear magnetic resonance spectroscopy, differential scanning calorimetry, and thermogravimetric analysis. The chapter "Flame Retardancy of Bioepoxy Polymers, Their Blends, and Composites" highlights the recent developments in flame-retardant bio-based epoxy resins. The chapter "Water Sorption and Solvent Sorption of Bioepoxy Polymers, Their Blends, and Composites" focuses on the water absorption properties of bioepoxy systems. The chapter "Bio-Based Epoxy: Applications in Mendable and Reprocessable Thermosets, Thermosetting Foams, and Pressure-Sensitive Adhesives" give an overview of the recent developments in self-healing bioepoxy systems and bio-based epoxy foams.

The editors are thankful to the contributors of all chapters and the Wiley editorial and publishing team for their kind support. The editors are also thankful to Ms. Junjiraporn Thongprasit (Baifern) for her involvement at the initial screening process of chapters.

20 August 2019

Dr. Jyotishkumar Parameswaranpillai (Thailand)
Dr. Sanjay Mavinkere Rangappa (Thailand)
Prof. Dr.-Ing. habil. Suchart Siengchin (Thailand)
Dr. Seno Jose (India)

About the Authors

Dr. Jyotishkumar Parameswaranpillai is currently working as a research professor at the Center of Innovation in Design and Engineering for Manufacturing, King Mongkut's University of Technology North Bangkok. He received his Ph.D. in Polymer Science and Technology (Chemistry) from Mahatma Gandhi University. He has research experience in various international laboratories such as Leibniz Institute of Polymer Research Dresden (IPF), Germany, Catholic University of Leuven, Belgium, and University of Potsdam, Germany. He has published more than 100 papers in high-quality international peer-reviewed journals on polymer nanocomposites, polymer blends and alloys, and biopolymers and has edited five books. He received numerous awards and recognitions including the prestigious Kerala State Award for the Best Young Scientist 2016, INSPIRE Faculty Award 2011, Best Researcher Award 2019 from King Mongkut's University of Technology North Bangkok.

https://scholar.google.co.in/citations?user=MWeOvlQAAAAJ&hl=en

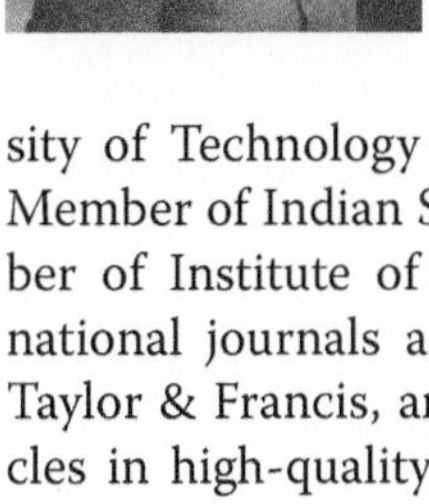

Dr. Sanjay Mavinkere Rangappa, received his B.E degree (Mechanical Engineering) from Visvesvaraya Technological University, Belagavi, India, in the year 2010, M. Tech degree (Computational Analysis in Mechanical Sciences) from VTU Extension Centre, GEC, Hassan, in the year 2013, Ph.D (Faculty of Mechanical Engineering Science) from Visvesvaraya Technological University, Belagavi, India, in the year 2017, and Postdoctorate from King Mongkut's University of Technology North Bangkok, Thailand, in the year 2019. He is a Life Member of Indian Society for Technical Education (ISTE) and Associate Member of Institute of Engineers (India). He has reviewed more than 50 international journals and international conferences (for Elsevier, Springer, Sage, Taylor & Francis, and Wiley). In addition, he has published more than 85 articles in high-quality international peer-reviewed journals, more than 20 book chapters, 1 book, and 15 books as an editor and also presented research papers at national/international conferences. His current research areas include natural

fiber composites, polymer composites, and advanced material technology. He is a recipient of DAAD Academic exchange-PPP Programme (Project-related Personnel Exchange) between Thailand and Germany to the Institute of Composite Materials, University of Kaiserslautern, Germany. He has received a Top Peer Reviewer 2019 Award, Global Peer Review Awards, Powdered by Publons, Web of Science Group.
https://scholar.google.com/citations?user=al91CasAAAAJ&hl=en

Prof. Dr.-Ing. habil. Suchart Siengchin is the president of King Mongkut's University of Technology North Bangkok (KMUTNB), Thailand. He received his Dipl.-Ing. in Mechanical Engineering from the University of Applied Sciences Giessen/Friedberg, Hessen, Germany, in 1999, M.Sc. in Polymer Technology from the University of Applied Sciences Aalen, Baden-Wuerttemberg, Germany, in 2002, M.Sc. in Material Science at the Erlangen-Nürnberg University, Bayern, Germany, in 2004, Doctor of Philosophy in Engineering (Dr.-Ing.) from the Institute for Composite Materials, University of Kaiserslautern, Rheinland-Pfalz, Germany, in 2008, and Postdoctoral Research from Kaiserslautern University and School of Materials Engineering, Purdue University, USA. In 2016, he received the habilitation at the Chemnitz University in Sachen, Germany. He worked as a lecturer for Production and Material Engineering Department at The Sirindhorn International Thai-German Graduate School of Engineering (TGGS), KMUTNB. He has been a full professor at KMUTNB and became the president of KMUTNB. He won the Outstanding Researcher Award in 2010, 2012, and 2013 at KMUTNB. His research interests are in polymer processing and composite material. He is the editor-in-chief of KMUTNB International Journal of Applied Science and Technology and the author of more than 150 peer-reviewed journal articles. He has participated with presentations in more than 39 international and national conferences with respect to materials science and engineering topics.
https://scholar.google.com/citations?user=BNZEC7cAAAAJ&hl=en

Dr. Seno Jose, a native of Kerala, India, is an assistant professor of Chemistry at Government College Kottayam. He did his masters in Chemistry in Mahatma Gandhi University. He has availed DST/DAAD fellowship and worked as a visiting researcher at the Institute for Composite Materials Ltd., Germany. He received his Ph.D. in Chemistry from Mahatma Gandhi University in 2007. He has coauthored over 40 peer-reviewed publications. His research interests include polymer blends, polymer nanocomposites, and shape memory polymeric materials.
https://scholar.google.com/citations?user=uoJt2ckAAAAJ&hl=en

1

Synthesis of Bio-Based Epoxy Resins

Piotr Czub and Anna Sienkiewicz

Cracow University of Technology, Department of Chemistry and Technology of Polymers, ul. Warszawska 24, 31-155 Cracow, Poland

1.1 Introduction

The term "epoxy resin" is understood to mean compounds containing at least one active epoxy group in their structure and which are capable of forming a cross-linked three-dimensional structure in the curing process involving these groups. Naturally, epoxy rings are found only in vernonia oil. However, epoxy functionality can be easily introduced into the compound structure, even by the oxidation of unsaturated bonds to oxirane rings. This is a typical method of obtaining cycloaliphatic resins, applied on a large scale in electronics to encapsulate electronic systems. The second method is the use of epichlorohydrin, which is commonly applied to prepare epoxy compounds via the reaction with polyalcohols or polyphenols. Epichlorohydrin together with bisphenols (mainly bisphenol A or F, and S) are the main raw materials used in industrial methods for the synthesis of epoxy resins most often produced and used on a large scale. All these compounds are of petrochemical origin. There are three main reasons for the search of new raw materials of natural origin for the synthesis of epoxy resins. The first is the need to replace petrochemical raw materials. The volatility of oil and gas prices and their strong connections with the changing political situation in various regions of the world, as well as the inevitable prospect of imminent exhaustion of their sources, and ecological considerations are the main reasons for the search of alternative sources of raw materials. Moreover, potential toxicological and endocrine disrupting properties of bisphenol A are discussed and emphasized, especially in recent years. The second reason is the need to solve the problem of annually increasing amount of postconsumer plastic waste. Epoxy resins belong to the category of polymeric materials practically not biodegradable. The application of bio-based raw materials can enable and facilitate their decomposition under the influence of biological factors. Epoxy resins are widely used as coating materials in products intended for contact with food or even storage of food (e.g. can-coating or paints for securing ship hold walls). Therefore, the third reason is the need to limit the penetration of harmful

Bio-Based Epoxy Polymers, Blends, and Composites: Synthesis, Properties, Characterization, and Applications, First Edition. Edited by Jyotishkumar Parameswaranpillai, Sanjay Mavinkere Rangappa, Suchart Siengchin, and Seno Jose.

substances such as bisphenol A into food from the coating material, preferably by eliminating them already at the stage of synthesis.

While searching for new bio-based resources for the synthesis of epoxy resins, particularly bisphenol substitutes, the crucial issue must be remembered. One of the most important challenges is to provide new bio-based resins with comparable performance properties to the currently manufactured and applied petrochemical-based commercial products, i.e. primarily high mechanical strength, thermal stability, and chemical resistance. The mentioned properties are characteristic of the resins based on bisphenol A (or other bisphenols), thanks to which these materials are produced on a large scale for many applications. Therefore, this chapter presents the most promising raw materials whose structure can provide the desired final properties of the epoxy system after cross-linking. At the same time, they must be raw materials easily available in large quantities from renewable sources, nontoxic and cheap to obtain and in preparation.

1.2 Plant Oil Bio-Based Epoxy Resins

Vegetable oils, as a material of natural origin and from renewable sources, are the subject of numerous studies aimed at their application for the synthesis or modification of various polymers [1]. Soybean, castor, linseed, rapeseed, sunflower, cotton, peanut, and palm oils are primarily used on a larger scale depending on the type of oil produced in a given region [2]. From the chemical point of view, plant-based oils are a mixture of esters derived from glycerol and free fatty acids, mainly unsaturated acids (primarily oleic, linoleic, linolenic, ricinoleic, and erucic acids) and in a small amount of saturated acids (stearic and palmitic acids) (Figure 1.1), depending on the type of oil.

When choosing vegetable oil for use in the synthesis of polymers, first of all, its structure should be taken into account: the presence of unsaturated bonds and possibly other functional groups (e.g. hydroxyl in castor oil or epoxy in vernonia oil), the amount of unsaturated bonds present in the molecule (referred as the oil functionality), and chain length alkyl derived from fatty acids (Table 1.1).

The functionality of oils (understood as the content of unsaturated bonds) primarily determines the cross-linking density of oil-based chemosetting polymers or polymers obtained by free radical polymerization as well as oil-modified polymeric materials. In turn, the final polymer properties such as mechanical strength, thermal stability, and chemical resistance strongly depend on the

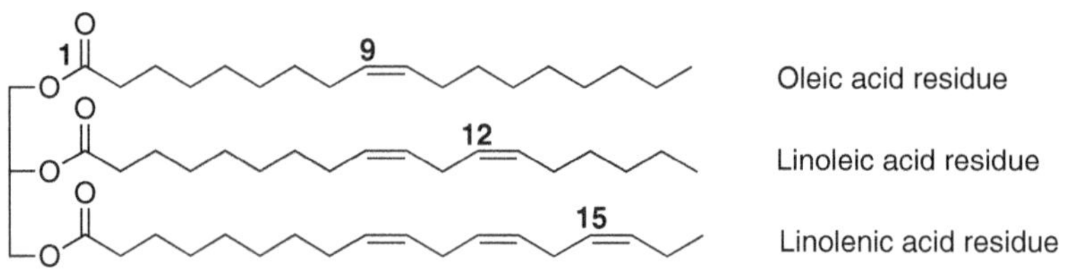

Figure 1.1 Schematic structure of triglycerides.

Table 1.1 The content of various fatty acids in selected vegetable oils.

Fatty acid	Vegetable oil, content of individual acids (wt%)					
	Soybean	Rapeseed	Linseed	Sunflower	Castor	Palm
Palmitic	12	4	5	6	1.5	39
Stearic	4	2	4	4	0.5	5
Oleic	24	56	22	42	5	45
Linoleic	53	26	17	47	4	9
Linolenic	7	10	52	1	0.5	—
Castor	—	—	—	—	87.5	—
Other	—	2	—	—	—	2
Functionality	4.6	3.8	6.6	4.6	2.8	1.8

cross-linking density. The elasticity of the polymers with the addition of vegetable oil or based on them depends on the length of the alkyl chains in the oil molecule derived from fatty acids.

Vegetable oils can be easily and efficiently converted into epoxy derivatives by oxidizing unsaturated bonds present in fatty acid residues. Several methods of double bond oxidation in triglyceride molecules are known and commonly used [3]: the method based on the Prilezhaev reaction, the radical oxidation, the Wacker-type oxidation, dihydroxylation of oils and fats, and enzyme–catalyst oxidation. The Prilezhaev reaction is the most often used method for natural oil epoxidation, commonly applied in the industry. In this method, the process of epoxidation of natural fatty acids and triglycerols is carried out in the system consisting of hydrogen peroxide, an aliphatic carboxylic acid, and an acidic catalyst. The organic peracid formed *in situ* by the reaction of acid with hydrogen peroxide is the real oxidizing agent in this method (Figure 1.2).

Carboxylic acids with one to seven carbon atoms are the most commonly used (in practice, mainly acetic acid). Inorganic or organic acids and their salts, as well as acidic esters, can be used as catalysts; however, sulfuric and phosphoric acid are the most often used in industrial practice. A promising method is oxidation in the presence of enzymes [4], heteropolyacids [5], and even ion exchange resins [6] as catalysts. The most commonly used oxidizing agent is hydrogen peroxide in the form of solution with a concentration of 35–90% (usually 50%). Epoxidation of plant oils in ionic liquids, as well as in supercritical carbon dioxide, is also described [7].

The earliest epoxidized esters of higher fatty acids have found wide applications as both plasticizers and stabilizers for thermoplastics, mostly poly(vinyl chloride), poly(vinylidene chloride), their copolymers, and poly(vinyl acetate) and chlorinated rubber [8, 9]. Epoxidized fatty acids containing oleic acid are used as a valuable intermediate in the production of lubricants and textile oils [10, 11]. It seems that epoxidized vegetable oils could also be used as hydraulic liquids [12]. However, primarily, they can also act as reactive diluents of bisphenol-based epoxy resins [13], which are usually highly viscous. They have oxirane groups, although

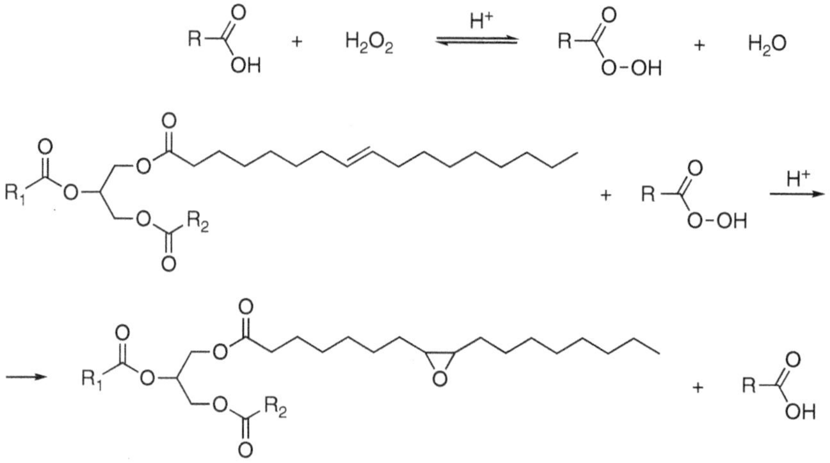

Figure 1.2 The reaction of triglyceride epoxidation with organic peracids.

less reactive because of their central location in triglyceride chains (compared to terminal glycidyl groups), but also capable of reacting with polyamines or carboxylic anhydrides. By building into the structure of the cured resin in the process of co-cross-linking with it, they affect its final properties – improving flexibility and impact strength. In this way, embedded triglycerides not only facilitate the processing of resins with high intrinsic viscosity but also allow limiting their typical disadvantages (high brittleness, low impact strength, and flexibility) resulting from the rigid structure they owe because of the structure of bisphenols [14].

However, the first, logically implied possibility of use of epoxidized vegetable oils is their application as stand-alone materials: the networks cross-linked with bifunctional compounds such as dicarboxylic acids or aliphatic and aromatic diamines, which are typically used as hardeners for the epoxy resins. Because of the content of more than one epoxy group in the molecule, epoxidized triglycerides may, according to the generally accepted definition, be treated as epoxy resins. However, curing of, e.g. epoxidized soybean oil [15] or vernonia oil (natural epoxidized oil mainly obtained from plant *Vernonia galamensis*), dicarboxylic acids [16] resulted in obtaining only soft elastomers. Materials with higher mechanical strength are synthesized by reacting epoxidized oils first with polyhydric alcohols (e.g. resorcinol) and phenols or bisphenols and then cross-linking the obtained modified oil with partially reacted epoxidized rings [17]. Finally, curing by photopolymerization or polymerization with latent initiators allows to obtain from modified vegetable oils, without the addition of the bisphenol-based or cycloaliphatic resins, coating materials with satisfactory mechanical properties. It was found [18] that the properties of hardened vegetable oils also depend on the type of used thermal latent initiator. The properties of epoxidized castor oil cross-linked with *N*-benzylpyrazine (BPH) and *N*-benzylquinoxaline (BQH) were studied (Figure 1.3).

It was found that materials characterized by a higher glass transition temperature, a higher value of the coefficient of thermal expansion, and greater

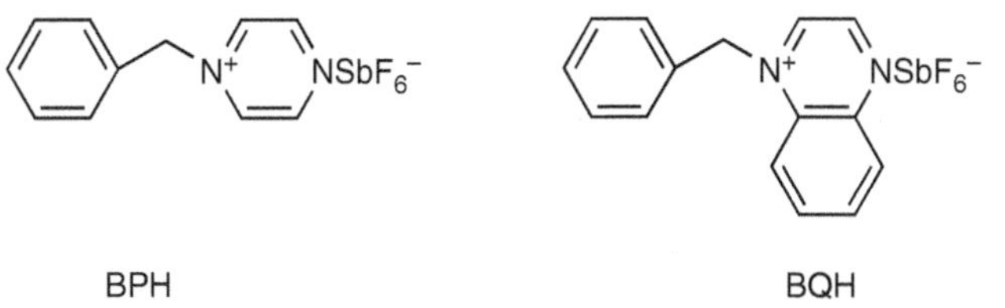

Figure 1.3 Chemical structure of cationic photoinitiators.

thermal stability are obtained using BPH as a photoinitiator. Nevertheless, the composition cross-linked with BQH is characterized by better mechanical properties. Most likely, the better final properties result from the higher cross-linking density of cured with BPH composition. Anhydrides of various carboxylic acids are used to cure epoxidized linseed oil [19], and cross-linking reactions are catalyzed by various tertiary amines and imidazoles. The materials obtained with phthalic anhydride and methylendomethylenetetrahydrophthalic anhydride hardeners exhibit a lower cross-linking density than those obtained with *cis*-1,2,3,6-tetrahydrophthalic anhydride. It was found that a greater degree of oil–anhydride conversion and thus higher cross-linking density and greater rigidity of the cured material are obtained using imidazoles. The best properties are achieved for the composition cured with *cis*-1,2,3,6-tetrahydrophthalic anhydride as the hardener and 2-methylimidazole as the catalyst.

High-molecular-weight epoxies are a special group of very important epoxy resins commonly used as coating materials, especially for powder, can and coil coatings mainly in automotive industry. Theoretically, they can be obtained in the traditional way in the Taffy process with epichlorohydrin and bisphenol. However, even the use of a slight excess of epichlorohydrin does not provide high-molecular-weight solid resins. Therefore, industrially, they are synthesized from low or moderate molecular weight epoxy resins and bisphenol A by the epoxy fusion process. It is the method of polyaddition carried out in bulk, in the molten state of reagents, and without the use of solvents. In this way, it is possible to obtain resins with a softening temperature of 100–150 °C, characterized by an epoxy value of 0.020–0.150 mol/100 g, and an average molecular weight of 1.5–10 thousands of Daltons. The application of epoxidized vegetable oils in place of low/medium molar mass resins, as well as hydroxylated oils in place of bisphenols, in the epoxy fusion process with bisphenol A (BPA) or BPA-based epoxy resin was proposed [20, 21]. Hydroxylated oils are obtained from epoxidized oils in the reaction of opening of oxirane rings using diols and the most often glycols. Depending on the type of starting oil, catalyst used, and reaction time, the products of the epoxy fusion process using modified oils (Figure 1.4) contain a large amount of hydroxyl groups (hydroxyl value 120–160 mg KOH/g), some free epoxy groups (epoxy value 0.050–0.150 mol/100 g), and are characterized by weight-average molecular weight even above 30 000 g/mol.

Therefore, for the cross-linking of these products, diisocyanates or blocked diisocyanates can be applied (Figure 1.5).

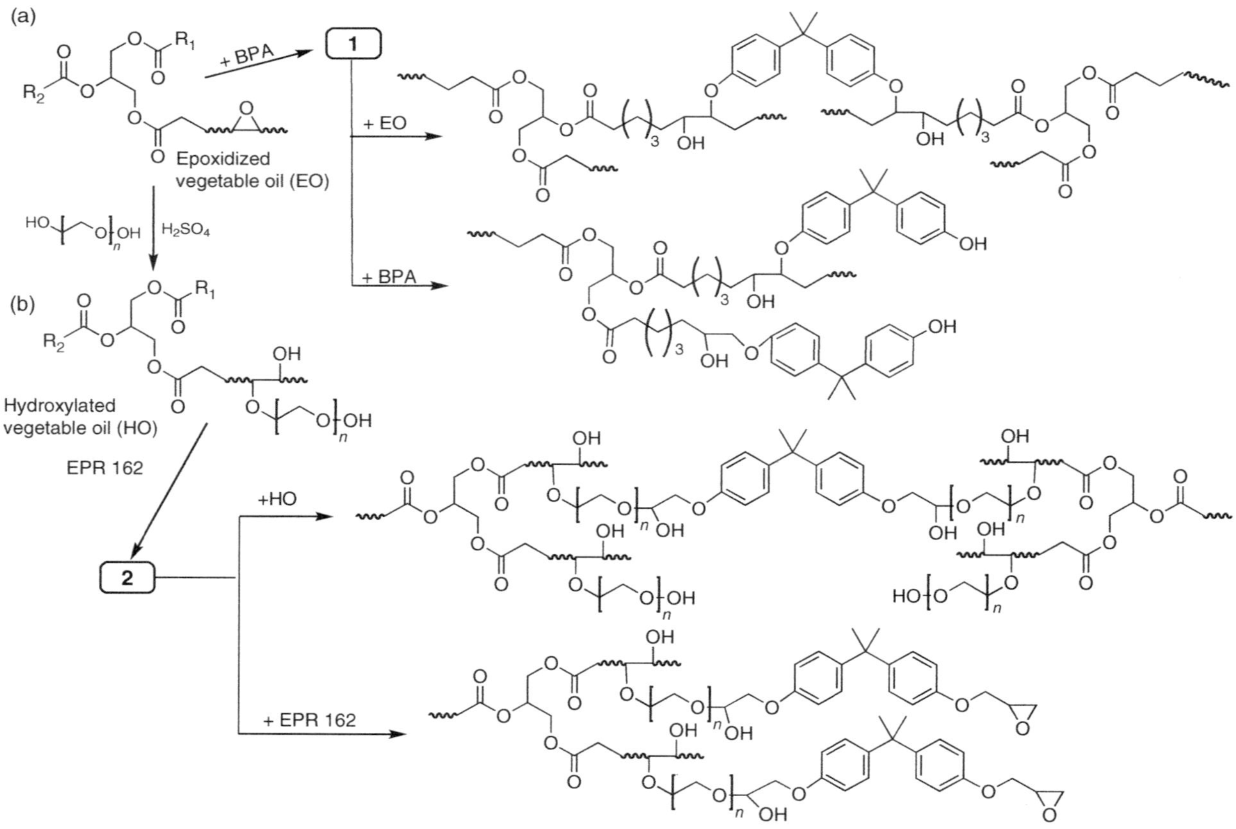

Figure 1.4 The synthesis of high-molecular-weight epoxy resins based on modified vegetable oil: (a) epoxidized or (b) hydroxylated soybean oil.

1 =

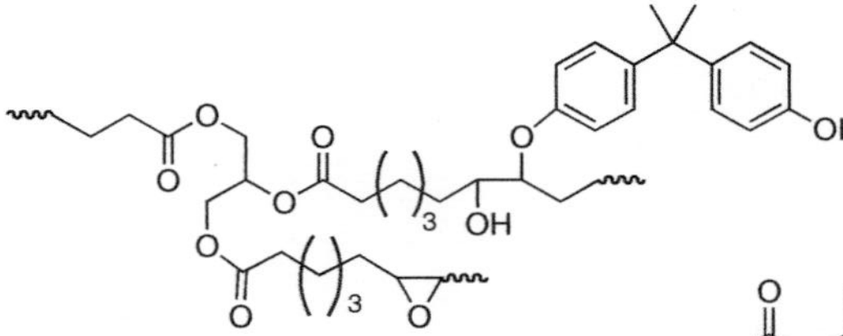

2 =

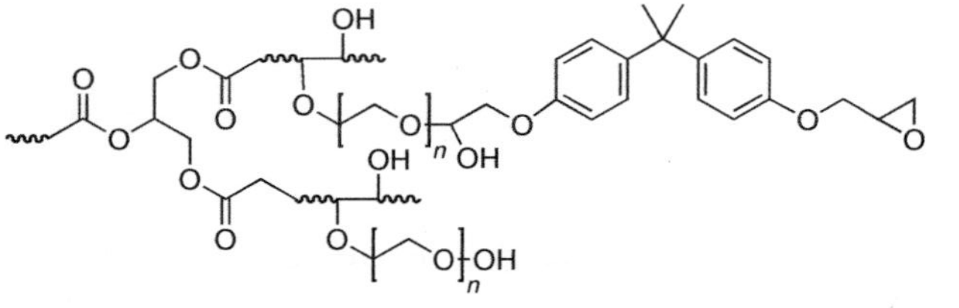

Figure 1.4 (*Continued*)

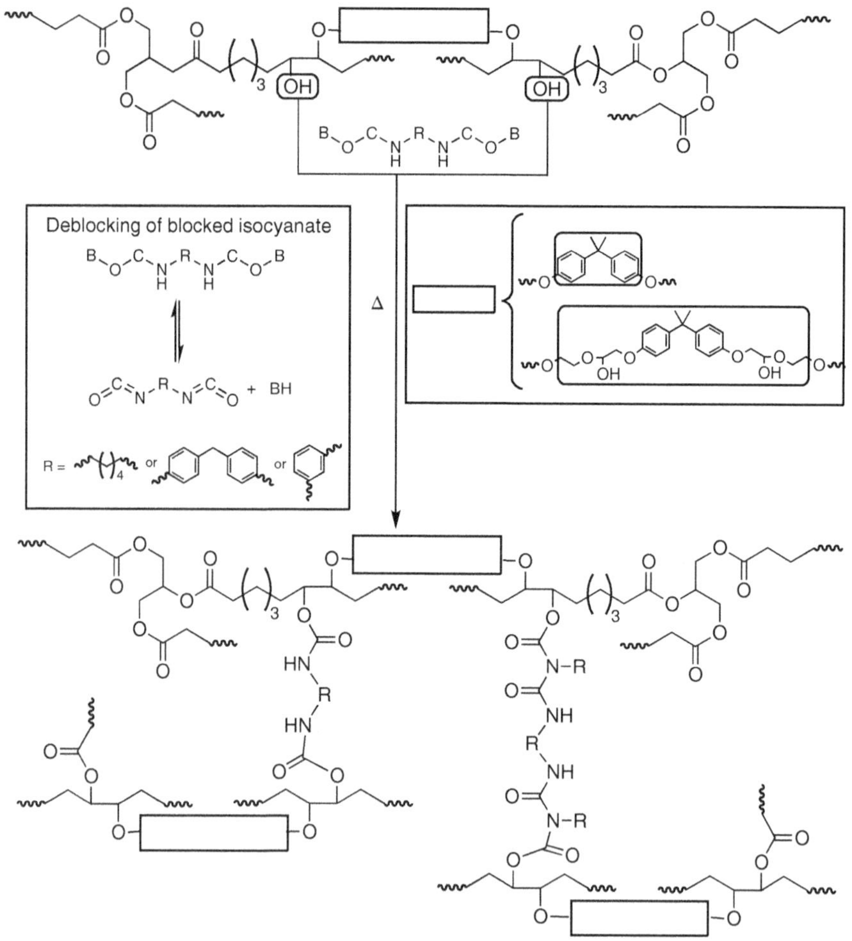

Figure 1.5 Cross-linking reactions of epoxy fusion process products.

The resins cross-linked with polyisocyanates are characterized by differential mechanical properties, which depend on the type of used isocyanate, and are higher than the one of the low-molecular-weight bisphenol A-based resin crosslinked with methyl-tetrahydrophthalic anhydride, however lower while cured with isophoronediamine [22]. Moreover, the presence of epoxy groups in the polyaddition products can be used to obtain two-layer materials [23], in which one layer is cured with polyamine epoxy resin and the other is a polyaddition product cross-linked with diisocyanate. The reaction of the amine hardener with the free epoxy groups that are present within the polyaddition product ensures a very good interlayer bonding.

Because of the usually unsatisfactory properties of the oils cured with amines or acid anhydrides, epoxidized vegetable oils began to be used as one of the components of epoxy compositions [24]. The compositions consisting of epoxidized esters of higher fatty acids obtained by the transesterification of various vegetable oils and natural or hydrocarbon resin acids can be used as an ingredient in, among others, epoxy adhesives with reduced crystallization tendency [25]. Compositions of modified vegetable oils with epoxy resins based on various bisphenols can generally be prepared via two methods. One of them is the homogenization of the components of the composition and their simultaneous co-cross-linking. In this way, compositions of bisphenol F diglycidyl ether with epoxidized linseed oil are prepared [26] and cured with methyltetrahydrophthalic anhydride in the presence of 1-methylimidazole or polyoxypropylenetriamine [27]. It turned out that with an increase in the content of epoxidized linseed oil in the anhydride-cured compositions, the storage modulus, glass transition temperature, and heat resistance under load decrease, while the impact strength measured by the Izod method does not change, but above 70 wt% of oil content increases the cross-linking density. In contrast, compositions cured with the use of amine are characterized by an almost fivefold increase in impact strength at the oil content of 30% by weight. Other discussed cured parameters change in the same way as in the case of anhydride cross-linked materials. In turn, comparison [28] of the properties of the composition with epoxidized linseed oil and soybean oil shows significant differences between the materials based on both oils. It was found that, due to the greater compatibility of linseed oil with the epoxy resin and better oil solubility in the resin (resulting from greater polarity and functionality and lower molecular weight), linseed oil does not tend to form a separate phase. However, the two-phase structure, observed in the case of epoxidized soybean oil, is responsible for improving the impact strength and fracture toughness of the epoxy resin composition. A decrease in cross-linking density is also observed in the compositions of 4,4′-tetraglycidyldiamino-diphenylmethane with epoxidized soybean oil cured with diaminodiphenylmethane [29]. Also in this case, besides the improvement in impact strength, as the effect of reducing the cross-linking density, a decrease in the heat resistance and the glass transition temperature is observed. Using the example of a bisphenol-based epoxy resin compositions with different contents of epoxidized castor [30] or soybean oil [31], cured with thermal latent initiator BPH, it was proven that the final properties of cross-linked materials are determined not only by the polarity, functionality, or structure of the used oil but also by its content, ensuring the optimal amount of flexible fragments embedded in the rigid epoxy resin structure, and the most favorable phase composition of the material.

Bisphenol-based epoxy resin compositions with modified vegetable oils might also be prepared in the two-step method. The first stage is the initial cross-linking of oil so that free functional groups capable of co-cross-linking with the epoxy resin remain in it. In this way, a prepolymer or, as it is called in some publications, a rubber is obtained from the modified oil. Only then, the prepared prepolymer is mixed in appropriate proportions with epoxy resin, and finally, the composition is cured. The cross-linked composition is characterized by a two-phase structure, analogous to that of epoxy resins modified with liquid acrylonitrile butadiene copolymers with reactive carboxyl or amine end groups and acrylic elastomers. The two-phase structure of the composition determines their postcuring properties. Using the two-step method, composition of diethylene epoxy resin with epoxidized soybean oil was prepared [32]. Initially, both the oil and then the composition with the epoxy resin were cross-linked with 2,4,6-*tri*(*N*,*N*-dimethylaminomethyl)phenol. The soybean prepolymer, cross-linked for 12–84 hours, is a highly viscous liquid that mixes well with the epoxy resin [33]. The formation of the two-phase structure of the cured composition was confirmed by DSC and DMA analyses. The adhesive joint prepared with the use of the tested composition shows a significant improvement in the impact strength and the strength. It has been found that the properties of the composition depend on both the content of soybean prepolymer and the time of its pre-cross-linking. The best results are obtained using the addition of 20 wt% of prepolimer pre-cross-linked for 60 hours. Compositions characterized by greater cross-linking density and mechanical strength than the networks with epoxidized soybean oil were obtained using methyl and allyl esters, synthesized by the transesterification of soybean oil [34]. The esters were epoxidated and then precured with *p*-aminocyclohexylmethane, which showed the highest reactivity to soybean oil derivatives among the tested polyamines. The curing conditions were selected in such a way that cross-linking of both esters and epoxidized oil, which was chosen for the comparison purposes, terminates at the gelation stage. The bisphenol-based epoxy resin compositions, with the content of prepolymers of 10–30 wt%, were cured using various polyamines, and their mechanical properties were compared with those of the samples of analogous composition but obtained via the one-step method. Generally, the mixed compositions with various soybean oil derivatives obtained by the two-stage method are characterized by the best strength parameters, definitely better than the networks synthesized only with epoxidized oil. In particular, the addition of epoxidized allyl ester increases the glass transition temperature and provides greater rigidity and mechanical strength of the composition.

Additionally [35], the process of cross-linking of the above-described materials with acid anhydride (the commercial product called Lindride LS 56V produced by Lindau Chemicals, USA) was studied. Based on the results of DSC and viscometric measurements, models describing the course of curing reactions have been developed, which might be applied in the industrial processing of the described compositions. The DMTA analysis showed [36] that the conservative modulus of elasticity and glass transition temperature increase with an increase in the content of epoxidized allyl ester in the case of anhydride cross-linking while decrease for polyamine-cured materials. Moreover, the value of the loss factor decreases in the case of anhydride cross-linking, but it is definitely higher for polyamine-cured compositions. That kind of formation of dynamic mechanical properties results from a greater degree of cross-linking of anhydride-cured compositions. The epoxidized palm oil was prepolymerized in a reaction with isophorone diamine [37]. The resulting palm oil derivatives were used as modifiers of a bisphenol A-based low molecular weight epoxy resin. The prepared compositions and the pure unmodified epoxy resin were cured with isophorone diamine. It was found that the palm oil derivatives led to a decrease in the mechanical strength of the resin, but on the other hand, they contributed to an increase in relative elongation at break and significant improvement (even twice) in impact strength of the cross-linked products. A two-phase structure of

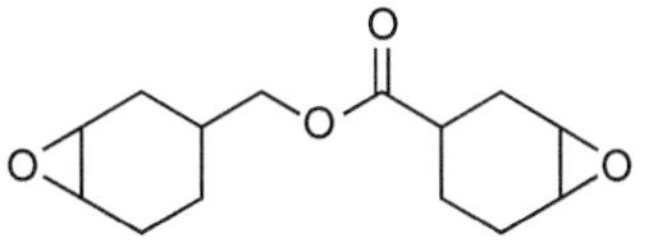

Figure 1.6 Chemical structure of the cycloaliphatic resin (3,4-epoxycyclohexylmethyl-3′,4′-epoxycyclohexane carboxylate).

the compositions studied, responsible for the increase of their impact strength, was observed.

One of the most important areas of application of epoxidized vegetable oils are compositions with epoxy resins, capable of cross-linking with UV or visible light. Photoinitiated polymerization is a commonly used industrial method of cross-linking of coating materials. Throughout this method, the cured coating is obtained in a short time and above all at the room temperature. Modified natural oils are a very interesting alternative to acrylic monomers, commonly used to obtain photosetting coatings, and starting from the first reports [38] are the subject of the research performed by scientific teams around the world. Compositions consisting of vernonia oil or epoxidized soybean oil and cycloaliphatic epoxy resin were tested [39] (Figure 1.6).

The compositions were cross-linked by photopolymerization using a cationic UV initiator, which was a mixture of triarylsulfonium salts of hexafluoroantimone with a trifunctional primary triol based on ε-caprolactone. Coatings with the addition of epoxidized vegetable oils are characterized by excellent adhesion to the surface, high impact strength, UV stability, corrosion resistance, and long-lasting shine. It was also found that pencil hardness and tensile strength of coating films decrease with increasing oil content. Similarly, the glycidyl castor oil derivative [40], added in an amount of up to 60 wt% to the same cycloaliphatic resin and cross-linked with it using triarylsulfonium salts as cationic initiators (Figure 1.7), leads to a significant improvement of epoxy coating properties: increasing its elasticity and gloss as well as reducing water absorption.

Flexible coatings characterized by high tensile strength and hardness are obtained by adding epoxidized palm oil to cycloaliphatic resin (Figure 1.6) [41]. Additionally, in the described research, the possibility of photopolymerization of prepared compositions with UV light in the presence of various initiators, radical, cationic, and hybrid ones, was tested. Because of the low solubility of triarylsulfonium salts in oil, divinyl ethers of various structures were also added to the composition, which, as it was found in the course of the study, did not affect the mechanical properties of cured coatings. The photocuring process of highly branched resins obtained from modified vernonia oil was also investigated [42]. For the reason that the final properties of cross-linked compositions with

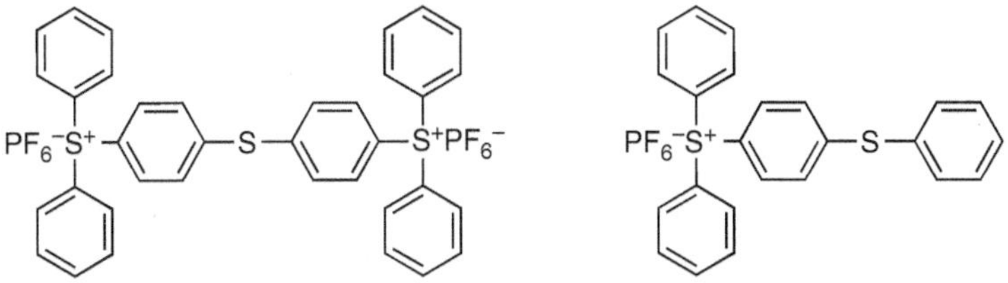

Figure 1.7 Triarylsulfonium salts applied as the cationic photoinitiators.

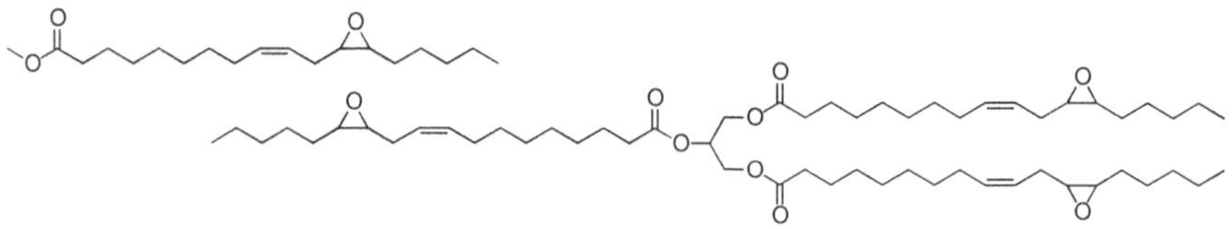

Figure 1.8 Structure of vernolic acid methyl ester and product of its reaction with trimethylol propane.

modified vegetable oils depend not only on the amount of oil but also on their structure, the authors decided to study the photopolymerization of epoxy resin with a strictly defined composition and structure. For this purpose, obtained in the oil transesterification reaction of *Euphorbia lagascae*, methyl vernolate was reacted with trimethylol propane to give the compounds depicted in Figure 1.8.

The obtained derivatives, including the hyperbranched polyether, were used to prepare compositions with different contents of individual components, with methyl vernolate acting as a reactive diluent. The compositions were polymerized with a cationic photoinitiator (octyloxydiphenyliodine hexafluoroantimonate). The application of methyl vernolate reduces the viscosity of the polyether as well as significantly decreases the glass transition temperature. An interesting example of the synthesis of epoxy resin based on vegetable oil, hardened later by the photopolymerization, is the attachment of bicyclo[2.2.1]heptane to linseed oil [43]. The derivative, which is shown in Figure 1.9, was obtained by the Diels–Alder reaction of cyclopentadiene with linseed oil, carried out at the temperature of 240 °C and a pressure of 1.4 MPa.

Compositions consisting of a epoxidized derivative, the addition of various divinyl ethers of epoxidized linseed oil, and cycloaliphatic epoxy resin (Figure 1.6) have been cured using the already mentioned triarylsulfonium salts. Divinyl monomers fulfilled the role of reactive diluents and compatibilizers primarily of oil and photoinitiator. It is also known that the presence of this type of monomers accelerates photocuring of cycloaliphatic epoxy resins. It has been observed that the cross-linking of the cycloaliphatic linseed oil derivative proceeds at a lower rate than the cycloaliphatic epoxy resin, but with a higher rate than epoxidized oil. The addition of divinyl monomers

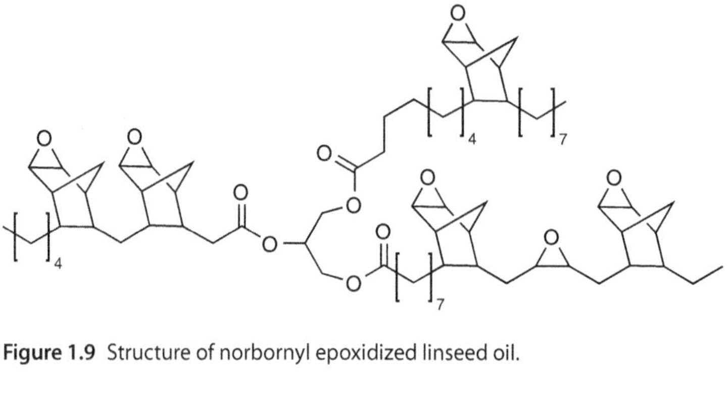

Figure 1.9 Structure of norbornyl epoxidized linseed oil.

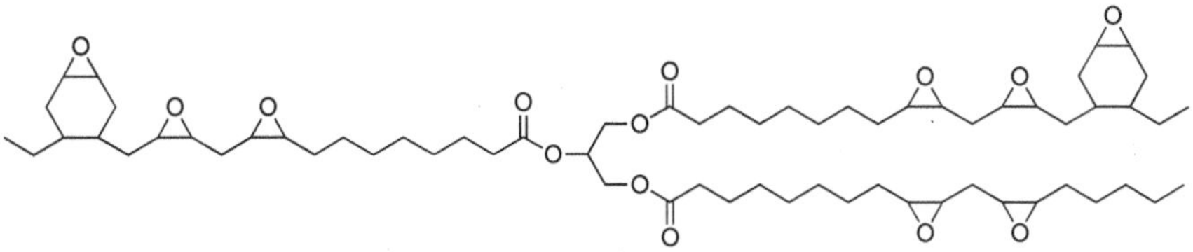

Figure 1.10 Structure of epoxidized cyclohexene-derivatized linseed oil.

accelerates the speed of curing and increases the elasticity of the cured materials. A similar relationship was also observed during kinetic studies of the cationic photopolymerization reaction of a cycloaliphatic linseed oil derivative [44]. It was found that the photo-cross-linking rate is controlled by the diffusion of active macromolecules, which depends on the viscosity of the environment. The different reactivity of the cycloaliphatic and epoxidized oil derivative in the main chains results from the differences in the diffusion of the molecules of both compounds and depends on the presence of divinyl monomers in the reaction environment. The improvement of the final properties of the described compositions was obtained by adding up to 20 wt% of tetraethyl orthosilane (TEOS) [45]. The organic–inorganic hybrid materials obtained in this way, containing the optimum amount of TEOS oligomers, amounting to about 10 wt%, were characterized by the highest value of the elastic modulus, the highest glass transition temperature, and the highest cross-linking density. Although the incorporation of TEOS oligomers in the structure of a cured cycloaliphatic linseed oil derivative simultaneously reduces the relative elongation at break and fracture toughness, it should be remembered that the biggest disadvantages of modified vegetable oils as materials susceptible to photocuring are low glass transition temperature and low speed of cross-linking. Another example of a cycloaliphatic linseed oil derivative, also intended for photocuring, is the product of a Diels–Alder reaction of linseed oil with 1,3-butadiene [46] (Figure 1.10).

Compositions based on modified vegetable oils, hardened by photopolymerization, are mainly intended for coating materials. However, it has been shown that it is also possible to use epoxidized soybean and linseed oils together with cycloaliphatic epoxy resin as binders for glass fiber-reinforced composites and cross-linked with visible or UV light [47].

1.3 Substitutes for Bisphenol A Replacement

1.3.1 Lignin-Based Phenols

Lignin (Figure 1.11) is the relatively large-volume renewable aromatic feedstock. Next to heteropolysaccharides, it is one of the most abundant biopolymer on Earth, which is found in most global plants [48, 49]. It is deposited in the cell walls and the middle lamella.

Lignin, whose concentration systematically decreases from the outer layer to the inner layer of the cell wall, is generally responsible for reinforcing the plant structure. It is described as a water sealant in the stems, playing an important

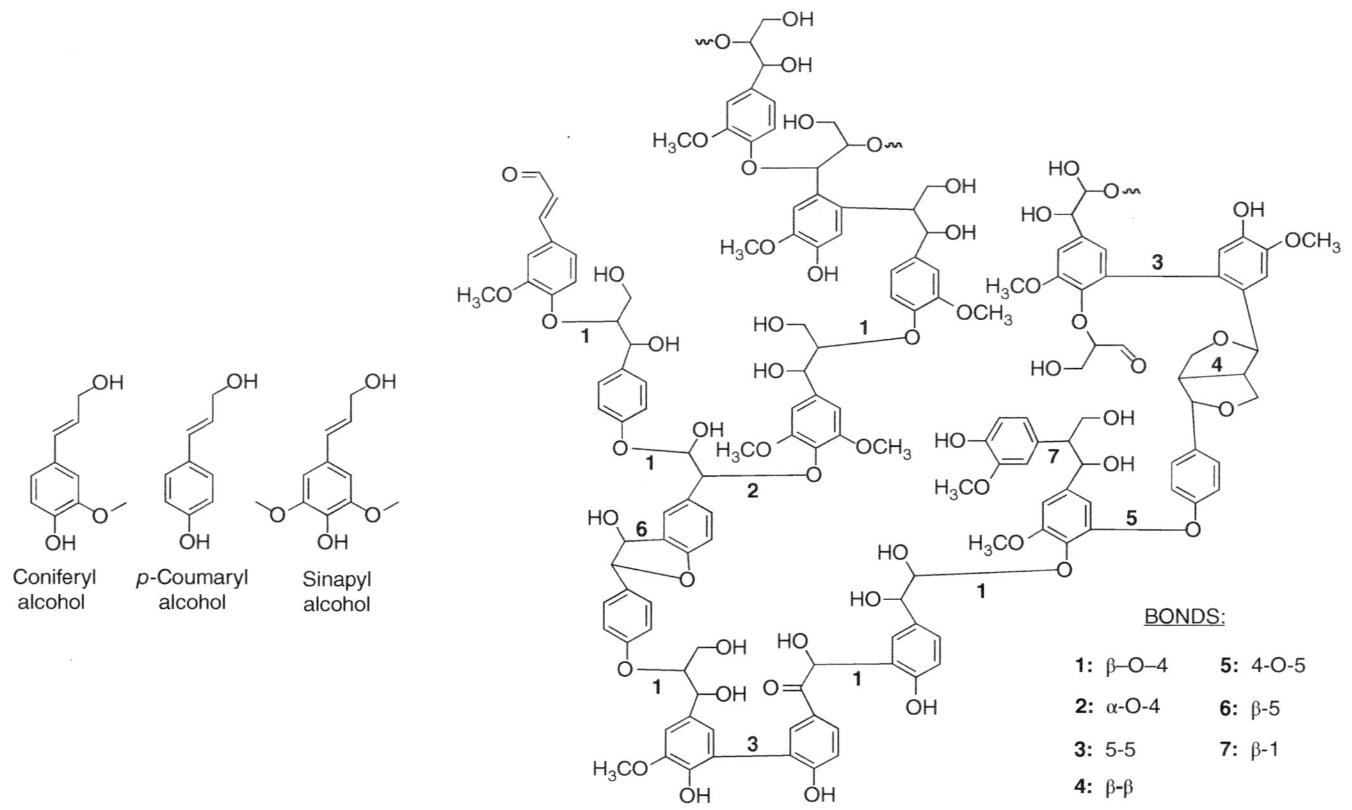

Figure 1.11 Simplified structure of softwood lignin (including three monolignols, the building blocks of lignin) [48, 49].

role in controlling water transport throughout the cell wall. Additionally, lignin of outer layers acts as a binding agent, holding the adjoining cells together, whereas the lignin within the cell walls gives rigidity by the chemical bonding with hemicellulose and cellulose microfibrils [50]. Moreover, it protects plants against decay and biological attacks [51].

Lignin is a complex and amorphous, three-dimensional network of hydroxylated phenylpropane units. Its contents vary with different types of plants, and overall, it is about 15–40% of the dry weight of lignocellulosic biomass [52]. Lignin is cross-linked with cellulose and hemicellulose through covalent and hydrogen bonds [53]. Generally, because of the complex structure and variety of possible degrees of polymerization, lignin is called by the term "lignins," which refers to the complex and diverse chemical composition and structure [54]. Mentioned properties, along with amorphous and hydrophobic nature, have an influence on difficult process of the isolation of lignin in unaltered form [55]. That is why, ball-milled wood lignin (MWL), isolated from finely powdered wood via the application of mild, neutral solvents, is considered to be the closest to *in vivo* lignin. In general, lignins contain a variety of alkyl- or aryl-ether interunit linkages (~60–70%), carbon–carbon (~25–35%), and small amounts of ester bonds, which include β-O-4, β-5, β-β, β-1, β-5, β-6, α-β, α-O-4, α-O-γ, γ-O-γ, 1-O-4, 4-O-5, 1-5, 5-5, and 6-5 (Figure 1.11) [56, 57]. Respectively, β-O-4-aryl ether (β-O-4), β-O-4-aryl ether (β-O-4), 4-O-5-diaryl ether (4-O-5), β-5-phenylcoumaran (β-5), 5-5-biphenyl (5-5), β-1-(1,2-diarylpropane) (β-1), and β-β (resinol) are major linkages, which are present within lignin macromolecules.

The lignin content is usually higher in softwoods (27–33%) than in hardwoods (18–25%), and herbaceous plants such as grasses (17–24%) have the lowest lignin contents [51]. Moreover, lignin originated from softwood and hardwood has different contents of methoxyl groups. Softwood lignin is composed of guaiacyl units, resulted from a polymerization of coniferyl alcohol (one methoxyl group per phenylpropane unit), whereas hardwood lignin is a copolymer of coniferyl and sinapyl alcohols (two methoxyl groups per phenylpropane unit). Additionally, hardwood lignin, on the one hand, contains fewer free phenolic hydroxyl groups but, on the other hand, contains more free benzyl alcohol groups than softwood lignin (Table 1.2).

Based on the literature [58], there is also a third type of lignin, which is formed by the polymerization of *p*-coumaryl alcohol. However, the resulting *p*-hydroxyphenyl lignin, is usually found in the form of a copolymer with guaiacyl lignin only in certain trees and tissues.

Even though lignin is one of the most abundant natural polymers, its industrial applications are rather limited. That is why, recently, in the era of greater ecological awareness, as well as unstable and diminishing petrochemical resources, the intensive research is being performed on application of lignin and its valuable compounds. However, the strong chemical bonding of lignin with hemicellulose and cellulose microfibrils makes it hard to isolate for effective utilization. Hence, great effort is being put on the development of pretreatment methods for more effective separation of lignin. Generally, the isolation of lignin is performed by its extraction using different methods, such as Kraft, soda, lignosulfate,

Table 1.2 Proportions of interunit linkages in softwood and hardwood [50].

Structure (%)	Lignocellulosic material	
	Softwood	Hardwood
Phenylpropane unit		
– Coumaryl	—	—
– Coniferyl	90–95	50
– Sinapyl	5–10	50
C_9-O-C_9		
– β-O-4	46	60
– α-O-4	6–8	6–8
– 4-O-5	3.5–4	6.5
C_9-C_9		
– β-5	9–12	6
– β-1	7	7
– β-β	2	3
– 5-5	9.5–11	4.5

organosolv [59–61], hydrolysis, enzymatic, ionic liquids [62], and ultrafiltration by membrane technology. Because all the mentioned isolation methods require specific conditions such as pH, temperature, pressure, reagents, time, and variety of different solvents, the isolated lignin is characterized by diverse structural and chemical properties [63]. Utilization of lignin might be performed: (i) without its chemical modification (via the incorporation of lignin into matrix to give new or improved properties) and (ii) with the chemical modification to prepare a large number of smaller chemicals, which might be used to obtain other chemical compounds including polymers. The chemical modification of lignin (Figure 1.12) is performed by (i) fragmentation or lignin depolymerization to use lignin as a carbon source or to split the structure of lignin into aromatic macromers; (ii) creation of new chemical active sites, and (iii) chemical modification of hydroxyl groups.

Nearly 90% of epoxy polymers are obtained from bisphenol A (BPA). However, there is a growing demand to develop renewable aromatic compounds to replace the petroleum-based BPA. Conducted studies are concentrated on maintenance in bio-based materials the desirable thermomechanical properties, provided by aromatic rings of BPA-based epoxy resins, which are linked to the rigidity provided by aromatic rings of BPA. That is why, efforts in the synthesis of novel epoxy resins are mainly directed toward renewable phenolic compounds derived from biomass feedstocks. Lignin is one of the most promising natural resource for replacement of bisphenol A because of the presence of aromatic structure with hydroxyl, carboxylic acid, and phenolic functional groups, which are able

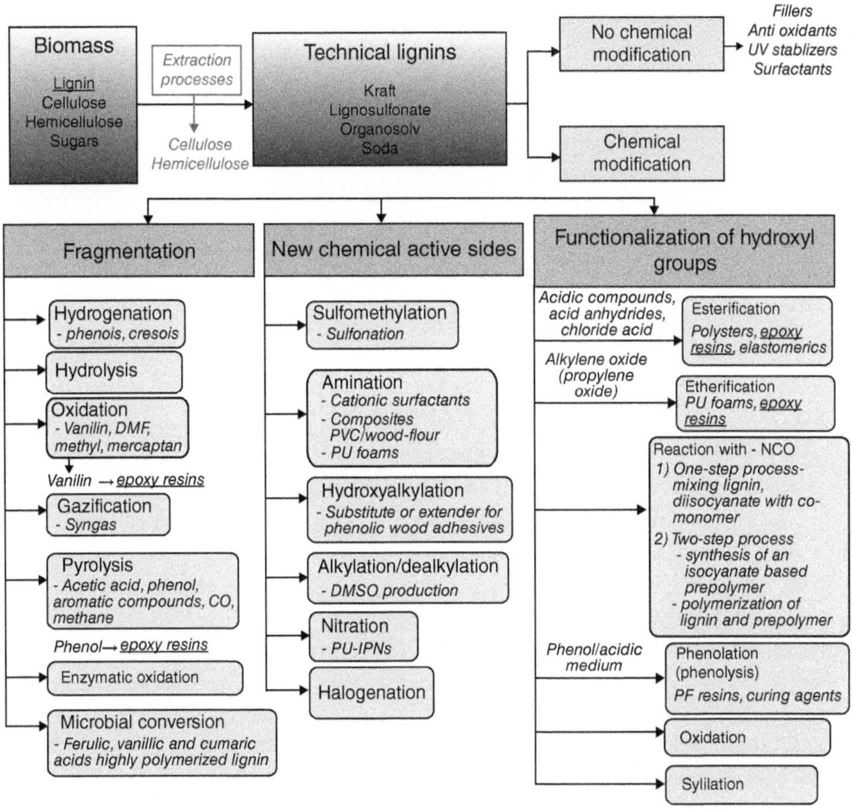

Figure 1.12 Summary of the main strategies for lignin conversion [49, 55].

to react with epichlorohydrin to form bio-based epoxy resins. The phenolic and alcohol hydroxyls, which are present within the lignin macromolecule, have found application in numerous research on incorporating that biopolymer into thermosetting resins, either as a component during the synthesis of epoxy monomers or as a reactive additive [64]. In general, the process of preparation of lignin-based epoxy resins might be conducted by (i) direct blending of lignin with epoxy resin [65], (ii) modifying lignin derivatives by the glycidylation [66], or (iii) modifying lignin derivatives to improve their reactivity, followed by the glycidylation [67]. Recently, one of the most common approaches to obtain lignin-derived polyols is the lignin depolymerization to lower molecular weight compounds, such as vanillin, vanillyl alcohol derivatives [68], phenol [69], isoeugenol [70], syringaresinol, and compounds based on propyl guaiacol and its demethylated product.

Bio-based epoxy resins are, for instance, synthesized from derivatives obtained on the course of lignin hydrogenolysis [71]. Lignin from pine wood is depolymerized by mild hydrogenolysis to give an oil product, containing aromatic polyols: dihydroconiferyl alcohol (DCA, 4-(3-hydroxypropyl)-2-methoxyphenol) and 4-propyl guaiacol (PG, 4-propyl-2-methoxyphenol), along with their dimers and

Table 1.3 Thermal analysis data for epoxy resin blends containing DGEBA and cured with DETA [68].

Resin	T_g a) (°C)	$T_{5\%}$ b) (°C)	T_s c) (°C)
DGEBA	117	328	169
LHEP/DGEBA 1 : 1	80	289	161
LHEP/DGEBA 2 : 1	70	270	156
LHEP/DGEBA 3 : 1	68	258	151
LHEP	53	236	144

a) T_g – the glass transition temperature.
b) $T_{5\%}$ – the initial decomposition temperature.
c) T_s – the statistic heat-resistant index temperature.

oligomers. Then, the obtained oil product is dissolved in refluxing epichlorohydrin in the presence of solution of NaOH to give epoxy prepolymer (LHEP) (Figure 1.13), which is then blended with bisphenol A diglycidyl ether (DGEBA) in mass ratios of LHEP : DGEBA up to 3 : 1. Epoxy composition might be cross-linked with diethylenetriamine (DETA).

Cured epoxy material LHEP/DGEBA is less thermally stable than the DGEBA resin (Table 1.3) because of the presence of methoxy groups on the aromatic ring. The initial decomposition temperature ($T_{5\%}$) and the statistic heat-resistant index temperature (T_s) are the lowest for samples containing the highest proportion of LHEP. On the other hand, the presence of methoxy groups in the lignin hydrogenolysis products is likely to contribute to the superior mechanical properties of the cured LHEP/BADGE blends. Values of flexural modulus and strength of bio-based materials are 52% and 28%, respectively, greater than DGEBA alone. Additionally, it is worth to mention here the research on the influence of the presence of lignin on the thermal performance and thermal decomposition kinetics of lignin-based epoxy resins [72]. The presence of lignin-based epoxy resin (depolymerized Kraft lignin, DKL-epoxy resin, and depolymerized organosolv lignin, DOL-epoxy resin, respectively) in epoxy composites, prepared by mixing the DGEBA and a desired amount (25, 50, and 75 wt%) of DKL-epoxy resin and DOL-epoxy resin at 80 °C, then cured with 4,4′-diaminodiphenylmethane (DDM), results in a significant effect on the activation energy of the decomposition process, in particular, at the early and the final stage of decomposition (Table 1.4).

The increase in the percentage value of lignin-based epoxy resins in the composites reduces the initial activation energy of the system. Additionally, the obtained bio-based materials exhibit higher limiting oxygen index (LOI) than that of the conventional BPA-based epoxy resin, which might indicate that the lignin-based epoxy composites are more effective fire retardants than the conventional BPA-based epoxy resin.

An interesting example of a novel approach to finding new epoxy application for bio-based derivatives from lignin is a conversion of lignin to epoxy compounds throughout the reaction of epichlorohydrin with partially depolymerized

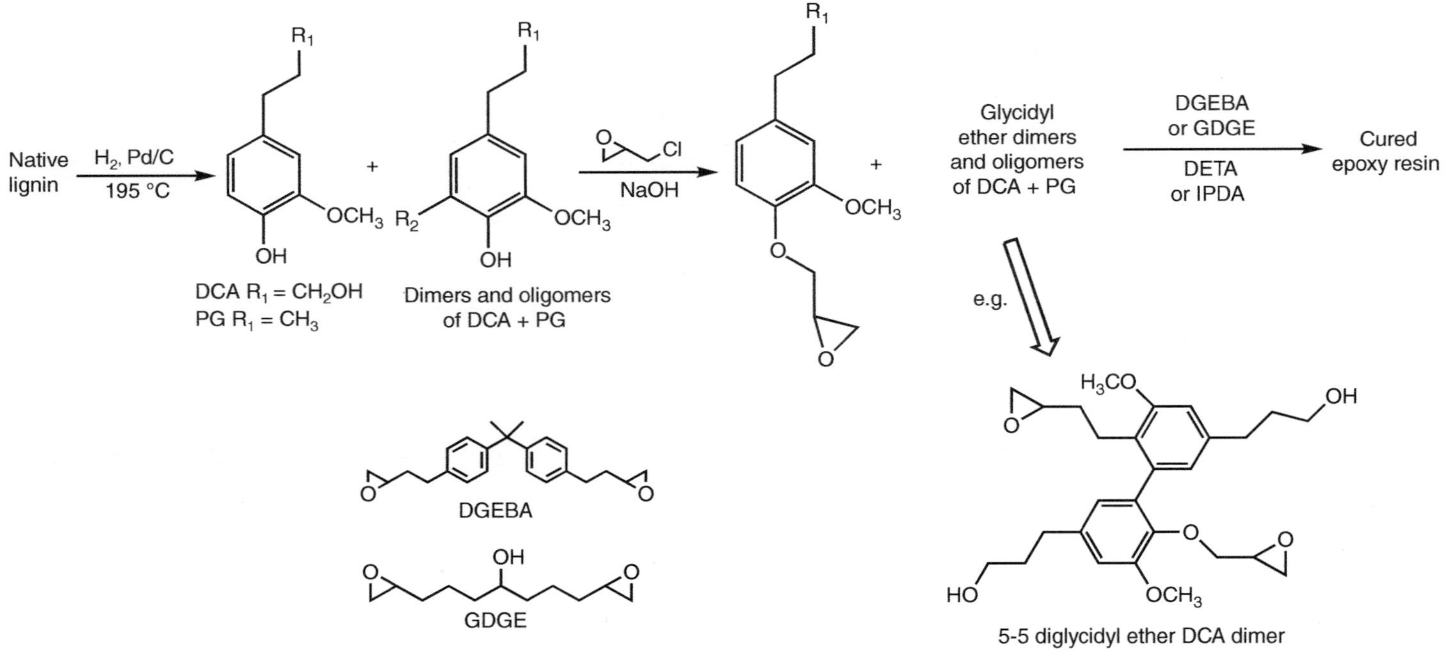

Figure 1.13 Cured epoxy resins from lignin hydrogenolysis products.

Table 1.4 Thermal decomposition of BPA-based epoxy resin and the DGEBA/lignin-based epoxy resin [72].

Sample	IDT (°C)	T_{max} (°C)		$Char_{800}$ (%)	LOI
		Shoulder peak	Main peak		
DGEBA-DDM	370	—	405	12.5	22.5
25% DKL-DDM	359	—	397	18	24.7
50% DKL-DDM	330	—	396	25	27.5
75% DKL-DDM	300	335	407	32	30.3
100% DKL-DDM	290	325	416	38	32.7
25% DOL-DDM	383	—	404	17	24.3
50% DOL-DDM	352	—	399	21	25.9
75% DOL-DDM	338	—	398	24	26.7
100% DOL-DDM	335	—	397	29	29.1

lignin (PDL) in the presence of benzyltriethylammonium chloride and dimethyl sulfoxide (Figure 1.14) [67].

The lignin-based epoxy material is characterized by comparable thermal and mechanical properties to those of BPA-based epoxy resin (DER332) cured with the same bio-based curing agent (the Diels–Alder adduct of methyl esters of eleostearic acid, a major tung oil fatty acid, and maleic anhydride [MMY]). The obtained bio-based epoxy product might be applied as a modifier for asphalt applications in the same manner as petroleum-based (and mostly BPA-based) epoxy resins, which are currently used for asphalt modification to improve its temperature performance. The PDL epoxy asphalt, in the same way as DER332-asphalt, exhibits significant improvement on the viscoelastic properties, especially at elevated temperatures.

The research on utilization of lignin derivatives toward the synthesis of the epoxy system is ongoing for several years; thus, there are numerous methods described in the literature. Among them, it is worth to mention the synthesis of epoxies by (i) direct epoxidation of the phenolic hydroxyl group in the technical lignin with epichlorohydrin and (ii) obtaining bisguaiacyl structure via the reaction using ketone compound and then the epoxidation (Figure 1.15a) [66].

The other route of lignin utilization toward the synthesis of epoxy system is the cleavage of lignin intermolecular bond and creating the phenolic hydroxyl group in the molecule (Figure 1.15c). The process is usually done by treating the Kraft lignin with acid (hydrochloric or sulfuric acid) and phenol derivatives. The obtained phenolic hydroxyl group is epoxidized with epichlorohydrin, resulting in the lignin-based epoxy resin, which in the next step is cross-linked using DETA or phthalic anhydride. The phenol derivative within the lignin structure might also be obtained on the course of the lignin phenolization with bisphenol A in the presence of hydrochloric acid and BF_3-ethyl etherate as catalysts (Figure 1.15b) [73]. The obtained product is soluble in organic solvent such as acetone because of the contribution of bisphenol A.

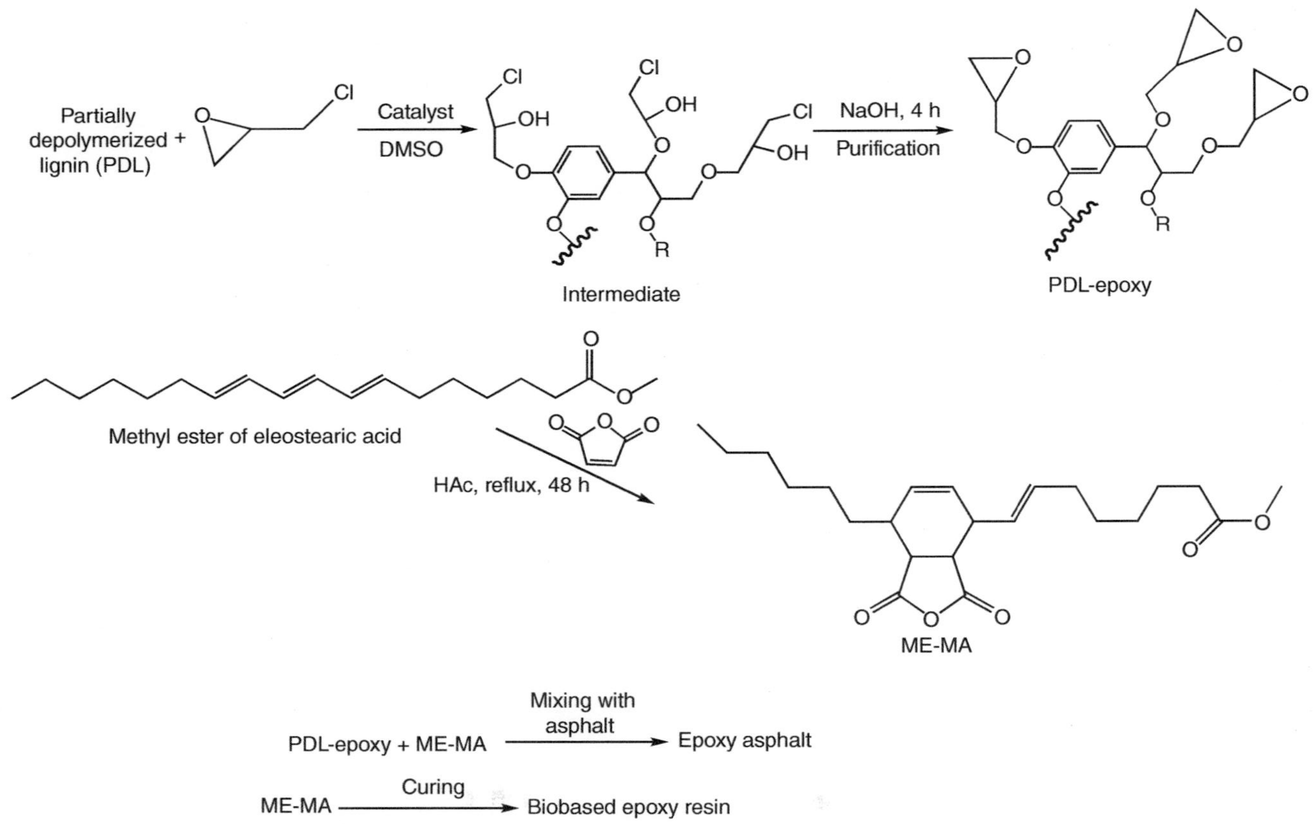

Figure 1.14 Synthesis of lignin-based epoxy and epoxy asphalt.

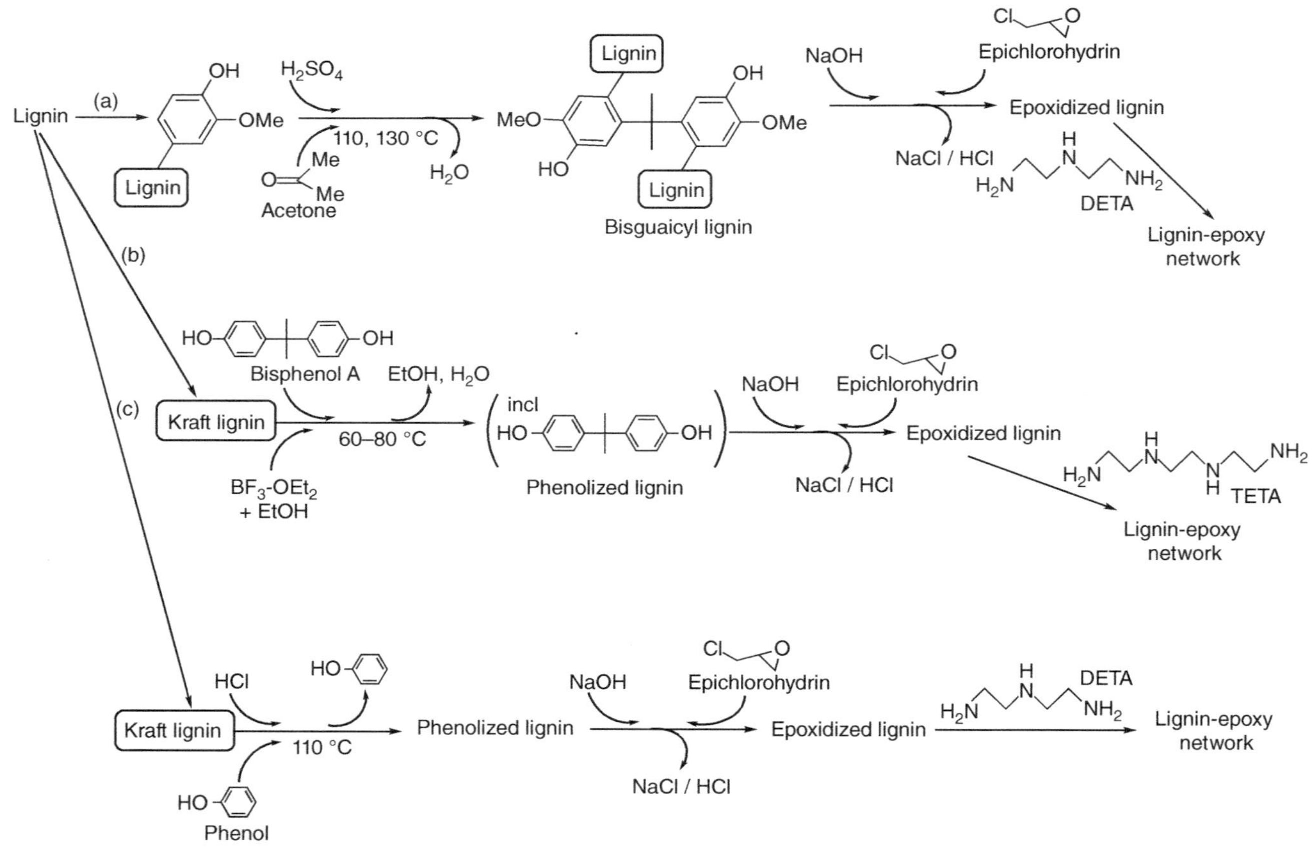

Figure 1.15 Schematic routes of lignin modification and crosslinking: (a) epoxidation with bisguaiacyl structure stage, (b) lignin' phenolization with bisphenol A and (c) direct epoxidation of the phenolic hydroxyl group in the technical lignin with epichlorohydrin.

It is worth noting that a substantial amount of lignin decomposing aromatics is characterized by the structure of phenol substituted by inert methoxy and alkyl groups (structures such as guaiacol or creosol), making polycondensation or radical polymerization especially difficult. Thus, there are numerous studies on (i) introducing the reactive groups, which are promoting further polymerization reactions [74], (ii) utilization of the reactive *ortho*- and *para*-sites of phenol for hydroxymethylation or obtaining novolac or resol-type resin using formaldehyde chemistry [75] otherwise, (iii) connecting lignin-derived compounds to make oligomers with additional functional groups [76]. Bimetallic Zn/Pd/C catalytic method for converting lignin via the selective hydrodeoxygenation ($T = 150\,°C$ and 20 bar H_2, using methanol as a solvent) directly into two methoxyphenol products has been reported [77]. The compound characterized by the increased content of hydroxyl groups might be obtained using the above method, via the reaction of *o*-demethylation of 2-methoxy-4-propylphenol and aqueous HBr. In the next step, propylcatechol is glycidylated to epoxy monomer (Figure 1.16).

Other techniques described in the literature involve the ozone oxidation of Kraft lignin toward splitting its aromatic rings and generation of the muconic acid derivatives. The ozonized lignin (Figure 1.17a) might then be dissolved in an alkali water solution and cross-linked with the water-soluble epoxy resin (glycerol polyglycidylether).

Another interesting synthesis described in the literature begins from the dissolution of alcoholysis lignin or lignin sulfuric acid in ethylene glycol and/or glycerin (Figure 1.17b) [78]. Next, the hydroxyl group in the lignin molecule is reacted with succinic acid to convert the lignin into multiple carboxylic acid derivatives. In the last step, the resulting products react with epoxy compound (ethylene glycol diglycidyl ether [EGDGE]) in the presence of dimethylbenzyl amine as a catalyst to provide the cross-linked epoxidized lignin resin. In the obtained curried epoxy material, lignin acts as a hard segment (increasing value of T_g with increasing lignin derivatives). Additionally, a slight decrease of T_d with increasing content of biocomponent in epoxy resin suggests that the thermal stability of obtained epoxy system is not affected by the presence of lignin derivatives.

Based on numerous studies, one can conclude that lignin is a very promising natural resource for replacement of bisphenol A in the synthesis of epoxy resins, as it has aromatic structure with hydroxyl, carboxylic acid, and phenolic functional groups, which can react with epichlorohydrin to form bio-based epoxy resins. One of the biggest problems for commercial application of lignin's derivatives, because of its complex and multifunctional nature, is isolation and the synthesis of monomers.

1.3.2 Vanilin

Vanillin (4-hydroxy-3-methoxybenzaldehyde) is an organic compound consisting of a benzene ring substituted with three functional groups: aldehyde –CHO, hydroxyl –OH, and methoxy –O–CH_3 (Figure 1.18a).

It is a naturally occurring compound (in the form of its β-D-glucoside, Figure 1.18b) that can be directly obtained in the extraction process from the bean or seed pods of *Vanilla planifolia*, the tropical orchid presently cultivated

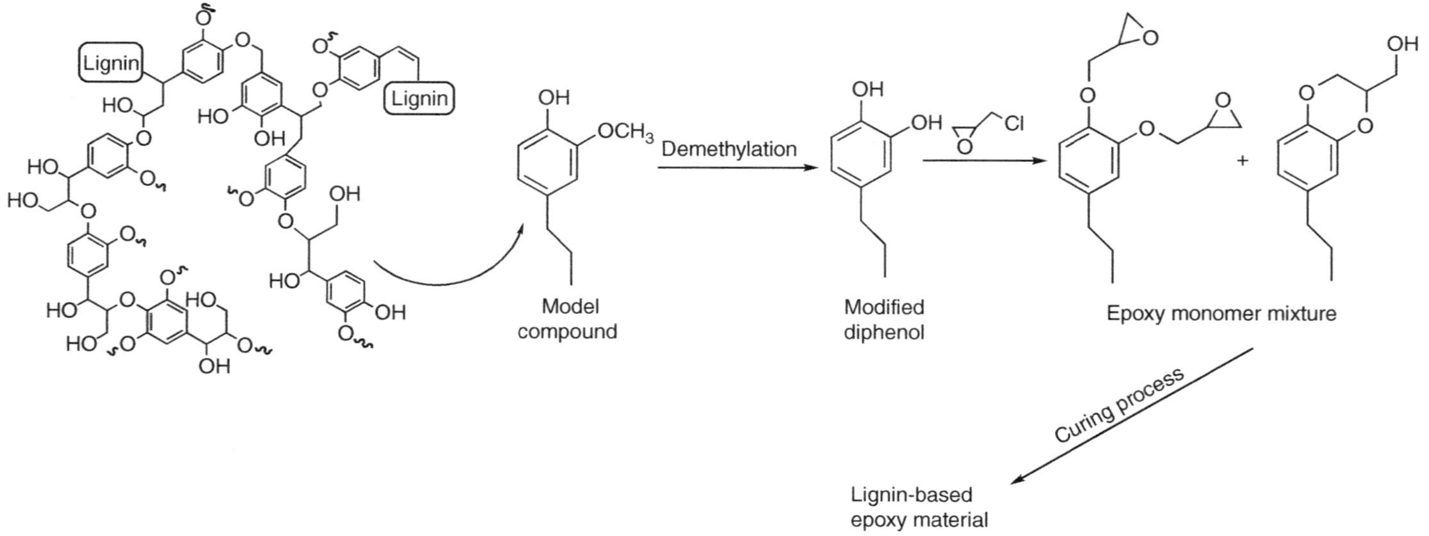

Figure 1.16 Route of the synthesis epoxy monomers from selectively hydrodeoxygenated lignin.

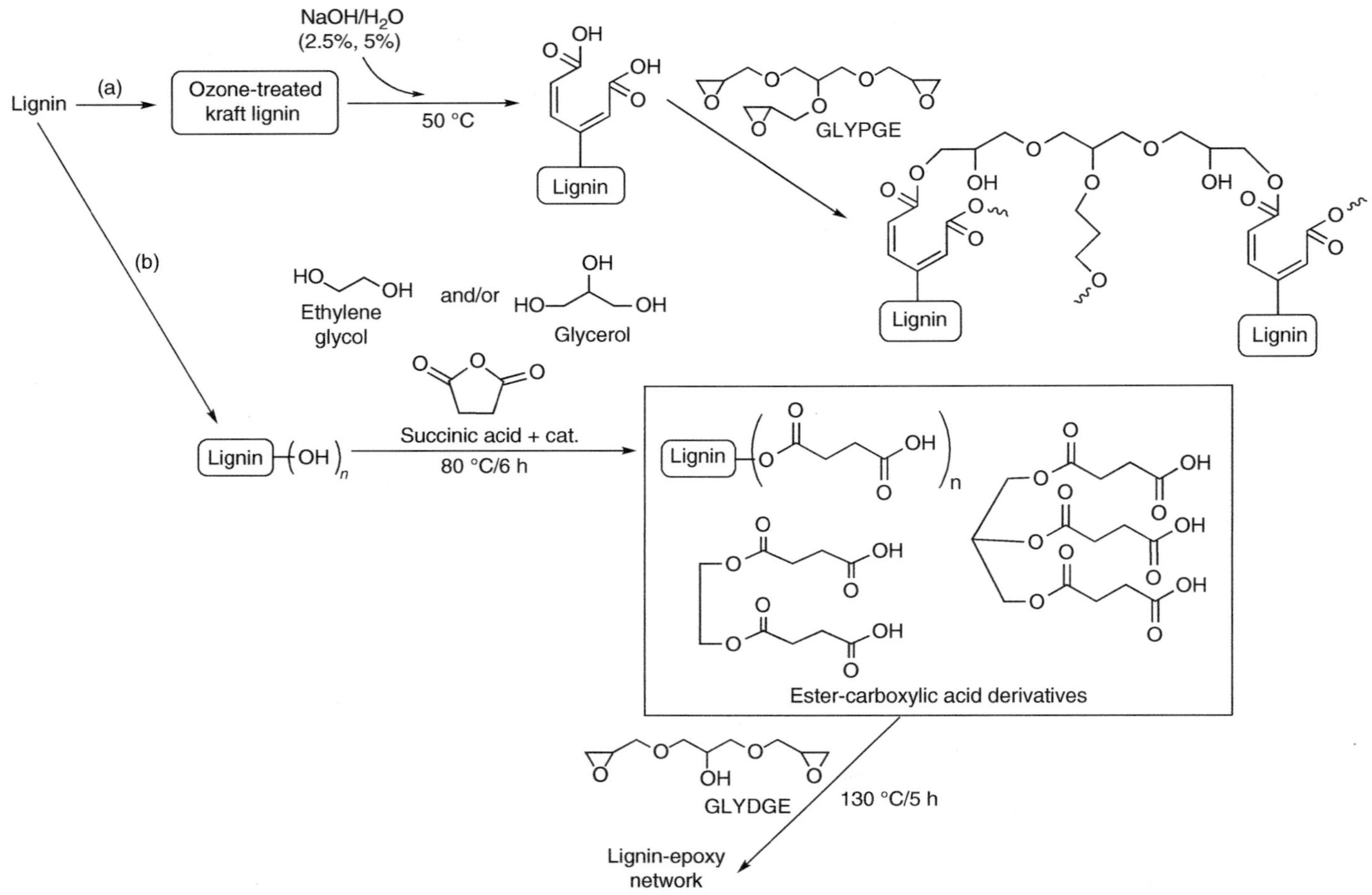

Figure 1.17 Lignin modification and cross-linking: (a) ozone oxidation of Kraft lignin and (b) synthesis of multiple carboxylic acid derivatives.

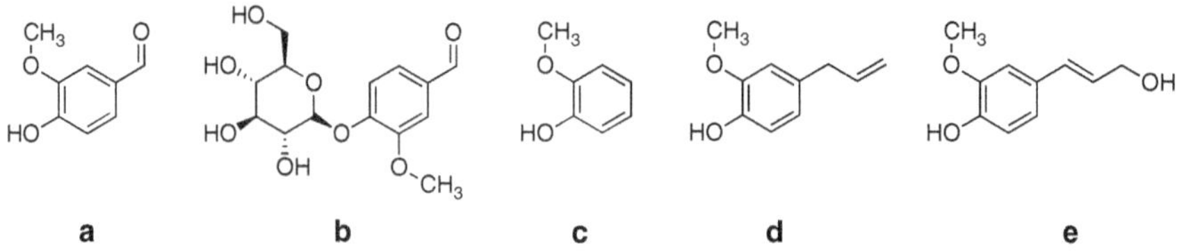

Figure 1.18 Chemical structures of (a) vanillin and its naturally occurring precursors: (b) vanillin glucoside, (c) guaiacol, (d) eugenol, and (e) coniferyl alcohol.

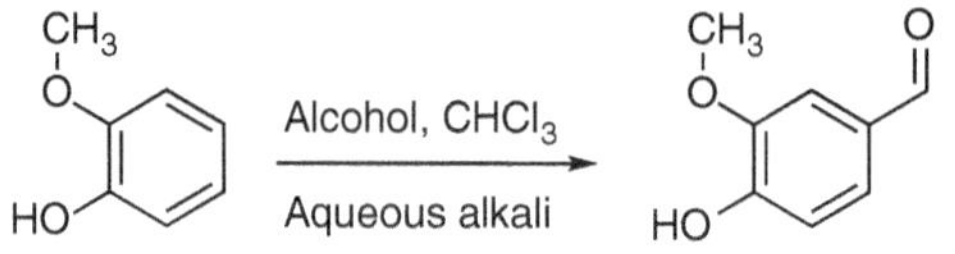

Figure 1.19 Synthesis of vanillin from guaiacol.

in a number of tropical countries. Although this method has been known for centuries and it is still used, actually less than 1% of vanilla produced in the world is obtained in such a way [79]. Almost all vanillin is now synthesized much more cheaply through chemical processes. Synthetic vanillin is commercially available and is commonly used in both food and nonfood applications, in fragrances, as a flavoring in pharmaceutical preparations, as an intermediate in the chemical and pharmaceutical industries for the production of herbicides, antifoaming agents or drugs, and in household products, such as air fresheners and floor polishes. Synthetic or semisynthetic vanillin can be derived from two compounds: guaiacol and eugenol, both available from petrochemical sources or of natural origin.

The first one, guaiacol (2-methoxyphenol) (Figure 1.18c), is a naturally occurring organic compound present in an aromatic oil from flowering plants *Guaiacum*. Guaiacol can also be gained from creosotes formed by distillation of various tars and pyrolysis of plant-derived material, such as wood. Semisynthetic vanillin can be obtained from guaiacol through the Reimer–Tiemann reaction of phenols formylation (Figure 1.19) [80].

The reaction is carried out using chloroform deprotonated by a strong base (hydroxide typically) to form the chloroform carbanion and finally the dichlorocarbene, which is the principal reactive specie in nucleophilic substitution also occurred in deprotonated phenol. Another method of vanillin synthesis from guaiacol is its reaction with glyoxylic acid (Figure 1.20a), leading to the formation of 2-hydroxy-2-(4-hydroxy-3-methoxyphenyl)-acetic acid (Figure 1.20b) [81]. The obtained vanillylmandelic acid is converted via 2-(4-hydroxy-3-methoxyphenyl)-2-oxoacetic acid (Figure 1.20c) to vanillin by the oxidative decarboxylation [82].

Eugenol (2-methoxy-4-(prop-2-en-1-yl)phenol) (Figure 1.18d) present in an essential oil extracted from the clove plant *Syzygium aromaticum* is the next important natural raw material for the vanillin synthesis (Figure 1.21).

The synthesis consists of two steps: the basic isomerization of the double bond in eugenol leading to the formation of isoeugenol and oxidation of the rearranged

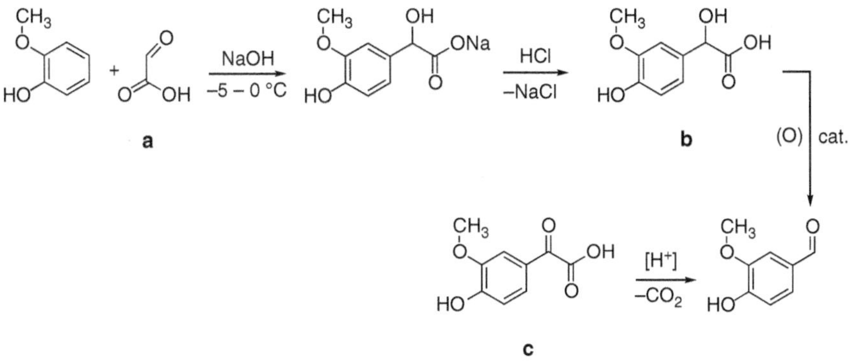

Figure 1.20 Synthesis of vanillin from guaiacol using glyoxylic acid.

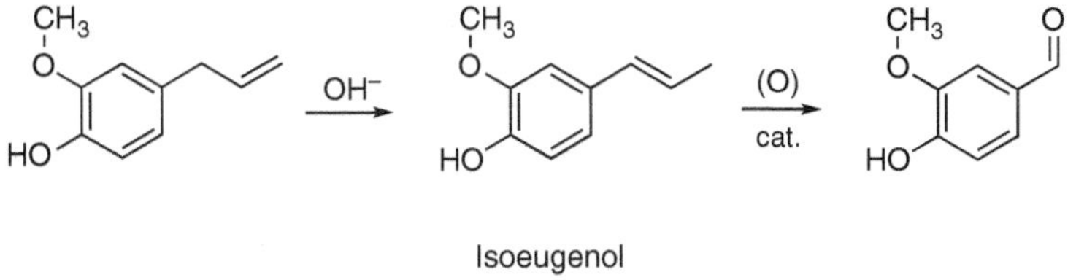

Figure 1.21 Synthesis of vanillin from eugenol.

double bond to vanillin [83, 84]. The process can be carried out with or without isolation of the intermediate product which is isoeugenol [85].

Lignin from softwood is still one of the most important sources of raw materials for the synthesis of vanillin. Three-dimensional network structures of lignin are composed of three types of monolignols: *p*-coumaryl alcohol, coniferyl alcohol, and sinapyl alcohol. Coniferyl alcohol (Figure 1.18e) is the main intermediate in the pathways to vanillin from a softwood (coniferous) lignin, as well as the precursor for eugenol in its biosynthesis. Vanillin can be produced from the lignin-containing waste manufactured by the sulfite pulping process for preparing wood pulp for the paper industry [86]. This first developed method of vanillin synthesis from lignin lost its relevance primarily for environmental reasons (the need to safely get rid of alkaline-based liquid waste). However, thanks to the results of work on the process optimization [87], it was possible to achieve an increase in the yield of vanillin and a reduction in waste stream volume. Therefore, the volume of vanillin production from lignin is still estimated at around 15% of the total world vanillin production.

There are also known for years [88] and still developed biotechnological processes of vanillin synthesis [89]. They are promising for the industrial-scale production of vanillin due to the use of natural raw materials, renewable and readily available in large quantities, such as rice bran [90] or corn sugar [91]. However, actually bio-synthesized vanillin is still very expensive [92] and its high cost of production are justified only for specific applications in the food, cosmetics, and pharmaceutical industries. Nowadays, the biotechnological

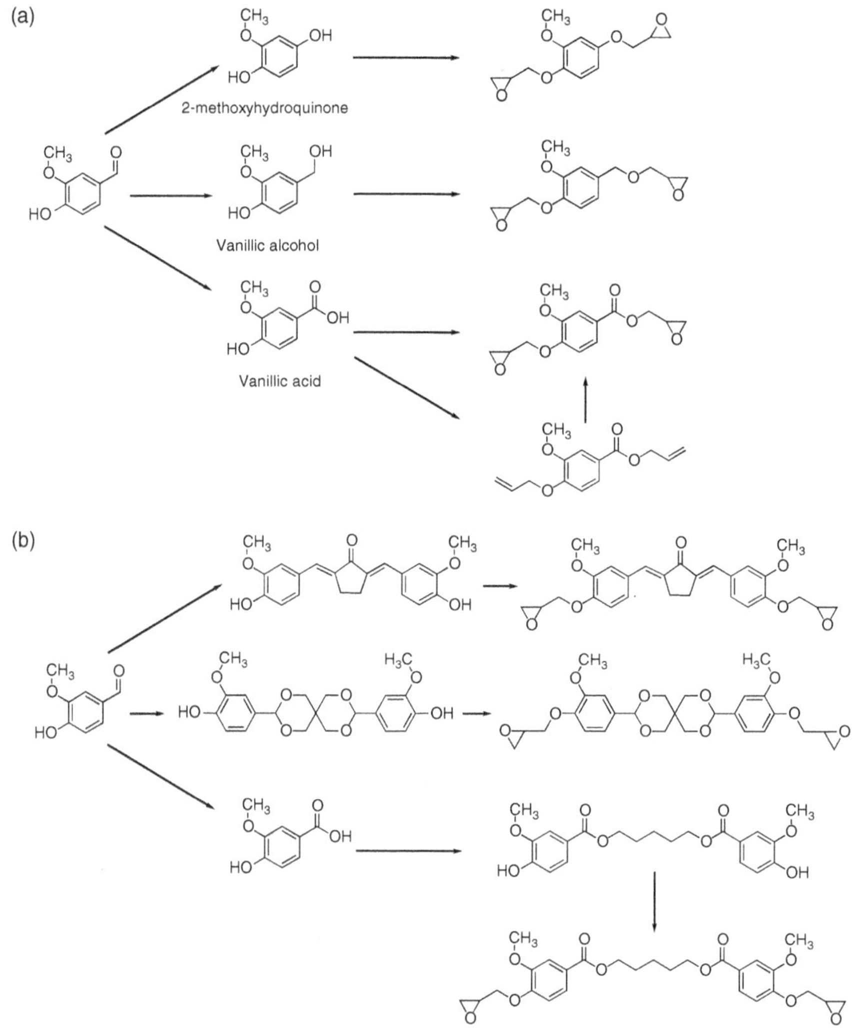

Figure 1.22 Schematic illustration of possible synthesis pathways (a) and (b) for vanillin-based epoxy resins.

production is not suitable and profitable source of vanillin for the synthesis of polymeric materials.

Due to its chemical structure as a phenolic compound, vanillin is the promising raw material that could replace bisphenol A (or other commonly used bisphenols) providing epoxy resins with adequate mechanical strength and thermal stability. However, as the trifunctional compound, but also only a monoalcohol, vanillin must be modified in order to serve as a substitute for bisphenols in the synthesis of epoxy resins. Figure 1.22 shows schematically the possible pathways of vanillin modification described in the literature that lead to obtaining epoxy resins.

Generally, two strategies for the synthesis of vanillin-based epoxy resins are mainly being investigated. The first one (Figure 1.22a) assumes converting

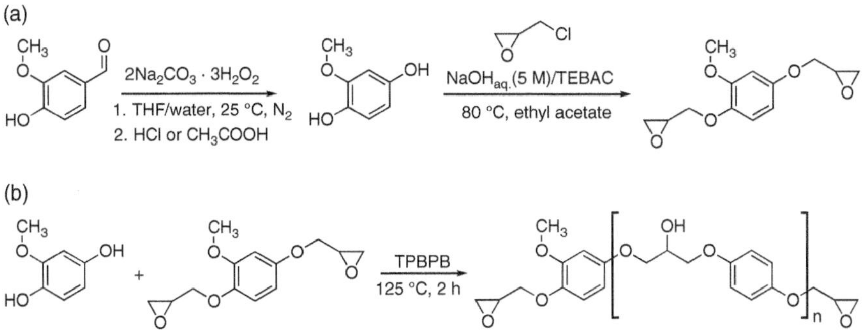

Figure 1.23 Synthesis of 2-methoxyhydroquinone and its epoxy derivatives - strategies (a) and (b).

vanillin into derivatives also containing, in addition to the phenol group already present in the vanillin molecule, a second functional group through which the epoxy functionality could be introduced. The second strategy (Figure 1.22b) involves coupling of two vanillin (or its derivative) molecules using another chemical compound, resulting in a product containing at least two phenolic or other groups through which the epoxy group can also be introduced. In both strategies, commonly used methods for introducing epoxy functionality have been applied: the oxidation of double bonds and the reaction with epichlorohydrin.

According to the first strategy, the Dakin oxidation can be applied to convert aldehyde group in vanillin to hydroxyl group [93] (Figure 1.23).

Synthesized 2-methoxyhydroquinone can be reacted with the large excess (10-fold) of epichlorohydrin under the typical phase-transfer catalysis conditions in the presence of triethylbenzylammonium chloride (TEBAC). The resulting product mainly contains diglycidyl ether of 2-methoxyhydroquinone (Figure 1.23a), which can be used together with 2-methoxyhydroquinone to obtain an epoxy resin (with an epoxy value of 0.060–0.340 mol/100 g) via the fusion process [94] in the presence of triphenylbutylphosphonium bromide (TPBPB) (Figure 1.23b). Such epoxy resin with an epoxy value of 0.404 mol/100 g [95] could be successfully cross-linked with the cycloaliphatic amine curing agent (commercial product Epikure F205), preferably in the presence of calcium nitrate as an accelerator. The vanillin-based epoxy resin cured using 2 wt% of the inorganic accelerator exhibits the tensile strength and the Izod impact strength higher than those for liquid diglycidyl ether of bisphenol A (epoxy resin Epon 828 with an epoxy value of 0.541 mol/100 g) used for comparison. As an aldehyde, vanillin can be easily oxidated to vanillic acid, as well as reduced to vanillic alcohol (Figure 1.22a). Under analogous conditions as in the case of 2-methoxyhydroquinone, the diglycidyl monomers (Figure 1.22a) can be obtained from both vanillic acid and alcohol [96]. After cross-linking with isophorone diamine bio-based epoxy resins derived from them are characterized by the high glass transition temperature (132 and 152 °C, respectively) and the storage modulus comparable with the value determined for diglycidyl ether of bisphenol A. They also exhibit high thermal stability, typical for epoxy resins

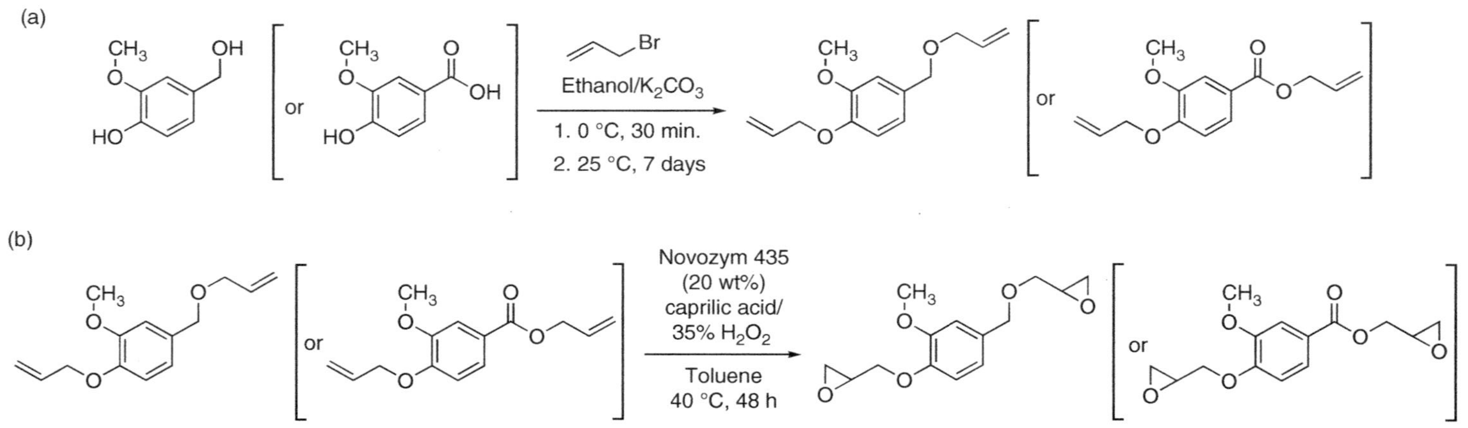

Figure 1.24 The *O*-alkylation of vanillin derivatives (a), followed by the epoxidation of the resulting double bonds (b).

based on bisphenol A. 2-Methoxyhydroquinone as well as vanillic alcohol and acid could be reacted with allyl bromide giving derivatives (Figures 1.22a and 1.24a) with terminal unsaturated bond [97], which can be e.g. enzymatically oxidized to oxirane rings using percaprylic acid as an oxygen carrier and immobilized lipase B from *Candida antarctica* (Novozym 435) as a biocatalyst [98] (Figure 1.24b).

This is another interesting reaction pathway for the synthesis of above-mentioned diglycidyl monomers without using bisphenol A and epichlorohydrin, and under mild conditions. Moreover, the other interesting epoxy compound derived from two coupled vanillic acid molecules (Figure 1.25) could also be prepared throughout this way.

However, obtaining the completely epoxidized products and the formation of various regioisomers still remain challenging.

According to the second strategy, the dimerization of vanillin is possible [99] by the selective enzymatic oxidative coupling (Figure 1.26a). After the reduction of aldehyde groups, a divanillin alcohol is obtained, which can be then reacted with epichlorohydrin (Figure 1.26b) [100].

The vanillin-based epoxy compounds are obtained as a mixture of glycidyl derivatives at different ratios, which can be fractionated by flash chromatography. The content of individual glycidyl derivatives in the product mixture can be controlled primarily by the sodium hydroxide content, as well as the duration of the second step reaction with epichlorohydrin (adding a base at room temperature in order to perform the ring closure of intermediate halohydrin species). For example, with a NaOH/OH ratio equal to 10, the tetraglycidyl compound is mainly obtained with about 90% yield. In contrast, the diglycidyl derivative is mainly created (80% yield) at lower NaOH/OH ratios. Separated vanillin-based epoxy compounds cross-linked with isophorone diamine, characterized with the glass transition temperature in terms of 138–198 °C, exhibit similar Young modulus and thermal stability values to the bisphenol A-based epoxy thermoset, but lower elongation at break.

The other possibility of the vanillin dimerization is the electrochemical synthesis of *meso*-hydrovanilloin (Figure 1.27) [101]. The symmetrical compound divanillin (two molecules of vanillin coupled by aromatic rings) can be easily prepared by the oxidative dimerization of vanillin catalyzed by $FeCl_3$ or heme iron enzymes [102]. However, vanillin can also be reductively dimerized at the aldehyde function using low-valent titanium generated via $TiCl_4$-Mn or by the electrochemical method [103]. The electrochemical coupling is a highly stereoselective reaction giving *meso*-hydrovanilloin. The vanillin electrolytic reduction in alkaline solution at a metallic cathode gives interesting bisphenol compound, which can be used as a direct substitute for bisphenol A [104].

The obtained *meso*-hydrovanilloin-based epoxy resin cured using long-chain aliphatic diamine (1,6-diaminohexane) and cycloaliphatic amine (isophorone diamine) showed the glass transition temperature and Shore hardness (D-type) values comparable with commercial diamine-cured bisphenol A-based epoxy resins.

Two molecules of vanillin can be coupled by the crossed aldol condensation (Figure 1.28) with cyclopentanone [105].

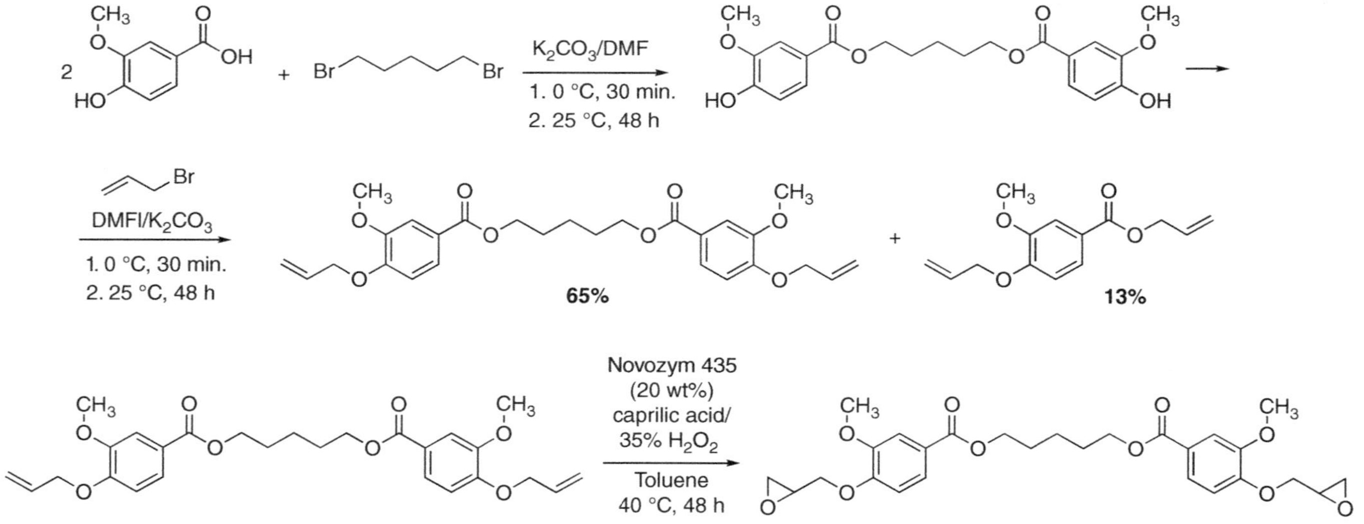

Figure 1.25 Esterification of vanillic acid, followed by the *O*-alkylation and subsequently by the epoxidation of the allylic double bonds.

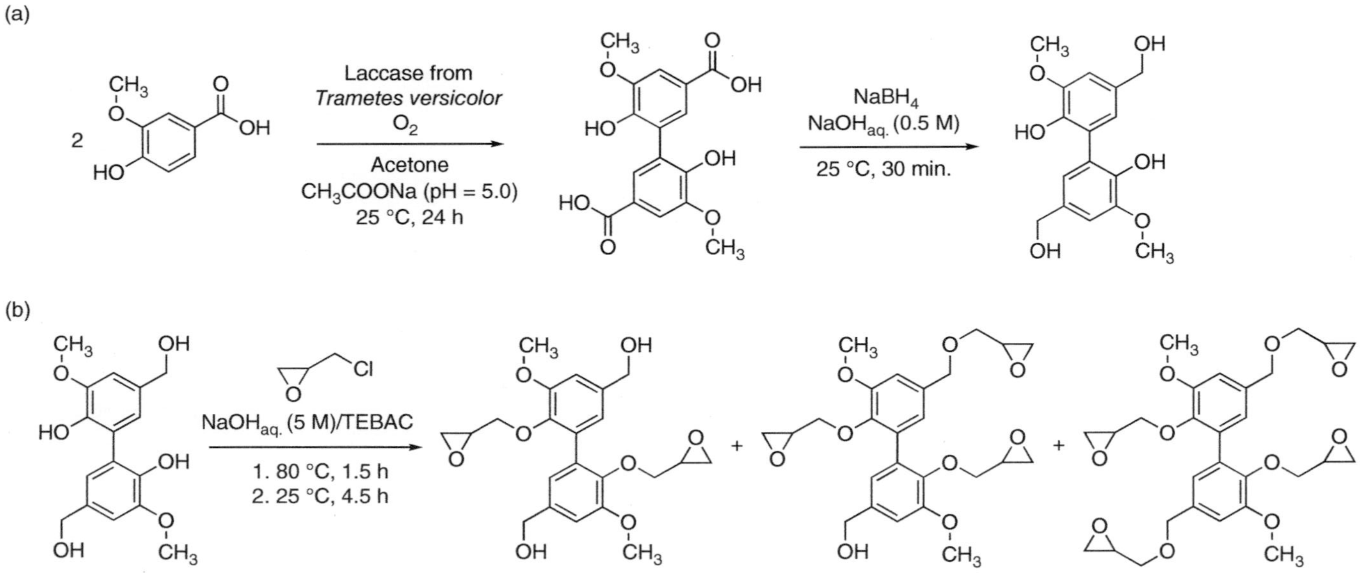

Figure 1.26 Synthesis of glycidyl derivatives (b) based on the product of vanillin dimerization (a).

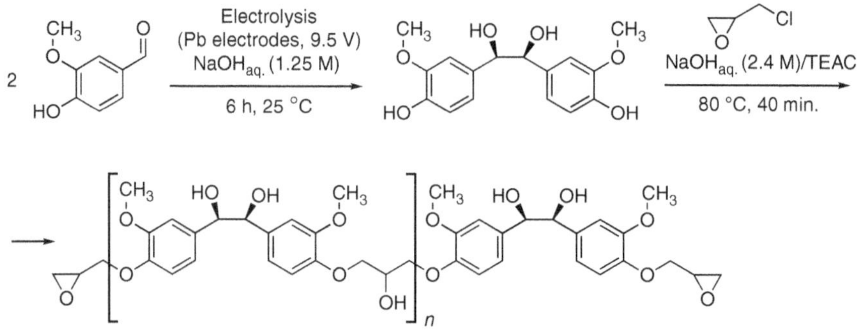

Figure 1.27 Synthesis of hydrovanilloin and the epoxy resin based on this vanillin dimer.

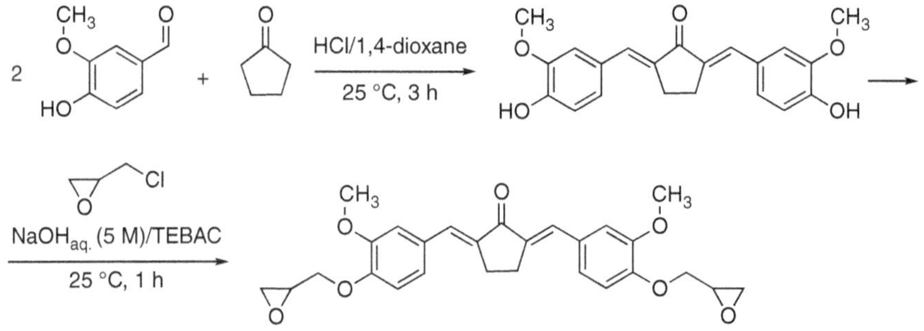

Figure 1.28 Synthesis of 2,5-bis(4-hydroxy-3-methoxybenzylidene)cyclopentanone and its diglycidyl derivative.

The thermal and mechanical properties of the synthesized resin cured with bio-based (quercetin and guaiacol novolac) hardeners and a petroleum-based hardener (phenol novolac) are comparable with those of the bisphenol A-based resins cross-linked with the same hardeners.

A diamine can also be used to couple two vanillin molecules [106]. Vanillin coupled with aromatic diamines and diethyl phosphite, followed by the reaction with epichlorohydrin, yields high-performance biorenewable and environment-friendly flame-retardant epoxy resins (Figure 1.29).

The coupling product with 4,4-diaminodiphenylmethane (DDM) or *p*-phenylenediamine (PDA) is synthesized (Figure 1.29a) through Schiff base condensation, and the generated Schiff base is further reacted with diethyl phosphite by the phosphorus–hydrogen addition reaction to yield phosphorus-containing vanillin-based bisphenols. The resulted bisphenol can be converted into diglycidyl derivative via the above-described reaction with an excess of epichlorohydrin, preferable under PTC conditions. The reactivity of the epoxy resins synthesized in this way is similar to the bisphenol A-based epoxy resin. After curing with a stoichiometric amount of 4,4-diaminodiphenylmethane, both resins showed excellent flame retardancy with UL-94 V0 rating and high LOI value of 31.4% (coupling with DDM) and 32.8% (coupling with PDA), due to their outstanding intumescent and dense

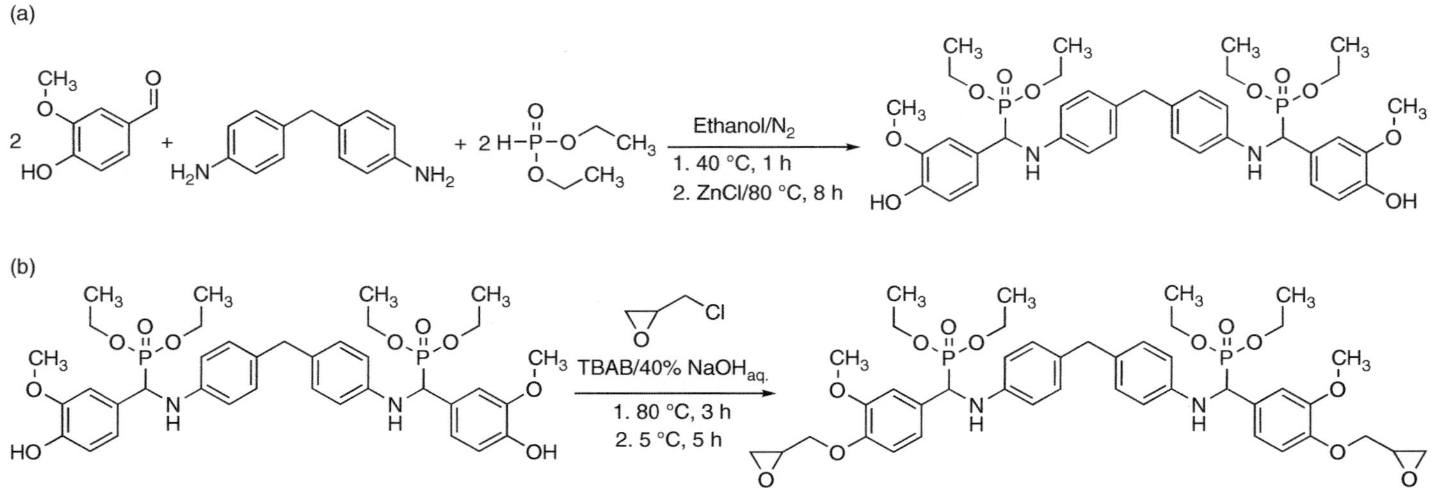

Figure 1.29 Synthesis of the vanillin coupling product (a) and the flame-retardant epoxy resin based on it (b).

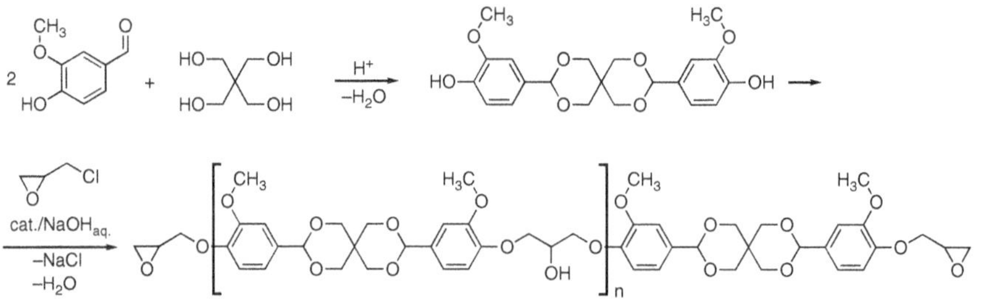

Figure 1.30 The coupling of vanillin with pentaerythritol and synthesis of the epoxy resins containing spiro-ring structure.

char formation ability. They also exhibit high glass transition temperature value of 183 °C (DDM) and 214 °C (PDA), the tensile strength of 80.3 MPa (DDM) and 60.6 MPa (PDA), and the tensile modulus of 2114 MPa (DDM) and 2709 MPa (PDA), much higher than the cured bisphenol A-based epoxy resin with a T_g of 166 °C, a tensile strength of 76.4 MPa, and a tensile modulus of 1893 MPa, respectively.

Two molecules of vanillin can also be coupled through the dehydration condensation with pentaerythritol, leading to obtain the bisphenol with the specific spiro-ring structure (Figure 1.30) [66], which can be further reacted with epichlorohydrin to give the epoxy resin.

This vanillin-based resin exhibits very interesting properties [107]. This solid resin with an epoxy value of 0.355 mol/100 g, cross-linked with diamine hardeners, DDM or 3,9-bis(3-aminopropyl)-2,4,8,10-tetroxaspiro(5,5)undecane, has several relaxations. The first is the β-relaxation, caused by the micro-Brownian motion of the aromatic methoxy group, observed from 50 to 100 °C for the spiro-ring-type resin systems in both mechanical and dielectric measurements. The peak height and the activation energy of this relaxation are independent of the degree of curing. The second one is the relaxation caused by the hydrogen bonding between the methoxy and the hydroxyl groups at around 0 °C [108]. This relaxation behavior is expected to have a positive effect on the damping characteristics. Moreover, the fracture toughness of the spiro-ring-type epoxide resin with methoxy branches is considerably greater above the temperature region of the β-relaxation than that of the bisphenol A type resin [109].

1.3.3 Cardanol

Cardanol is extracted from the shell of the cashew nut. Cashew nut comes from the cashew tree, *Anacardium occidentale*, mostly grown in India, East Africa, and Brazil [110]. The nut has a shell of about 1/8 in. thickness inside, which is characterized by a soft honey comb structure containing a dark brown viscous liquid, called cashew nut shell liquid (CNSL). Nut shells, depending on the extraction method used, contain about 30 wt% CNSL. The world production of CNSL is about one million tonnes annually [111]. CNSL is extracted from nuts using hot oil process; roasting process using solvents such as benzene, toluene, and

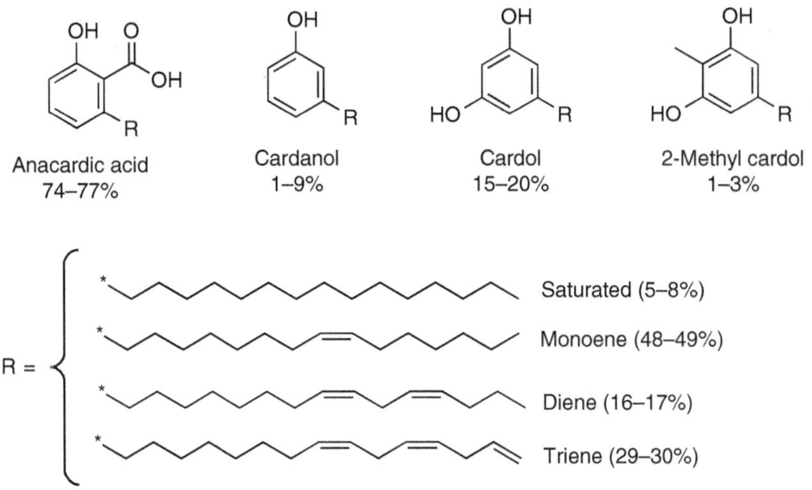

Figure 1.31 Schematic illustration of components of CNSL.

petroleum hydrocarbon; or supercritical extraction of oil using a mixture of CO_2 and isopropyl alcohol [112].

CNSL is a large and relatively cheap source of naturally occurring phenols [110]. The crude CNSL contains different long-chain phenols such as anacardic acid (3-*n*-pentadecylsalicylicacid), cardanol (3-*n*-pentadecylphenol), cardol (5-*n*-pentadecylresorcinol), and 2-methylcardol (2-methyl-5-*n*-pentadecylresorcinol) (Figure 1.31) [113].

There are various methods of purifying technical CNSL. Among them, it is worth mentioning two methods (i) column chromatography of CNSL and (ii) the method based on the formation of an amine-cardol and distillation of cardanol under high vacuum [114]. Cardanol of industrial grade is obtained throughout the thermal treatment of CNSL, followed by distillation. During that process, the decarboxylation of anacardic acid occurs, resulting in cardanol (about 90% purity) and a small quantity of cardol and methylcardol [114]. The diepoxidized cardanol (NC-514, Cardolite Corporation), obtained in a two-step process (Figure 1.32) of phenolation of aliphatic chain and then the reaction of phenol hydroxyl groups with epichlorohydrin in basic conditions, with $ZnCl_2$, at 95 °C, is an example of commercially available cardanol [115, 116].

Cardanol is a nonedible by-product of CNSL industry, and it is known as a promising aromatic renewable source, which is available in large scale. It is a yellow liquid composed of four meta-alkyl phenols differing by the unsaturation degree of aliphatic chain: 8.4% saturated, 48.5% monoolefinic (8), 16.8% diolefinic (8,11), and 29.3% triolefinic (8,11,14).

The presence of long aliphatic alkyl chain in the *meta*-position of the phenolic ring within the cardanol molecule is responsible for attractive properties such as good processability and high solubility in organic solvents, but it also influences many chemical transformations (Figures 1.33 and 1.34).

Cardanol is used as an interesting substitute to bisphenol A. For instance, cardanol-based epoxy resin is synthesized and tested as a replacement of

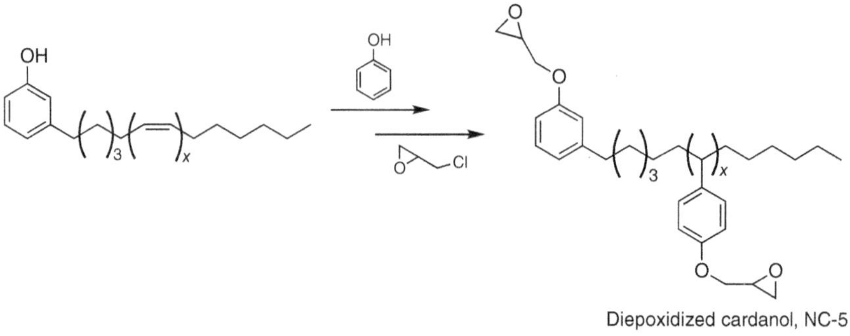

Figure 1.32 Synthesis of NC-514.

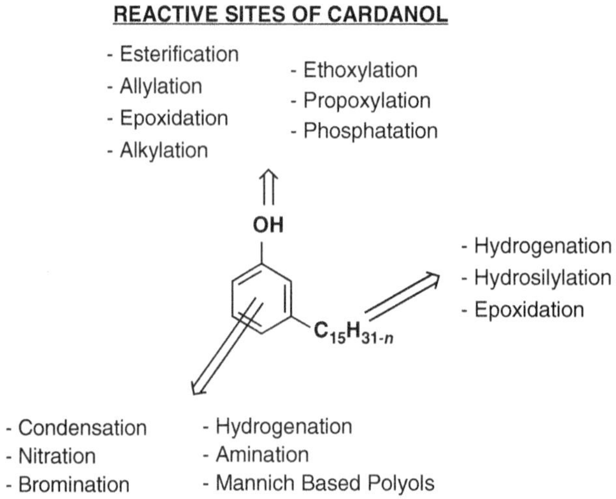

Figure 1.33 Reactive sides of cardanol.

BPA-based epoxy in coating applications [117]. The triglycidyl resin (TGC) is obtained from acetyl cardanol via a two-step reaction (Figure 1.35).

In the first step, the acetyl cardanol molecule is modified with maleic anhydride, followed by hydrolysis of the anhydride ring to prepare carboxyl functional cardanol (CFC). In the second step, the CFC reacts with excess of epichlorohydrin to prepare TGC. During that synthesis, various reactions occur: (i) grafting of maleic anhydride moiety on the aliphatic chain facilitated by hydrogen transfer, (ii) thermal rearrangements of the nonconjugated double bonds, resulting in the formation of conjugated double bonds, (iii) Diels–Alder reaction of newly created conjugated double bonds with maleic anhydride forming a Chroman ring, (iv) addition reaction by proton transfer mechanism of cardanol double bonds, and (v) radical polymerization of the double bonds resulting in a polymeric structure.

Further, the prepared TGC is used as a binder in combination with the BPA-based epoxy at various weight ratios (formulations containing 40–90% of TGC on mass basis), and it is cured with different amine hardeners: isophorone diamine (cycloaliphatic amine), 4,9-dioxadodecane-1,12-diamine (long chain

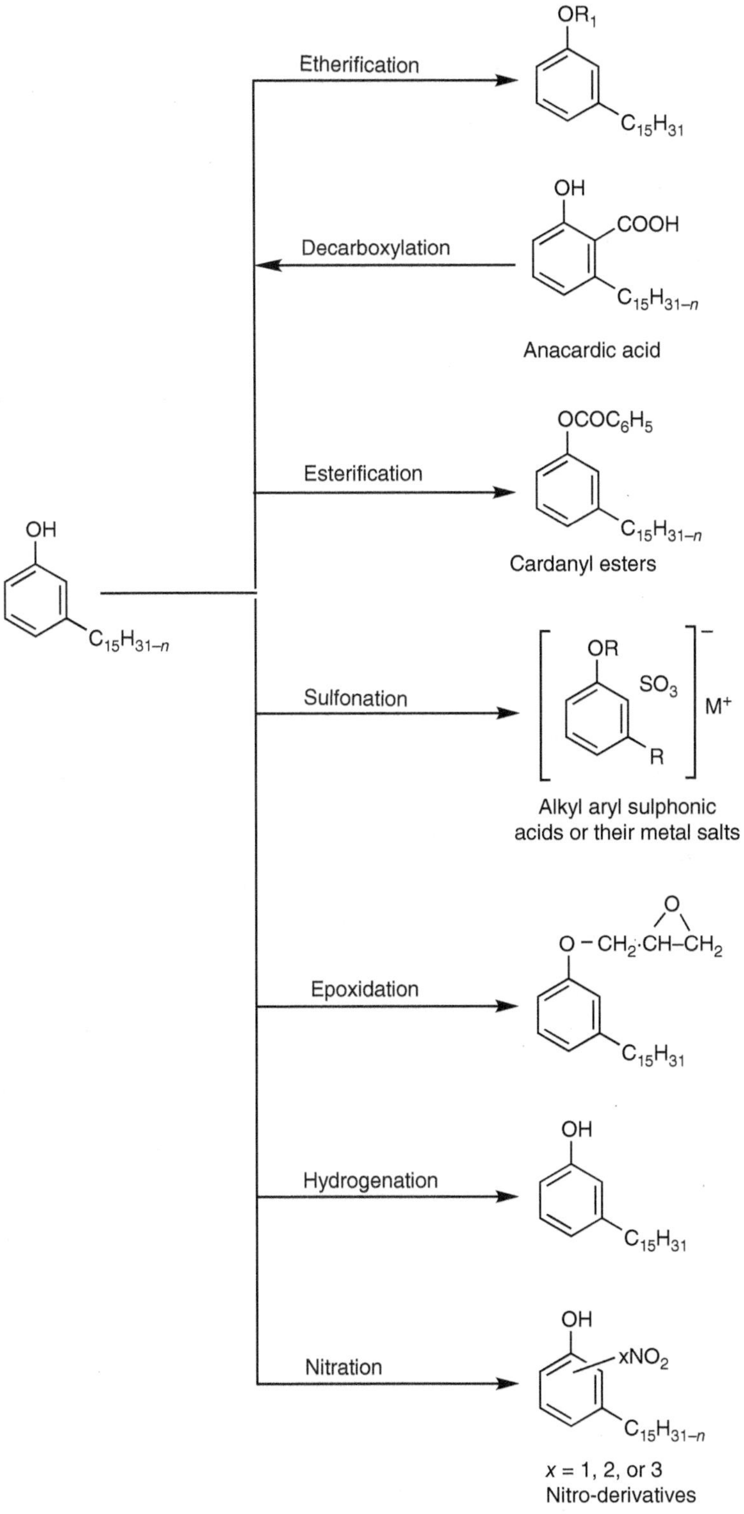

Figure 1.34 Chemical transformation of cardanol.

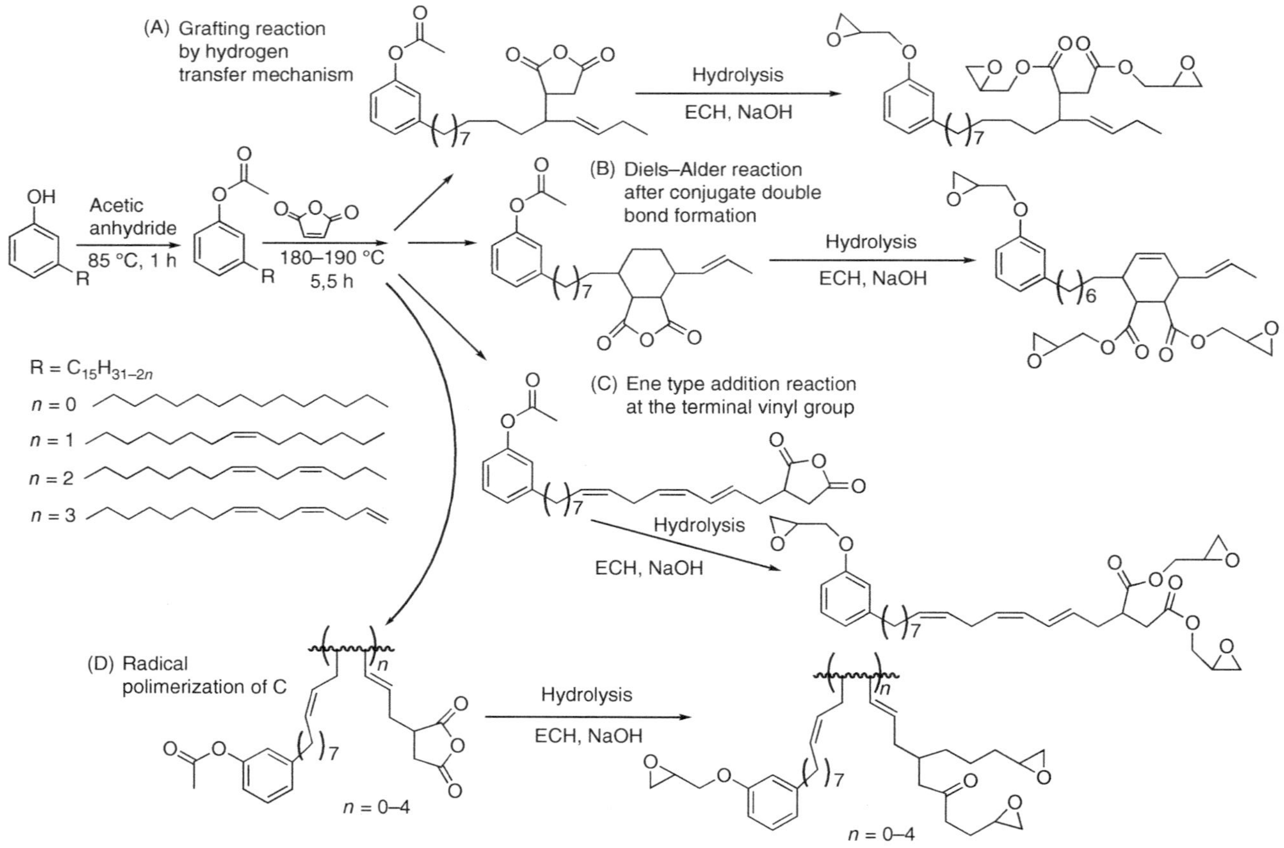

Figure 1.35 Synthesis of carboxyl functional cardanol and structures of the glycidyl products formed via the reaction of CFC and epichlorohydrin.

amine), and Replamide 325 (polymeric amine). Samples containing 40–60 wt% of DGEBA are characterized by comparable properties to that of completely DGEBA-based system, resulting from a good balance between soft and hard segments in the cross-linked structure. Additionally, the coatings with TGC are characterized by slightly lower thermal stability at higher temperatures than DGEBA-based system ($T_{10\%}$ = 340 °C, while for the coatings containing TGC, $T_{10\%}$ = 280 °C). The aromatic content in the case of DGEBA is higher than TGC and hence gives relatively higher thermal stability of tested samples.

NC-514, a mix of polymers of epoxidized cardanol (Figure 1.36), is used to obtain cardanol-based materials [114].

In order to synthesize epoxy networks, epoxidized cardanol is reacted with isophorone diamine and Jeffamine D400 diamine. Mass loss temperatures of cardanol networks are slightly lower than the DGEBA-based material. Additionally, the polymers synthesized from epoxidized cardanol NC-514 and IPDA exhibit lower T_g values, between 41 and 50 °C (T_g = 158 °C for DGEBA/IPDA 1 : 1, respectively). Moreover, the cross-linking density of NC-514/IPDA and NC-514/Jeff400 is almost five times lower than the cross-linking density of DGEBA/IPDA. The decrease of cross-linking density, in comparison to DGEBA, is correlated with the presence of long aliphatic chain within the structure of NC-514, which increases the distance between the epoxy groups.

In order to replace bisphenol A in epoxy networks, the application of bio-based saccharides (sorbitol and isosorbide) as epoxidized reactants in epoxidized cardanol-based resins is an interesting approach for achieving higher thermomechanical properties of cardanol-based materials [118]. Several mixtures of diglycidyl ether of cardanol NC-514 (EEW = 400 g/eq.) at different ratios (100, 75, 50, and 25 wt%) with various epoxy reactants, polyglycidyl ether of sorbitol Denacol EX622 (EEW = 188 g/eq.), and two epoxy reactants of isosorbide, Denacol GSR100 (EEW = 155 g/eq.) and Denacol GSR102 (EEW = 158 g/eq.) (Figure 1.37), are cured at room temperature with two commercial amines: trifunctional polyetheramine (Jeffamine T403) and isophorone diamine (Aradur 42BD).

Based on the results, it was stated that even if epoxidized cardanol-based networks cannot fully replace DGEBA materials, the combinations of epoxy cardanol with other bio-based epoxy reactants can alter the thermomechanical properties of cardanol-based networks. The obtained materials are characterized by properties comparable to bisphenol A-based materials. T_g and hardness in Shore A scale of samples of epoxidized sorbitol-cardanol at a 1 : 1 weight ratio are higher than that of the same ratio of epoxidized isosorbide (60 °C and 93 as compared to 48 °C and 91). Material based on epoxidized isosorbide-cardanol (3 : 1) exhibits T_g of 83 °C (isosorbide 100, with EEW = 155 g/eq.) and 62 °C (isosorbide 102 with EEW = 158 g/eq.), respectively.

Moreover, cardanol derivative obtained throughout the reaction of cardanol with 9,10-dihydro-9-oxa-10-phosphaphenanthrene-10-oxide (DOPO) (Figure 1.38) might be used as the flame-retardant agent [119].

A triscardanyl phosphate (PTCP) containing phosphaphenanthrene groups is synthesized via dehydrochlorination, epoxidation, and ring opening reaction of cardanol (Figure 1.39).

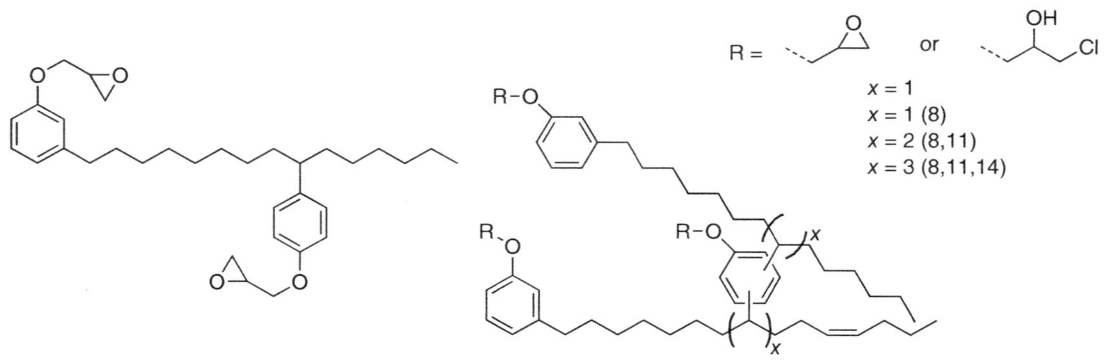

Figure 1.36 Chemical structure of cardanol NC-514.

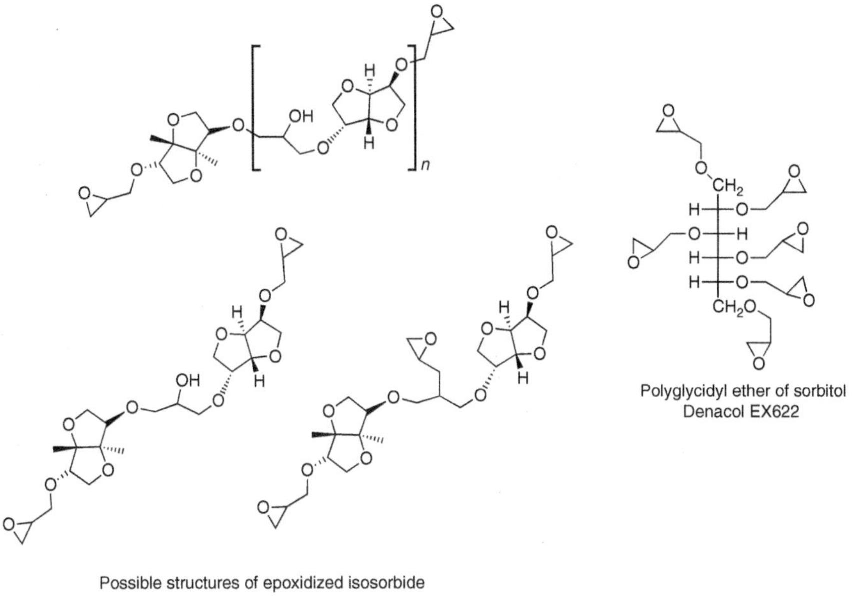

Figure 1.37 Epoxy reactants: epoxidized isosorbide and polyglycidyl ether of sorbitol.

Figure 1.38 Chemical structure of DOPO.

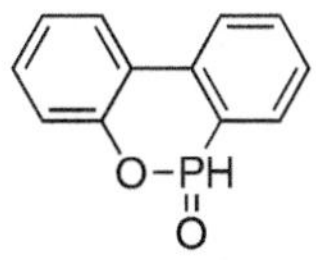

Based on the performed studies (Table 1.5), it was observed that the incorporation of PTCP into DGEBA epoxy resin accelerates the thermal degradation process and improves the char yield.

The same higher char yield is valuable toward improving the flame-retardant properties of epoxy resins. The LOI value of EP/PTCP-30% increases from 23.0% (neat epoxy material) to 30.5%. Moreover, compared with neat epoxy resin, the impact strength of EP/PTCP-30% increases by 29%.

Cardanol is an interesting nonedible by-product of CNSL industry. Because of the presence of phenolic hydroxyl group, olefinic linkages in the alkyl chain, and aromatic ring, it is a promising nonharmful renewable substituent to BPA. The presence of long aliphatic chain of cardanol in bio-based epoxy resins results in lower T_g values of obtained materials than those with DGEBA. At the same time, these materials are characterized by very interesting thermal stabilities. Moreover, the mechanical properties of cardanol-based epoxies are lower than DGEBA epoxies. However, numerous, described in the literature, studies tested coating applications of epoxies obtained with the cardanol derivatives.

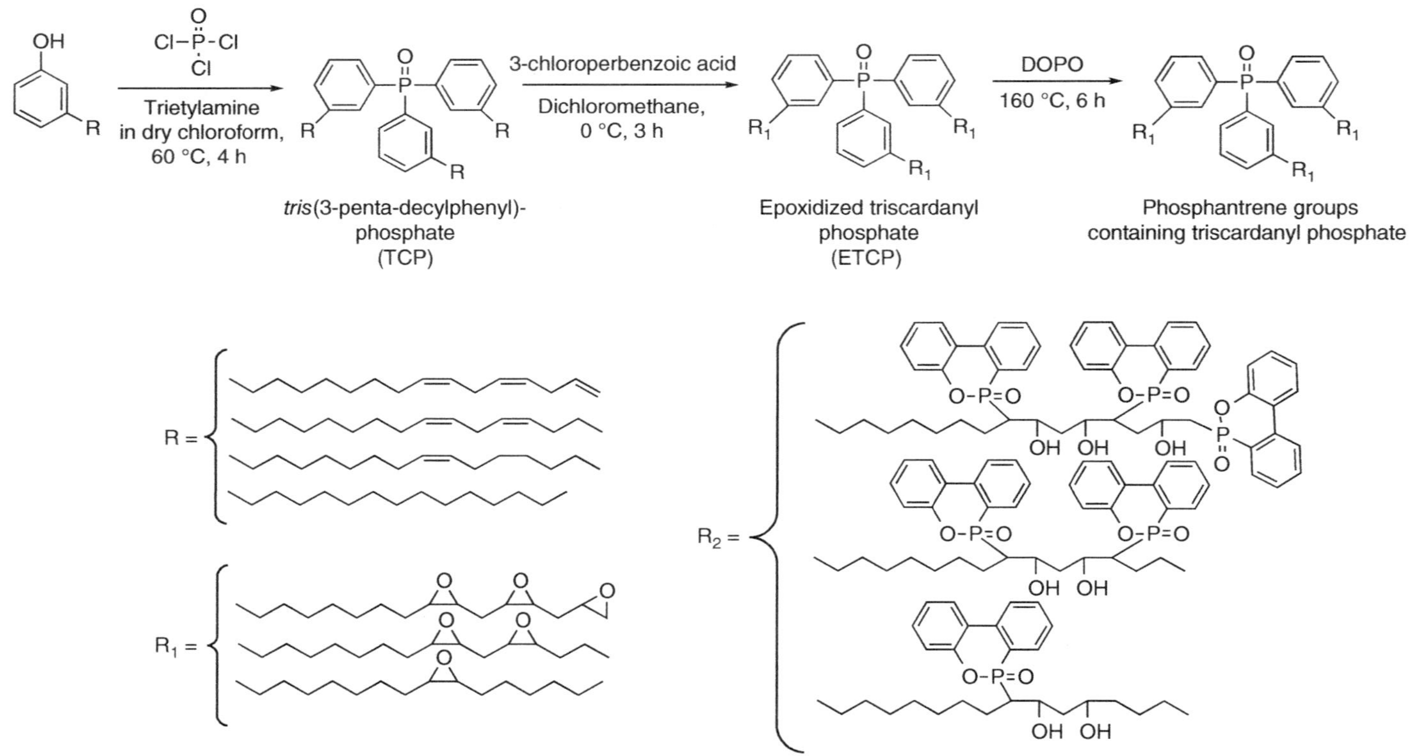

Figure 1.39 Synthesis of PTCP.

Table 1.5 Thermal and mechanical properties of PTCP, neat epoxy, and EP/PTCP samples.

Sample	Thermal properties					Mechanical properties			
	$T_{10\%}$ (°C)	T_{max1} (°C)	T_{max2} (°C)	Char residue (%)	LOI (%)	Impact strength (kJ/m^2)	Tensile modulus (GPa)	Tensile strength (MPa)	Elongation at break (%)
PTCP	286	322	500	2.8	—	—	—	—	—
EP	361	367	554	1.4	23.0	14.9 ± 1.1	1.56 ± 0.10	40.6 ± 2.5	2.2 ± 0.8
EP/PTCP-10%	336	344	554	4.2	26.5	16.8 ± 1.0	1.35 ± 0.00	46.3 ± 3.4	5.5 ± 01
EP/PTCP-20%	328	344	575	5.7	28.0	18.1 ± 1.4	1.46 ± 0.03	60.8 ± 4.4	7.7 ± 1.8
EP/PTCP-30%	311	335	577	8.3	30.5	19.1 ± 0.5	1.09 ± 0.09	46.7 ± 1.1	8.2 ± 0.4

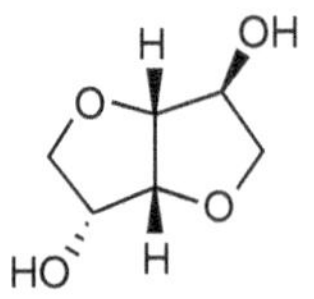

Figure 1.40 Chemical structures of isosorbide.

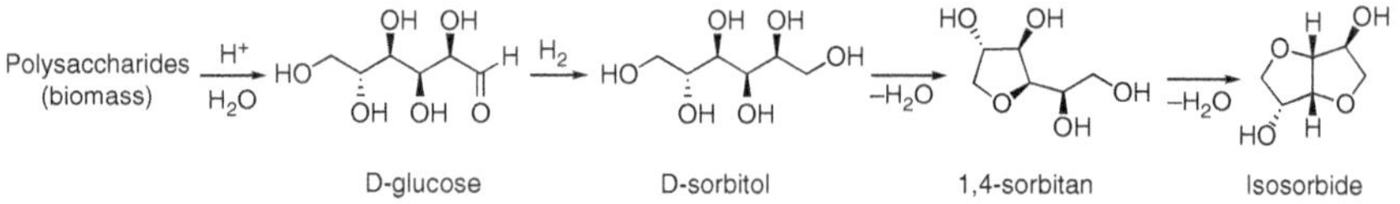

Figure 1.41 Schematic reaction pathway for the production of bio-based isosorbide.

1.3.4 Isosorbide

Isosorbide (1,4:3,6-dianhydro-D-glucitol) is an organic oxygen-containing heterocyclic compound composed of two fused furan rings (Figure 1.40).

Isosorbide does not occur naturally. It can be obtained from various raw materials by organic synthesis through different reaction pathways. However, nowadays, isosorbide is produced from diverse polysaccharides through the multistep process that includes several intermediates [120] (Figure 1.41).

The synthesis of isosorbide can start from polysaccharides (mainly starch or cellulose biomass) or directly from all of the intermediate compounds because they are currently commercially available on a large scale. Lignocellulosic biomass is considered as one of the best resource due to its abundancy, versatility, and price [121]. The synthetic process of isosorbide production from polysaccharides consists of the following stages: acid-catalyzed hydrolysis of the glycosidic bonds in the polymeric carbohydrates, hydrogenation of obtained glucose to sorbitol and further dehydration to sorbitan, and finally dehydration of sorbitan to isosorbide. Moreover, the different side reactions (such as degradation or polymerization) can occur in this complex process. Therefore, different new synthesis strategies (including the one-step synthesis from glucose and cellulose) and catalysts are still elaborated and proposed [122].

Isosorbide can be derivatized by functionalization or substitution of the two secondary hydroxyl groups present in its molecule. A certain difficulty is the different reactivity and steric hindrance of the hydroxyl groups (against, due to these, a selective monoderivatization is also possible). Nevertheless, isosorbide is currently converted to valuable derivatives for use as pharmaceuticals, detergents, emulsifiers for cosmetics, surfactants, stabilizers, or plasticizers. In addition to the above-mentioned applications, for a long time, isosorbide is considered and intensively studied as a potential monomer building block for the biopolymeric materials (e.g. polycarbonates, polyurethanes, or polyesters). As the bio-based diol with bicyclic rigid structure isosorbide can also be applied as a substitute for bisphenol A replacement, there are generally two ways for introducing an epoxy functionality into the isosorbide molecule (Figure 1.42): conversion to the diallylic derivative, followed by oxidation to oxirane rings (the first pathway) and reaction with epichlorohydrin (the second and third pathways).

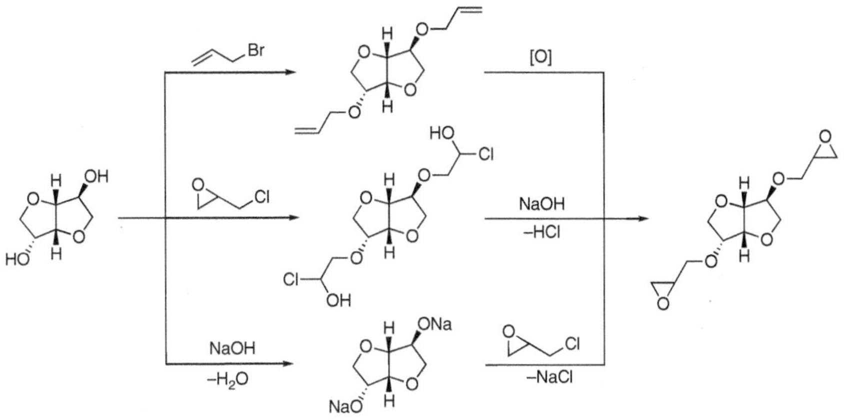

Figure 1.42 Possible reaction pathways for the synthesis of the diepoxide derivative of isosorbide.

Figure 1.43 Structure of isosorbide-based epoxy resins.

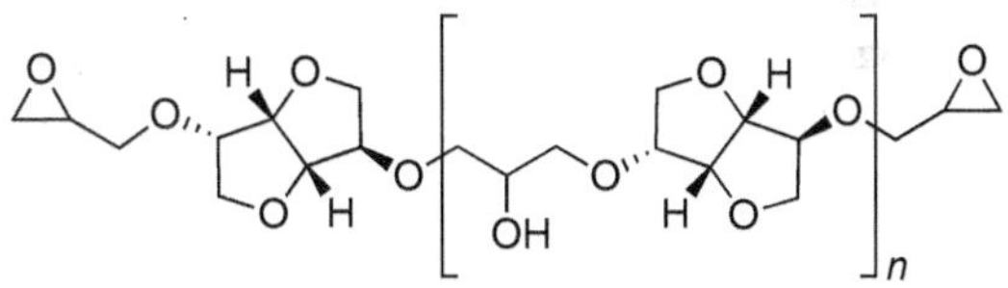

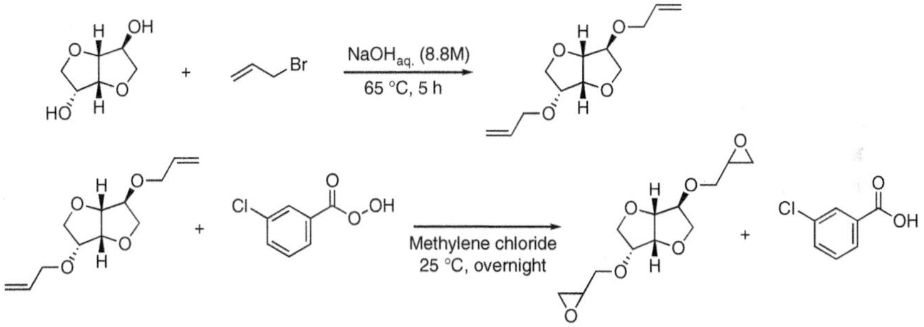

Figure 1.44 Synthetic route of diallyl isosorbide and isosorbide diglycidyl ether.

In fact, the chemical structure of the diglycidyl isosorbide derivative shown in Figure 1.42 is valid only for the first reaction pathway. Only by the reaction with allyl bromide, followed by oxidation of the terminal unsaturated bonds, it is possible to obtain the product containing only one isosorbide molecule (the monomer). For the remaining reaction pathways, even the use of a large excess of epichlorohydrin leads to the formation of the products with an oligomer structure (Figure 1.43).

According to the first reaction pathway, the diallyl ether can be prepared by heating the isosorbide with allyl bromide in sodium hydroxide solution (Figure 1.44) [123].

Oxidation of diallyl isosorbide ether to isosorbide diglycidyl ether is carried out in the reaction with *meta*-chloroperbenzoic acid as an oxygen carrier. It is also possible to use 4-allyoxybenzoly chloride to introduce the diallyl functionality (Figure 1.45) [124].

The next stage of the reaction is carried out using the *meta*-chloroperbenzoic acid, as described above.

The synthesis of isosorbide-based epoxy resins using epichlorohydrin (Figure 1.43) can be accomplished in a different manner. Isosorbide can be reacted with a large excess (even a 10-fold) of epichlorohydrin in the presence of strong alkali–sodium hydroxide (40% aq. solution) in a single-stage reactor with continuous removal of water [125]. The reaction is carried out at a temperature of 109–115 °C for eight hours and the product with an epoxy value of 0.451–0.467 mol/100 g is obtained. In another method, sodium hydride as a base in diglyme is used to prepare the disodium salt of isosorbide (reaction time about six hours at a temperature of 43–48 °C and then one hour at a temperature of 85 °C) [126], which is then reacted with nearly 20-fold excess of epichlorohydrin (six hours of dropwise addition, leaving overnight at room temperature, and heating for 2.5 hours at a temperature of 55 °C). The resulted product is obtained with an epoxy value of 0.359 mol/100 g. Sodium hydroxide (50% aq. solution) can be used instead of sodium hydride [127], also in the two-step method. The disodium salt of isosorbide synthesis is catalyzed by trimethylcetylammonium bromide. In the second step, the disodium salt is reacted with the 10-fold excess of epichlorohydrin in the presence of another phase transfer catalyst – tetrabutylammonium bromide (heating at a temperature of 115 °C for about three hours is carried out). The obtained isosorbide-based epoxy resin has an epoxy value of 0.518 mol/100 g.

It is also possible to react isosorbide with epichlorohydrin in a two-stage process under different conditions [128]: using an aqueous alkali (mostly sodium/potassium hydroxide) and an organic solvent (toluene and dimethyl acetamide) or the first stage is carried out in the presence of the Lewis acid catalyst (stannous fluoride).

The cross-linked bio-based epoxy resin (an epoxy value of 0.440 mol/100 g), synthesized in the one-step reaction from isosorbide and epichlorohydrin [129] in the presence of 50% aqueous NaOH, exhibits properties comparable to those of the commercial bisphenol A-based resin Epidian 5 (an epoxy value of 0.510 mol/100 g). Depending on the cross-linking agent used (triethylenetetramine, isophoronediamine, tetrahydrophthalic, and phthalic anhydrides), the selected mechanical properties of the isosorbide-based resin are in some cases even better (the flexural and compression strengths and the Brinell hardness). Also, the Izod impact strength of this resin is usually better than that of the cross-linked resin Epidian 5 (even more than four times). Water sorption of the isosorbide-based resin is much higher than that of the bisphenol-A-based resin (for the sample cross-linked with triethylenetetramine, even their disintegration is observed), and as a result, the chemical resistance is less than that for the resin Epidian 5.

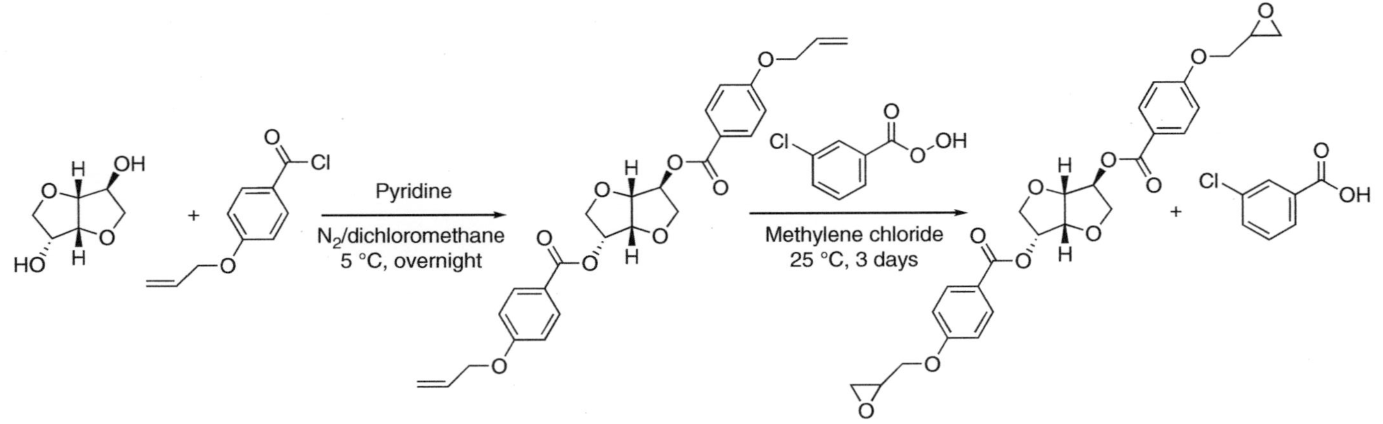

Figure 1.45 Synthesis of isosorbide diglycidyl ether using 4-allyoxybenzoly chloride.

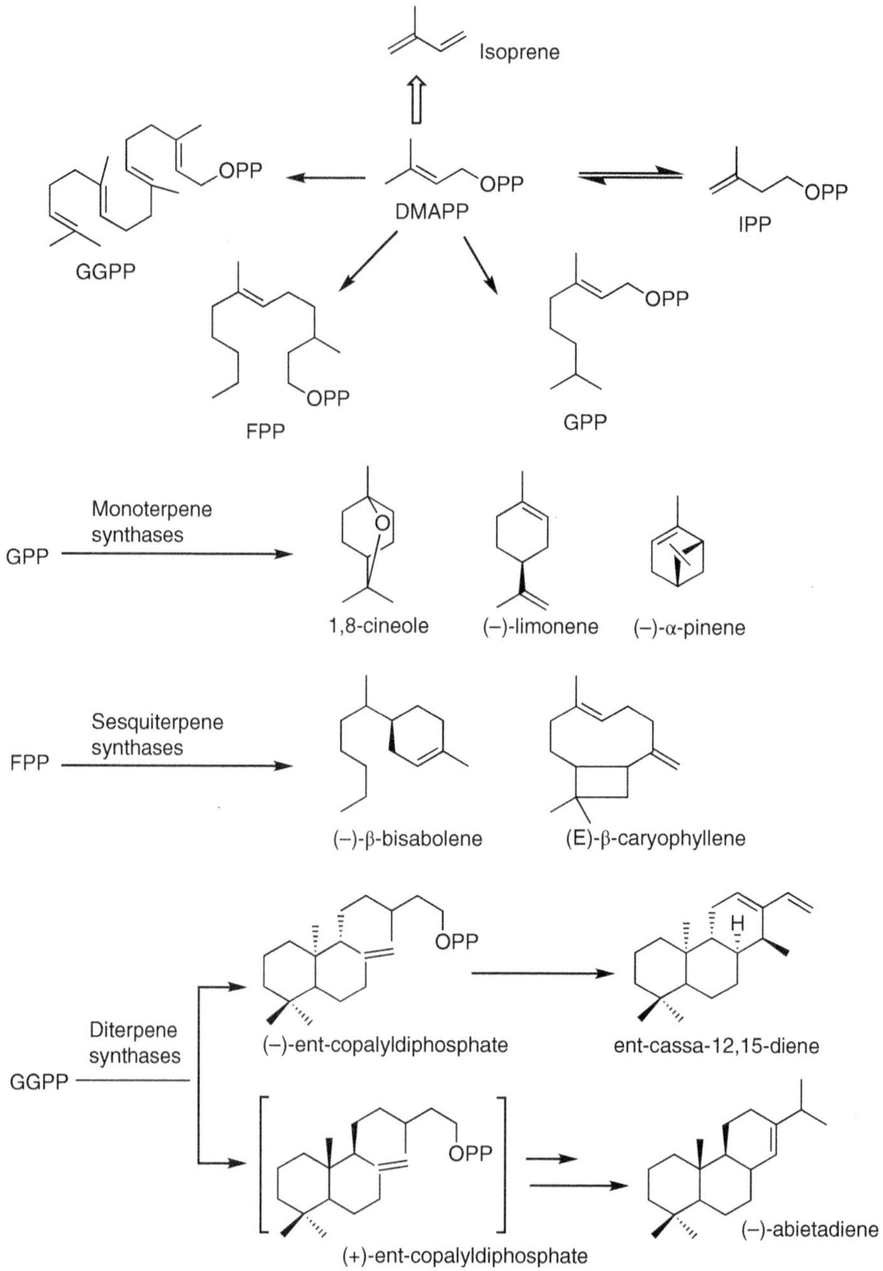

Figure 1.46 Formation of plant terpenes.

1.3.5 Terpene Derivatives

Terpenes and terpenoids are an interesting group of natural resources, with relatively large potentials as substrates for the synthesis of various polymers. They are unsaturated aliphatic structures, predominantly derived from turpentine, the volatile fraction of resins exuded from conifers [130]. The C5-units of isopentenyl diphosphate (IPP) and its isomer dimethylallyl diphosphate (DMAPP) are the initial substrates for the synthesis of terpenes. It is worth highlighting here that the linear prenyl diphosphates: geranyl diphosphate (GPP, C10), farnesyl diphosphate (FPP, C15), and geranylgeranyl diphosphate (GGPP, C20) are precursors of terpenes obtained via the biosynthesis in the presence of prenyltransferases (Figure 1.46). Due to the fact that terpenes consist of multiple isoprene units (C_5H_8, 2-methyl-1,4-butadiene), they are categorized based on the number of units into hemiterpene (one isoprene unit, C5), monoterpene (two units, C10), sesquiterpene (two units, C15), diterpene (four units, C20), and so on [131] (Figure 1.46) [132].

Terpenes are built from a wide range of cycloaliphatic and hydrocarbon chain structures with repeating isoprenyl double bonds. Terpenoids can be viewed as modified terpenes with added, missing, or shifted methyl and oxygenated functional groups (Figure 1.47).

The global turpentine production is more than 300 000 tons per year, and it includes α-pinene (45–97%) and β-pinene (0.5–28%), with smaller amounts of other monoterpenes [130]. The pinenes and limonenes are natural chemical precursors to a wide variety of compounds used in the pharmaceutical, fragrance, and flavor industry. Pinenes, additionally, are a source of other less common terpenes (Figure 1.48).

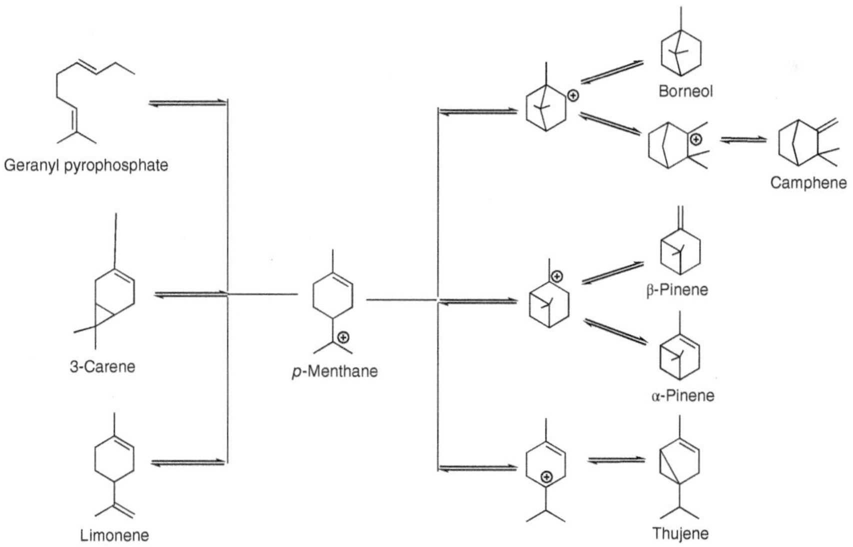

Figure 1.47 Outline of the biosynthetic pathway leading to the major monoterpenes from a single precursor.

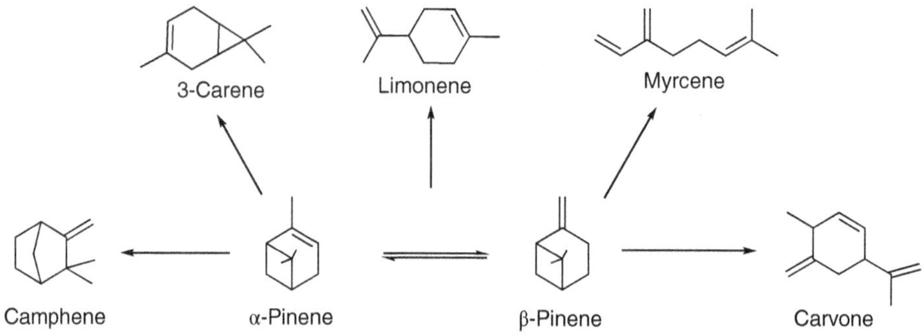

Figure 1.48 Isomerization and oxidation processes for converting pinenes into other terpenes and a terpenoid.

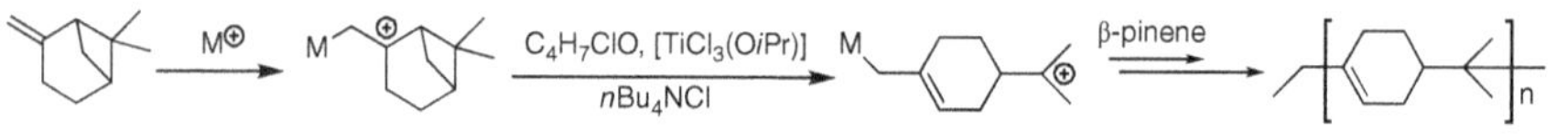

Figure 1.49 Mechanism of the cationic polymerization of β-pinene.

A methyl group or other electron-donating groups on the double bond present within the structure of compounds obtained via the isomerization and oxidation of pinenes makes them susceptible for cationic polymerization. However, because of the presence of highly reactive *exo*-methylene double bond within the β-pinene structure, most of the polymerization reactions involve β-pinene, not α-pinene (Figure 1.49).

Some cationic polymerization of terpenes leads to oligomers or low molecular weight polymers, which along with other terpene monomers might be used for the synthesis of epoxy resins. One of the interesting applications of terpenes is the synthesis of hydrogenated terpinene-maleic ester type epoxy resin [133]. A bio-based epoxy resin (denoted as TME) and a waterborne dispersion of TME (denoted as WTME) obtained from the turpentine (Figure 1.50) are nontoxic alicyclic structure epoxy resins without BPA.

The resulting film with good thermal stability and antifouling properties is transparent and flexible. However, because the obtained cured terpene-based products are characterized by worse mechanical properties than those of BPA-based epoxy resin, two different approaches had also been studied: (i) incorporation of cellulose nanowhiskers (CNWs) suspension, hydrolyzed from microcrystalline cellulose [134], and (ii) in combination with polyurethane [133]. In the first modification of the synthesis, the incorporation of CNWs in the WTME matrix (0.5–8 wt%) results in the increase of the storage modulus at 150 °C (from 0.8834 to 4.756 MPa), Young's modulus (from 295.6 to 800.1 MPa), and tensile strength (from 7.08 to 15.2 MPa) compared to unmodified WTME. Noted improved properties might be attributed to the formation and increase of interfacial interaction by hydrogen bonds between CNWs nanofiller and the WTME matrix. On the other hand, in the second approach (Figure 1.51), an anionic polyol (T-PABA) dispersion is prepared by modifying terpene-based epoxy resin with *para*-aminobenzoic acid and then cross-linked

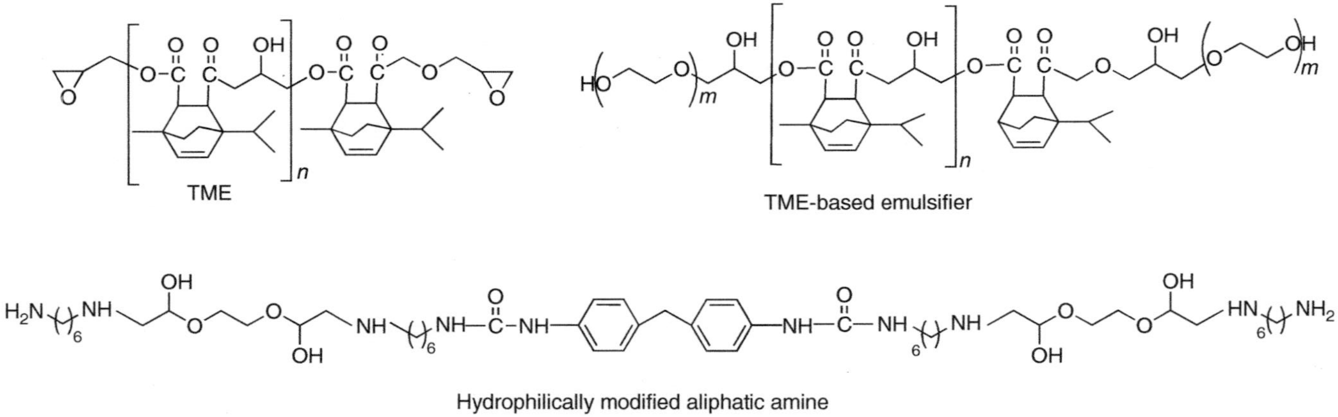

Figure 1.50 Chemical structures of TME, TME-based emulsifier, and aliphatic amine.

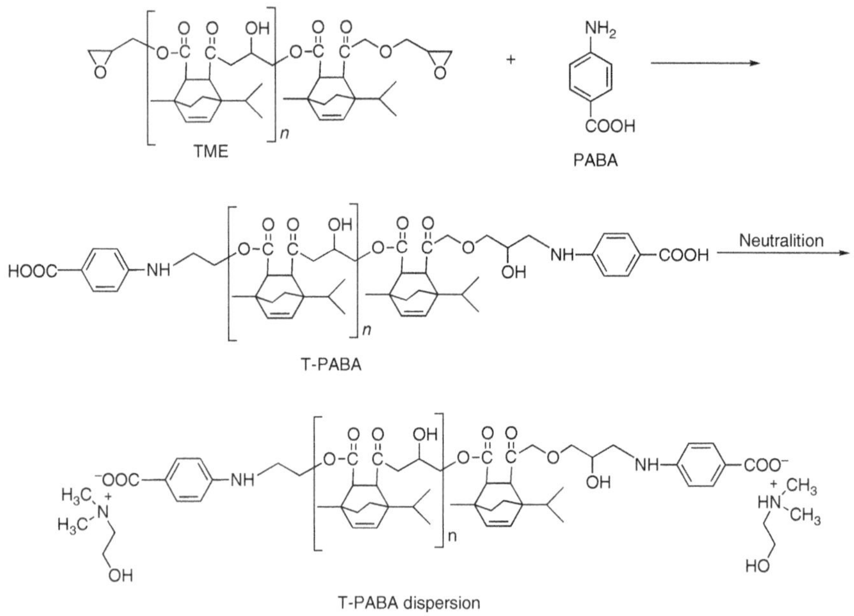

Figure 1.51 Preparation of T-PABA and T-PABA dispersion.

with a hexamethylene diisocyanate (HDI) tripolymer to prepare waterborne polyurethane/epoxy resin composite coating. These new cross-linked products combine the rigidity and weatherability of the saturated terpinene alicyclic epoxy resin with the flexibility and tenacity of the polyurethane.

On the other hand, the oxidation of limonene, the terpene derivative, which within the structure contains two double bonds, (i) vinylene group allocated in the ring and (ii) vinylidene side group, results in limonene monoepoxide and diepoxide – compounds commercially applied as reactive diluents in epoxy applications [135]. Additionally, D-limonene might also be used in the synthesis of epoxy resins as a bio-based replacement for conventional DGEBA resins [136]. Hybrid epoxy resin, including naphthalene and limonene moieties, is obtained on the course of the three-step reaction (Figure 1.52) consisting of (i) alkylation of naphthol with limonene in the presence of Friedel–Crafts catalyst, (ii) introduction of methylene linkage between naphthalene rings to obtain product **2** with higher molecular weight, and subsequently (iii) epoxidation of **2** with epichlorohydrin in the presence of sodium hydroxide and polyethylene glycol to give epoxy resin **3**.

The obtained epoxy resin is mixed with dicyanodiamide and a bisphenol A formaldehyde novolac resin used as curing agents in a molar stoichiometric ratio of 1 : 1 and in the presence of 2-methylimidazole as an accelerator. Compared to DGEBA resins, the cured D-limonene/naphthol-based epoxy resins are characterized by higher T_g (by 75 °C) and thermal stability with higher temperatures of maximum rate of weight loss in air by about 40 °C.

Epoxy resins might also be synthesized from rosin, produced by heating fresh tree resin to remove the volatile liquid terpenes [137]. Because of the presence of

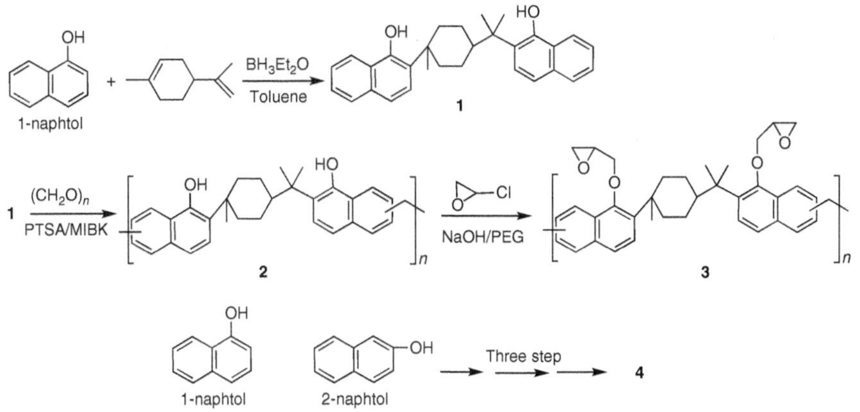

Figure 1.52 Synthesis of epoxy resins starting from naphthol and limonene.

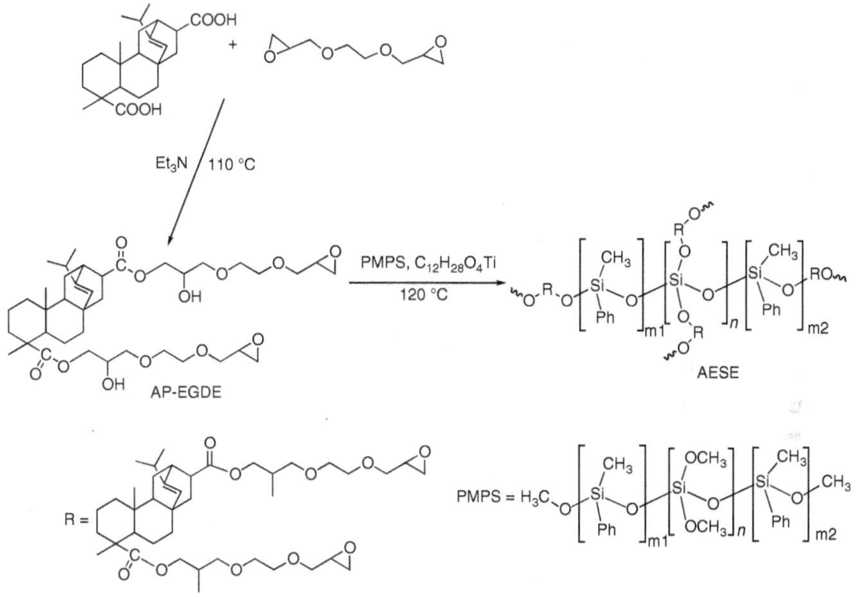

Figure 1.53 Synthetic route to a rosin-based siloxane epoxy monomer (AESE).

rigid hydrogenated phenanthrene ring in the molecular structure, rosin is suitable as an alternative to DGEBA [138, 139]. The interesting examples are the flame-retardant rosin-based epoxy thermosets, such as a rosin-based siloxane epoxy monomer (AESE), which is prepared by the reaction of ethylene glycol diglycidyl ether modified acrylpimaric acid (AP-EGDE) with polymethylphenylsiloxane (PMPS) (Figure 1.53) [140].

The incorporation of PMPS results in the improvement in the LOI value compared to the AP-EGDE/MHHPA thermosets. The highest LOI value of 30.2% is noted for the cured product containing 30 wt% of AESE (AESE30/MHHPA). Because of the presence of the flexible chains of the PMPS, all the AESE/MHHPA thermosets display a relatively low tensile strength (<15 MPa) but on the other hand much larger elongation at break (>50%).

Terpenes are an interesting group of raw materials that might be used in the synthesis of epoxy resins. Even though not all terpenoids contain aromatic and/or phenolic moieties, these requirements can be reached via different synthesis steps (e.g. carvacrol, for instance, might be obtained from other turpentine components, limonene, for instance, throughout an oxidation, followed by an isomerization process).

1.4 Bio-Based Epoxy Curing Agents

Cross-linking of epoxy resins takes place with the participation of oxirane rings present in them. The strained structure of these rings is the reason for their high reactivity and facilitates their opening under the influence of very different factors. Epoxy resin curing generally takes place under the influence and with the participation of multifunctional chemical compounds with active protons. These are mainly polyamines (aliphatic, aromatic, and cycloaliphatic) and carboxylic compounds (acids and anhydrides). Cross-linking of the epoxy resins can also be accomplished by a mechanism of ring-opening polymerization using suitable catalysts. This is an important method of cross-linking, especially coating materials. However, the search for compounds of natural origin, which could replace petrochemical hardeners, is the most important in the case of polyamines and carboxyl compounds that are used for cross-linking in stoichiometric ratios to the content of epoxy groups in the resin. The consumption of ring-opening polymerization catalysts is insignificant compared to the polyamine and carboxylic hardeners. Usually, they are used in an amount of up to several percents by weight relative to the weight of the cross-linked resin. Therefore, new biohardeners should be sought first and foremost among the compounds of natural origin with amine or carboxylic functionalities. The appropriate reactivity and adequate miscibility with epoxy resins are the conditions, which are limiting the use of these new compounds.

Considering the possibilities of using plant oils as raw materials for the synthesis or modification of epoxy resins, an interesting solution would be just to give the plant oils proper functionality so that they can also act as natural hardeners. The use of modified vegetable oils in this role is favored by their very good miscibility with the epoxidized vegetable oils and good miscibility with the epoxy resins based on bisphenol A (despite the differences in the polarity of both groups of materials). For example, hydrolyzed castor oil can serve as a source of dehydrated fatty acids from which reactive polyamides are obtained by reaction with acrylic acid [141] (Figure 1.54) and then with various polyamines (diethylenetriamine, triethylenetetramine, and tetraethylenepentamine) [17].

The obtained polyamides (with amine values from 310 to 389 mg KOH/g) can be used to cross-link the epoxy resins based on bisphenol A giving the materials with good coating properties.

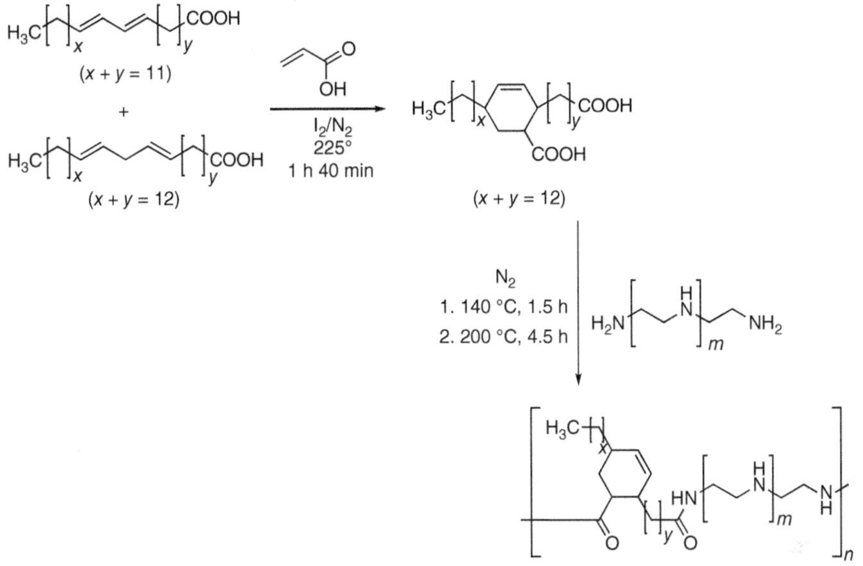

Figure 1.54 Synthesis of C_{21} cycloaliphatic dicarboxylic acids and reactive polyamides from them.

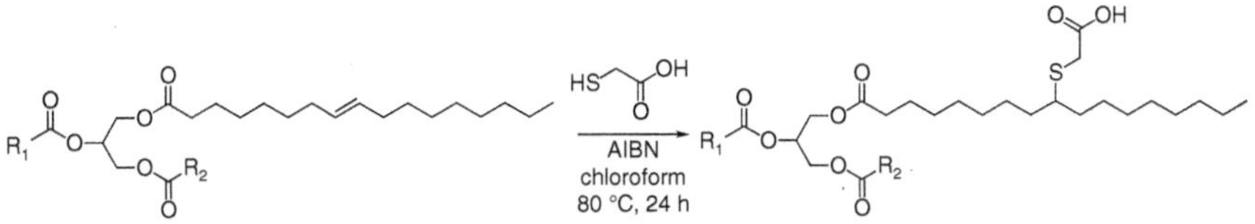

Figure 1.55 Scheme of soybean oil functionalization with thioglycolic acid.

Vegetable oil-based curing agents can be obtained through direct oil modification. The novel bio-based polyacid hardener is synthesized by the thiol-ene coupling of soybean oil with thioglycolic acid (Figure 1.55) [142].

The synthesized soybean oil-based polyacid exhibit a functionality of 3.3 acid functions per triglyceride molecule. This polyacid curing agent is characterized by the high reactivity toward epoxy groups. The low molecular weight bisphenol A-based epoxy resin cross-linked with the bio-polyacid exhibits interesting properties for coating and binders (the Shore A hardness of 52 and $E' = 0.59$ MPa). A commercially available mercaptanized soybean oil (prepared by direct addition of H_2S to soybean oil) reacted with allylamine (Figure 1.56) or its salts can also be potentially used as a cross-linking agent for epoxy resins [143].

Modified lignin derivatives are also applied as curing agents for the epoxy resins. Commercially, various curing agents for epoxy resins are available. The commonly used petroleum-based curing agents include amines, amides, hydroxyls, acid anhydrides, phenols, and polyphenols. However, recently, a great effort has been put on obtaining new bio-based hardeners for epoxy systems. Generally, lignin-based curing agents are prepared by two different methods: (i) the reaction of lignin with ozone in the presence of NaOH to give lignin with

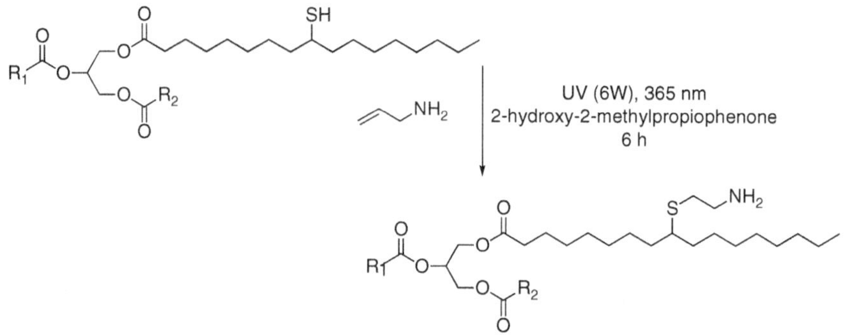

Figure 1.56 Synthesis of a polyamine cross-linking agent via thiol-ene reaction of mercaptanized soybean oil with allylamine.

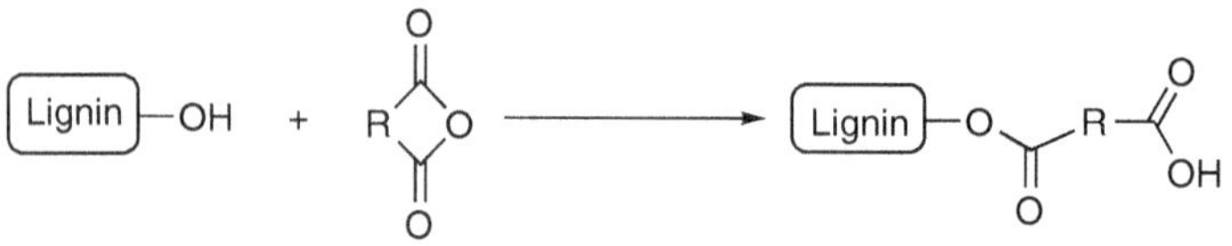

Figure 1.57 Synthesis of carboxylic acid from lignin.

unsaturated carboxyl groups (Figure 1.57) or (ii) throughout the reaction of modified lignin (partially depolymerized lignin or polyol solutions of alcoholysis lignin) with anhydrides or trimellitic anhydride chloride [144].

Another interesting approach is using aminated lignin (black powder, amine value: 180–200 mg KOH/g) as a cross-linker of bisphenol A-based epoxy resin (epoxy value: 0.48–0.54 mol/100 g) [145]. The obtained aminated lignin contains a large number of primary and secondary amine groups, which successfully cure the epoxy network (Figure 1.58).

The application of aminated lignin has the positive effect at the initial degradation stage of the epoxy resin. Because lignin itself has a good thermal–mechanical performance, samples prepared with its higher content presented accordingly improved properties (Table 1.6).

TGA and DMA tests reveal improved thermal–mechanical properties of the bio-based epoxy resin. In comparison to the epoxy resin cross-linked with the commercial curing agent based on modified isophorone diamine (W93 curing agent: amine value: 550–600 mg KOH/g), the recorded value of T_{10}, for the material cured with the hardener containing 20 wt% of aminated lignin, increased nearly by 50 °C, while the highest value of T_{10} was achieved for the epoxy system cured by 100% aminated lignin (T_{10} = 266 °C for 100% of petrochemical-based hardener W93, T_{10} = 315 °C and T_{10} = 332 °C for aminated lignin hardener, 80%W93 + 20%AL and 100%AL, respectively). Additionally, the mass loss before 300 °C of the epoxy resin cured by W93 is four times higher than the one recorded for the aminated lignin. Moreover, the obtained materials are characterized by improved values of the glass transition temperature and thermal deformation temperature. T_g and T_d of epoxy resin sample cured with

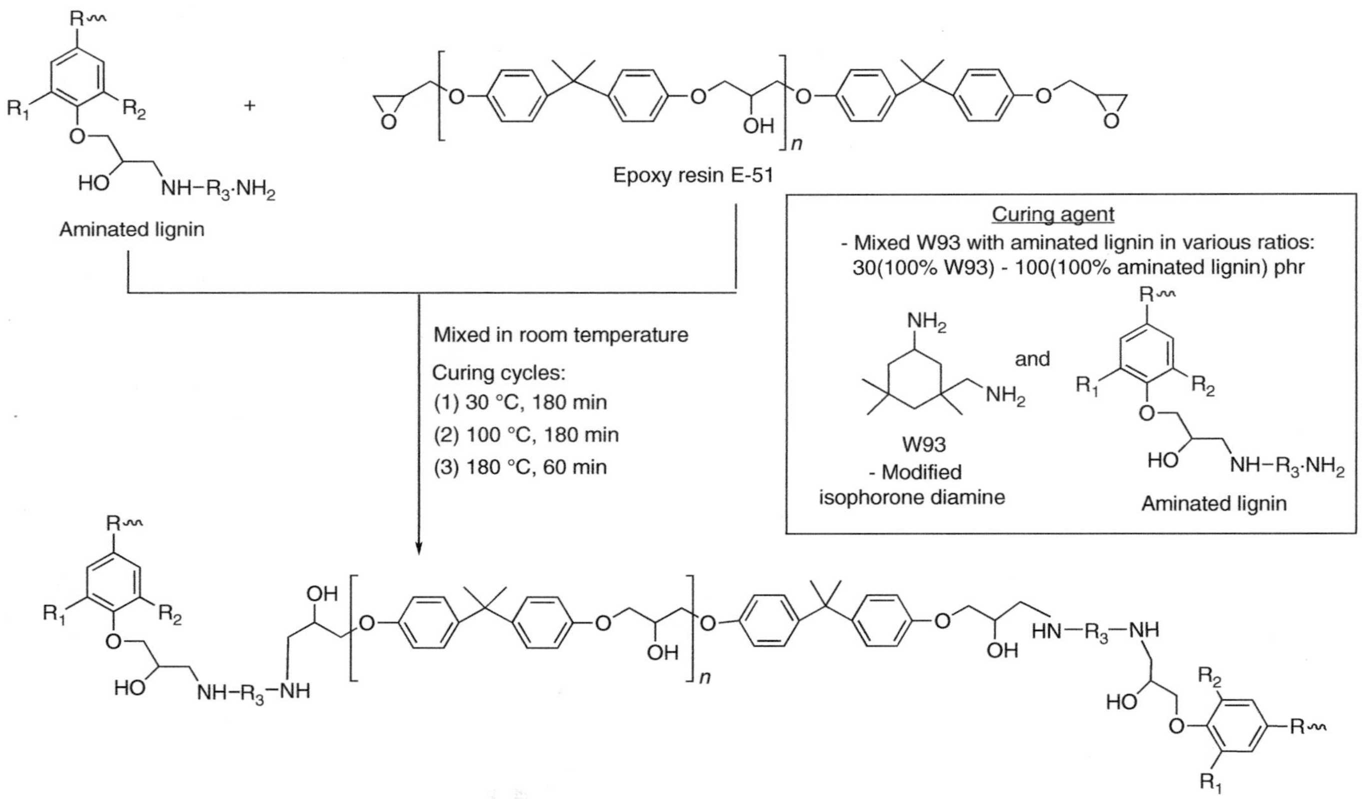

Figure 1.58 The curing of epoxy resin using aminated lignin as a curing agent.

Table 1.6 TGA, T_g, and T_d values of epoxy resin samples cured with different contents of lignin in the curing agent [145].

Epoxy resin cured by different hardeners	Total mass loss before 300 °C (%)	T_{10} (°C)	T_{50} (°C)	T_g (°C)	T_d (°C)
100%W93	11.1	266	362	79	70
80%W93 + 20%AL	8.7	315	370	86	74
70%W93 + 30%AL	—	—	—	89	76
60%W93 + 40%AL	7.2	325	364	92	79
50%W93 + 50%AL	—	—	—	93	84
40%W93 + 60%AL	8.4	317	371	—	—
20%W93 + 80%AL	5.5	327	363	—	—
100%AL	3.7	332	372	—	—

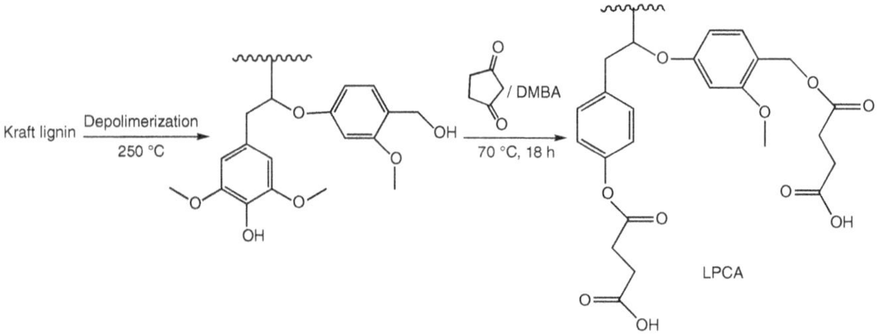

Figure 1.59 Preparation of partially depolymerized lignin (PDL) and lignin polycarboxylic acid LPCA) from Kraft lignin.

the hardener containing 50 wt% of lignin increased by 14 °C compared with the one without lignin.

Another interesting utilization of lignin-based compounds for the curing purposes of epoxy resin is application of partially depolymerized Kraft lignin [146]. In order to increase its solubility in organic solvents, lignin is subjected to the base-catalyzed depolymerization in supercritical methanol. The resulting partially depolymerized lignin (PDL) is then converted to lignin-based polycarboxylic acid (LPCA) by reacting with succinic anhydride (Figure 1.59).

LPCA might be applied as a curing or co-curing agent for epoxy resins. The curing of a commercial epoxy (DER353, epoxy value: 0.500–0.526 mol/100 g) using LPCA is conducted in the presence of 1 wt% of ethyl-4-methyl-imidazole as a catalyst and at the similar temperature range to the commercial hexahydrophthalic anhydride (HHPA). The obtained cured material exhibits a moderate T_g and comparable storage modulus to that cross-linked with a commercial anhydride curing agent. Additionally, linear succinate monoester, used in the synthesis of LPCA, enhances the flexibility of the lignin molecules. Therefore, increasing the content of bio-curing agent in the formulation tends to reduce the T_g of cured resins. For composition of DER353/LPCA with equivalent ratio 1/0.6, 1/0.8, to 1/1, the T_g of the cured resins decreases from 78.5 and 69.4 to 62.3 °C, while the storage moduli at room temperature is comparable (2.4–2.7 GPa). Based on the studies on the application of the solid LPCA together with other liquid curing agents, such as glycerol *tris*(succinate monoester) and commercial hexahydrophthalic anhydride to cure epoxies, it was observed that using a mixture of LPCA and a liquid curing agent not only adjusts the viscosity of the resin system but also significantly regulates the dynamic mechanical properties and thermal stability of the obtained epoxy materials.

Moreover, interesting two different types of novel cross-linked epoxy resins from lignin and glycerol are reported [147]. The first one is obtained by mixing the product of sodium lignosulfonate (LS) and glycerol (LSGLYPA, where the content of LSGLYPA varies at 0, 20, 40, 60, 80, and 100%) with polyacid of sodium lignosulfonate and ethylene glycol (LSEGPA) and ethylene glycol diglycidyl ether (EGDGE) at 80 °C (Figure 1.60).

The second one is by mixing LSGLYPA with a mixture of EGDGE/GLYDGE (the content of EGDGE/GLYDGE mixture was 0, 20, 40, 60, 80, and 100%) under similar reaction conditions (Figure 1.61).

The glass transition temperature of the cross-linked epoxy resins increases with increasing LSGLYPA and GLYDGE contents. The increase of T_g for the product of the first synthesis is due to the increase in cross-linking density. In turn, for the second sample, an increase in the GLYDGE content increases T_g of the cured epoxy resins (hydrogen bonding became the dominant factor).

Vanillin can be converted into the diamine derivatives (Figure 1.62) [148].

Starting from diglycidyl ethers of 2-methoxyhydroquinone and vanillyl alcohol through an epoxy ring opening with ammonia, it is possible to obtain two primary amines, which could be applied as epoxy resin cross-linking agents. Because of β-hydroxyl groups present in the molecules of diamines, they exhibit the autocatalytic effect on the epoxy–amine reactions. These new β-hydroxylamines can be used for cross-linking of diglycidyl ether of methoxyhydroquinone and diglycidyl ether of vanillyl alcohol giving fully bio-based epoxy systems with good thermo-mechanical properties and high thermal stability.

Cardanol derivatives are classified as the phenolic curing agents that are cross-linked with epoxy groups via the phenolic hydroxyl group. Novolac resins (Nov-I and Nov-II), containing an amount of unreacted cardanol of 35 and 20 wt%, respectively, are synthesized by the condensation reaction of cardanol and paraformaldehyde using oxalic acid as a catalyst (Figure 1.63) [149].

The cardanol-based novolacs might be used as curing agents of commercial (the diglycidyl ether of bisphenol A) epoxy resin in the presence of 2-ethyl-4-methyl-imidazole as a catalyst. The higher cross-linking density is observed with higher amounts of epoxy resin. Moreover, the resin cured with Nov-II is characterized by higher T_g and better mechanical properties than Nov-I-resin. On the other hand, because of the higher molecular weight and lower unreacted cardanol content, the similar thermal degradation properties are observed for both tested materials.

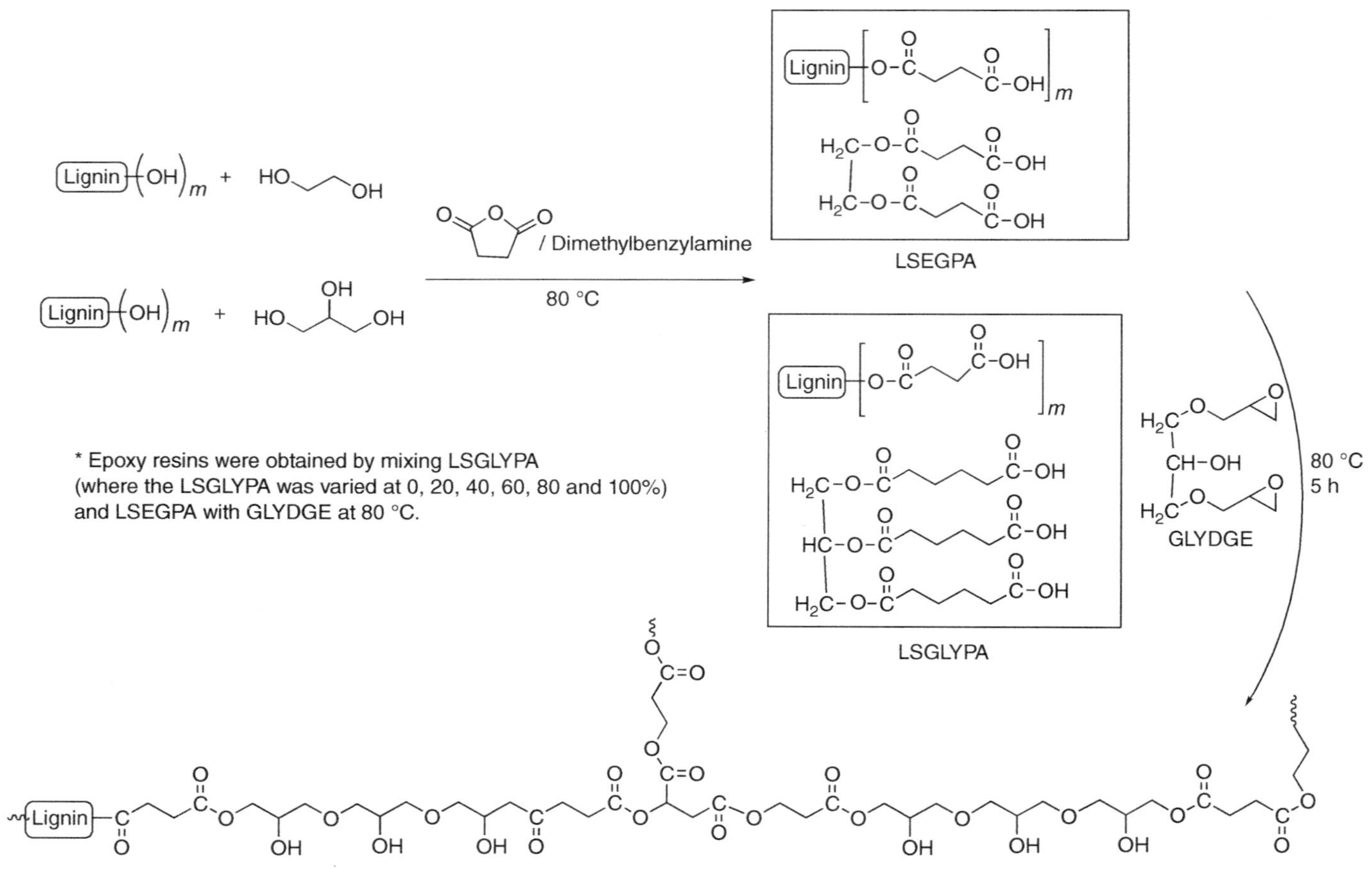

Figure 1.60 Scheme of preparation cross-linked epoxy resin.

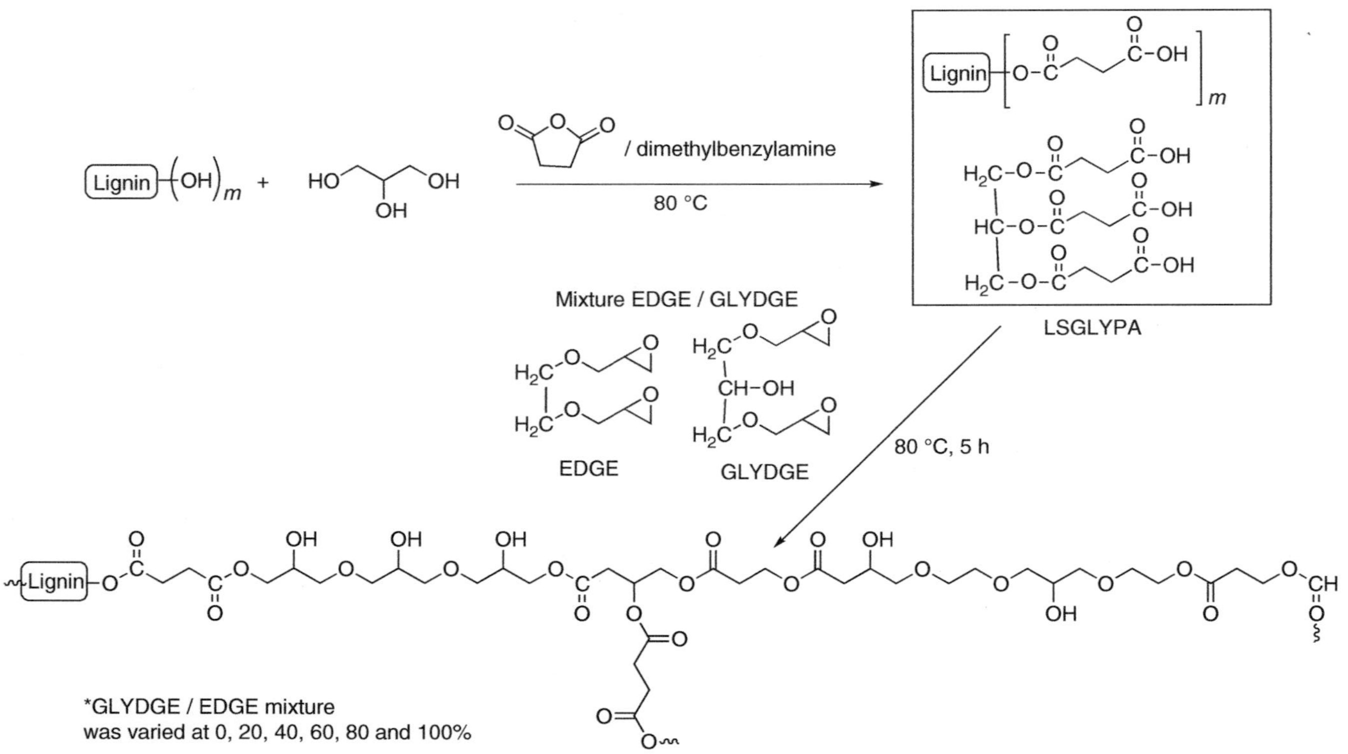

Figure 1.61 Scheme of epoxy resin by cross-linking LSGLYPA with a mixture of EGDGE/GLYDGE.

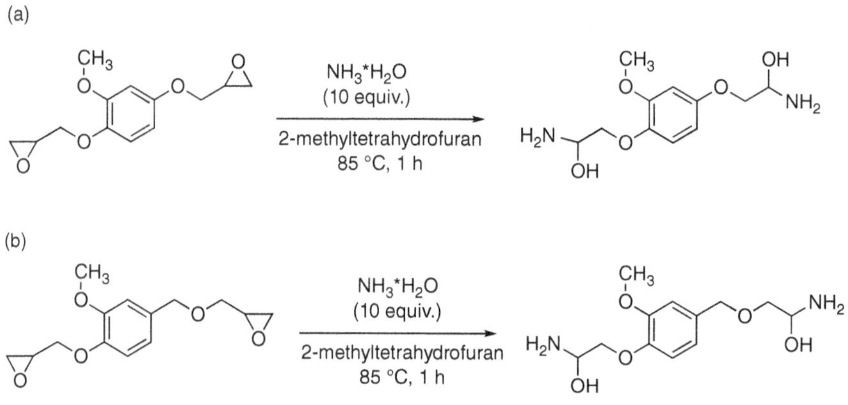

Figure 1.62 Synthesis of dihydroxyaminopropane of 2-methoxyhydroquinone (a) and (b) vanillyl alcohol.

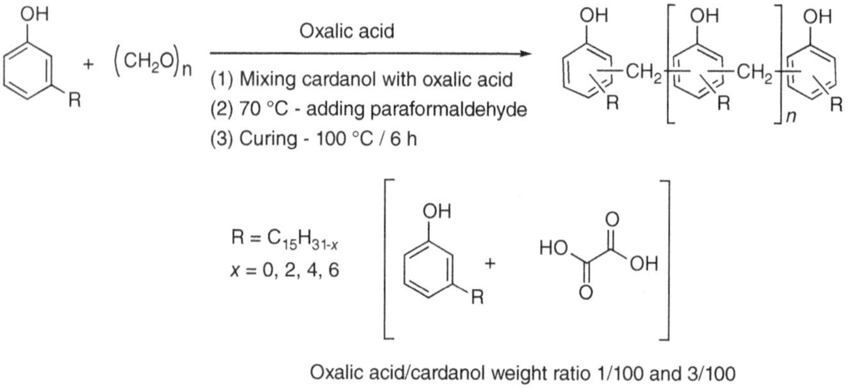

Figure 1.63 Synthesis of the cardanol-based novolacs.

The interesting cross-linking agent is possible to obtain as the diamine derivative of isosorbide (Figure 1.64) and it could be used for curing of diglycidyl ether of isosorbide giving the completely bio-based epoxy system [150].

The diamine derivative of isosorbide is obtained using microwave assistant thiol-ene coupling reaction in the aqueous media and with the water-soluble initiator $(NH_4)_2S_2O_8$, as the alternative to AIBN. The cured isosorbide-based resin has good shape fixity, good shape recovery, and satisfied thermal stability. This fully bio-based resin shows great potential to be used as a candidate for shape memory material. Another synthetic method for obtaining the diamine derivative of isosorbide is the reaction of cyanoethylation of isosorbide, followed by the hydrogenation of di-cyanoethylated product (Figure 1.65) [151].

Terpene derivatives might also be applied as bio-based curing agents for epoxy resins. For example, a novel terpene-based curing agent prepared as an adduct of myrcene, monoterpene containing three double bonds including conjugated diene, and maleic anhydride (MMY) is used to cure bisphenol A-based epoxy resin [152]. The obtained cured material is characterized by a tensile strength

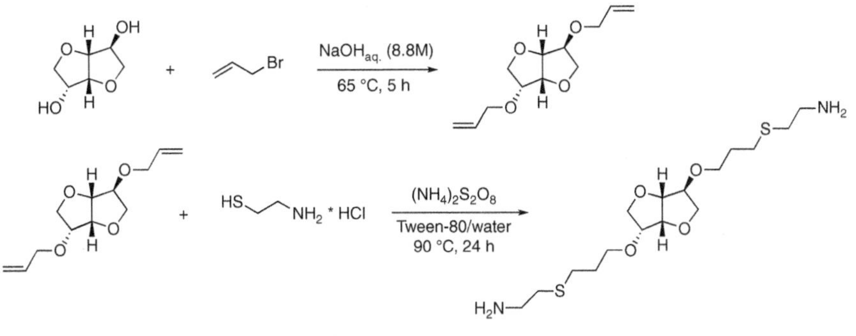

Figure 1.64 Synthesis of the isosorbide-based cross-linking agent.

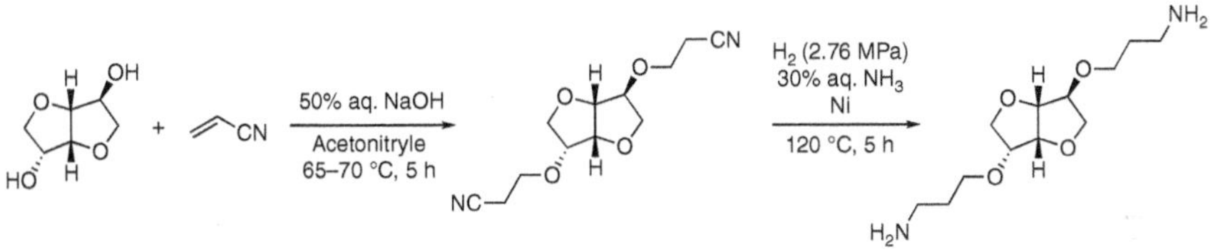

Figure 1.65 Synthesis of the isosorbide-based cross-linking agent via cyanoethylation step.

Table 1.7 Properties of epoxy resin networks cured with bio-based curing agents.

Cured samples	T_g (°C)	E at $T_{g+30°C}$ (MPa)	Tensile strength (MPa)	Elongation at break (%)	Impact strength (kJ/m^2)
MMY100	61.59	19.06	48.74	7.5	8.55
MMY75/CMMY25	58.03	19.00	42.11	6.5	13.87
MMY50/CMMY50	54.03	4.18	35.55	6.6	17.29
MMY25/CMMY75	45.09	2.72	11.11	259.4	62.51
CMMY100	15.14	1.11	0.43	565.8	Unbroken

of 48.74 MPa and a storage modulus of 19.06 MPa, but a poor impact property of 8.55 kJ/m^2 and a very low elongation at break (7.54%). Improvement of properties of cured bio-based epoxy resins is performed by application of mixture (mixed by weight ratios of 100/0, 75/25, 50/50, 25/75, and 0/100, respectively) of two curing agents: MMY and castor oil-modified adduct of myrcene and maleic anhydride (CMMY) (Figure 1.66).

Based on the obtained results (Table 1.7), with the increase of CMMY weight ratio, the tensile strength and T_g of the cross-linked resin decreases, but the elongation at break and the impact strength increase.

As can be seen from the examples presented above, it is possible to obtain not only epoxy resins from raw materials of natural origin but also cross-linking agents for them. Interestingly, it is also possible to synthesize completely bio-derived epoxy systems. Both the epoxy resins based on bisphenol A, as well

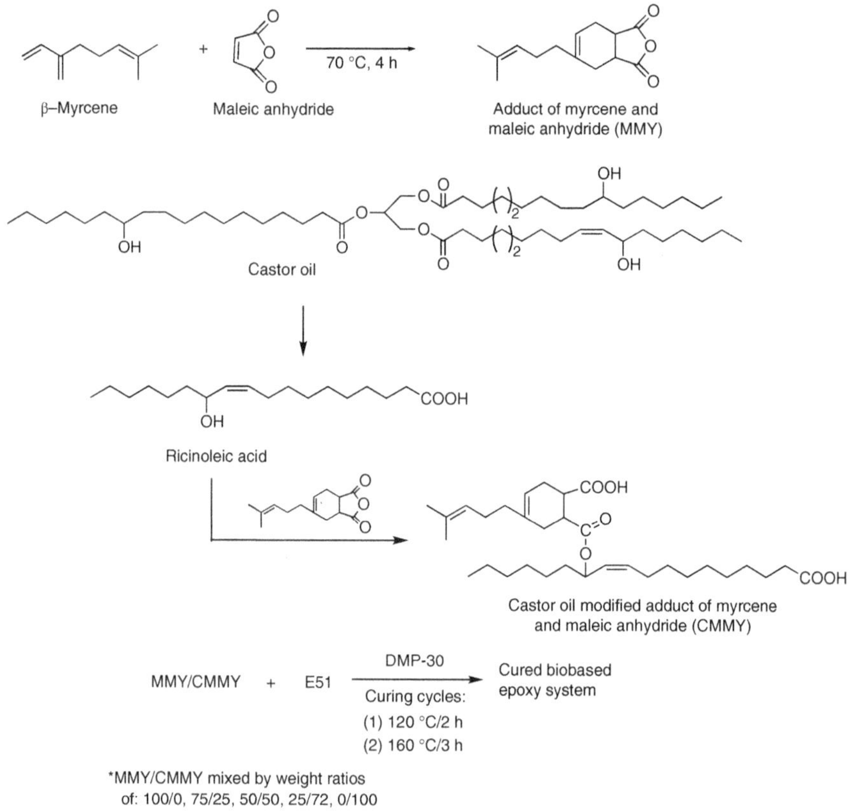

Figure 1.66 Synthesis of bio-based epoxy curing agent derived from myrcene and castor oil.

as cross-linked with curing agents on the basis of raw materials of natural origin, and the fully epoxy biosystems are characterized by very good final properties.

References

1 Derksen, J.T.P., Cuperus, F.P., and Kolser, P. (1996). *Prog. Org. Coat.* 27 (1–4): 45–53.
2 Sharma, V. and Kundu, P.P. (2006). *Prog. Polym. Sci.* 31 (11): 983–1008.
3 Köckritz, A. and Martin, A. (2008). *Eur. J. Lipid Sci. Technol.* 110: 812–824.
4 Hilker, I., Bothe, D., Prüss, J., and Warnecke, H.-J. (2001). *Chem. Eng. Sci.* 56: 427–432.
5 Chen, J., de Liedekerke Beaufort, M., Gyurik, L. et al. (2019). *Green Chem.* 21: 2436–2447.
6 Sienkiewicz, A.M. and Czub, P. (2016). *Ind. Crops Prod.* 83: 755–773.
7 Chua, S.Ch., Xu, X., and Guo, Z. (2012). *Process Biochem.* 47: 1439–1451.
8 Gan, L.H., Goh, S.H., and Ooi, K.S. (1992). *J. Am. Oil Chem. Soc.* 69 (4): 347–351.

9 Faria-Machado, A.F., Altenhofen da Silva, M., Vieira, M.G.A., and Beppu, M.M. (2013). *J. Appl. Polym. Sci.* 127 (5): 3543–3549.
10 Walton, H.W. and Nevin, Ch.S. (1965). Reaction of half esters of an alpha, beta-ethylenically unsaturated-alpha, beta-dicarboxylic acid with a vicinal epoxy compound and products thereof. US Patent 3, 190, 899, assigned to A.E. Staley Manufacturing Company, June 22, 1965.
11 Panchal, T.M., Patel, A., Chauhan, D.D. et al. (2017). *Renewable Sustainable Energy Rev.* 70: 65–70.
12 Soni, S. and Agarwal, M. (2014). *J. Green Chem. Lett. Rev.* 7 (4): 359–382.
13 Czub, P. (2006). *Macromol. Symp.* 242 (1): 60–64.
14 Czub, P. (2006). *Macromol. Symp.* 245–246 (1): 533–538.
15 Wear, R.L. (1960). Aromatic epoxidized polyester and method of making. US Patent 2, 944, 035, assigned to Minnesota Mining & Manufacturing Company, July 5, 1960.
16 Afolabi, O.A., Aluko, M.E., Wang, G.C. et al. (1989). *J. Am. Oil Chem. Soc.* 66 (7): 983–985.
17 Vijayalakshmi, P., Rao, T.Ch., Kale, V. et al. (1992). *Polymer* 33 (15): 3252–3256.
18 Park, S.J., Jin, F.L., Lee, J.R., and Shin, J.S. (2005). *Eur. Polym. J.* 41: 231–237.
19 Boquillon, N. and Fringant, C. (2000). *Polymer* 41 (24): 8603–8613.
20 Czub, P. (2009). *Polym. Adv. Technol.* 20 (3): 194–208.
21 Czub, P. (2009). *Macromol. Symp.* 277 (1): 162–170.
22 Sienkiewicz, A. and Czub, P. (2017). *eXPRESS Polym. Lett.* 11 (4): 308–319.
23 Sienkiewicz, A. and Czub, P. (2019). *eXPRESS Polym. Lett.* 13 (7): 642–655.
24 Lewis, M. and Rohrer, J.F. (1977). Cured epoxy resins. US Patent 4, 040, 994, assigned to Unitech Chemical Inc., August 9, 1977.
25 Eicken, U., Gorzinski, M., Birnbrich, P., and Tamcke, T., Modified resins made from the reaction of epoxidized esters and resin acids. US Patent 5, 770, 662, assigned to Henkel KGaA, June 23, 1998.
26 Miyagawa, H., Mohanty, A.K., Misra, M., and Drzal, L.T. (2004). *Macromol. Mater. Eng.* 289 (7): 629–635.
27 Miyagawa, H., Mohanty, A.K., Misra, M., and Drzal, L.T. (2004). *Macromol. Mater. Eng.* 289 (7): 636–641.
28 Miyagawa, H., Misra, M., Drzal, L.T., and Mohanty, A.K. (2005). *Polym. Eng. Sci.* 45 (4): 487–495.
29 Park, S.J., Jin, F.L., and Lee, J.R. (2004). *Mater. Sci. Eng., A* 374, 1–2: 109–114.
30 Park, S.J., Jin, F.L., and Lee, J.R. (2004). *Macromol. Chem. Phys.* 205 (15): 2048–2054.
31 Jin, F.L. and Park, S.J. (2008). *Mater. Sci. Eng., A* 478, 1-2: 402–405.
32 Ratna, D. and Banthia, A.K. (2000). *J. Adhes. Sci. Technol.* 14 (1): 15–25.
33 Kar, S. and Banthia, A.K. (2004). *Mater. Manuf. Processes* 19 (3): 459–474.
34 Zhu, J., Chandrashekhara, K., Flanigan, V., and Kapila, S. (2004). *J. Appl. Polym. Sci.* 91 (6): 3513–3518.
35 Liang, G. and Chandrashekhara, K. (2006). *J. Appl. Polym. Sci.* 102: 3168.
36 Shabeer, A., Garg, A., Sundararaman, S. et al. (2005). *J. Appl. Polym. Sci.* 98 (4): 1772–1780.

37 Czub, P. and Franek, I. (2013). *Polimery* 58 (2): 135–139.
38 Crivello, J.V. and Narayan, R. (1992). *Chem. Mater.* 4 (3): 692–699.
39 Thames, S.F. and Yu, H. (1999). *Surf. Coat. Technol.* 115 (2–3): 208–214.
40 Thames, S.F., Yu, H., and Subramanian, R. (2000). *J. Appl. Polym. Sci.* 77 (1): 8–13.
41 Wan Rosli, W.D., Kumar, R.N., Mek Zah, S., and Hilmi, M.M. (2003). *Eur. Polym. J.* 39 (3): 593–600.
42 Samuelsson, J., Sundell, P.E., and Johansson, M. (2004). *Prog. Org. Coat.* 50 (3): 193–198.
43 Chen, J., Soucek, M.D., Simonsick, W.J., and Celikay, R.W. (2002). *Polymer* 43 (20): 5379–5389.
44 Zong, Z., Soucek, M.D., Liu, Y., and Hu, J. (2003). *J. Polym. Sci, Part A: Polym. Chem.* 41 (21): 3440–3456.
45 Zong, Z., He, J., and Soucek, M.D. (2005). *Prog. Org. Coat.* 53 (2): 83–90.
46 Zou, K. and Soucek, M.D. (2005). *Macromol. Chem. Phys.* 206 (9): 967–975.
47 Crivello, J.V., Narayan, R., and Sternstein, S.S. (1997). *J. Appl. Polym. Sci.* 64 (11): 2073–2087.
48 Achyuthan, K.E., Achyuthan, A.M., Adams, P.D. et al. (2010). *Molecules* 15 (12): 8641–8688.
49 Laurichesse, S. and Avérous, L. (2014). *Prog. Polym. Sci.* 39 (7): 1266–1290.
50 Azadi, P., Inderwildi, O.R., Farnood, R., and King, D.A. (2013). *Renewable Sustainable Energy Rev.* 21: 506–523.
51 Ghaffar, S.H. and Fan, M. (2014). *Int. J. Adhes. Adhes.* 48: 92–101.
52 Cao, L., Zhang, C., Chen, H. et al. (2017). *Bioresour. Technol.* 245: 1184–1193.
53 Gillet, S., Petitjean, L., Aguedo, M. et al. (2017). *Bioresour. Technol.* 233: 216–226.
54 Morreel, K., Dima, O., Kim, H. et al. (2010). *Physiology* 153 (4): 1464–1478.
55 Wang, H., Pu, Y., Ragauskas, A., and Yang, B. (2019). *Bioresour. Technol.* 271: 449–461.
56 Ponnusamy, V.K., Nguyen, D.D., Dharmaraja, J. et al. (2019). *Bioresour. Technol.* 271: 462–472.
57 Agarwal, A., Rana, M., and Park, J.H. (2018). *Fuel Process. Technol.* 181: 115–132.
58 Zakzeski, J., Bruijnincx, P.C., Jongerius, A.L., and Weckhuysen, B.M. (2010). *Chem. Rev.* 110 (6): 3552–3599.
59 Watkins, D., Nuruddin, M., Hosur, M. et al. (2015). *J. Mater. Res. Technol.* 4 (1): 26–32.
60 Zhang, J., Deng, H., Lin, L. et al. (2010). *Bioresour. Technol.* 101 (7): 2311–2316.
61 Li, M.F., Sun, S.N., Xu, F., and Sun, R.C. (2012). *Sep. Purif. Technol.* 101: 18–25.
62 Binder, J.B., Gray, M.J., White, J.F. et al. (2009). *Biomass Bioenergy* 33 (9): 1122–1130.
63 Lourençon, T.V., Hansel, F.A., da Silva, T.A. et al. (2015). *Sep. Purif. Technol.* 154: 82–88.
64 Sun, G., Sun, H., Liu, Y. et al. (2007). *Polymer* 48 (1): 330–337.

65 Upton, B.M. and Kasko, A.M. (2015). *Chem. Rev.* 116 (4): 2275–2306.
66 Koike, T. (2012). *Polym. Eng. Sci.* 52 (4): 701–717.
67 Xin, J., Li, M., Li, R. et al. (2016). *ACS Sustainable Chem. Eng.* 4 (5): 2754–2761.
68 Hernandez, E.D., Bassett, A.W., Sadler, J.M. et al. (2016). *ACS Sustainable Chem. Eng.* 4 (8): 4328–4339.
69 Ye, Y., Zhang, Y., Fan, J., and Chang, J. (2011). *Ind. Eng. Chem. Res.* 51 (1): 103–110.
70 Salanti, A., Orlandi, M., Tolppa, E.L., and Zoia, L. (2010). *Int. J. Mol. Sci.* 11 (3): 912–926.
71 van de Pas, D.J. and Torr, K.M. (2017). *Biomacromolecules* 18 (8): 2640–2648.
72 Ferdosian, F., Yuan, Z., Anderson, M., and Xu, C.C. (2016). *J. Anal. Appl. Pyrolysis* 119: 124–132.
73 Shiraishi, N. (1989). *ACS Symp. Ser.* 397: 488–495.
74 Stanzione, J.F. III, Giangiulio, P.A., Sadler, J.M. et al. (2013). *ACS Sustainable Chem. Eng.* 1 (4): 419–426.
75 Enjoji, M., Yamamoto, A., and Shibata, M. (2015). *J. Appl. Polym. Sci.* 132 (4): 41347–41347.
76 Kaya, İ., Doğan, F., and Gül, M. (2011). *J. Appl. Polym. Sci.* 121 (6): 3211–3222.
77 Parsell, T., Yohe, S., Degenstein, J. et al. (2015). *Green Chem.* 17 (3): 1492–1499.
78 Hirose, S., Hatakeyama, T., and Hatakeyama, H. (2003). *Macromol. Symp.* 197 (1): 157–170.
79 Walton, N.J., Mayer, M.J., and Narbad, A. (2003). *Phytochemistry* 63: 505–515.
80 Wynberg, H. (1960). *Chem. Rev.* 60 (2): 169–184.
81 Fatiadi, A. and Schaffer, R. (1974). *J. Res. Natl. Bur. Stand. Sect. A* 78A (3): 411–412.
82 Kamlet, J. (1953). Manufacture of vanillin and its homologues. US Patent 2, 640, 083, assigned to Mathieson Chemical Corporation and Olin Corporation, May 26, 1953.
83 Bots, R.H. (1927). Process of manufacturing vvanillin. US Patent 1, 643, 804, September 27, 1927.
84 Fiecchi, A., Nano, G.M., Cabella, P., and Cicognani, G. (1970). Method of preparing vanillin from eugenol. US Patent 3, 544, 621, assigned to Collins Chemical Corporation Incorporation, December 1, 1970.
85 Lampman, G.M., Andrews, J., Bratz, W. et al. (1977). *J. Chem. Educ.* 54 (12): 776–778.
86 Hocking, M.B. (1997). *J. Chem. Educ.* 74 (9): 1055–1059.
87 Bjørsvik, H.R. and Minisci, F. (1999). *Org. Process Res. Dev.* 3 (5): 330–340.
88 Duffey, S.S., Aldrich, J.R., and Blum, M.S. (1977). *Comp. Biochem. Physiol. B: Biochem. Mol. Biol.* 56 (2): 101–102.
89 Furuya, T., Kuroiwa, M., and Kino, K. (2017). *J. Biotechnol.* 243 (10): 25–28.
90 Schmidt, C.G., Gonçalves, L.M., Prietto, L. et al. (2014). *Food Chem.* 146: 371–377.

91 Bomgardner, M.M. (2016). *Chem. Eng. News* 94 (36): 38–42.
92 Havkin-Frenkel, D. and Belanger, F.C. (2016). Biotechnological production of vanillin. In: *Biotechnology in Flavor Production*, 2e (eds. D. Havkin-Frenkel and F.C. Belanger), 83–103. Chichester: Wiley.
93 Fache, M., Darroman, E., Besse, V. et al. (2014). *Green Chem.* 16 (4): 1987–1998.
94 Fache, M., Viola, A., Auvergne, R. et al. (2015). *Eur. Polym. J.* 68: 526–535.
95 Nikafshar, S., Zabihi, O., Hamidi, S. et al. (2017). *RSC Adv.* 7: 8694–8701.
96 Fache, M., Auvergne, R., Boutevin, B., and Caillol, S. (2015). *Eur. Polym. J.* 67: 527–538.
97 Fache, M., Boutevin, B., and Caillol, S. (2015). *Eur. Polym. J.* 68: 488–502.
98 Aouf, Ch., Lecomte, J., Villeneuve, P. et al. (2012). *Green Chem.* 14 (8): 2328–2336.
99 Llevot, A., Grau, E., Carlotti, S. et al. (2015). *Polym. Chem.* 6 (33): 6058–6066.
100 Savonnet, E., Grau, E., Grelier, S. et al. (2018). *ACS Sustainable Chem. Eng.* 6: 11008–11017.
101 Pearl, I.A. (1952). *J. Am. Chem. Soc.* 74 (17): 4260–4262.
102 Nishimura, R.T., Giammanco, Ch.H., and Vosburg, D.A. (2010). *J. Chem. Educ.* 87 (5): 526–527.
103 Amarasekara, A.S., Wiredu, B., and Razzaq, A. (2012). *Green Chem.* 14 (9): 2395–2397.
104 Amarasekara, A.S., Garcia-Obergon, R., and Thompson, A.K. (2019). *J. Appl. Polym. Sci.* 36 (4): 47000. (1–6).
105 Shibata, M. and Ohkita, T. (2017). *Eur. Polym. J.* 92: 165–173.
106 Wang, Sh., Ma, S., Xu, Ch. et al. (2017). *Macromolecules* 50 (5): 1892–1901.
107 Ochi, M., Shimbo, M., Saga, M., and Takashima, N. (1986). *J. Polym. Sci., Part B: Polym. Phys.* 24 (10): 2185–2195.
108 Ochi, M., Yoshizumi, M., and Shimbo, M. (1987). *J. Polym. Sci., Part B: Polym. Phys.* 25 (9): 1817–1827.
109 Ochi, M., Shiba, T., Takeuchi, H. et al. (1989). *Polymer* 30 (6): 1079–1084.
110 Caillol, S. (2018). *Curr. Opin. Green Sustain. Chem.* 14: 26–32.
111 Dworakowska, S., Cornille, A., Bogdał, D. et al. (2015). *Eur. J. Lipid Sci. Technol.* 117 (11): 1893–1902.
112 Voirin, C., Caillol, S., Sadavarte, N.V. et al. (2014). *Polym. Chem.* 5 (9): 3142–3162.
113 Jaillet, F., Darroman, E., Ratsimihety, A. et al. (2014). *Eur. J. Lipid Sci. Technol.* 116 (1): 63–73.
114 Kumar, P.P., Paramashivappa, R., Vithayathil, P.J. et al. (2002). *J. Agric. Food. Chem.* 50 (16): 4705–4708.
115 Ionescu, M. and Petrovic, Z.S.J. (2011). *J. Serb. Chem. Soc.* 76 (4): 591–606.
116 Jaillet, F., Darroman, E., Boutevin, B., and Caillol, S. (2016). *Oilseeds Fats Crops Lipids* 23 (5): D511.
117 Kathalewar, M. and Sabnis, A. (2014). *J. Coat. Technol. Res.* 11 (4): 601–618.
118 Darroman, E., Durand, N., Boutevin, B., and Caillol, S. (2015). *Prog. Org. Coat.* 83: 47–54.

119 Wang, X., Zhou, S., Guo, W.W. et al. (2017). *ACS Sustainable Chem. Eng.* 5 (4): 3409–3416.

120 Rose, M. and Palkovits, R. (2012). *Chem. Sustain. Energy Mater.* 5 (1): 167–176.

121 Keskiväli, J., Rautiainen, S., Heikkilä, M. et al. (2017). *Green Chem.* 19: 4563–4570.

122 Dussenne, C., Delaunay, T., Wiatz, V. et al. (2017). *Green Chem.* 19: 5332–5344.

123 Feng, X., East, A.J., Hammond, W.B. et al. (2011). *Polym. Adv. Technol.* 22 (1): 139–150.

124 Hammond, W., East, A., Jaffe, M., and Feng, X. (2017). Isosorbide-derived epoxy resins and methods of making same. US Patent 9, 605, 108, assigned to New Jersey Institute of Technology, March 28, 2017.

125 Morrison, J.G. (1962). Polyglycidyl ethers of ether anhydro hexitols, method of production, and aqueous solutions thereof. US Patent 3, 041, 300, assigned to Martin Marietta Corporation, June 26, 1962.

126 Zech, J.D. and Maistre, J.L.W. (1966). Bisglycidyl ethers of isohexides. US Patent 3, 272, 845, assigned to Atlas Chemical Industries, September 13, 1966.

127 Hong, J., Radojčić, D., Ionescu, M. et al. (2014). *Polym. Chem.* 5 (18): 5360–5368.

128 East, A., Jaffe, M., Zhang, Y., and Catalani, L.H. (2009). Thermoset epoxy polymers from renewable resources. US Patent 7, 619, 056, assigned to New Jersey Institute of Technology, November 17, 2009.

129 Łukaszczyk, J., Janicki, B., and Kaczmarek, M. (2011). *Eur. Polym. J.* 47 (8): 1601–1606.

130 Gandini, A. and Lacerda, T.M. (2015). *Prog. Polym. Sci.* 48: 1–39.

131 John, G., Nagarajan, S., Vemula, P.K. et al. (2019). *Prog. Polym. Sci.* 92: 158–209.

132 Tholl, D. (2006). *Curr. Opin. Plant Biol.* 9 (3): 297–304.

133 Wu, G.-M., Kong, Z.-W., Chen, J. et al. (2014). *Prog. Org. Coat.* 77: 315–321.

134 Wu, G.-M., Liu, D., Liu, G.-F. et al. (2015). *Carbohydr. Polym.* 127: 229–235.

135 Bähr, M., Bitto, A., and Mülhaupt, R. (2012). *Green Chem.* 14 (5): 1447–1454.

136 Xu, K., Chen, M., Zhang, K., and Hu, J. (2004). *Polymer* 45: 1133–1140.

137 Ng, F., Couture, G., Philippe, C. et al. (2017). *Molecules* 22 (1): 149–197.

138 Deng, L., Ha, C., Sun, C. et al. (2013). *Ind. Eng. Chem. Res.* 52 (37): 13233–13240.

139 Atta, A.M., Mansour, R., Abdou, M.I., and Sayed, A.M. (2004). *Polym. Adv. Technol.* 15 (9): 514–522.

140 Deng, L., Shen, M., Yu, J. et al. (2012). *Ind. Eng. Chem. Res.* 51 (24): 8178–8184.

141 Vijayalakshmi, P., Subbarao, R., and Lakshminarayana, G. (1988). *J. Am. Oil Chem. Soc.* 65 (6): 939–941.

142 Jaillet, F., Desroches, M., Auvergne, R. et al. (2013). *Eur. J. Lipid Sci. Technol.* 115 (6): 698–708.

143 Ionescu, M., Radojčić, D., Wana, X. et al. (2015). *Eur. Polym. J.* 67: 439–448.

144 Ding, C. and Matharu, A.S. (2014). *ACS Sustainable Chem. Eng.* 2 (10): 2217–2236.
145 Pan, H., Sun, G., Zhao, T., and Wang, G. (2015). *Polym. Eng. Sci.* 55 (4): 924–932.
146 Qin, J., Wolcott, M., and Zhang, J. (2013). *ACS Sustainable Chem. Eng.* 2 (2): 188–193.
147 Ismail, T.N.M.T., Hassan, H.A., Hirose, S. et al. (2010). *Polym. Int.* 59 (2): 181–186.
148 Mora, A.S., Tayouo, R., Boutevin, B. et al. (2018). *Green Chem.* 20 (17): 4075–4084.
149 Campaner, P., D'Amico, D., Longo, L. et al. (2009). *J. Appl. Polym. Sci.* 114 (6): 3585–3591.
150 Li, C., Dai, J., Liu, X. et al. (2016). *Macromol. Chem. Phys.* 217 (13): 1439–1447.
151 Kumar, S., Samal, S.K., Mohanty, S., and Nayak, S.K. (2018). *Polym. Plast. Technol. Eng.* 57 (3): 133–155.
152 Yang, X., Wang, C., Li, S. et al. (2017). *RSC Adv.* 7 (1): 238–247.

2

Natural/Synthetic Fiber-Reinforced Bioepoxy Composites

Bo Wang[1], *Silu Huang*[1], *and Libo Yan*[1,2]

[1] *Technische Universität Braunschweig, Institute for Building Materials, Solid Construction and Fire Protection, Department of Organic and Wood-Based Construction Materials, Hopfengarten 20, 38102 Braunschweig, Germany*
[2] *Fraunhofer Institute for Wood Research Wilhelm-Klauditz-Institut WKI, Centre for Light and Environmentally-Friendly Structures, Bienroder Weg 54E, 38108 Braunschweig, Germany*

2.1 Introduction

Fiber-reinforced polymer (FRP) is a composite material that consists of two main components, i.e. fiber and polymer matrix [1]. Fibers as reinforcement materials are added in the polymer matrix to contribute to tensile strength, tensile stiffness, and toughness of the composite material. The polymer matrix can, on the one hand, bind the fibers together and transfer the stress through adhesion or friction to the fiber. On the other hand, it provides a protection of the embedded fibers against chemical and mechanical attacks, such as corrosion and impact [2]. Conventional synthetic fibers for FRP composites are carbon, glass, and aramid fibers. The polymer matrix can be either thermoplastics or thermosets. The commonly applied thermoplastics are polypropylene (PP), polyethylene (PE), and polystyrol (PS). The common thermosets are epoxy (EP), polyester (PET), and vinyl ester resin (VE) [1]. Recently, bioepoxy has gained its popularity and has been considered as an alternative to petroleum-based polymer because of its sustainability and renewability. Nowadays, natural and synthetic fiber-reinforced bioepoxy composites are utilized in many fields such as ski/skateboard [3], furniture, and automotive [4]. This chapter provides an overview of the classification and recent advancements on natural fibers, synthetic fibers, bioepoxy, and their fiber-reinforced bioepoxy composites.

2.2 Synthetic and Natural Fibers

Fiber materials can be mainly classified into synthetic and natural fibers based on whether the fiber is produced from natural sources or by artificial synthesization. The classification of fibers is shown in detail in Figure 2.1.

Bio-Based Epoxy Polymers, Blends, and Composites: Synthesis, Properties, Characterization, and Applications, First Edition. Edited by Jyotishkumar Parameswaranpillai, Sanjay Mavinkere Rangappa, Suchart Siengchin, and Seno Jose.

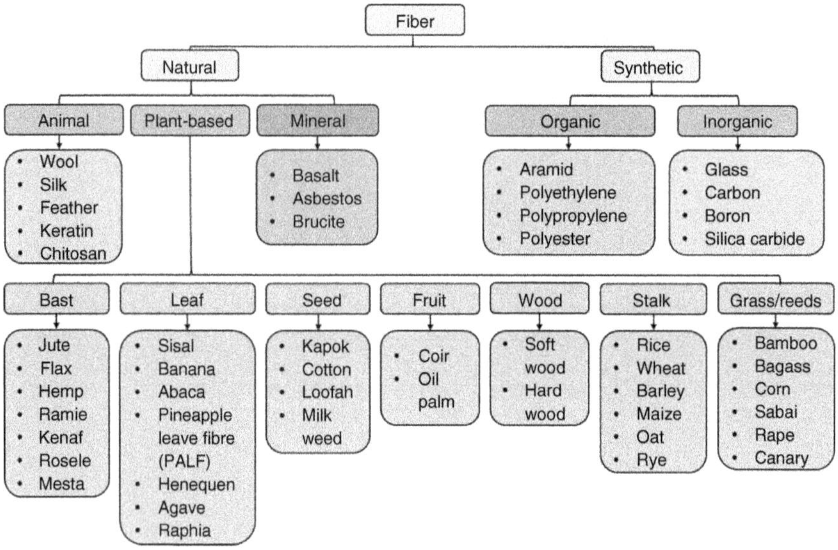

Figure 2.1 The classification of fibers [5].

2.2.1 Synthetic Fibers

As shown in Figure 2.1, synthetic fibers are further classified into organic and inorganic synthetic fibers according to the main chemical compositions of the fibers.

2.2.1.1 Organic Synthetic Fibers

Aramid fibers are one of the important organic fibers. They are made by carboxylic acid halide group reacting with amine group [6]. Aramid fibers are mainly divided into two types according to the position of the amide linkage on the aromatic ring: *meta*-aramid (*m*-aramid) and *para*-aramid (*p*-aramid) fibers, as shown in Figure 2.2. The *m*-aramid fibers were first applied commercially in 1960s, and the *p*-aramid fibers were then developed in the early 1970s [7]. *m*-Aramid fibers are generally manufactured by wet spinning process. They are

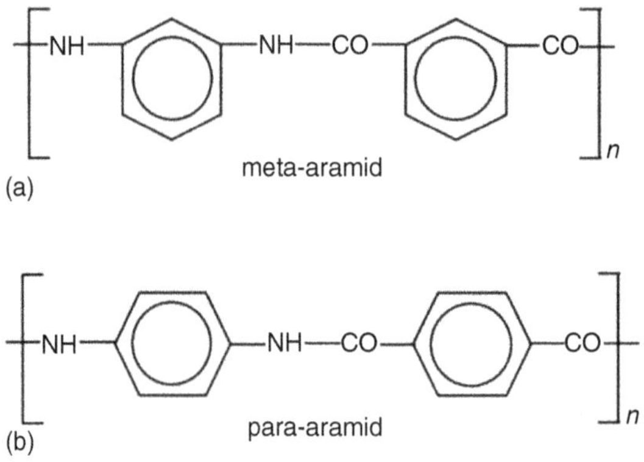

Figure 2.2 Chemical structure of (a) *meta*-aramid and (b) *para*-aramid fibers.

Table 2.1 Some commercial aramid fibers [9].

Trade name	Producer	Remarks
Aramica®	Asahi	*p*-Aramid film
Armos®	Chimvolokno JSC	
Heracron®	Kolon Industries, Inc.	
Hydlar®	A.L. Hyde Co.	Reinforced aramid fiber
Ixef®	Solvay Advanced Polymers	
Kevlar®	Dupont	*p*-Aramid fiber
Mictron®	Toray Industries	*p*-Aramid film
Nomex®	DuPont	*m*-Aramid fiber
Rusar®	Termotex Co., Mytishchi	
Sulfron®	Teijin Chemicals	Sulfur-modified aramid
Technora®	Teijin Chemicals	
Teijinconex®	Teijin Chemicals	*m*-Aramid fiber
Thermatex®	Difco Performance Fabrics, Inc.	Aramid and aramid blend fabrics
Twaron®	Teijin Twaron B.V.	

semicrystalline fibers. Their structure is the result of the incomplete extension of the molecular chain, which is hindered by the amide linkage on the aromatic ring [8]. *p*-Aramid fibers are produced by the dry jet wet spinning process. In *p*-aramid fibers, the position of the amide linkage is beneficial for extending the molecular chain. Based on this, *p*-aramid fibers obtain crystallinity and orientation at a high level [8]. Therefore, *p*-aramid fibers usually show higher tensile strength and modulus than *m*-aramid fibers. Some commercial aramid fibers are presented in Table 2.1. Aramid fibers are lightweight and have good resistance to fatigue and impact loading. Therefore, they are widely applied for producing body armor [10]. Besides, aramid fibers also show superior resistance to heat and low flammability. Therefore, they can be used in firefighter apparel as well [11]. The aramid FRP composites are usually applied in aircraft and military vehicles because of their high impact resistance and vibration damping properties [8]. According to the comparison among several aerospace materials, the vibration damping loss factor of aramid epoxy is approximately 10 times than the loss factor of carbon epoxy and is nearly 200 times higher than that of aluminum [12]. However, aramid fibers are sensitive to ultraviolet (UV) light, and the unprotected ones will discolor and lose strength with prolonged exposure [10]. Besides, because of the lack of chemically active groups, there is usually a poor adhesion between the aramid fiber and the matrix. Therefore, some surface treatments can be used to overcome these drawbacks. For example, tetrabutyl titanate as a sol–gel precursor of a nanosized TiO_2 coating of aramid fibers can increase the photostability of the aramid fibers [13]. Gamma-ray radiation on the surface of aramid fibers was effective to improve the interfacial bonding strength between aramid fibers and epoxy resin [14].

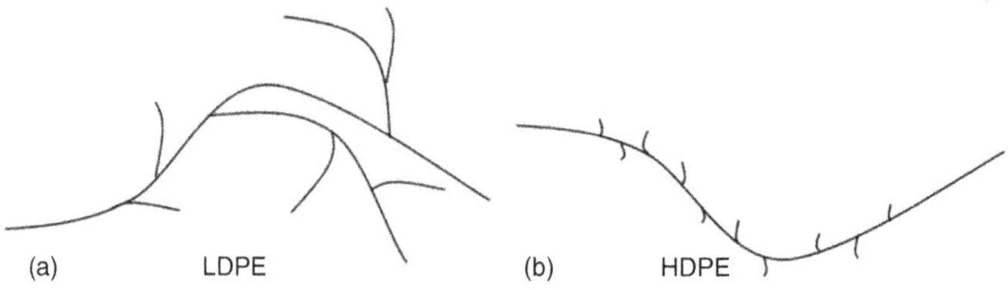

Figure 2.3 (a) Structure of LDPE and (b) HDPE [16].

PE fibers are made of polymerized PE units. PE has a semicrystalline structure and its molecular structure is linear [15]. They are produced from PE by the melt spinning process [15]. The mechanical properties of PE are highly dependent on the branch, crystal structure, and molecular weight. Therefore, PE is generally classified by density and molecular weight, such as low-density PE (LDPE), linear low-density PE (LLDPE), medium-density PE (MDPE), high-density PE (HDPE), and ultra-high-molecular-weight PE (UHMWPE). Figure 2.3a shows the structure of the LDPE, which has relatively long branches on the main molecular. This structure prevents the molecules from packing closely together and results in the flexibility and low tensile strength. In the structure of HDPE, as shown in Figure 2.3b, there are fewer major branches on long chains. This results in a higher crystallinity and higher tensile properties compared with LDPE [16].

UHMWPE is produced by the gel spinning process. In this method, a highly oriented fibrous structure with high crystallinity (95–99%) can be formed [17]. Therefore, it presents the highest tensile strength and modulus in the PE family. Besides, UHMWPE also has higher abrasion resistance and wear resistance than HDPE. This is because UHMWPE has extremely long chains, with a molecular mass between 3.5 and 7.5 million [18]. These long chains can transfer load more effectively to the polymer backbone because the intermolecular interactions can be enhanced [15]. This results in a very tough material, with the highest impact strength of any thermoplastics presently used [15]. The melting temperature, however, is a limitation for the application of PE fibers. As presented in Table 2.2, the maximum temperature limitation of UHMWPE is around 155 °C. The conventional forms of UHMWPE are yarns and fabrics.

HDPE and UHMWPE fibers have been used in producing FRP composites. Incorporating chopped HDPE fiber into acrylic denture base resin can reduce the water sorption, dimensional changes and increase the stiffness and impact strength [20, 21]. The wear behavior of UHMWPE fiber-reinforced ethylene–butene copolymers was researched to extend the lifetime of artificial

Table 2.2 Typical melting temperature ranges of PE [17–19].

	UHMWPE	HDPE	MDPE	LLDPE	LDPE
Density (g/cm^3)	0.93–0.949	0.94–0.97	0.93–0.94	0.91–0.93	0.92–0.93
Melt temperature (°C)	144–155	128–136	120–130	120–130	105–115

joints made by UHMWPE fiber-reinforced composites. The thickness of the neat matrix surface layer has more influence on the running-in wear rates than the intrinsic material properties [22].

Polypropylene (PP) fiber is formed from a PP by the traditional melt spinning process. It has a linear structure based on the monomer C_nH_{2n}. The crystallinity of PP fiber is generally between 50% and 65%, which is higher than that of LDPE fiber but lower than that of HDPE fiber. PP fiber has a low density of 0.91 g/cm^3, which is the lowest of all synthetic fibers [23]. PP fiber also displays good heat-insulating properties, and it is highly resistant to acids, alkalis, and organic solvents [24]. The application of PP fiber depends on the form of fiber configuration. The PP fibers in filament and monofilament forms are used to manufacture floating cables and are woven to nets and filter fabrics. Staple PP fiber is used for producing blankets, outerwear fabrics, and knitwear. Some filter fabrics are also manufactured by staple fibers [25]. Because of the lighter weight and higher corrosion resistance when compared with steel fiber, PP fiber can even be used as a reinforcement material in concrete as the replacement of steel fiber. In FRP composites, PP is mainly used as the matrix. The common fiber-reinforced PP composites include carbon fiber-reinforced PP [26], glass fiber-reinforced PP [27], and plant-based natural fiber-reinforced PP composites [28].

Polyester fiber is made from long-chain synthetic polymer. At least 85 wt% of the polymer is an ester of dihydric alcohol (HOROH) and terephthalic acid [29]. Polyethylene terephthalate (PET) predominates the polyester fiber production because PET has good end-use properties and low cost of production. Besides, it is easy to modify PET through physical and chemical techniques, which can weaken the negative properties (e.g. hydrophobicity and pilling tendency) and enhance the positive properties (e.g. tenacity or resistance to UV light) of PET fibers [30]. In PET fiber-reinforced composites, grafted copolymers (such as glycidyl methacrylate and maleic anhydride) are usually used to graft PET fiber to obtain a better interfacial adhesion [31]. PET fiber-reinforced composite has also been studied to retrofit reinforced concrete (RC) square columns in order to improve the axial compressive strength and the ductility of the columns [32]. In the study of Ref. [32], specimens were under a constant axial compression load and cyclic lateral loads of increasing displacement. The elastic modulus and tensile strength of PET–FRP were 18 times and 2.9 times lower than those of the corresponding high-modulus aramid FRP (HMA–FRP), respectively. However, the ultimate tensile strain of PET–FRP was 3.2 times higher than that of HMA–FRP. The large strain capacity provided the ductility of the jacketed RC square columns, and PET–FRP ruptured after the ultimate displacement was reached.

2.2.1.2 Inorganic Synthetic Fibers

Inorganic fibers are constituted mainly by inorganic natural elements such as carbon, silicon, and boron. Conventional inorganic synthetic fibers include carbon, glass, boron, and silica carbide (SiC) fibers.

Carbon fiber is described as the fiber that contains at least 92% carbon content. When the carbon content exceeds 99%, the fiber is also named as graphite fiber. In the atomic structure of carbon fiber, carbon atoms arrange in a regular hexagonal pattern. Layer planes in carbon fibers may be either turbostratic, graphitic, or a turbostratic–graphitic hybrid structure [33]. In turbostratic carbon fiber, which has high tensile strength, the sheets of carbon atoms are together haphazardly folded or crumpled. The graphitic fiber has high Young's modulus and thermal conductivity. According to the tensile properties, carbon fibers can be roughly classified into ultra-high-modulus (>450 GPa), high-modulus (350–450 GPa), intermediate modulus (200–350 GPa), low-modulus (<100 GPa), and high-strength (>4.5 GPa) carbon fibers [15, 33]. Additionally, carbon fibers can also be categorized by the type of the precursor fibers, such as polyacrylonitrile (PAN)-based carbon fiber, pitch-based carbon fiber, mesophase pitch-based carbon fiber, isotropic pitch-based carbon fiber, rayon-based carbon fiber, and gas-phase-grown carbon fiber. Most PAN-based carbon fibers are high-strength fibers. Mesophase-pitch-based carbon fibers produced by spinning of anisotropic pitches are high-modulus fibers. Isotropic pitch and rayon-based carbon fibers have very high surface areas (>1500 m^2/g) and are likely to dominate as materials for liquid as well as gaseous adsorption and environmental protection [34]. The characteristic of gas-phase-grown carbon fiber is its high graphitizability, which indicates high conductivity.

Because of the high tensile strength and modulus, high resistance to creep, fatigue, wear, and corrosion [35], carbon fiber has been used to reinforce polymers, ceramics, and alloys. The forms of carbon fiber include short carbon fibers, carbon yarn, carbon fabric (Figure 2.4), and recycled carbon fiber mats. Carbon fabric-reinforced polymer is one of the most commonly used FRP composites. It has been widely applied in many industries including aerospace, automotive, wind turbines, sport and leisure, and civil engineering (Figure 2.5). Carbon fibers are also used to reinforce ceramics and alloys, such as carbon

Figure 2.4 Carbon fabric [35].

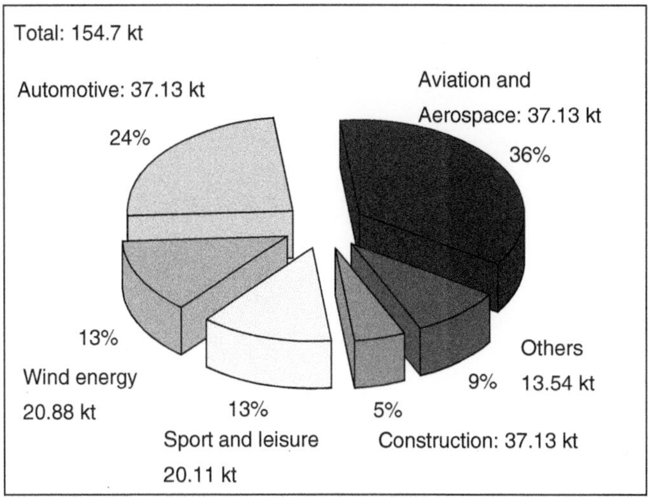

Figure 2.5 The global annual required quantity for carbon composites by application in 2018 (kt = kilo tonnes) [36].

short fiber-reinforced C/SiC composites [37] and carbon short fiber-reinforced Al–12Si alloys [38]. With the development of recycling technology of recycled carbon fiber, which can reduce environmental impacts compared to virgin carbon fiber production, it has gained more interest from researchers. The variation of recycled carbon fiber, however, is an obstacle in their application. In the research of Varna [39], the recycled carbon fibers used in the study were produced by milling of carbon fiber composites from the production waste of the aircraft industry. The length varied from 1 to 20 mm in recycled carbon fiber/PP composite, and elastic modulus was in the range from 6 to 10 GPa [39].

Glass fibers are extremely fine fibers of silicate glasses, which are produced from limestone, borosilicate, and other raw materials by melting, drawing, and cooling [40]. Glass fibers can be categorized in the form of continuous fiber, roving, woven fabric (Figure 2.6), and chopped strand mat. Another classification of glass fiber is based on its functions. Table 2.3 presents the composition for commercial glass fibers with different functions. Glass fibers are used without matrix as fibrous blankets for thermal and acoustical insulation as well as filters [43]. Glass fiber-reinforced polymer (GFRP) composites have a wide range of application such as civil, automobile, marine, and aeronautical engineering. In addition, GFRP has also been applied in the manufacturing of sport and leisure goods. The proportion of its applications in Europe in 2018 is shown in Figure 2.7. Over 70% of GFRP composites were applied in transport and construction fields. For example, GFRP has been used to make car bonnet and bumper. Some existing bridge decks and construction beams have been strengthened by GFRP plates. Although the stiffness of glass fiber is lower than that of carbon fiber, the glass fiber has a high specific strength and, most of all, a very reasonable cost [44].

Boron fibers are made by a chemical vapor deposition process in which the boron vapors are deposited onto a fine tungsten or carbon filament [45]. Boron

Table 2.3 Composition ranges (%) and applications for commercial glass fibers [41, 42].

Types	Material	Application	Composition												
			SiO_2	Al_2O_3	B_2O_3	CaO	MgO	ZnO	BaO	Li_2O	$Na_2O + K_2O$	TiO_2	ZrO_2	Fe_2O_3	F_2
C-glass	Calcium borosilicate glasses	Used for their exceptional stability in acidic environments	64–68	3–5	4–6	11–15	2–4	—	0–1	—	7–10	—	—	0–0.8	—
D-glass	Borosilicate glasses	Radomes and other specialty applications requiring permeability to electromagnetic waves	72–75	0–1	21–24	0–1	—	—	—	—	0–4	—	—	0–0.3	—
E-glass	Alumina–calcium–borosilicate glasses	Strength and high electrical resistivity	52–56	12–16	5–10	16–25	0–5	—	—	—	0–2	0–1.5	—	0–0.8	0–1
ECR-glass	Calcium aluminosilicate glasses	Acid corrosion resistivity, strength, and electrical resistivity	54–62	9–15	—	17–25	0–4	2–5	—	—	0–2	0–4	—	0–0.8	—
R-glass	Calcium aluminosilicate glasses	Higher strength and temperature resistance	55–65	15–30	—	9–25	3–8	—	—	—	0–1	—	—	—	0–0.3
AR-glass	Alkali-resistant glasses	Used in cement and concrete	55–75	0–5	0–8	1–10	—	—	—	0–1.5	11–21	0–12	1–18	0–5	0–5
S-glass	Magnesium aluminosilicate glasses	High strength, modulus, and durability under conditions of extreme temperature of corrosive environments	64–66	24–25	—	0–0.18	9.5–10.2	—	—	—	0–0.2	—	—	0–0.1	—

Figure 2.6 Glass fabric [40].

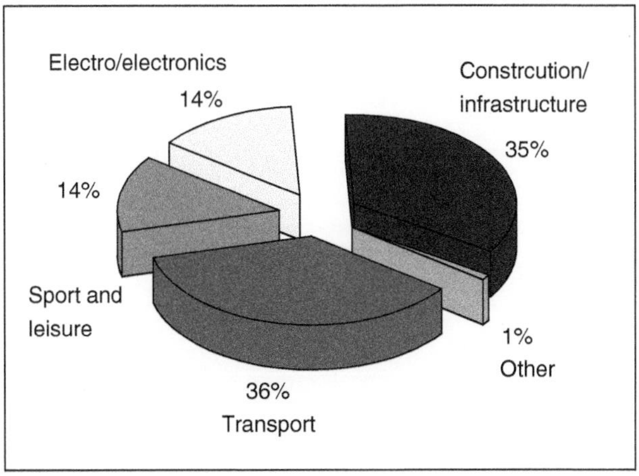

Figure 2.7 GFRP production in Europe by application industry in 2018 [36].

fibers are very stiff. They are five times stiffer than glass fibers [45]. In FRP composites, generally boron fibers are used to reinforce epoxy, polyimide, or phenolic resin [46]. The application of boron fiber is limited mostly in the field of aircraft and space because of its high cost [45]. For example, an equivalent quantity of boron/epoxy composite is roughly 12 times the price of carbon/epoxy composite [47]. Therefore, when boron fibers are applied in sport goods to make products thinner and stiffer, they are usually hybridized with carbon fibers to reduce the cost.

SiC fibers are inorganic fibers with β-silicon carbide structure. They are produced by spinning, carbonizing, or chemical vapor deposition from silicide. SiC fibers are extremely strong, oxidation-resistant fibers. Therefore, they are mostly applied for high-temperature applications. SiC can retain the structural stability and strength retention at 1000 °C. At 1200 °C, the passive oxidation begins, but the bulk form SiC can be used up to 1600 °C [48]. In FRP composites, mostly SiC is the filler in the form of a microparticle or nanoparticle/whisker. Adding

4 wt% SiC nanoparticles and nanowhiskers into epoxy can increase the wear and frictional properties 50% and 30%, respectively [49].

2.2.2 Natural Fibers

The use of natural fibers has a long history in textile products. In the FRP industry, however, synthetic fibers dominate the market because of their high mechanical properties and consistency. Recently, the concern of using plant-based natural fibers to replace synthetic fibers increases the popularity because of their benefits in economy, renewability, and biodegradability. There is a wide range of natural fibers as presented in Figure 2.1, which can be classified into plant-based, mineral-based, and animal-based natural fibers.

2.2.2.1 Plant-Based Natural Fibers

Plant-based natural fibers have many types that can be categorized by their origins. For example, jute, flax, hemp, ramie, and kenaf fibers are classified as bast fibers because they come from the bast of plants. Meanwhile, sisal and banana fibers are leaf fibers as they come from the leaf of their plants. Cotton fiber is a seed fiber, while coir and oil palm are fruit fibers. Wood fibers, such as soft wood or hard wood, are bored from the stem of the wood. Although plant-based fibers have different origins, they have a similar microstructure (Figure 2.8) and consist of similar chemical constituents including cellulose, hemicellulose, lignin, pectin, and wax in different concentrations.

In Figure 2.8, the microstructure of plant-based fiber cell is a multilayer structure that consists of two cell walls: the primary wall and the secondary cell wall.

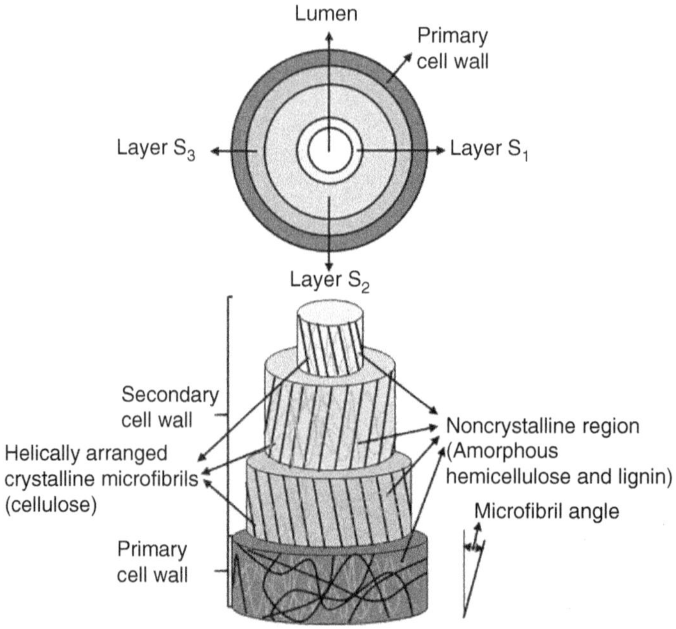

Figure 2.8 Structure of natural fibers [50].

In the primary wall, there are disorderly arranged crystalline cellulose microfibril networks [51]. The secondary cell wall, which is responsible for most of the strength of the fiber, consists of three cell walls (S_1, S_2, and S_3) and has helically arranged crystalline cellulose microfibril networks. Among three cell walls, S_2 layer is the thickest and has the most cellulose content. In S_2 layer, cellulose microfibrils are highly oriented with an angle (mircrofibril angle). Mircrofibril angle is the angle between microfibrils and the fiber axis. When the mircrofibril angle is smaller, the fiber has higher stiffness and strength [52]. When the mircrofibril angle is larger, the fiber will have higher ductility and elongation in tension [52]. In the secondary cell wall, these highly crystalline microfibrils are glued together by the amorphous hemicellulose [53]. Hemicellulose is a hydrophilic component that is responsible for the moisture absorption of plant-based fibers. In the amorphous region, there are also pectins and lignins. Pectin is a collective name for heteropolysaccharides. The flexibility of plants is mainly dependent on this amorphous hemicellulose/pectin matrix [54]. Lignin is a complex hydrocarbon polymer that is amorphous and hydrophobic. It cannot be broken down to monomeric units and is the compound that is responsible for the rigidity of plants [54]. The wax substances affect the fiber wettability and adhesion characteristics [2]. The properties of natural fibers are significantly dependent on the chemical composition and structure of the fiber, which are sensitive to fiber type, growing environments as well as the completely manufacturing procedure including harvesting, extracting, treating, and storage. [55]. The further detailed mechanical and physical properties as well as the chemical compositions of different plant-based fiber in comparison to E-glass fiber are listed in Table 2.4.

According to the report of world natural fiber production in 2018 (see Figure 2.9) [58], cotton fiber takes the mainstay in the production of plant-based fibers and animal-based fibers.

Cotton fiber comes from cotton plant (*Gossypium*), which is native to tropical and subtropical regions around the world [59]. The diameter is around 10–45 μm. China, the United States, Russia, and India are the four major producers of

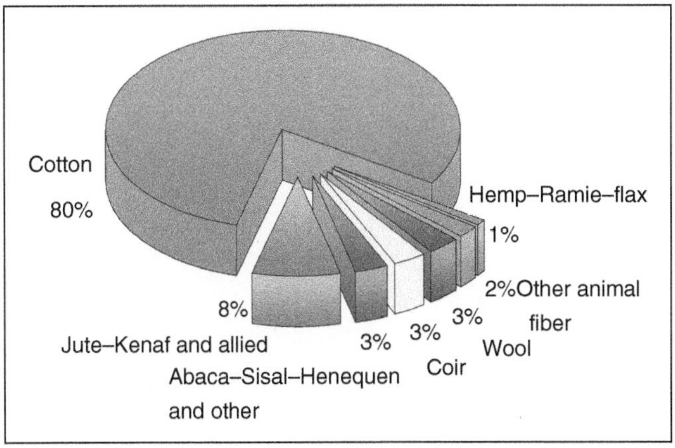

Figure 2.9 World natural fiber production: 110 million tons in 2018 [57].

Table 2.4 Compiled properties and chemical compositions of natural fibers [56].

Fiber type	Density (g/cm^3)	Diameter (μm)	Tensile strength (MPa)	Tensile modulus (GPa)	Specific modulus (approx.)	Elongation (%)	Cellulose (wt%)	Hemicellulose (wt%)	Lignin (wt%)	Pectin (wt%)	Waxes (wt%)	Microfibrillar angle (°)
E-glass	2.5–2.59	<17	2000–3500	70–76	29	1.8–4.8	—	—	—	—	—	—
Abaca	1.5	—	400–980	6.2–20	9	1.0–10	56–63	20–25	7–13	1	3	—
Alfa	0.89	—	35	22	25	5.8	45.4	38.5	14.9	-	2	—
Bagasse	1.25	10–34	222–290	17–27.1	18	1.1	32–55.2	16.8	19–25.3	—	—	—
Bamboo	0.6–1.1	25–40	140–800	11–32	25	2.5–3.7	26–65	30	5–31	—	—	—
Banana	1.35	12–30	500	12	9	1.5–9	63–67.6	10–19	5	—	—	—
Coir	1.15–1.46	10–460	95–230	2.8–6	4	15–51.4	32–43.8	0.15–20	40–45	3–4	—	30–49
Cotton	1.5–1.6	10–45	287–800	5.5–12.6	6	3–10	82.7–90	5.7	<2	0–1	0.6	—
Flax	1.4–1.5	12–600	343–2000	27.6–103	45	1.2–3.3	62–72	18.6–20.6	2–5	2.3	1.5–1.7	5–10
Hemp	1.4–1.5	25–500	270–900	23.5–90	40	1–3.5	68–74.4	15–22.4	3.7–10	0.9	0.8	2–6.2
Jute	1.3–1.49	20–200	320–800	8–78	30	1–1.8	59–71.5	13.6–20.4	11.8–13	0.2–0.4	0.5	8.0
Kenaf	1.4	—	223–930	14.5–53	24	1.5–2.7	31–72	20.3–21.5	8–19	3–5	—	—
Oil palm	0.7–1.55	150–500	80–248	0.5–3.2	2	17–25	60–65	—	11–29	—	—	42–46
Piassava	1.4	—	134–143	1.07–4.59	2	7.8–21.9	28.6	25.8	45	—	—	—
PALF	0.8–1.6	20–80	180–1627	1.44–82.5	35	1.6–14.5	70–83	—	5–12.7	—	—	14.0
Ramie	1.0–1.55	20–80	400–1000	24.5–128	60	1.2–4.0	68.6–85	13–16.7	0.5–0.7	1.9	0.3	7.5
Sisal	1.33–1.5	8–200	363–700	9.0–38	17	2.0–7.0	60–78	10.0–14.2	8.0–14	10.0	2.0	10–22

Figure 2.10 Flax fabric [2].

cotton. Most application of cotton is in textile industry. In FRP composite, cotton fiber has been investigated to reinforce epoxy [60], PP [61], polyester [62], and poly(lactic acid) [63]. In cotton fiber-reinforced PP composite, when the content of cotton fiber is 10 wt%, the tensile strength is lower than pure PP. When the content of cotton fiber increases to 20 and 30 wt%, the tensile strength increases because of entanglement of cotton fibers [61]. Besides, using cotton fibers to reinforce unsaturated polyester resin can improve the structural integrity under the sliding wear condition [62].

Flax belongs to the family *Linaceae* and grows in temperate climates. The diameter is 12–600 μm. China, France, Canada, and Belarus are the largest flax-producing countries [59]. The forms available in the application of flax fiber include short fiber, roving, nonwoven mat, and fabric (shown in Figure 2.10). These forms are also studied in FRP composites. Many researchers have investigated the mechanical properties (including tensile strength, modulus, bending strength, impact strength, fracture toughness, and fatigue behavior) of flax fiber-reinforced epoxy, PP, HDPE, and polyactides (PLA) [2]. Flax fibers are cost-effective materials and have the potential to replace glass fibers as the reinforcement material in composites [2]. Flax FRP composites have been applied in many fields such as sporting goods [64] and automobile industry [58].

Except for cotton and flax fibers, there are many other plant-based fibers that have been researched in FRP composites, such as hemp [65], jute [66], sisal [67], kenaf [68], banana [69], PALF [70], abaca [71], coir [72], oil palm [72], bamboo [73], and ramie [74] fibers. Like flax in FRP composites, these plant-based fibers have been investigated to reinforce various thermoplastic and thermoset polymers, such as PP [65, 74], PLA [75], and epoxy [76]. Recently, using plant-based fiber-reinforced biodegradable polymers is becoming more popular to obtain completely "green" composites. Examples of biodegradable polymers include lignin-based epoxy, soy-based resins, and epoxidized linseed and soybean oil [65]. The mechanical properties of plant-based FRP composites in short term are not only dependent on the nature of materials (i.e. fiber strength and fiber structure) but also on the volume fraction of fiber and the interfacial bond between fiber and resin. If plant-based fibers are applied in FRP in the form of fabric, the woven method and the fabric stacking sequence will also influence the properties especially to the bending and impact strength. For the mechanical properties of plant-based FRP composites after long-term exposure, because of the hydrophilic nature of plant-based fiber, the water absorption is crucial for the interfacial bond and will result in the reduction of mechanical properties. Therefore, many approaches have been conducted to reduce the water absorption and increase the interfacial bond, such as alkali treatment [67, 69] on the fibers and adding nanoparticles [77] into resins.

2.2.2.2 Animal-Based Natural Fibers

Animal-based fibers are the second most widely used natural fibers after plant-based fibers [78]. Animal-based fibers mainly include keratin, silk, and chitosan fibers.

The α-keratin fibers mainly come from wool, feathers, and hair of various animals. Wool is usually obtained from sheep, goat, rabbits, and other mammals. Its structure is formed by epidermis, cortex, and medullary tissue. The annual production of sheep wool is approximately 1.2 million metric tons and 90% of them are consumed by the textile industry [79]. Conzatti et al. investigated the tensile properties of short wool fiber-reinforced PP (Wool-PP) composites. The Wool-PP composites were manufactured by melt blending in an internal batch mixer (170 °C). The author found a critical length of the short fiber. Under this length, the tensile strength of Wool-PP increases as the fiber length increases. When the fiber length is beyond the critical length, the increasing length will reduce the tensile strength of the composites [80]. Chicken feather is a waste animal-based material from poultry industry. The low density and good thermal and acoustic insulation are the main advantages of chicken feather. When the sound frequency was about 4 kHz, the sound absorption of feather-reinforced PP composite was around twice than jute-reinforced PP composite [81]. Other keratin fibers, such as human hair, have also been investigated to reinforce concrete and PP. When the content of human hair fiber increases in the range of 3–5 wt%, hair fiber-reinforced PP composites have better flexural, impact strength but lower tensile strength than nonreinforced PP composites. However, when the fiber content increases, in the range of 10–15%, all the flexural, impact, and tensile strength were smaller than the nonreinforced PP composites [82]. This may be interpreted by the poor distribution of the random short fibers into the matrix.

Silk is another important animal-based fiber obtained from various insects (including butterfly species, spiders, and silkworm). Silk consists of highly structured proteins and has high tensile strength and high elongation [79]. Mulberry silk from silkworm and dragline silk from spider have been widely discussed and used [79, 83]. Their mechanical and physical properties are presented in Table 2.5. In FRP composite, silk has been studied to reinforce PP

Table 2.5 Mechanical and physical properties of two types of silk fibers [79, 83].

Fiber	Degree of fiber crystallinity (%)	Maximum use temperature (°C)	Thermal degradation (°C)	Tensile strength (GPa)	Extensibility (%)
Mulberry silk	38–66	170	250	0.6	18
Dragline silk	20–45	150	234	1.1	30

polymer. In Ref. [84], silk fiber-reinforced PP (Silk-PP) and jute fiber-reinforced PP (Jute-PP) had the same weight percent of fiber (20 wt%). The tensile strength, bending strength, Young's modulus, bending modulus, and impact strength of Silk-PP composites were 35%, 1.33%, 29%, 51%, and 42%, respectively, higher than those properties of Jute-PP composites [85]. After exposing in soil for 12 weeks, the reductions of tensile strength and Young's modulus of silk-PP were 10.1% and 28.94%, respectively. However, Jute-PP lost 32.2% and 56% of tensile strength and Young's modulus, respectively [85].

Chitosan is one of the available natural polysaccharides. It can be found in the exoskeletons of the arthropod (e.g. crab, shrimp, and butterfly) and some fungi [86]. Melt and wet spinning processes are the main methods to produce chitosan fibers. Chitosan textiles can be used in medical field because they are not adaptable for the bacterial growth and can accelerate wound healing [87]. In FRP composite, chitosan has been investigated to reinforce epoxy in the form of nanoparticle, and chitosan has also been used as a grafting agent [88]. Adding 4 wt% of chitosan, the tensile and bending strength of epoxy were increased by 80% and 47%, respectively. In addition, the hemolytic ratio (hemoglobin release ratio in blood) of chitosan/epoxy composite was less than 5% [89], which has only little effect on the erythrocyte necrosis. Therefore, the chitosan/epoxy composites are favorable biomaterials in medical field.

2.2.2.3 Mineral-Based Natural Fibers

The mineral-based fibers include basalt, asbestos, and brucite fibers. Asbestos fibers were widely used earlier, as they are excellent for fireproof and thermal insulation. Asbestos fibers have been banned because of their detriments to human health. Brucite fiber is an alkaline mineral that has excellent antialkaline property. It is researched to add into a cement mortar to enhance the flexural strength of concrete [90].

As an alternative mineral-based fiber to asbestos, basalt fiber has been widely utilized until now. Similar to asbestos fiber, basalt fibers also have good thermal stability because they are originated from the molten basalt, which is obtained by melting crushed basalt rocks at 1400 °C [91]. However, the high melting temperature makes basalt fiber a fiber with high-energy consumption. Basalt fiber has higher tensile strength and modulus than those of glass fiber, as shown in Table 2.6. In the aerospace and automotive applications, basalt fibers are used as textile (e.g. nonwoven mat) for fire prevention. In civil engineering, basalt fibers in the form of rebar, chopped

Table 2.6 Mechanical and physical properties of basalt fiber [92].

Fiber type	Diameter (μm)	Tensile strength (MPa)	Elastic modulus (GPa)	Breaking extension (%)	Temperature withstand (°C)
Continuous basalt	6–21	3000–4840	79.3–93.1	3.1	−260 to +700
E-glass	6–21	3100–3800	72.5–75.5	4.7	−50 to +380

Table 2.7 Price of several fibers [95].

Fiber type	PP fiber	Aramid fiber	Basalt fiber	Carbon fiber	Glass fiber	Plant-based fiber
Price (€/kg)	3.0–6.0	20–80	4.0–6.0	23–36	1.0–1.5	0.3–1.5

fibers, and mesh have been used to strengthen concrete. The conventional forms of basalt fiber are rebar, chopped fiber, mesh, woven fabric, and nonwoven mat. In FRP composites, basalt fibers have been investigated to reinforce thermoplastic, thermoset polymers, and biodegradable polymers [84]. According to the research by Lopresto et al., basalt FRP has higher Young's modulus than E-glass FRP when the fiber volume fractions were 51% and 46%, respectively, for basalt and glass FRP with the same thickness [93].

2.2.3 Hybrid Fiber Product

As discussed above, synthetic and natural fibers have different properties because of their chemical compositions and structures, which determine their main application fields. Their further applications, however, are limited by their own weaknesses, such as the high cost of carbon fiber and hygroscopicity of plant-based fiber. Fiber hybridization is a strategy in which two or more types of fibers are combined. The hybrid fiber composite can obtain a better balance in mechanical properties than composites with a single fiber type [94]. For example, from Table 2.7, from the economics point of view, it can be seen that the cost of carbon fiber is much higher than plant-based fibers. Hybridizing carbon fiber with plant-based fibers can reduce the cost while maintaining the high mechanical performance of carbon fiber. Various fiber forms (short fiber, yarn, fabric, and mat) also provide the possibility for hybridization to obtain designated properties. There are three main configurations of fiber hybridization, i.e. intrayarn, intralayer, and interlayer (Figure 2.11). On the fiber level, different types of fibers can be twisted together to the hybrid yarn, as depicted in Figure 2.11a. The intralayer hybrid composite is produced by weaving different fiber yarns in one fabric (Figure 2.11b). The interlayer hybrid composite is manufactured by stacking different fabrics (Figure 2.11c), which is the cheapest and simplest method to obtain hybrid composites [94].

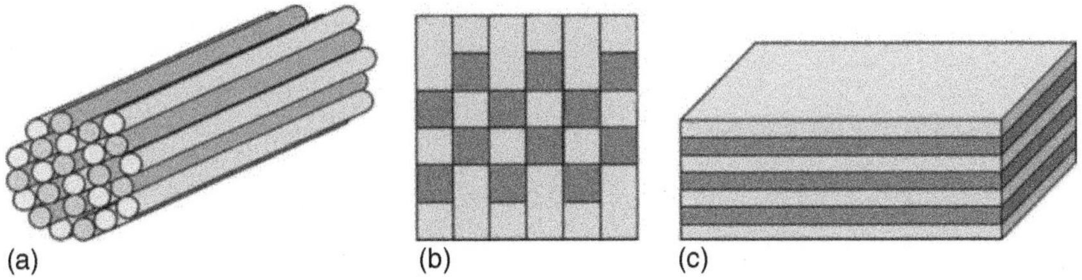

Figure 2.11 The three main hybrid configurations: (a) intrayarn or fiber-by-fiber, (b) intralayer or yarn-by-yarn, and (c) interlayer or layer-by-layer [94].

2.3 Bioepoxy

Thermosetting polymers have a broad application because of their high performance in strength, durability, and thermal and chemical resistances achieved by their high cross-linked structures [96]. The production of thermosetting polymers contributes about 20% of the global plastic production. Among them, approximately 70% are epoxy. Generally, epoxy resins are low molar mass prepolymer. They usually contain at least two epoxide groups (i.e. glycidyl or oxirane groups). The epoxy can be polymerized through its own anionic or cationic homopolymerization, or the reaction with coreactants, e.g. polyfunctional amines, acids, anhydrides, phenols, alcohols, and thiols [97]. However, in epoxy cross-linked polymer, more than 90% are produced with bisphenol A (BPA), which has been classified as carcinogen mutagen and reprotoxic (CMR) [98]. Besides, BPA may also have a negative health effect on the immune and reproductive systems [97]. Deriving from renewable biomass, the bioepoxy can be the alternative to the epoxies produced by BPA.

Based on the original source and molecular structure, according to Ramon et al. [99], the bioepoxy resins can be categorized into the following groups: (i) natural oil-based epoxy, (ii) isosorbide-based epoxy, (iii) furan-based epoxy, (iv) phenolic and polyphenolic epoxy, (v) epoxidized natural rubber, (vi) epoxy lignin derivatives, and (vii) rosin-based epoxy.

2.3.1 Natural Oil-Based Epoxy

The natural oil is a renewable resource that is most utilized in the chemical industry. It contains a plenty of triglycerides. The hydrolysis of a triglyceride will form one glycerol and three fatty acids that account for 95% of the total weight of the triglyceride. Most of the fatty acids have double bonds, hydroxyl groups, or possibly epoxy groups. The representative chemical structure of soybean oil can be found in Figure 2.12. There are three different ways for the transformation of vegetable oil to bioepoxy: (i) the direct polymerization of the double bonds; (ii) the transformation of double bonds into functional groups (e.g. oxidation process that oxidizes the double bonds to oxirane group) and then polymerization; and (iii) the simple fatty acids or diglycerides through hydrolysis and then they are used as base monomers for polymer syntheses [101]. The commonly utilized vegetable oils are soybean oil, linseed oil, canola oil, and karanja oil. The mechanical

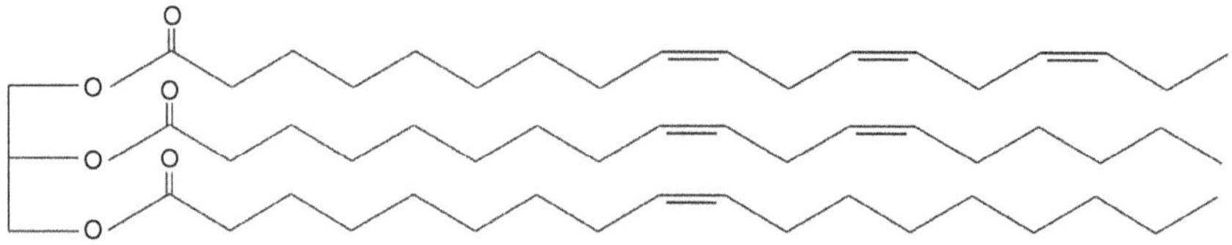

Figure 2.12 Representative chemical structures of soybean oil [100].

properties of bio-based polymer derived from some of the epoxidized vegetable oils are presented in Table 2.8 according to Zhu et al. [102] and Sudha et al. [103]. For the nomenclature in the table, ESO stands for the epoxidized soybean oil and EMS and EAS are modified soy-based resins, i.e. epoxidized methyl soyate and epoxidized allyl soyate, respectively. ECO stands for epoxidized canola oil. T_g stands for glass transition temperature. In the one-step curing, the epoxidized soyate resin was directly mixed with base Epon resin. In the two-step curing, a prepolymer was prepared by adding a highly reactive curing agent (*para*-amine cyclohexylmethane) firstly into the epoxidized soyate resin. Then, the prepolymer was mixed with Epon epoxy [102]. According to the table, the glass transition temperature, the tensile and flexural strength, and modulus of the blends decrease as more epoxidized vegetable oil was added. Exceptional are the blends with low bio-based polymer content from Zhu et al. [102]. The 10 wt% ESO and 10 wt% EAS in both curing methods as well as 20 wt% EAS and 10 wt% EMS in two-step curing exhibit higher tensile and flexural modulus and strength than Epon epoxy because of the higher reactivity of EAS and EMS compared to ESO. This gives EAS and EMS denser intermolecular cross-linking and in turn higher mechanical properties. The epoxy and curing agents can form the intermolecular bond easier and the functionalities and reactivity increase. This results in a more densely cross-linked networks in blends and leads to higher properties.

2.3.2 Isosorbide-Based Epoxy (IS-EPO)

Isosorbide is derived from starch. The starch is firstly hydrolyzed into D-glucose. Through hydrogenation, D-glucose will be transferred into sorbitol. After the multistep process, isosorbide can be obtained [99, 104, 105] (Figure 2.13). Besides biodegradability and bio-origin, isosorbide is a rigid organic diol and has a similar structure to BPA. However, it has no endocrine disrupting effect [106]. In isosorbide-based epoxy, the structure of diglycidyl ethers of isosorbide (DGEI) and the curing agent structure are the main factors on the final mechanical properties of cured epoxy resin. Table 2.9 exhibits the comparison of the bio-based DGEI with the petroleum-based diglycidyl ether of bisphenol A (DGEBA) with two different amines, i.e. diethylenetriamine (DETA, purity of 97%) and isosorbide-based diamine (ISODA) from Hong et al. [105]. In the test, two types of DGEI were investigated, i.e. DGEI monomer and DGEI polymeric. DGEI monomer has a monomer percentage of 96%, and its epoxy oxygen content (EOC) is 10.9%. The DGEI polymeric has a monomer percentage of 16% with EOC = 6.9%. The results show that the isosorbide-based epoxies have lower glass transition temperature than bisphenol-A-based epoxy. However, they have

Table 2.8 Mechanical properties of epoxidized natural oil.

Literature	Sample	T_g (°C)	Tensile modulus (MPa)	Tensile strength (MPa)	Flexural modulus (MPa)	Flexural strength (MPa)
Zhu et al. [102] (one-step curing)	Epon epoxy	74.8	3041	58	3021	110
	10 wt% ESO	69.1	3145	59	3117	106
	20 wt% ESO	61.2	2807	51	2841	93
	30 wt% ESO	54.2	2434	36	2641	79
	10 wt% EAS	72.5	3193	60	3214	112
	20 wt% EAS	63.0	2972	53	2959	100
	30 wt% EAS	58.4	2979	41	2579	89
	10 wt% EMS	64.0	2952	54	3310	107
	20 wt% EMS	54.2	2890	45	2917	88
	30 wt% EMS	47.1	2621	31	2510	70
Zhu et al. [102] (two-step curing)	Epon epoxy	74.8	3041	58	3021	110
	10 wt% ESO	72.3	3076	63	3234	119
	20 wt% ESO	67.0	3034	60	3090	111
	30 wt% ESO	61.9	2738	50	2910	99
	10 wt% EAS	75.1	3359	63	3503	127
	20 wt% EAS	69.2	3207	58	3359	123
	30 wt% EAS	65.0	2945	54	2979	103
	10 wt% EMS	68.0	3483	53	3214	115
	20 wt% EMS	63.3	3234	46	3083	110
	30 wt% EMS	55.3	3131	45	2841	98
Sudha et al. [103]	0 wt% ECO	96.6	3343	70	3358	96
	10 wt% ECO	91.4	3285	51	3509	83
	20 wt% ECO	71.4	2728	54	2507	100
	30 wt% ECO	47.4	2131	42	2485	82
	50 wt% ECO	39.2	901	18	1201	40

higher tensile strength, tensile modulus, and elongation. No obvious tendency can be found in flexural modulus. Łukaszczyk et al. [107] compared Epidian 5 DGEBA and isosorbide-based epoxy with four different hardeners, i.e. phthalic anhydride (PHA), tetrahydrophthalic anhydride (TPHA), triethylenetetramine (TETA), and isophoronediamine (IPHA). The glass transition temperature and the flexural and compressive strength are shown in Table 2.10. Similar results to the study of Hong et al. [105] can be found, such as the glass transition temperature of isosorbide-based epoxy is significantly lower than that of Epidian 5 DGEBA. Except for IPHA, the isosorbide-based epoxy has higher flexural strength than Epidian 5 DGEBA with all other types of hardener. The isosorbide-based epoxy with TETA or IPHA shows higher compressive strength than Epidian 5 DGEBA.

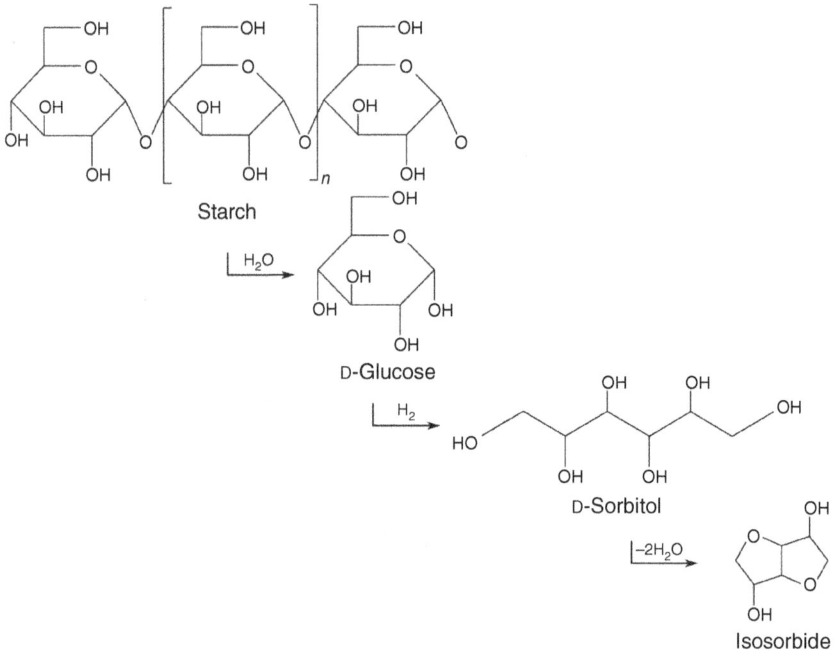

Figure 2.13 Chemical structure of D-isosorbide and the synthesis process of isosorbide [105].

Table 2.9 Comparison of the bio-based DGEI with the petroleum-based DGEBA with two different amines [105].

Composition (resin/hardener)	T_g [a)]	Tensile modulus (MPa)	Tensile strength (MPa)	Elongation (%)	Flexural modulus (MPa) [b)]
DGEBA/DETA	129 (134)	1389 (± 65.3)	26 (± 2.1)	3 (± 0.2)	3061 (± 49.0)
DGEBA/ISODA	74 (79)	1825 (± 82.1)	67 (± 2.3)	5 (± 0.7)	3364
DGEI (mono)/DETA	75 (76)	1798 (± 21.6)	62 (± 5.6)	6 (± 1.8)	4027
DGEI (mono)/ISODA	32 (43)	1532 (± 39.8)	41 (± 8.6)	5 (± 1.0)	1168
DGEI (polymeric)/ISODA	36 (43)	2461 (± 233.8)	52 (± 4.2)	3 (± 0.7)	3520
DGEI (polymeric)/DETA	48 (63)	1774 (± 141.9)	52 (± 9.4)	5 (± 0.3)	2747

a) Glass transition temperature; the numbers in parenthesis are T_g of the samples with additional curing at 150 °C for two hours.
b) The results without parenthesis are based on one specimen.

2.3.3 Furan-Based Epoxy

One of the first-generation furan derivatives of furan-based epoxy is furfural, which is the dehydration product of xylose. The xylose is a pentose sugar that has large quantities in the hemicellulose fraction of lignocellulosic biomass [108] such as corncobs and sugar cane bagasse [109]. To

Table 2.10 Comparison of Epidian 5 DGEBA and isosorbide-based epoxy with four different hardeners [107].

Composition (resin/hardener)	T_g	Flexural modulus (GPa)	Flexural strength (MPa)	Compression modulus (GPa)	Compressive strength (MPa)
Epidian 5/PHA	171	10.4	158.4	2.8	290.8
IS-EPO/PHA	108	17.4	225.5	2.1	254.1
Epidian 5/THPHA	172	9.8	27.9	2.5	122.2
IS-EPO/THPHA	95	15.1	100.5	2.2	88.8
Epidian 5/TETA	116	8.3	170.8	2.1	234.2
IS-EPO/TETA	49	5.5	228.3	2.1	311.6
Epidian 5/IPHA	141	10.4	175.4	2.3	193.9
IS-EPO/IPHA	73	14.6	158.5	2.0	318.1

Table 2.11 Tensile and fracture properties of furan-based epoxy compared with DGEBA [113]..

Sample	T_g (°C)	E modulus (GPa)	Ultimate tensile stress (MPa)	Failure strain (%)	Stress intensity factor (MPa m$^{1/2}$)
BOF/PACM	71	2.8 (± 0.1)	65.2 (± 1.1)	6.6 (± 2.6)	2.00
BOB/PACM	100	1.8 (± 0.1)	54.4 (± 1.7)	10.4 (± 3.5)	1.22
DGEBA/PACM	167	2.3 (± 0.1)	71.3 (± 5.1)	7.2 (± 1.7)	0.61

produce the furan-based epoxy, furfural will be transformed into furfuryl alcohol through hydrogenation. Furfuryl alcohol resin is one of the important furan-based epoxies, which is produced by self-polycondensation of furfuryl alcohol monomers [110]. During the synthesis process, one furfuryl alcohol monomer will react with the active a-hydrogen of another monomer in the presence of acid catalyst. Another furan derivative of furan-based epoxy is 5-hydroxymethylfurfural (HMF). The synthesis of HMF is based on the acid-catalyzed dehydration of hexoses [111]. Hu et al. [112] synthesized 2,5-bs[(2-oxiranylmethoxy)methyl]-furan (BOF) through etherification reaction between epichlorohydrin and 2,5-bis(hydroxymethyl)furan (b-HMF) catalyzed by tetrabutylammonium hydrogen sulfate and a ring-closing reaction due to the dehydrochlorination reaction with alkali. The authors also synthesized 1,4-bis[(2-oxiranylmethoxy)methyl]-benzene (BOB) through condensation reaction between epichlorohydrin and 1,4-benzenedimethanol. Besides, the tensile and fracture properties of BOF and BOB cured with 4,4′-methylene biscyclohexanamine (PACM) were compared with DGEBA/PACM system in another research of the author [113]. Their results are shown in Table 2.11. As can be seen, BOF/PACM exhibits highest elastic modulus, which may due to the existence of hydrogen bonding between hydroxyl groups and oxygen atoms according to the authors. Except for that, the cured furan-based epoxies all have lower T_g and ultimate tensile stress than that of DGEBA. However, the

Table 2.12 Flexural strength of polyphenolic epoxy compared with DGEBA [115].

Sample	T_g (°C)	Flexural strength (MPa)
GEC	178	63
GEHDGTE	155	56
GEFHDGTE	173	40
DGEBA	150	29

failure of furan-based epoxies shows higher fracture toughness than the ordinary DGEBA. The stress intensity factor of BOF/PACM and BOB/PACM are 2.00 and 1.22 MPa $m^{1/2}$, respectively. They are three and two times more than that of DGEBA/PACM (0.61 MPa $m^{1/2}$), respectively.

2.3.4 Polyphenolic Epoxy (Vegetable Tannins)

The vegetable tannins are the natural polyphenolic structures, which come from a multitude of trees and shrubs, such as black wattle or black mimosa bark (*Acacia mearnsii*), quebracho wood (*Schinopsis balansae* or *lorentzii*), oak bark (*Quercus* spp.), and chestnut wood (*Castanea sativa*). According to Pizzi [114], tannins can be divided into two groups, i.e. hydrolyzable tannins and condensed (polyflavonoid) tannins [114]. The main component in tannins is catechin, which is one of the most studied natural polyphenols for bioepoxy resin. For example, Basnet et al. [115] compared the glass transition temperature and flexural strength of the glycidyl ether of catechin (GEC), glycidyl ether of heat-dried green tea extract (GEHDGTE), and glycidyl ether of freeze-dried green tea extract (GEFDGTE) with commercial DGEBA epoxy resin (Table 2.12). Lignin derivative was used as a curing agent. Their results show that all these bio-based polyphenolic epoxies have higher temperature resistance and flexural strength than the DGEBA epoxy resin. This indicates that these materials have potential applicability to replace DGEBA epoxy resin in electronic applications. Tarzia et al. [116] epoxidized tri- and tetra-glycidyl ethers of gallic acid (GEGAs) with three different curing agents, i.e. isophorone diamine (IPDA), Jeffamine D-230 (DPG), and N,N-dimethylbenzylamine (BMDA, as an ionic initiator). Their mechanical properties were compared with DGEBA cured with the corresponding curing agent and are shown in Table 2.13. The results present that in comparison with IPDA, the cured polymers (both GEGA and DGEBA) all exhibit higher tensile strength and larger failure elongation. In comparison with the ordinary DGEBA epoxy resin, GEGA has inferior tensile properties (strength and failure elongation) but higher elastic modulus.

2.3.5 Epoxidized Natural Rubber (ENR)

The natural rubber is produced from the Brazilian rubber tree and can be epoxidized with peracid by the double bonds. The structure of ENR is shown in Figure 2.14 according to Chan [117].

Table 2.13 Comparison of mechanical properties of polyphenolic epoxy and DGEBA [116].

Sample	T_g (°C)	E modulus (GPa)	Tensile strength (MPa)	Elongation at break (%)
GEGA/IPDA	158	3.6 (± 0.3)	43.1 (± 13.1)	1.4 (± 0.3)
GEGA/DPG	98	3.5 (± 0.2)	70.6 (± 2.9)	6.1 (± 0.6)
GEGA/BDMA	136	3.2 (± 0.2)	31.2 (± 2.3)	1.1 (± 0.1)
DGEBA/IPDA	—	3.1 (± 0.8)	34.1 (± 2.3)	1.7 (± 0.2)
DGEBA/DPG	—	3.0 (± 0.2)	116.4 (± 7.0)	8.6 (± 0.3)

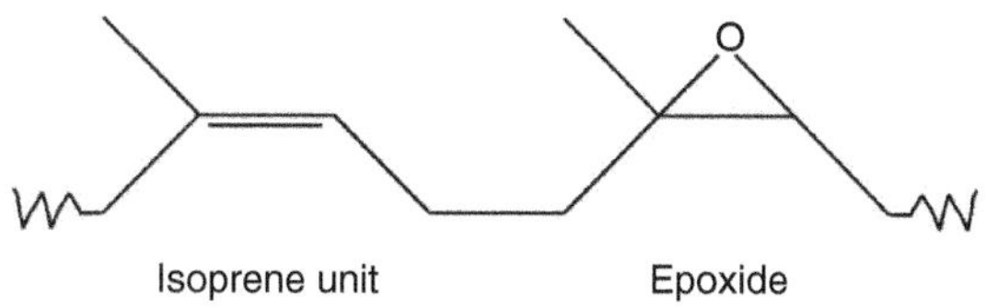

Figure 2.14 Chemical structure of epoxidized natural rubber [117].

Table 2.14 Impact properties of ENR-modified DGEBA [118].

Sample	T_g (°C)	Unnotched impact strength (J/m)	Notched impact strength (J/m)
Neat resin	118	6.87 (± 0.8)	1.85 (± 0.1)
5 wt% ENR	112	12.74 (± 0.7)	2.45 (± 0.05)
10 wt% ENR	112	20.98 (± 0.9)	2.93 (± 0.07)
15 wt% ENR	114	18.22 (± 0.7)	2.78 (± 0.05)
20 wt% ENR	109	16.59 (± 0.6)	2.55 (± 0.06)

Generally, ENR is used as a modifier for epoxy resin. For example, Mathew et al. [118] modified DGEBA epoxy resin with ENR. In their work, DGEBA epoxy resins modified by ENR of 5, 10, 15, and 20 wt% content are compared with neat epoxy polymer. Their results of T_g and impact properties are presented in Table 2.14.

In comparison to neat resin, the T_g value decreases when the resin is blended with ENR. The reduction is more significant when higher weight percentage of ENR is applied. The ENR-modified neat resin also shows higher impact resistance than the neat resin. Among all the percentages, the cured 10 wt% ENR-modified resin exhibits largest impact strength (unnotched: 20.98 ± 0.9 J/m and notched 2.93 ± 0.07 J/m) compared to that with other ENR content ratios. According to the author, the effective stress concentration and stress transfer behavior of the phase-separated, rubber-rich particles can be reached as the ENR content increases. Therefore, it amplified the plastic deformation to a certain extent. As ENR increases beyond 10 wt%, the aggregate size of rubber particles increases. These large particles act as deflection sites, which can lead to the catastrophic failure of the matrix.

Table 2.15 Thermal properties of epoxidized lignin and DEGBA [121].

Epoxy	Curing agent	Td_5 (°C)	Td_{10} (°C)	Lignin content (%)
DGEBA	Phenol novolac	361	378	0
DGEBA	Lignin (Cedar)	326	363	34.2
DGEBA	Lignin (Eucalyptus)	319	349	39.5
DGEBA	Lignin (Bamboo)	315	351	35.4
Epoxidized lignin (Cedar)	Phenol novolac	293	336	63.3
Epoxidized lignin (Eucalyptus)	Phenol novolac	275	311	62.8
Epoxidized lignin (bamboo)	Phenol novolac	266	313	64.7
Epoxidized lignin (Cedar)	Lignin (Cedar)	296	329	88.2
Epoxidized lignin (Eucalyptus)	Lignin (Eucalyptus)	274	307	86.1
Epoxidized lignin (bamboo)	Lignin (Bamboo)	259	298	87.3

2.3.6 Lignin-Based Epoxy

Lignin is one of the three main constituents (the other two are cellulose and hemicellulose) of land plant-based materials. Generally, lignin comprises about 15–40% [119] of the dry weight of land plants and can be extracted through different processes (e.g. kraft process and organosolv process), which will affect the final structure of the lignin (e.g. kraft lignin and organosolv lignin) [120]. According to Baroncini et al. [101], both kraft lignin and organosolv lignin have the potential to produce high-value products such as bitumen, carbon fibers, and vanillin. Asada et al. [121] compared epoxidized lignin (obtained from cedar, eucalyptus, and bamboo) cured with lignin (also obtained from the same plants) as a curing agent with DGEBA as an epoxy resin and commercial fossil resource-derived phenol novolac as a curing agent. Their thermal properties can be seen in Table 2.15. For the nomenclature in the table, Td_5 and Td_{10} stand for decomposition temperature at 5% weight loss and 10% weight loss, respectively.

It is suggested by the authors that higher lignin content can reduce the decomposition temperature. Although the decomposition temperature of cured lignin-based resin (ranging from 259 to 336 °C) is slightly lower than that of DGEBA cured with phenol novolac (361 °C for Td_5 and 378 °C for Td_{10}), the lignin-based polymer still satisfies the heat stability property of solder dip resistance (250–280 °C). This indicates that such kind of lignin-based matrix has great potential in the electronic applications. Another common industrially available lignin derivatives is vanillin. Wang et al. [122] synthetized two vanillin-derived epoxy resins (EP1 and EP2). The tensile properties are presented in Table 2.16. As can be seen, the tensile modulus of both EP1 (2.2 ± 0.1 GPa) and EP2 (2.7 ± 0.1 GPa) are significantly higher than that of DGEBA (1.9 ± 0.1 GPa). The reason is that both EP1 and EP2 have more rigid aromatic rings and polar N-H-originated intramolecular hydrogen bonds than cured DGEBA. However, only EP1 showed higher tensile strength than DGEBA. The reason

Table 2.16 Tensile properties of vanillin-derived epoxy [122].

Sample	T_g (°C)	Tensile strength (MPa)	Tensile modulus (GPa)	Elongation at break (%)
EP1	185	80.3 (± 5)	2.1 (± 0.1)	5.2 (± 0.4)
EP2	214	60.6 (± 3)	2.7 (± 0.1)	2.6 (± 0.2)
DGEBA	166	76.4 (± 6)	1.9 (± 0.1)	7.4 (± 0.5)

Table 2.17 Flexural properties of DGEDVCP [123].

Sample	Flexural strength (MPa)	Flexural modulus (GPa)	Flexural strain at break (%)
DGEDVCP-QC	75.0	2.61	2.89
DGEDVCP-GCN	105.9	2.93	3.77
DGEDVCP-PN	66.5	3.82	1.86
DGEBA-QC	71.4	2.79	2.57
DGEBA-GCN	102.5	2.54	3.85
DGEBA-PN	121.6	2.44	4.69

can be due to the internal stress of EP2, which has higher T_g than the EP1 and DGEBA.

Not only tensile properties but vanillin-derived epoxy resin also has better flexural properties than DGEBA. Shibata and Ohkita [123] synthesized a new bio-based epoxy resin (DGEDVCP) and compared its flexural properties with DGEBA by curing with three different curing agents, i.e. renewable quercetin (QC), guaiacol novolac (GCN), and petroleum-based phenol novolac (PN). The results of flexural properties are presented in Table 2.17. The DGEDVCP–GCN, for example, showed higher flexural strength (105.9 MPa) and flexural modulus (2.93 GPa) than that of DGEVA–GCN (flexural strength: 102.5 MPa and modulus: 2.54 GPa). This indicates that the lignin/vanillin-derived resin has the possibility to be an alternative to conventional DGEBA.

2.3.7 Rosin-Based Epoxy

Rosin, a resin obtained mainly from pine trees, has a large global annual production of about 1.2 million tons [124]. Isomerized acids ($C_{19}H_{29}COOH$) are the main components in rosin, which account for more than 90% of rosin [124]. Liu et al. [125] synthesized a bio-based epoxy with rosin-based resin and rosin-based curing agent, i.e. maleopimaric acid (MPA) and triglycidyl ester of MPA, respectively. The mechanical properties of this cured rosin-based epoxy and petroleum-based counterparts DEGBA obtained from the literature [123, 126–128] are listed in Table 2.18. Although all the flexural properties and impact strength of rosin-based epoxy are less than conventional DEGBA, the

Table 2.18 Flexural and impact properties of rosin-based resin [125–128].

Sample	Flexural modulus (GPa)	Flexural strength (MPa)	Flexural strain at break (%)	Impact strength (kJ/m^2)
Rosin-based epoxy	2.2 (± 0.0)	70 (± 1)	1.9 (± 0.3)	2.1 (± 0.2)
DEGBA	3.0 (± 0.2)	80 (± 3)	2.6	3.2

Table 2.19 Tensile properties of bioepoxy with different ratios of FPR and EGDE [129].

Sample	m (FPA) (%)	m (EGDE) (%)	T_g (°C)	Tensile strength (MPa)	Tensile modulus (GPa)	Elongation at break (%)
DGEBA			140	56.25	0.29	12.35
FPAE1C	100	0	167	48.54	0.471	13.37
FPEG1C	36	64	81	68.75	0.495	17.35
FPEG2C	32	68	79	58.18	0.300	20.54
FPEG3C	27	73	75	42.41	0.270	13.67

rosin-based epoxy still provides comparable properties with that of DEGBA, e.g. the flexural modulus and strength of rosin-based epoxy is 2.2 GPa and 70 MPa, respectively. Therefore, the authors concluded that this rosin-based epoxy provides a possibility for the synthesis of wholly bio-based epoxy with high performance, and it has a great potential to replace the petroleum-based epoxy.

Deng et al. [129] investigated the tensile properties of rosin-based epoxy, which is characterized with different ratios of fumaropimaric acid (FPR) and ethylene glycol diglycidyl ether (EGDE) and cured with 75.6 phr methylhexahydrophthalic anhydride (MeHHPA). Their ratio of FPR and EGDE, T_g, and tensile properties are listed in Table 2.19. Compared to DGEBA, FPAE1C, FPEG1C, and FPEG3C exhibit similar tensile strengths (48.54, 68.75, and 58.18 MPa, respectively) to that of DGEBA (56.25 MPa). Moreover, the tensile modulus and elongation at break of these three rosin-based epoxies are all larger than that of DGEBA, which indicates that the rosin-based epoxy is stiffer than DGEBA. According to the authors, the rigid ring in rosin provided improvement in the material strength; meanwhile, the strength can also be decreased because of the brittleness of rosin. Therefore, a moderate flexible chain should be introduced in rosin-based resin.

2.4 Fiber-Reinforced Bioepoxy Composites

2.4.1 Synthetic Fiber-Reinforced Bioepoxy Composites

Compared with other conventional FRP, e.g. synthetic/natural fiber-reinforced petroleum-based polymer or natural fiber-reinforced bioepoxy, the synthetic fiber-reinforced bioepoxy has its own benefits. The synthetic fiber can have unlimited length and it is homogeneous along the fiber direction (longitudinal direction). Therefore, the mechanical, physical, and chemical properties along the fiber direction is more stable than that of natural fibers, which usually contain several defects, e.g. knot for plant-based fibers, and have limited length. Besides, the synthetic fibers such as glass and carbon fibers have higher tensile properties than that of natural fibers. Generally, the tensile strengths of E-glass fiber, carbon fiber, and aramid fiber are about 2000–3500 MPa, 2500–5000 MPa, and 1400–1450 MPa [1], respectively, which are significantly higher than those of most of the natural fibers listed in Table 2.4. Because of that, the synthetic fibers are feasible to be used in the fields that require high strength such as in structural applications for concrete column confinement [130–132] or concrete beam reinforcement [133–135]. Nevertheless, as discussed in Section 2.3, some of the bioepoxies are reported to be possible as the replacement of petroleum-based DGEBA because of their comparable mechanical properties to DGEBA (tensile strength, modulus, flexural strength, and flexural modulus). Besides, the utilization of partially or fully bio-based epoxy has its benefit to increase the renewability and sustainability of the entire composites. Therefore, developing synthetic fiber-reinforced bioepoxy composites is of great practical significance.

Lu and Larock [136] reinforced soybean oil-based epoxy with continuous glass fiber and investigated the morphology, thermal stability, and mechanical properties of the composites. In this study, two kinds of soybean oils were used, i.e. Wesson soybean oil (SOY) and LoSatSoy oil (LSS) without purification. The resins were synthesized by cationic copolymerization of the oil with styrene (ST) and divinylbenzene (DVB) initiated by distilled-grade boron trifluoride diethyl etherate (BFE) initiators modified by Norway Pronova fish oil ethyl ester (NFO). The influence of the type of soybean oil, the ratio of DVB to ST and NFO, and the fiber content on the tensile properties was investigated. Among all the testing groups, the group of LSS50-ST20-DVB20-NFO5-BFE5 (50 wt% LSS, 20 wt% ST, 20 wt% DVB, 5 wt% NFO, and 5 wt% BFE) always has highest tensile modulus and tensile strength for the same glass fiber content. Its tensile strength and tensile modulus increase from 7.9 to 76 MPa and from 0.15 to 2.73 GPa, respectively, as glass fiber content increases from 0 to 52 wt%.

Thulasiraman et al. [137] investigated chlorinated soy epoxy (CSE)-modified commercial epoxy reinforced by glass fiber. The epoxidized soybean oil (ESBO) epoxy was prepared by dissolved soybean oil in $CHCl_3$ and mixed with *m*-chloroperbenzoic acid. CSE was prepared by blending chlorinated soypolyol and epichlorohydrin with the presence of 30% aq. NaOH and isopropanol. The tensile, flexural, and impact properties of glass fiber-reinforced epoxy modified with different CSE contents are listed in Table 2.20. An obvious tendency can be observed that all the tensile and flexural strength as well as modulus slightly decreases as more epoxy is replaced by CSE. However, the impact strength of the FRP increases until 15% of CSE replacement ratio, i.e. 38.6 J/cm compared to 34.2 J/cm of pure epoxy.

McIsaac [138] considered two variables for comparison of FRP, i.e. fiber materials (carbon fiber and glass fiber) and polymer matrices (conventional epoxy [E]

Table 2.20 Mechanical properties of glass fiber-reinforced CSE-modified epoxy [137].

Replacement ratio of CSE	Tensile strength (MPa)	Tensile modulus (GPa)	Flexural strength (MPa)	Flexural modulus (GPa)	Impact strength (J/cm)
0% CSE	299	3.4	379	7.8	34.2
5% CSE	288	3.2	370	7.6	36.8
10% CSE	271	2.9	363	7.1	37.9
15% CSE	255	2.8	359	6.8	38.6
20% CSE	248	2.4	346	6.3	29.7

Table 2.21 Tensile properties of carbon and glass fiber-reinforced bioepoxy [138].

Specimens	Bio-based resin replacement ratio (wt%)	Biocontent of total (wt%)	Tensile strength (MPa)	Tensile modulus (GPa)	Ultimate strain (%)
E-C	0	0	841.0 (± 61.9)	79.1 (± 15.6)	1.09 (± 0.16)
VW-C	100	41	762.0 (± 62.1)	80.9 (± 11.0)	0.96 (± 0.18)
CN20-C	20	20	522.4 (± 43.9)	100.8 (± 10.9)	0.48 (± 0.05)
CN30-C	30	30	845.6 (± 64.2)	79.2 (± 5.7)	1.07 (± 0.04)
CN40-C	40	40	561.3 (± 82.9)	108.0 (± 17.1)	0.66 (± 0.03)
ELO10-C	10	10	974.5 (± 58.5)	81.2 (± 2.9)	1.20 (± 0.08)
ELO20-C	20	20	848.4 (± 62.3)	81.5 (± 11.8)	1.06 (± 0.22)
ELO30-C	30	30	625.3 (± 53.9)	81.5 (± 26.9)	0.83 (± 0.25)
ELO40-C	40	40	364.5 (± 107.4)	84.4 (± 19.9)	0.56 (± 0.18)
E-G	0	0	491.7 (± 30.5)	24.7 (± 1.6)	2.00 (± 0.06)
VW-G	100	41	508.8 (± 84.8)	24.6 (± 1.6)	2.06 (± 0.25)
CN20-G	20	20	293.3 (± 28.9)	33.6 (± 12.6)	1.01 (± 0.51)
CN30-G	30	30	423.7 (± 41.3)	23.3 (± 2.1)	1.83 (± 0.24)
CN40-G	40	40	419.7 (± 53.1)	29.7 (± 9.2)	1.34 (± 0.23)
ELO10-G	10	10	394.6 (± 24.1)	25.1 (± 2.8)	1.58 (± 0.12)
ELO20-G	20	20	399.6 (± 15.6)	23.1 (± 2.4)	1.74 (± 0.16)
ELO30-G	30	30	425.7 (± 10.5)	24.3 (± 4.0)	1.79 (± 0.30)
ELO40-G	40	40	290.4 (± 14.8)	20.2 (± 3.2)	1.47 (± 0.13)

and epoxy replaced by bio-based resin blend[VW], cashew nut shell liquid [CN], or epoxidized linseed oil [ELO]). In the bio-based resin blend, 41 wt% of them are derived from by-products of wood and vegetables. The results of the tensile test are shown in Table 2.21. For nomenclature of the table, E, VW, CN, and ELO stand for epoxy, bio-based resin blend, cashew nut shell liquid, and ELO; the number after the resin represents the replacement ratio of other resin to epoxy in wt% of the composites; C and G stand for carbon and glass fibers, respectively. For carbon fiber, the sample with 30 wt% CN (CN30-C) and 10 wt% ELO exhibits higher tensile strength (845.6 MPa for CN30-C and 974.5 MPa for ELO10-C) and similar tensile modulus (79.2 GPa for CN30-C and 81.2 GPa for ELO10-C) to epoxy with carbon fiber. For glass fiber, the sample with the commercial bio-based epoxy (VW-G) exhibits a similar tensile strength (508.8 MPa) and tensile modulus (24.6 GPa) to that of E-G (tensile strength 491.7 MPa and modulus 24.6 GPa). Great difference in mechanical properties (e.g. tensile strength) can be found for different polymer matrixes with the same fiber material. The reason can be the different compatibility in the interfacial bond between the polymer matrix and fiber.

Additionally, to epoxidized plant oil, some researchers also investigated furan-based bioresin reinforced by synthetic fibers. For example, Fam et al. [139] applied glass fiber in furfuryl alcohol bioresin and compared its tensile properties with conventional glass fiber-reinforced epoxy. Two types of commercial furfuryl alcohol extracted from corncob were used for the investigation. The results show that the samples with a low-viscosity furfuryl alcohol type (Q2001) present similar tensile strength (520 MPa) and modulus (26.58 GPa) to that of composite with epoxy (tensile strength 525 MPa and modulus 26.78 MPa) as a polymer matrix. This result indicates that in consideration of environment impacts and mechanical properties, bioresin can replace the epoxy resin in GFRP when it is carefully selected.

Besides the mechanical properties of the synthetic fiber-reinforced bioepoxy, some researchers also studied such composites toward practical applications, e.g. strengthening of concrete structure. MeSwiggan and Fam [140] applied furfural alcohol resin and epoxidized pine oil with carbon and glass fiber as external strengthening of full-scale RC beams and compared them with the one with epoxy. The epoxidized pine oil resin provided similar results to epoxy for both carbon and glass fibers for full-scale RC beams. Both resins resulted in approximately 50% increase in yielding load and ultimate load with carbon fiber and about 23% in yielding load with glass fiber. The increases of ultimate load for epoxy and epoxidized pine oil with glass fiber are 43% and 32%, respectively. This is due to the failure mode in both situations. In the case of epoxy, concrete was crushed. Shear failure took place in the case of epoxidized pine oil resin. Based on this investigation, McIsaac [138] has performed a further investigation on the durability of the synthetic FRP with bioepoxy under freeze–thaw cycles (from +6 to −27 °C). After 300 cycles (seven months), no negative effect on tensile properties of FRP laminates can be observed. Besides, carbon fiber-reinforced epoxidized pine oil and furfuryl alcohol composites both increased the yield load and peak load of the RC beams. For epoxidized pine oil, the increases of yield load and peak load are 22% and 25%, respectively. For furfuryl alcohol, the increases of yield load and peak load are 25% and 31%, respectively. The yield load and peak load of the unconfined counterpart are 77.59 and 98.21 kN, respectively.

2.4.2 Natural Fiber-Reinforced Bioepoxy Composites

In comparison to synthetic fibers, natural fibers were also investigated because of the lower cost, comparable mechanical properties, and their renewability as well as degradability. Throughout the literature, plenty of investigations were

Table 2.22 Mechanical properties of flax fiber-reinforced flaxseed oil based bioresin [141].

Specimens	Tensile modulus (GPa)	Tensile strength (MPa)	Flexural modulus (GPa)	Flexural strength (MPa)
AEFO	0.37	29.8	2.84	53.5
AEFO + 2%FF	0.43	30.2	2.88	58
AEFO + 5%FF	0.47	30.9	2.92	61
AEFO + 10%FF	0.52	31.4	2.98	64.5
AEFO10 + 10%FF	0.57	31.9	3.00	63.5
AEFO20 + 10%FF	0.62	32.1	3.05	63.0
AEFO30 + 10%FF	0.66	33.3	3.11	62.4
AEFO40 + 10%FF	0.70	35.6	3.16	62.3
AEFO50 + 10%FF	0.73	37.9	3.22	61.6

Table 2.23 Comparison of hemp, ramie, and flax for cardanol-modified epoxy [142].

Specimens	Fiber content (wt%)	Flexural strength (MPa)	Flexural modulus (GPa)	Elongation at break (%)
Hemp	13	>20 (± 4)	1.75 (± 0.15)	>3.5 (± 0.5)
Ramie	15	22 (± 1.5)	1.79 (± 0.12)	1.5 (± 0.15)
Flax	15	>5 (± 0.9)	0.42 (± 0.03)	>3.15 (± 0.5)

conducted on the natural fiber-reinforced bioepoxy in order to create a 100% bioproduct. Among all the natural fibers, flax fiber is the most investigated fiber for bioepoxy, as flax fiber has comparable tensile properties to E-glass fiber according to Table 2.4. Rana and Evitts [141] investigated the mechanical properties of flax fiber-reinforced flaxseed oil based bioresin. The results can be found in Table 2.22. For nomenclature, AEFO stands for acrylated epoxidized flaxseed oil; FF stands for flax fiber, the number after AEFO stands for the ratio of styrene to the entire matrix, e.g. the ratio of styrene in AEFO10 is 10 wt%. Generally, all the mechanical properties (i.e. tensile strength and modulus and flexural modulus) increase as the content of flax fiber or styrene increases. However, the flexural strength decreases as the styrene content increase. The flax fiber (10%)-reinforced 100% AEFO exhibits proper tensile properties and flexural modulus compared to other groups, i.e. tensile strength is 31.4 ± 1.2 MPa, tensile modulus is 0.52 ± 0.03 GPa, and flexural modulus is 2.98 ± 0.12 GPa, and it has the highest flexural strength of 64.5 ± 2.3 MPa.

Maffezzoli et al. [142] compared short hemp, ramie, and flax in cardanol-modified epoxy, which has a cardanol content of 40 wt%. Their flexural mechanical properties are shown in Table 2.23. Although flax-reinforced bioepoxy exhibits lower flexural strength and modulus to that with hemp or ramie, the low variation in flexural properties still makes flax favorable in application.

Similar experimental work on flax can also be found in other research studies such as by Fahimian et al. [143] for canola oil-based resin and by O'Donnell et al. [144] for acrylated ESBO.

Besides plant-based natural fibers, other natural fibers such as mineral-based and animal-based natural fibers were also considered to reinforce bioepoxy. Samper et al. [145] applied basalt fiber in epoxidized vegetable oils. Three types of basalt fibers, i.e. basalt fiber without surface modification (Basalt TT), basalt fiber modified with [3-(2-aminoethylamine) propyl]-trimethoxysilane (Basalt A), and basalt fiber modified with glycidyl trimethoxysilane [2-(7-oxabicyclo [4.1.0] hept-3-yl) ethyl] silane (Basalt B), and two kinds of bioepoxy, i.e. ELO and ESBO, are considered as experimental parameters in the study. Generally, the samples with ELO exhibit better properties (i.e. tensile strength and flexural strength) than that with ESBO, which according to the authors is due to the higher cross-linking density of ELO. Besides, both modification methods significantly improve the tensile and flexural properties of the composites because of the improvement of wetting properties in the fiber–matrix interaction. Hong et al. [105] investigated animal-based hollow keratin fiber in acrylated ESBO. Although vacuum-assisted resin transfer molding (VARTM) process was applied, only 5% of the hollow keratin fiber was filled by resin. The dielectric constant k of the composite is between 1.7 and 2.7, and its thermal expansion is 67.4 ppm/°C. This makes the composites to be possible for the electronic applications. Besides, the authors also pointed out that the chicken feathers can be a potential material for lightweight, high-performance, and low cost composites.

2.4.3 Natural–Synthetic Hybrid Fiber-Reinforced Bioepoxy Composites

In the investigation of fiber-reinforced bioepoxy composites, plant-based natural fibers are prior to other synthetic fibers because of the better compatibility with bio-based resin and the renewability. At the present stage, however, there is a significant difference in mechanical properties, durability, and consistency between plant-based natural fibers and other stronger fibers. Hybridizing different fibers into one bio-based resin can balance the renewability, load carrying capabilities, and durability.

Scalici [146] investigated the aging resistance of jute–basalt hybrid fiber-reinforced bioepoxy composites. Composites were exposed to cyclic conditions of hygrothermal stress and UV radiation. The exposure lasted for a period of 84 days. The hybrid composites exhibited higher impact energy, flexural properties, and greater aging resistance than jute fiber-reinforced bioepoxy. Cicala et al. [147] researched a recyclable epoxy system that consists of bioepoxy-based monomers, a cure inhibitor, and a cleavable amine. Hybrid flax/carbon fibers were used to reinforce this epoxy system. Hybrid FRP can provide a good balance of tensile properties (strength and modulus) and loss factor (the ratio of loss modulus to residual modulus), when all the carbon fibers were placed on the external layers. The cured composites can be recycled partially by using mild acetic acid aqueous solutions to obtain clean fiber and create a reusable thermoplastic. Morye and Wool [148] researched glass/flax hybrid fiber-reinforced-modified soybean oil matrix composites. Mechanical tests (including tensile, compressive, bending, and impact tests) were performed to investigate the influence of glass/flax ratio and fiber arrangement on the mechanical properties. According to the research, when the glass/flax ratio was the same, the composites with unsymmetrical fiber arrangement had lower tensile and compressive strength than the symmetric composites. The tensile modulus of the composites with these two types of fiber arrangement were, however, very similar. The strength and failure modes of bending and impact tests were sensitive to fiber arrangements. When the flax was in the tension side, glass fiber was in the compressive side, flax FRP broke firstly, and then the composite was damaged. When both the tension and compressive sides were glass fibers, the bending strength was increased. Based on the results of the impact test, the sequence from high-energy absorption to low energy absorption is as follows: hybrid composite > glass composite > flax composite. Besides, the hybrid composite, in which the impact was loaded on flax fiber face, had the highest energy absorption.

2.5 Future Perspectives

The experimental results (e.g. Table 2.21) show that for certain fiber materials, the compatibility varies under the different types of bioepoxy. This is suspected to be the results of different bond behaviors between different fibers and bioepoxies. In order to improve the interfacial bond between fiber and resin, many treatments have been investigated in the previous studies on fiber/fabric-reinforced petroleum-based polymer, such as corona treatment [149], grafting coupling agent [150], adding nanoparticles [151], and alkali [152] and silane treatment [153]. These methods can enhance the bond behavior through increasing the roughness of fiber or having a bridge effect by coupling agent. It is expected that these methods can be used to improve interfacial bond between various fibers and bioepoxies. In the future research, more studies should be focused on the surface treatment of fiber materials and bioepoxy to improve their interfacial bond and in turn the resulting fiber-reinforced bioepoxy composites.

Although FRP composites with bioepoxy have been already applied in practices such as ski/skateboard, furniture, or automobile, it is still seldom in massive applications such as structural applications in civil engineering as the potential demand of using fiber-reinforced bioepoxy composites in civil engineering is extremely high. The reason is that most of the current investigations on fiber-reinforced bioepoxy composites focus on the short-term behavior of the composites. Only very few studies have considered the long-term durability of those composites. The lack of knowledge on the long-term performance of fiber-reinforced bioepoxy composites hinders the acceptance of those composites for civil engineering applications. According to Diez-Pascual [154], the biopolymer usually has low resistance to fatigue and chemicals as well as inferior long-term durability in comparison to petroleum-based polymer. Therefore, it is expected that the long-term durability of fiber/fabric-reinforced bioepoxy is also inferior to that with petroleum-based epoxy. Based on the potential application orientation, the long-term behaviors of fiber/fabric-reinforced bioepoxy composites under, e.g. hygrothermal environments [155], alkali [156], and salt conditions [157], should be further investigated. Similar to fiber/fabric-reinforced petroleum-based polymer, the one with bioepoxy should have the potential in structural applications such as confinement of concrete column, strengthening element for timber or concrete beams, or as postrepairing elements for damaged structures.

Although bioepoxy derives from sustainable and renewable resources, this does not make the bioepoxy biodegradable. The disposal or recycle methods for fiber-reinforced bioepoxy will be a critical issue in the future at the end of the life cycle of those composites.

2.6 Conclusions

This chapter introduces the fibers (natural and synthetic fibers) that are applied for fiber-reinforced bioepoxy composites and the bioepoxies including natural oil-based epoxy, isosorbide-based epoxy, furan-based epoxy, polyphenolic epoxy, epoxidized natural rubber, lignin-based epoxy, and rosin-based epoxy. Besides, this chapter also presents the up-to-date investigations on fiber-reinforced bioepoxy composites. Literature studies show that bioepoxy has the potential to replace petroleum-based polymer in FRP laminates, no matter which kinds of fiber materials (synthetic, natural, or hybrid fiber) are used. This is due to not only their sustainability and renewability but also the comparable mechanical properties (e.g. tensile strength and modulus) to DEGBA. The interfacial bonding between fiber/fabric and bioepoxy is one of the most critical issues in this composite material. The modification methods for improving the bonding behavior can be, e.g., alkali, silane, and corona treatments as well as the addition of nanoparticles; further studies should be considered.

The fiber/fabric-reinforced bioepoxy has the potential to be the new generation of composite materials as they are renewable, sustainable, and have comparable mechanical properties (e.g. tensile strength and tensile modulus) in comparison with fiber/fabric-reinforced petroleum-based polymer. Therefore, they can be a possible replacement of fiber/fabric-reinforced petroleum-based polymer in structural applications. Future work on fiber/fabric-reinforced bioepoxy should be focusing on the long-term durability under such as hygrothermal environments, alkali or salt condition. Moreover, the disposal and recycling methods for the end-of-life of this type of composite should also be investigated.

Acknowledgments

The authors acknowledge the support by Fachagentur Nachwachsende Rohstoffe e.V. (FNR, Agency for Renewable Resources) founded by Bundesministerium für

Ernährung und Landwirtschaft (BMEL, The Federal Ministry of Food and Agriculture of Germany), under the grant award 22011617 and by Bundesministerium für Bildung und Forschung (BMBF, Federal Ministry of Education and Research of Germany) (grant no. 01DS18023). The first two authors also acknowledge the PhD scholarship awarded by the Chinese Scholarship Council (CSC).

References

1 Zoghi, M. (2013). *The International Handbook of FRP Composites in Civil Engineering*. CRC Press.
2 Yan, L., Chouw, N., and Jayaraman, K. (2014). Flax fibre and its composites – a review. *Composites Part B: Eng.* 56: 296–317. https://doi.org/10.1016/j.compositesb.2013.08.014.
3 Sicomin Epoxy Systems (2019). Greenpoxy Bio Resins Facilitate Sustainable Production Practices for Zag Skis. http://www.sicomin.com/documents/zag-casestudy-77.pdf (accessed 9 July 2020).
4 TFC Biomass Based Chemicals (2019). Biocomposites and gree technology products. https://www.polyfurfurylalcohol.com/automotive-furniture (accessed 9 July 2020).
5 Jawaid, M. and Abdul Khalil, H.P.S. (2011). Cellulosic/synthetic fibre reinforced polymer hybrid composites: a review. *Carbohydr. Polym.* 86 (1): 1–18. https://doi.org/10.1016/j.carbpol.2011.04.043.
6 Sudarsan, V. *Materials for Hostile Chemical Environments*, 129–158. https://doi.org/10.1016/B978-0-12-801300-7.00004-8.
7 Bhatnagar, N. and Asija, N. *Durability of High-Performance Ballistic Composites*, 231–283. https://doi.org/10.1016/B978-0-08-100406-7.00008-8.
8 Deopura, B.L. and Padaki, N.V. (2015). Synthetic textile fibres. In: *Textiles and fashion: Materials, design and technology/edited by Rose Sinclair* (ed. R. Sinclair), 97–114. Cambridge: Elsevier/Woodhead Publishing.
9 Fink, J.K. (2014). Aramids. In: *High Performance Polymers*, Chapter 13 (ed. J.K. Fink), 301–320. Oxford: William Andrew.
10 Gong, R.H. and Chen, X. *Technical Yarns*, 43–62. https://doi.org/10.1016/B978-1-78242-458-1.00003-0.
11 Ash, R.A. (2016). Vehicle armor. In: *Lightweight Ballistic Composites: Military and Law-Enforcement Applications*, Chapter 9 (ed. A. Bhatnagar), 285–309. Duxford: Woodhead Publishing is an imprint of Elsevier.
12 Mouritz, A.P. (ed.) (2012). *Introduction to Aerospace Materials*. Reston, VA, Cambridge: American Institute of Aeronautics and Astronautics.
13 Xing, Y. and Ding, X. (2007). UV photo-stabilization of tetrabutyl titanate for aramid fibers via sol–gel surface modification. *J. Appl. Polym. Sci.* 103 (5): 3113–3119.
14 Zhang, Y., Huang, Y., Liu, L., and Wu, L. (2007). Surface modification of aramid fibers with γ-ray radiation for improving interfacial bonding strength with epoxy resin. *J. Appl. Polym. Sci.* 106 (4): 2251–2262. https://doi.org/10.1002/app.26887.

15 Park, S.-J. and Seo, M.-K. *Element and Processing*, vol. 18, 431–499. https://doi.org/10.1016/B978-0-12-375049-5.00006-2.

16 Khanam, P.N. and AlMaadeed, M.A.A. (2015). Processing and characterization of polyethylene-based composites. *Adv. Manuf. Polym. Compos. Sci.* 1 (2): 63–79. https://doi.org/10.1179/2055035915Y.0000000002.

17 Yan, Y. *Developments in Fibers for Technical Nonwovens*, 19–96. https://doi.org/10.1016/B978-0-08-100575-0.00002-4.

18 Kurtz, S.M. *A Primer on UHMWPE*, 1–12. https://doi.org/10.1016/B978-012429851-4/50002-4.

19 UHMW (Ultra-High-Molecular-Weight Polyethylene). https://dielectricmfg.com/knowledge-base/uhmw/ (accessed 9 July 2020).

20 Chow, T.W., Cheng, Y.Y., and Ladizesky, N.H. (1993). Polyethylene fibre reinforced poly(methylmethacrylate)—water sorption and dimensional changes during immersion. *J. Dent.* 21 (6): 367–372. https://doi.org/10.1016/0300-5712(93)90014-H.

21 Ladizesky, N.H., Cheng, Y.Y., Chow, T.W., and Ward, I.M. (1993). Acrylic resin reinforced with chopped high performance polyethylene fiber-properties and denture construction. *Dent. Mater.* 9 (2): 128–135. https://doi.org/10.1016/0109-5641(93)90089-9.

22 Jacobs, O., Kazanci, M., Cohn, D., and Marom, G. (2002). Creep and wear behaviour of ethylene–butene copolymers reinforced by ultra-high molecular weight polyethylene fibres. *Wear* 253 (5-6): 618–625. https://doi.org/10.1016/S0043-1648(02)00110-2.

23 McIntyre, J.E. (2005). *Synthetic Fibres: Nylon, Polyester, Acrylic, Polyolefin.* Boca Raton, FL, Cambridge: CRC Press.

24 Sastri, V.R. (2010). Commodity thermoplastics: polyvinyl chloride, polyolefins, and polystyrene. In: *Plastics in Medical Devices: Properties, Requirements and Applications*, Chapter 6 (ed. V.S. Sastri), 73–119. Norwich, NY: Elsevier/William Andrew.

25 Rivera-Gómez, C. and Galán-Marín, C. (2017). Biodegradable fiber-reinforced polymer composites for construction applications. In: *Natural Fiber-Reinforced Biodegradable and Bioresorbable Polymer Composites*, Chapter 4 (eds. A.K.T. Lau and A.P. Yan Hung), 51–72. Oxford: Woodhead Publishing.

26 Rezaei, F., Yunus, R., and Ibrahim, N.A. (2009). Effect of fiber length on thermomechanical properties of short carbon fiber reinforced polypropylene composites. *Mater. Des.* 30 (2): 260–263.

27 Wang, K., Guo, M., Zhao, D. et al. (2006). Facilitating transcrystallization of polypropylene/glass fiber composites by imposed shear during injection molding. *Polymer* 47 (25): 8374–8379. https://doi.org/10.1016/j.polymer.2006.10.003.

28 Mohanty, S., Verma, S.K., Nayak, S.K., and Tripathy, S.S. (2004). Influence of fiber treatment on the performance of sisal-polypropylene composites. *J. Appl. Polym. Sci.* 94 (3): 1336–1345. https://doi.org/10.1002/app.21161.

29 Vigneswaran, C., Ananthasubramanian, M., and Kandhavadivu, P. *Bioprocessing of Synthetic Fibres*, 189–250. https://doi.org/10.1016/B978-93-80308-42-5.50004-4.

30 Militky, J. *The Chemistry, Manufacture and Tensile Behaviour of Polyester Fibers*, 223–314. https://doi.org/10.1533/9781845696801.2.223.

31 Asgari, M. and Masoomi, M. (2012). Thermal and impact study of PP/PET fibre composites compatibilized with glycidyl methacrylate and maleic anhydride. *Composites Part B* 43 (3): 1164–1170. https://doi.org/10.1016/j.compositesb.2011.11.035.

32 Dai, J.-G., Lam, L., and Ueda, T. (2012). Seismic retrofit of square RC columns with polyethylene terephthalate (PET) fibre reinforced polymer composites. *Constr. Build. Mater.* 27 (1): 206–217. https://doi.org/10.1016/j.conbuildmat.2011.07.058.

33 Huang, X. (2009). Fabrication and properties of carbon fibers. *Materials* 2 (4): 2369–2403.

34 Inagaki, M. (2000). Carbon fibers. In: *New Carbons: Control of Structure and Functions*, Chapter 4, 1e (ed. M. Inagaki), 82–123. New York: Elsevier Science.

35 Manocha, L.M. (2001). Carbon fibers. In: *Encyclopedia of materials: Science and technology* (eds. K.H.J. Buschow, R.W. Cahn, M.C. Flemings, et al.), 906–916. Amsterdam, London: Elsevier.

36 Witten, E., Mathes, V., Sauer, M., and Kühnel, M. Composites Market Report 2018-Market-developments, trends, outlooks and challeng. https://www.avk-tv.de/files/20181115_avk_ccev_market_report_2018_final.pdf (accessed 9 July 2020).

37 Raether, F., Meinhardt, J., and Kienzle, A. (2007). Oxidation behaviour of carbon short fibre reinforced C/SiC composites. *J. Eur. Ceram. Soc.* 27 (2–3): 1217–1221. https://doi.org/10.1016/j.jeurceramsoc.2006.04.019.

38 Yao-hui, L., Du Jun, S.-r.Y., and Wei, W. (2004). High temperature friction and wear behaviour of Al_2O_3 and/or carbon short fibre reinforced Al–12Si alloy composites. *Wear* 256 (3–4): 275–285. https://doi.org/10.1016/S0043-1648(03)00387-9.

39 Giannadakis, K., Szpieg, M., and Varna, J. (2011). Mechanical performance of a recycled carbon fibre/PP composite. *Exp. Mech.* 51 (5): 767–777. https://doi.org/10.1007/s11340-010-9369-8.

40 Jones, F.R. (2001). Glass fibres. In: *High Performance Fibres* (ed. H. JWS), 191–238. Cambridge: Woodhead Publishing.

41 Hartman, D.R., Greenwood, M.E., and Miller, D.M. (1996). High Strength Glass Fibers Inc. Technical paper Ref. 1-Pl-19025-A.

42 Hu, H. and Liu, Y. *High Modulus, High Tenacity Yarns*, 329–386. https://doi.org/10.1533/9781845699475.2.329.

43 Shioya, M. and Kikutani, T. (2015). Synthetic textile fibres: non-polymer fibres. In: *Textiles and Fashion: Materials, Design and Technology*, Chapter 7 (ed. R. Sinclair), 139–155. Cambridge: Elsevier/Woodhead Publishing.

44 Chawla, K.K. (2015). Glass fibers. In: *Reference Module in Materials Science and Materials Engineering* (ed. S. Hashmi), 3541–3545. Amsterdam: Elsevier.

45 Meola, C. and Boccardi, S. Carlomagno Gm. *Compos. Mater. Aeron. Ind.*: 1–24. https://doi.org/10.1016/B978-1-78242-171-9.00001-2.

46 Biron, M. *Composites*, 343–463. https://doi.org/10.1016/B978-185617411-4/50008-5.
47 Dutton, S., Kelly, D., and Baker, A. (2004). *Composite Materials for Aircraft Structures*, 2e. Reston, VA: American Institute of Aeronautics and Astronautics.
48 Bunsell, A.R. *CVD Monofilaments*, 1953–1957. https://doi.org/10.1016/B0-08-043152-6/00354-5.
49 Naeimirad, M., Zadhoush, A., and Neisiany, R.E. (2016). Fabrication and characterization of silicon carbide/epoxy nanocomposite using silicon carbide nanowhisker and nanoparticle reinforcements. *J. Compos. Mater.* 50 (4): 435–446. https://doi.org/10.1177/0021998315576378.
50 Mukhtar, I., Leman, Z., Ishak, M.R., and Zainudin, E.S. (2016). Sugar palm fibre and its composites: a review of recent developments. *BioResources* 11 (4) https://doi.org/10.15376/biores.11.4.10756-10782.
51 Liu, K., Takagi, H., Osugi, R., and Yang, Z. (2012). Effect of physicochemical structure of natural fiber on transverse thermal conductivity of unidirectional abaca/bamboo fiber composites. *Composites Part A* 43 (8): 1234–1241. https://doi.org/10.1016/j.compositesa.2012.02.020.
52 Djafari Petroudy, S.R. *Physical and Mechanical Properties of Natural Fibers*, 59–83. https://doi.org/10.1016/B978-0-08-100411-1.00003-0.
53 Baley, C. (2002). Analysis of the flax fibres tensile behaviour and analysis of the tensile stiffness increase. *Composites Part A* 33 (7): 939–948. https://doi.org/10.1016/S1359-835X(02)00040-4.
54 John, M.J. and Anandjiwala, R.D. (2008). Recent developments in chemical modification and characterization of natural fiber-reinforced composites. *Polym. Compos.* 29 (2): 187–207. https://doi.org/10.1002/pc.20461.
55 Pickering, K.L., Efendy, M.A., and Le, T.M. (2016). A review of recent developments in natural fibre composites and their mechanical performance. *Composites Part A* 83: 98–112. https://doi.org/10.1016/j.compositesa.2015.08.038.
56 Dittenber, D.B. and GangaRao, H.V.S. (2012). Critical review of recent publications on use of natural composites in infrastructure. *Composites Part A* 43 (8): 1419–1429. https://doi.org/10.1016/j.compositesa.2011.11.019.
57 Townsend, T. (2019). Natural Fibres and the World Economy July 2019. https://dnfi.org/coir/natural-fibres-and-the-world-economy-july-2019_18043/ (accessed 9 July 2020).
58 Midani, M. Natural fiber composites: what's holding them back? https://www.compositesworld.com/blog/post/natural-fiber-composites-whats-holding-them-back (accessed 9 July 2020).
59 Rowell, R.M. Natural Fibres: Types and Properties, 3–66. https://doi.org/10.1533/9781845694593.1.3.
60 Mahdi, E., Hamouda, A.M.S., Sahari, B.B., and Khalid, Y.A. (2003). On the collapse of cotton/epoxy tubes under axial static loading. *Appl. Compos. Mater.* 10 (2): 67–84. https://doi.org/10.1023/A:1022890104728.
61 Kim, S.-J., Moon, J.-B., Kim, G.-H., and Ha, C.-S. (2008). Mechanical properties of polypropylene/natural fiber composites: comparison of wood fiber

and cotton fiber. *Polym. Test.* 27 (7): 801–806. https://doi.org/10.1016/j.polymertesting.2008.06.002.

62 Hashmi, S.A.R., Dwivedi, U.K., and Chand, N. (2007). Graphite modified cotton fibre reinforced polyester composites under sliding wear conditions. *Wear* 262 (11–12): 1426–1432. https://doi.org/10.1016/j.wear.2007.01.014.

63 Graupner, N. (2008). Application of lignin as natural adhesion promoter in cotton fibre-reinforced poly(lactic acid) (PLA) composites. *J. Mater. Sci.* 43 (15): 5222–5229. https://doi.org/10.1007/s10853-008-2762-3.

64 Pil, L., Bensadoun, F., Pariset, J., and Verpoest, I. (2016). Why are designers fascinated by flax and hemp fibre composites? *Composites Part A* 83: 193–205. https://doi.org/10.1016/j.compositesa.2015.11.004.

65 Shahzad, A. (2012). Hemp fiber and its composites – a review. *J. Compos. Mater.* 46 (8): 973–986. https://doi.org/10.1177/0021998311413623.

66 Gogna, E., Kumar, R., Anurag, S.A.K., and Panda, A. (2019). A comprehensive review on jute fiber reinforced composites. In: *Advances in Industrial and Production Engineering* (eds. K. Shanker, R. Shankar and R. Sindhwani), 459–467. Singapore: Springer.

67 Ibrahim, I.D., Jamiru, T., Sadiku, E.R. et al. (2016). Mechanical properties of sisal fibre-reinforced polymer composites: a review. *Compos. Interfaces* 23 (1): 15–36. https://doi.org/10.1080/09276440.2016.1087247.

68 Akil, H.M., Omar, M.F., Mazuki, A.A.M. et al. (2011). Kenaf fiber reinforced composites: a review. *Mater. Des.* 32 (8-9): 4107–4121. https://doi.org/10.1016/j.matdes.2011.04.008.

69 Venkateshwaran, N., Elaya Perumal, A., and Arunsundaranayagam, D. (2013). Fiber surface treatment and its effect on mechanical and visco-elastic behaviour of banana/epoxy composite. *Mater. Des.* 47: 151–159. https://doi.org/10.1016/j.matdes.2012.12.001.

70 Todkar, S.S. and Patil, S.A. (2019). Review on mechanical properties evaluation of pineapple leaf fibre (PALF) reinforced polymer composites. *Composites Part B* 174: 106927. https://doi.org/10.1016/j.compositesb.2019.106927.

71 Delicano, J.A. (2018). A review on abaca fiber reinforced composites. *Compos. Interfaces* 25 (12): 1039–1066. https://doi.org/10.1080/09276440.2018.1464856.

72 Hill, C.A.S. and Khalil, H.P.S.A. (2000). The effect of environmental exposure upon the mechanical properties of coir or oil palm fiber reinforced composites. *J. Appl. Polym. Sci.* 77 (6): 1322–1330. https://doi.org/10.1002/1097-4628(20000808)77:6<1322:AID-APP17>3.0.CO;2-R.

73 Muhammad, A., Rahman, M.R., Hamdan, S., and Sanaullah, K. (2019). Recent developments in bamboo fiber-based composites: a review. *Polym. Bull.* 76 (5): 2655–2682. https://doi.org/10.1007/s00289-018-2493-9.

74 He, L., Li, W., Chen, D. et al. (2015). Effects of amino silicone oil modification on properties of ramie fiber and ramie fiber/polypropylene composites. *Mater. Des.* 77: 142–148. https://doi.org/10.1016/j.matdes.2015.03.051.

75 Hinchcliffe, S.A., Hess, K.M., and Srubar, W.V. (2016). Experimental and theoretical investigation of prestressed natural fiber-reinforced polylactic

acid (PLA) composite materials. *Composites Part B* 95: 346–354. https://doi.org/10.1016/j.compositesb.2016.03.089.

76 Mittal, V., Saini, R., and Sinha, S. (2016). Natural fiber-mediated epoxy composites – a review. *Composites Part B* 99: 425–435. https://doi.org/10.1016/j.compositesb.2016.06.051.

77 Prasad, V., Joseph, M.A., and Sekar, K. (2018). Investigation of mechanical, thermal and water absorption properties of flax fibre reinforced epoxy composite with nano TiO_2 addition. *Composites Part A* 115: 360–370. https://doi.org/10.1016/j.compositesa.2018.09.031.

78 Puttegowda, M., Rangappa, S.M., Jawaid, M. et al. *Potential of Natural/Synthetic Hybrid Composites for Aerospace Applications*, 315–351. https://doi.org/10.1016/B978-0-08-102131-6.00021-9.

79 Ramamoorthy, S.K., Skrifvars, M., and Persson, A. (2015). A review of natural fibers used in biocomposites: plant, animal and regenerated cellulose fibers. *Polym. Rev.* 55 (1): 107–162. https://doi.org/10.1080/15583724.2014.971124.

80 Conzatti, L., Giunco, F., Stagnaro, P. et al. (2013). Composites based on polypropylene and short wool fibres. *Composites Part A* 47: 165–171. https://doi.org/10.1016/j.compositesa.2013.01.002.

81 Reddy, N. and Yang, Y. (2010). Light-weight polypropylene composites reinforced with whole chicken feathers. *J. Appl. Polym. Sci.* 36 https://doi.org/10.1002/app.31931.

82 Verma, A. and Singh, V.K. (2016). Human hair: a biodegradable composite fiber–a review. *Int. J. Waste Resour* 6 (2) https://doi.org/10.4172/2252-5211.1000206.

83 Müssig, J. (2010). *Industrial Applications of Natural Fibres.* Chichester: Wiley.

84 Fiore, V., Scalici, T., Di Bella, G., and Valenza, A. (2015). A review on basalt fibre and its composites. *Composites Part B* 74: 74–94. https://doi.org/10.1016/j.compositesb.2014.12.034.

85 Shubhra, Q.T.H., Alam, A.K.M.M., Gafur, M.A. et al. (2010). Characterization of plant and animal based natural fibers reinforced polypropylene composites and their comparative study. *Fibers Polym.* 11 (5): 725–731. https://doi.org/10.1007/s12221-010-0725-1.

86 Khan, A., Khan, R.A., Salmieri, S. et al. (2012). Mechanical and barrier properties of nanocrystalline cellulose reinforced chitosan based nanocomposite films. *Carbohydr. Polym.* 90 (4): 1601–1608. https://doi.org/10.1016/j.carbpol.2012.07.037.

87 Luo, X. and Li, L. Chitosan's Wide Profile from Fibre to Fabrics: An Overview. https://doi.org/10.5772/intechopen.76196.

88 Sheik, S., Sheik, S., Nairy, R. et al. (2018). Study on the morphological and biocompatible properties of chitosan grafted silk fibre reinforced PVA films for tissue engineering applications. *Int. J. Biol. Macromol.* 116: 45–53. https://doi.org/10.1016/j.ijbiomac.2018.05.019.

89 Soundhar, A. and Jayakrishna, K. (2019). Investigations on mechanical and morphological characterization of chitosan reinforced polymer

nanocomposites. *Mater. Res. Express* 6 (7): 75301. https://doi.org/10.1088/2053-1591/ab1288.
90 Guan, B.W., Chen, S.F., and Liu, K.P. (2013). Mechanical property of brucite fiber reinforced composite material. *AMM* 357–360: 1019–1022. https://doi.org/10.4028/www.scientific.net/AMM.357-360.1019.
91 Maxineasa, S.G. and Taranu, N. *Life Cycle Analysis of Strengthening Concrete Beams with FRP*, 673–721. https://doi.org/10.1016/B978-0-08-102181-1.00024-1.
92 Singha, K. A short review on basalt fiber. http://article.sapub.org/pdf/10.5923.j.textile.20120104.02.pdf (accessed 9 July 2020).
93 Lopresto, V., Leone, C., and de Iorio, I. (2011). Mechanical characterisation of basalt fibre reinforced plastic. *Composites Part B* 42 (4): 717–723. https://doi.org/10.1016/j.compositesb.2011.01.030.
94 Swolfs, Y., Gorbatikh, L., and Verpoest, I. (2014). Fibre hybridisation in polymer composites: a review. *Composites Part A* 67: 181–200. https://doi.org/10.1016/j.compositesa.2014.08.027.
95 Unterweger, C., Brüggemann, O., and Fürst, C. (2014). Synthetic fibers and thermoplastic short-fiber-reinforced polymers: properties and characterization. *Polym. Compos.* 35 (2): 227–236. https://doi.org/10.1002/pc.22654.
96 Pascault, J.-P. (2002). *Thermosetting Polymers*. New York: Marcel Dekker.
97 Auvergne, R., Caillol, S., David, G. et al. (2014). Biobased thermosetting epoxy: present and future. *Chem. Rev.* 114 (2): 1082–1115. https://doi.org/10.1021/cr3001274.
98 Couture, G., Granado, L., Fanget, F. et al. (2018). Limonene-based epoxy: anhydride thermoset reaction study. *Molecules*: 23(11). https://doi.org/10.3390/molecules23112739.
99 Ramon, E., Sguazzo, C., and Moreira, P. (2018). A review of recent research on bio-based epoxy systems for engineering applications and potentialities in the aviation sector. *Aerospace* 5 (4): 110. https://doi.org/10.3390/aerospace5040110.
100 Costa, C.S.M.F., Fonseca, A.C., Moniz, J. et al. (2016). Soybean and coconut oil based unsaturated polyester resins: thermomechanical characterization. *Ind. Crops Prod.* 85: 403–411. https://doi.org/10.1016/j.indcrop.2016.01.030.
101 Baroncini, E.A., Kumar Yadav, S., Palmese, G.R., and Stanzione, J.F. (2016). Recent advances in bio-based epoxy resins and bio-based epoxy curing agents. *J. Appl. Polym. Sci.* 133 (45): 1473. https://doi.org/10.1002/app.44103.
102 Zhu, J., Chandrashekhara, K., Flanigan, V., and Kapila, S. (2004). Curing and mechanical characterization of a soy-based epoxy resin system. *J. Appl. Polym. Sci.* 91 (6): 3513–3518. https://doi.org/10.1002/app.13571.
103 Sudha, G.S., Kalita, H., Mohanty, S., and Nayak, S.K. (2017). Biobased epoxy blends from epoxidized castor oil: effect on mechanical, thermal, and morphological properties. *Macromol. Res.* 25 (5): 420–430. https://doi.org/10.1007/s13233-017-5063-3.
104 Flèche, G. and Huchette, M. (1986). Isosorbide. Preparation, properties and chemistry. *Starch/Stärke* 38 (1): 26–30. https://doi.org/10.1002/star.19860380107.

105 Hong, J., Radojčić, D., Ionescu, M. et al. (2014). Advanced materials from corn: isosorbide-based epoxy resins. *Polym. Chem.* 5 (18): 5360–5368. https://doi.org/10.1039/C4PY00514G.

106 Korugic-Karasz, L.S. (ed.) (2010). *Contemporary Science of Polymeric Materials: A Symposium in Honor of Professor Frank E. Karasz on the Occasion of his 75th Birthday*, Valletta, Malta, (February 28–March 2, 2009). Washington, DC: American Chemical Soc.

107 Łukaszczyk, J., Janicki, B., and Kaczmarek, M. (2011). Synthesis and properties of isosorbide based epoxy resin. *Eur. Polym. J.* 47 (8): 1601–1606. https://doi.org/10.1016/j.eurpolymj.2011.05.009.

108 Mathew, A.K., Abraham, A., Mallapureddy, K.K., and Sukumaran, R.K. (2018). Lignocellulosic biorefinery wastes, or resources? In: *Waste Biorefinery: Potential and Perspectives*, Chapter 9 (eds. T. Bhaskar, A. Pandey, S.V. Mohan, et al.), 267–297. Amsterdam, Netherlands: Elsevier.

109 Lamminpää, K., Ahola, J., and Tanskanen, J. (2012). Kinetics of xylose dehydration into furfural in formic acid. *Ind. Eng. Chem. Res.* 51 (18): 6297–6303. https://doi.org/10.1021/ie2018367.

110 Wang, R.-M., Zheng, S.-R., and Zheng, Y.-P. (2011). 3 - Matrix materials. In: *Polymer Matrix Composites and Technology* (eds. R.-M. Wang, S.-R. Zheng and Y.-P. Zheng), 101–548. Beijing, Oxford: SP, Science Press; WP, Woodhead Publ.

111 Torres, A.I., Daoutidis, P., and Tsapatsis, M. (2011). Biomass to chemicals: Design of an extractive reaction process for the production of 5-hydroxymethylfurfural. In: *Computer Aided Chemical Engineering 21 European Symposium on Computer Aided Process Engineering* (eds. E.N. Pistikopoulos, M.C. Georgiadis and A.C. Kokossis), 236–240. Elsevier.

112 Hu, F., La Scala, J.J., Sadler, J.M., and Palmese, G.R. (2014). Synthesis and characterization of thermosetting furan-based epoxy systems. *Macromolecules* 47 (10): 3332–3342. https://doi.org/10.1021/ma500687t.

113 Hu, F., Yadav, S.K., Sharifi, M. et al. (2016). Characterization of furanyl thermosetting polymers with superior mechanical properties and high-temperature char yield. In: *Proceedings of the International SAMPE Technical Conference, Long Beach, CA, USA*, 23–26.

114 Pizzi, A. (2009). Tannins: major sources, properties and applications. In: *Monomers, Polymers and Composites from Renewable Resources*, Chapter 8, cop. 2008 (eds. M.N. Belgacem and A. Gandini), 179–199. Amsterdam: Elsevier.

115 Basnet, S., Otsuka, M., Sasak, C. et al. (2015). Functionalization of the active ingredients of Japanese green tea (*Camellia sinensis*) for the synthesis of bio-based epoxy resin. *Ind. Crops Prod.* 73: 63–72. https://doi.org/10.1016/j.indcrop.2015.03.091.

116 Tarzia, A., Montanaro, J., Casiello, M. et al. (2018). Synthesis, curing, and properties of an epoxy resin derived from gallic acid. *BioResources* 13 (1): 632–645.

117 Chan, B.L. (2016). Potential application of ENR/EPDM Blends. *AJSTD* 33 (1): 48. https://doi.org/10.29037/ajstd.5.

118 Mathew, V.S., George, S.C., Parameswaranpillai, J., and Thomas, S. (2014). Epoxidized natural rubber/epoxy blends: Phase morphology and thermomechanical properties. *J. Appl. Polym. Sci.* 131 (4) https://doi.org/10.1002/app.39906.

119 Faruk, O. and Sain, M. (2015). *Lignin in Polymer Composites.* William Andrew.

120 Glasser, W.G., Northey, R.A., and Schultz, T.P. (eds.) (2000). *Lignin: Historical, Biological, and Materials Perspectives.* Washington, DC: American Chemical Society.

121 Asada, C., Basnet, S., Otsuka, M. et al. (2015). Epoxy resin synthesis using low molecular weight lignin separated from various lignocellulosic materials. *Int. J. Biol. Macromol.* 74: 413–419. https://doi.org/10.1016/j.ijbiomac.2014.12.039.

122 Wang, S., Ma, S., Xu, C. et al. (2017). Vanillin-derived high-performance flame retardant epoxy resins: facile synthesis and properties. *Macromolecules* 50 (5): 1892–1901. https://doi.org/10.1021/acs.macromol.7b00097.

123 Shibata, M. and Ohkita, T. (2017). Fully biobased epoxy resin systems composed of a vanillin-derived epoxy resin and renewable phenolic hardeners. *Eur. Polym. J.* 92: 165–173. https://doi.org/10.1016/j.eurpolymj.2017.05.007.

124 Ma, S., Li, T., Liu, X., and Zhu, J. (2016). Research progress on bio-based thermosetting resins. *Polym. Int.* 65 (2): 164–173. https://doi.org/10.1002/pi.5027.

125 Liu, X.Q., Huang, W., Jiang, Y.H. et al. (2012). Preparation of a bio-based epoxy with comparable properties to those of petroleum-based counterparts. *Express Polym. Lett.* 6 (4).

126 Liu, F., Wang, Z., Wang, Y., and Zhang, B. (2010). Copolymer networks from carboxyl-containing polyaryletherketone and diglycidyl ether of bisphenol-A: preparation and properties. *J. Polym. Sci., Part B: Polym. Phys.* 48 (23): 2424–2431. https://doi.org/10.1002/polb.22141.

127 Morell, M., Ramis, X., Ferrando, F. et al. (2009). New improved thermosets obtained from DGEBA and a hyperbranched poly(ester-amide). *Polymer* 50 (23): 5374–5383. https://doi.org/10.1016/j.polymer.2009.09.024.

128 Jin, F.-L. and Park, S.-J. (2008). Thermomechanical behavior of epoxy resins modified with epoxidized vegetable oils. *Polym. Int.* 57 (4): 577–583. https://doi.org/10.1002/pi.2280.

129 Deng, L., Ha, C., Sun, C. et al. (2013). Properties of bio-based epoxy resins from rosin with different flexible chains. *Ind. Eng. Chem. Res.* 52 (37): 13233–13240. https://doi.org/10.1021/ie4005223.

130 Fam, A. and Rizkalla, S.H. (2001). Behavior of axially loaded concrete-filled circular FRP tubes. *Aci. Struct. J.* 98: 280–289.

131 Wang, L.-M. and Wu, Y.-F. (2008). Effect of corner radius on the performance of CFRP-confined square concrete columns: test. *Eng. Struct.* 30 (2): 493–505. https://doi.org/10.1016/j.engstruct.2007.04.016.

132 Shehata, I.A.E.M., Carneiro, L.A.V., and Shehata, L.C.D. (2002). Strength of short concrete columns confined with CFRP sheets. *Mater. Struct.* 35 (1): 50–58. https://doi.org/10.1007/BF02482090.

133 Täljsten, B. (2003). Strengthening concrete beams for shear with CFRP sheets. *Constr. Build. Mater.* 17 (1): 15–26. https://doi.org/10.1016/S0950-0618(02)00088-0.

134 Barros, J.A.O. and Fortes, A.S. (2005). Flexural strengthening of concrete beams with CFRP laminates bonded into slits. *Cem. Concr. Compos.* 27 (4): 471–480. https://doi.org/10.1016/j.cemconcomp.2004.07.004.

135 Saadatmanesh, H., Ehsani, M.R., and Beams, R.C. (1991). Strengthened with GFRP plates. I: Experimental study. *J. Struct. Eng.* 117 (11): 3417–3433. https://doi.org/10.1061/(ASCE)0733-9445(1991)117:11(3417).

136 Lu, Y. and Larock, R.C. (2007). Fabrication, morphology and properties of soybean oil-based composites reinforced with continuous glass fibers. *Macromol. Mater. Eng.* 292 (10–11): 1085–1094. https://doi.org/10.1002/mame.200700150.

137 Thulasiraman, V., Rakesh, S., and Sarojadevi, M. (2009). Synthesis and characterization of chlorinated soy oil based epoxy resin/glass fiber composites. *Polym. Compos.* 30 (1): 49–58. https://doi.org/10.1002/pc.20532.

138 McIsaac, A. The effects of bio content and type of resin on FRP wet layup tensile and bond strength to concrete and durability under freeze-thaw cycles. MSc. Thesis. 2017. Queen's University, Kingston, Ontario, Canada.

139 Fam, A., Eldridge, A., and Misra, M. (2014). Mechanical characteristics of glass fibre reinforced polymer made of furfuryl alcohol bio-resin. *Mater. Struct.* 47 (7): 1195–1204. https://doi.org/10.1617/s11527-013-0122-5.

140 McSwiggan, C. and Fam, A. (2017). Bio-based resins for flexural strengthening of reinforced concrete beams with FRP sheets. *Constr. Build. Mater.* 131: 618–629. https://doi.org/10.1016/j.conbuildmat.2016.11.110.

141 Rana, A. and Evitts, R.W. (2015). Development and characterization of flax fiber reinforced biocomposite using flaxseed oil-based bio-resin. *J. Appl. Polym. Sci.* 132 (15) https://doi.org/10.1002/app.41807.

142 Maffezzoli, A., Calo, E., Zurlo, S. et al. (2004). Cardanol based matrix biocomposites reinforced with natural fibres. *Compos. Sci. Technol.* 64 (6): 839–845. https://doi.org/10.1016/j.compscitech.2003.09.010.

143 Fahimian, M., Adhikari, D., Raghavan, J., and Wool, R.P. (2008). Biocomposites from canola oil based resins and hemp and flax fibers. In: *13th European Conference on Composite Materials*, 2–5 June, Stockholm, Sweden.

144 O'Donnell, A., Dweib, M.A., and Wool, R.P. (2004). Natural fiber composites with plant oil-based resin. *Compos. Sci. Technol.* 64 (9): 1135–1145. https://doi.org/10.1016/j.compscitech.2003.09.024.

145 Samper, M.D., Petrucci, R., Sanchez-Nacher, L. et al. (2015). Properties of composite laminates based on basalt fibers with epoxidized vegetable oils. *Materials & Design* 72: 9–15. https://doi.org/10.1016/j.matdes.2015.02.002.

146 Fiore, V., Scalici, T., Badagliacco, D. et al. (2017). Aging resistance of bio-epoxy jute-basalt hybrid composites as novel multilayer structures for cladding. *Compos. Struct.* 160: 1319–1328. https://doi.org/10.1016/j.compstruct.2016.11.025.

147 Cicala, G., Pergolizzi, E., Piscopo, F. et al. (2018). Hybrid composites manufactured by resin infusion with a fully recyclable bioepoxy resin. *Composites Part B* 132: 69–76. https://doi.org/10.1016/j.compositesb.2017.08.015.

148 Morye, S.S. and Wool, R.P. (2005). Mechanical properties of glass/flax hybrid composites based on a novel modified soybean oil matrix material. *Polym. Compos.* 26 (4): 407–416. https://doi.org/10.1002/pc.20099.

149 Pizzi, A., Kueny, R., Lecoanet, F. et al. (2009). High resin content natural matrix–natural fibre biocomposites. *Ind. Crops Prod.* 30 (2): 235–240. doi: https://doi.org/10.1016/j.indcrop.2009.03.013.

150 Yu, T., Jiang, N., and Li, Y. (2014). Study on short ramie fiber/poly(lactic acid) composites compatibilized by maleic anhydride. *Composites Part A* 64: 139–146. https://doi.org/10.1016/j.compositesa.2014.05.008.

151 Wang, A., Xian, G., and Li, H. (2019). Effects of fiber surface grafting with nano-clay on the hydrothermal ageing behaviors of flax fiber/epoxy composite plates. *Polymers (Basel)* 11 (8) https://doi.org/10.3390/polym11081278.

152 Yan, L., Chouw, N., and Yuan, X. (2012). Improving the mechanical properties of natural fibre fabric reinforced epoxy composites by alkali treatment. *J. Reinf. Plast. Compos.* 31 (6): 425–437. https://doi.org/10.1177/0731684412439494.

153 Xie, Y., Hill, C.A.S., Xiao, Z. et al. (2010). Silane coupling agents used for natural fiber/polymer composites: a review. *Composites Part A* 41 (7): 806–819. https://doi.org/10.1016/j.compositesa.2010.03.005.

154 Díez-Pascual, A.M. (2019). Synthesis and applications of biopolymer composites. *Int. J. Mol. Sci.* 20 (9) https://doi.org/10.3390/ijms20092321.

155 Cadu, T., van Schoors, L., Sicot, O. et al. (2019). Cyclic hygrothermal ageing of flax fibers' bundles and unidirectional flax/epoxy composite. Are bio-based reinforced composites so sensitive? *Ind. Crops Prod.* 141: 111730. https://doi.org/10.1016/j.indcrop.2019.111730.

156 Hong, B., Xian, G., and Li, H. (2018). Effects of water or alkali solution immersion on the water uptake and physicomechanical properties of polyurethane. *Polym. Eng. Sci.* 58 (12): 2276–2287. https://doi.org/10.1002/pen.24848.

157 Fiore, V., Calabrese, L., Di Bella, G. et al. (2016). Effects of aging in salt spray conditions on flax and flax/basalt reinforced composites: wettability and dynamic mechanical properties. *Composites Part B* 93: 35–42. https://doi.org/10.1016/j.compositesb.2016.02.057.

3

Polymer Blends Based on Bioepoxy Polymers

Sudheer Kumar[1] *and Sukhila Krishnan*[2]

[1] *Central Institute of Plastics Engineering & Technology (CIPET), School for Advanced Research in Polymers (SARP), Laboratory for Advanced Research in Polymeric Materials (LARPM), B/25, CNI Complex, Patia, Bhubaneswar 751024, Odisha, India*

[2] *Sahrdaya College of Engineering and Technology, Department of Applied Science and Humanities, Kodakara, Thrissur 680684, Kerala, India*

3.1 Introduction

Recently, researchers and polymer industries play a very significant role in the preparation of new biobased materials and their blends with the help of renewable resources. Until now, epoxy resins have been exploited in various industrial applications such as painting, coatings, adhesives, electronics, and aerospace industry [1–5]. On the other hand, the depletion of petroleum feedstocks and utilization of sustainable resources as the initial materials of blending components in the development of bioepoxy blend systems is attracting great attention within the academicians and industries. The bioepoxy resin partially or completely replaces petroleum-based resins and enhances the physical and mechanical properties of newly developed materials [1, 6, 7]. However, after the addition of this type of materials, bioepoxy resin reduces global warming and carbon dioxide (CO_2) emission and causes no toxic effect on human health and the environment.

Further, epoxy resins demonstrated several outstanding properties, such as low cost, higher mechanical strength, good chemical stability, strong adhesion, and low shrinkage. However, they also exhibit low impact strength, poor toughness, and brittle nature after curing, which limit their applications in certain areas [8–11]. For that reason, bioepoxies were used as binders or diluents to enhance the performance of the newly developed materials and replace the existing materials. The properties of new materials are better or similar to those of the petroleum-based resin. Currently, several kinds of bioepoxies have been synthesized, such as rosin, lignin, furan, carbohydrate, tannin, itaconic acid, CNSL, and protein [12, 13].

Recently, the bioepoxy polymer obtained from these resources has been the main area of interest in the research and is a better substitute for the petroleum-based epoxy resin for sustainable development of biobased thermoset polymers. Nevertheless, a few epoxies reveal the major issues, such as limited

Bio-Based Epoxy Polymers, Blends, and Composites: Synthesis, Properties, Characterization, and Applications, First Edition. Edited by Jyotishkumar Parameswaranpillai, Sanjay Mavinkere Rangappa, Suchart Siengchin, and Seno Jose.

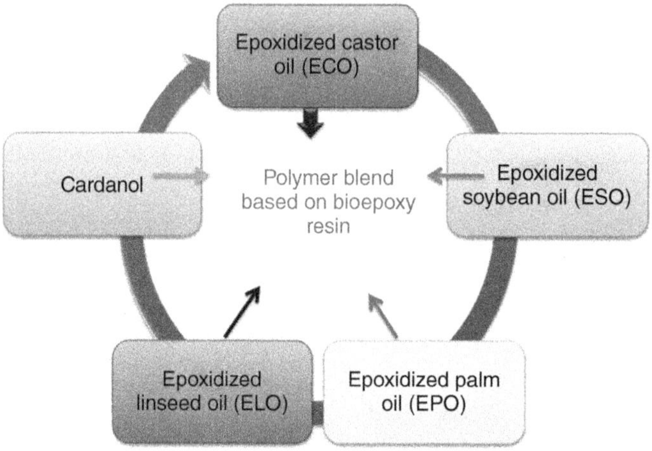

Figure 3.1 PO-based bioepoxy resin blend with DGEBA epoxy resin.

manufacturing, complicated structure, structural control deficiency, hydrophobicity, low purity, brittleness, and toxicity. However, the production of useful and low-cost bioepoxy with higher characteristics has remained a challenge [14, 15]. In this perspective, among all sustainable resource-based bioepoxies, the vegetable plant oil-based bioepoxy is extensively employed as a pioneer to prepare biobased polymer blends because plant oil (PO)-based bioepoxy shows numerous advantages such as easy availability, low cost, renewability, environment-friendliness, and the presence of more reactive sites for modification [16]. Plant oils are tri-esters of glycerol that contain long saturated and unsaturated aliphatic chains of fatty acids (with one or more double bonds). Further, the plant oil-based triglyceride bioepoxy with a higher content of oxirane rings and higher flexibility owing to their aliphatic chains enhances the mechanical (impact strength and fracture toughness) properties of the blend systems.

In the current year, POs as castor oil, soybean oil, linseed, palm, and cottonseed oil have stimulated the researchers for the development of bioepoxy polymers and their biobased blends that can serve as substitutes to petroleum-based materials, as shown in Figure 3.1. In this chapter, we had briefed about the synthesis of bioepoxies and their biobased thermoset blends used for various applications.

3.2 Plant Oils

3.2.1 Chemical and Physical Properties of Plant Oils

Renewable resources based on plant or vegetable oils can be utilized as consistent preliminary materials to obtain innovative products in a broad range of structural and functional varieties. Plant oils, such as soybean oil [17–19], linseed oil [20, 21], canola oil [22], castor oil [23], Karanja oil [24, 25], hemp oil [26, 27], cottonseed oil [28], rapeseed oil [29, 30], and palm oil [31], are broadly used

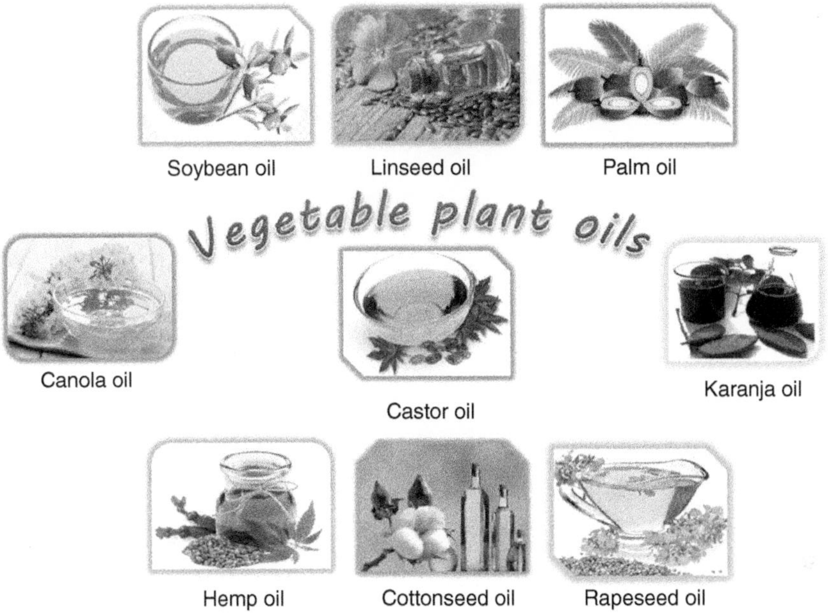

Figure 3.2 Schematic representation of vegetable plant oils used for various industrial applications. Source: Sudheer Kumar.

as the most important renewable feedstocks processed in the chemical industry for the preparation of bioepoxies and other biobased polymers as depicted in Figure 3.2.

The life cycle of PO-based polymer materials is described in Figure 3.3. The cycle demonstrated that there is no depletion of energy or reserves after the usage of biomass-derived plant oil-based polymers [32, 33]. Further, the waste can also be employed to generate the source of oils for a second time.

Various researchers had reported that the vegetable oil-based bioepoxy resins and their advanced biobased thermoset blend materials have been developed with the different methods because of the presence of unsaturated acid content. The main constituents of vegetable oils are fatty acids, triglycerides, or glycerol. The quantities of the various fatty acids describe the oil type. The plant oil generally contains 14–22 carbon in length and 1–3 double bond fatty acids. The fatty acid distribution of numerous common plant oils are revealed in Table 3.1, and the structures are illustrated in Figure 3.4.

Plant oils contain highly functionalizable triglyceride molecules with the presence of high content of unsaturated double bonds, and they have been employed to synthesize new polymeric materials. This is due to the presence of reactive sites or unsaturated C=C double bonds in the plant oils that polymerize and form prepolymers or polymers. The instant functional groups in triglycerides, such as hydroxyl (–OH) and epoxide group, can be cross-linked by employing different polymerization processes. However, the most appropriate and useful strategy to attain high-performance polymers is the modification (chemical) of oils before undergoing the polymerization process. The incorporation of simple

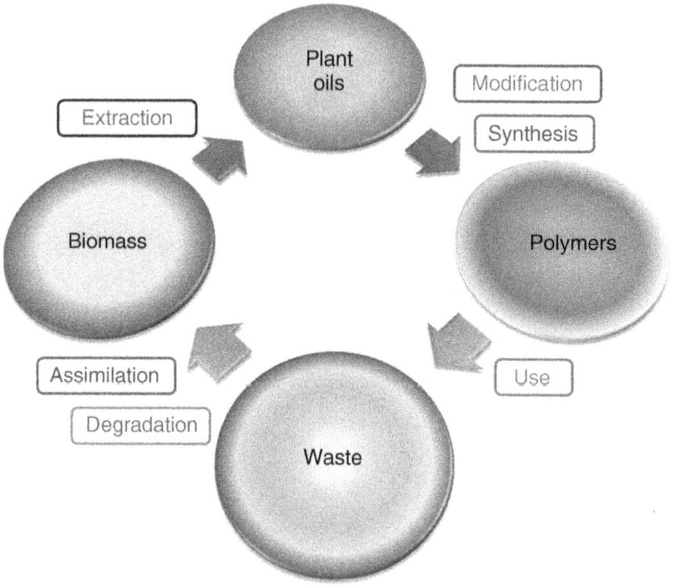

Figure 3.3 The life cycle of plant oil (PO)-based polymer materials.

polymerizable functional groups into the plant oil is possible but found to be very difficult because of the low reactivity of triglycerides. Currently, the industries are exploring the utilization of plant oils.

3.2.2 Chemical Modification of Plant Oils

Various researchers have described the chemical modification of plant oils (triglycerides) by transesterification [35], epoxidation [4, 36], and acrylation [37]. Triglycerides (triglycerols or tri-ester of glycerol) are the esterified products of glycerol with three acids as depicted in Figure 3.5.

The unmodified plant cannot be employed as the main matrix in the development of blends and nanocomposites. Because the unsaturated double bonds or vinyl groups present in the triglyceride chain are not enough to provide the stiffness and strength to the new materials [38–41], they are unable to undergo some polymerization reaction apart from direct polymerization such as cationic/radical polymerization [42, 43]. As a result, modification of these groups is necessary for enhancing the reactivity of polymerization of triglycerides to synthesize new polymeric materials. After the introduction of these groups into large aliphatic glyceride chains, they can provide better capability during the polymerization [27, 44–48].

Various researchers already reported the most commonly used chemical modification method of the double bond in plant oils to procure high characteristic polymer materials after introducing the different functional groups, for instance, –OH, acrylate, epoxy, or –CO moieties [48–53]. Further, different reactions were also utilized for the functionalization of triglycerides, which include transesterification, epoxidation, acrylation, methoxylation, and maleation. Among these

Table 3.1 The fatty acid distribution of numerous common plant oils [34].

Plant-based natural oils	Average annual production (10^6 tons)	Fatty acid (%)					Double bonds	Iodine value (mg/100 g)
		Palmitic	Stearic	Oleic	Linoleic	Linolenic		
Soybean	26.52	11.0	4.0	23.4	53.3	7.8	4.6	117–143
Castor	0.56	9	0.5	5.0	4.0	1.0	3.9	82–88
Linseed	0.83	5.5	3.5	19.1	15.3	56.6	6.6	168–204
Palm	23.53	42.8	4.2	40.5	10.1	—	1.7	44–58
Rapeseed/ canola	15.29	4/4.1	4.2	56/60.9	26/21.0	10/8.8	3.8/3.9	94–120/11–126
Sunflower	15.29	5.2	2.7	37.2	53.8	1.0	4.7	110–143
Cottonseed	4.49	21.6	2.6	18.6	54.4	0.7	3.9	90–106
Groundnut	5.03	11.4	2.4	48.3	31.9	—	3.4	80–106
Coconut	3.74	9.8	3.0	6.9	2.2	—	—	6–11
Palm kernel	2.95	8.8	2.4	13.6	1.1	—	—	14–24
Corn	2.30	10.9	2.0	24.4	59.6	1.2	4.5	102–130
Olive	2.52	13.7	2.5	71.1	10.0	0.6	2.8	75–94
Sesame	0.76	9	6	41	43	1.0	3.9	103–116

methods, the epoxidation of the oil is most commonly used owing to its flexibility in forming a variety of prepolymers.

Triglycerides are highly functionalized molecules and therefore can be used in the synthesis of cross-linked polymers. The internal C=C bonds present in the triglycerides can act as a reactive site for the alteration of fatty acids of POs. Epoxidation is the most important functionalization reaction connecting C=C bonds of fatty acid [54]. The epoxidation process is catalyzed by H_2O_2 through chemical or enzymatic oxidation of the C=C bonds [28, 46, 47]. According to another approach, the formation of methyl ester groups reduced the viscosity and enhances the reactivity of resins to obtain a high-performance polymer. Conversely, lots of other strategies are also employed for the modification of plant oils through ring-opening, free radical polymerization, metathesis, and polycondensation polymerization reactions to make prepolymers. Recently, different synthetic methods have been employed for the transformation of plant oils to functional polymers by the greener approach.

3.3 Preparation of Bioepoxy Polymer Blends with Epoxy Resins

Recently, plant oils (POs) have gained eminence as a substitute to biobased platform for numerous industrial applications because of their characteristics such as renewability, low cost, and easy availability globally. The POs are a

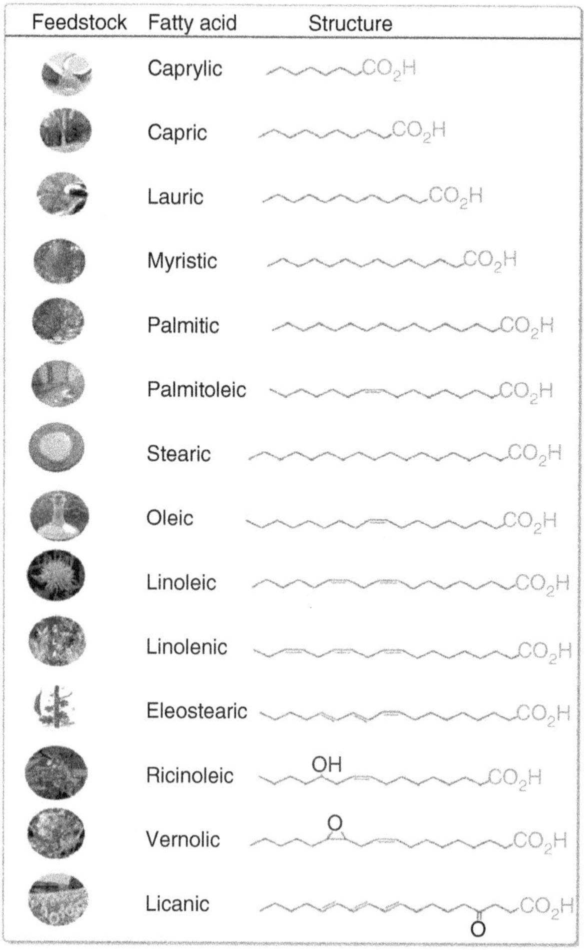

Figure 3.4 Structure of fatty acid in plant oil-based triglycerides. Source: Sudheer Kumar.

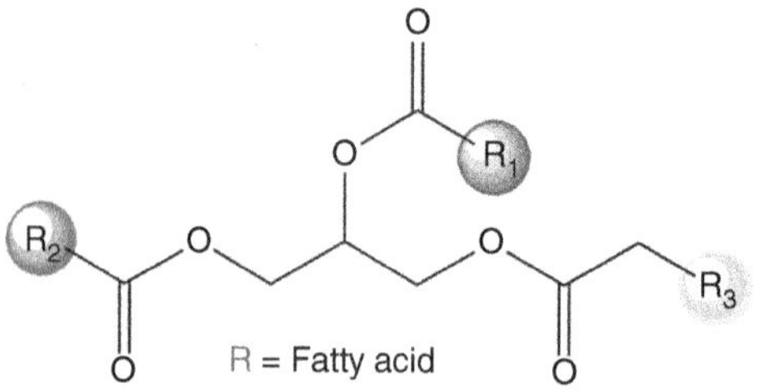

Figure 3.5 Structure of triglyceride.

major component for the preparation of green materials in the near future (e.g. adhesives, plasticizers, and bioepoxies) [55–58]. The PO-based bioepoxy resin demonstrated structural characteristics of the triglyceride such as a long aliphatic chain that directly influence the properties of the materials. Further, the synthesized resin exhibited lower mechanical properties as compared to

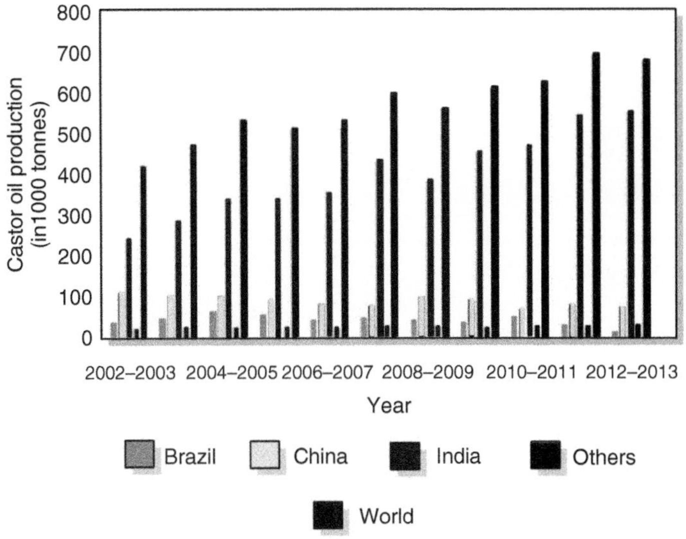

Figure 3.6 Representation of castor oil production [60].

commercial petroleum-based epoxy resin. Because of this, we need to prepare the partially biobased blends of epoxidized plant oils (EPOs) with DGEBA that have been widely investigated to generate bioepoxy and epoxy blend materials with enhanced mechanical properties [59].

3.3.1 Castor Oil-Based Bioepoxy Polymer Blend

India is the biggest exporter of castor oil (CO) in the world compared to other countries, as shown in Figure 3.6 [60]. CO is a nonedible oil that exhibits C=C and hydroxyl group in the triglyceride fatty acid chains. Per triglyceride chain, there is a 2.7 hydroxyl group, and 90% fatty acids in the CO are ricinoleic acids. The CO exhibits chemical versatility because of its trifunctional nature.

Sudha et al. [61] synthesized an epoxidized castor oil (ECO)-based bioresin by in situ polymerization process and further prepared the biobased epoxy blend with diglycidyl bisphenol A (DGEBA) epoxy resin at different wt% of ECO by employing triethylenetetramine as the curing agent, as shown in Figure 3.7. The results demonstrated that the 20 wt% ECO blend system exhibits a higher flexural and impact strength of 100 and 25.3 MPa, respectively. Further, the fracture toughness parameter such as the K_{IC} values of the epoxy/ECO blends improved up to 20% ECO. On the other hand, the G_{IC} values increased up to the addition of 50% ECO content. The results indicate that after inclusion of ECO content into the DGEBA matrix, the toughness of the biobased blend systems enhances. Thermogravimetric analysis also confirmed the improvement of the thermal stability of DGEBA and its biobased blend systems.

Sahoo et al. [62] synthesized the epoxy methyl ricinoleate (EMR) bioepoxy with ECO, and blends with various ratios such as 10–30 wt% of epoxy resin are compared with those of DGEBA/ECO (10, 20, and 30 wt%) blend systems.

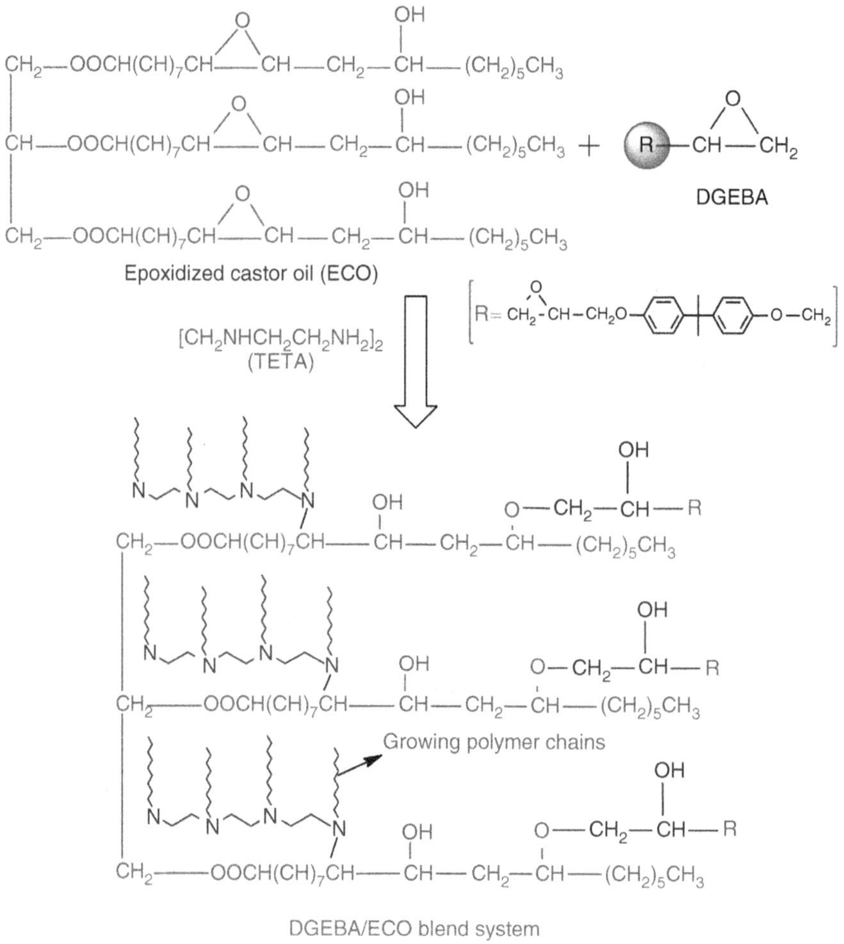

Figure 3.7 Schematic representations of epoxy/ECO bioepoxy blend systems.

Both systems were cured with phenalkamine (PKA) curing agent, as shown in Figure 3.8. The results revealed that the EMR blend system has higher tensile strength (TS) and elongation (%) as compared to ECO blend systems but exhibits a reduction in its modulus. However, 20% EMR and 30% EMR blend systems demonstrate lower modulus than ECO blend system owing to lower epoxide value and less entanglement. According to the elongation (%) investigation, it was confirmed that 20 wt% of ECO and EMR shows effective toughening of epoxy and moderate strength and modulus suitable for the specific application.

The toughening and impact properties of DGEBA epoxy resin get highly influenced by the presence of bioepoxies such as ECO and EMR. The impact strength of the virgin DGEBA enhanced after incorporation of ECO and EMR bioepoxies owing to the alternate of the rigid DGEBA epoxy groups by aliphatic segment and methyl ester chain correspondingly. Finally, in all the systems, 30% EMR exhibited a highest impact strength of 14.5 kJ/m^2 as compared to 20% ECO

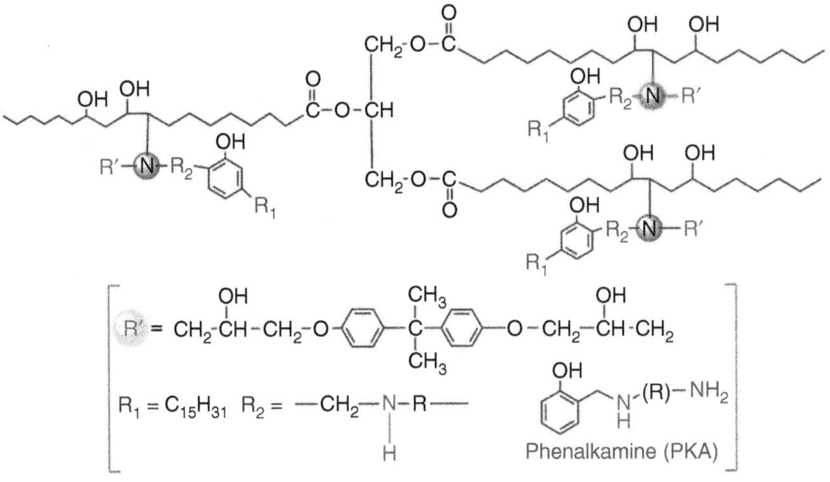

Figure 3.8 The representation of PKA-cured epoxy/ECO blend cross-linked network.

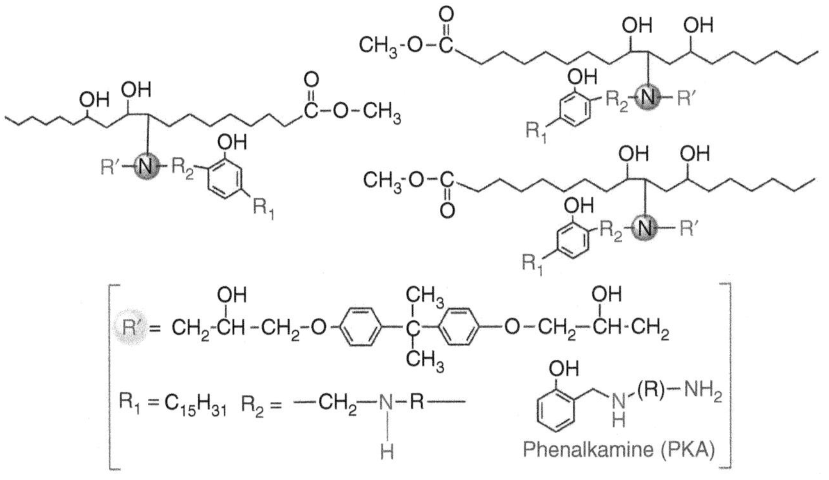

Figure 3.9 The representation of PKA-cured epoxy/EMR blend cross-linked network.

(7.45 kJ/m^2) and 315% higher than pure epoxy (3.49 kJ/m^2). TGA analysis also indicates that 10% ECO blend system corroborated higher thermal stability ($T_{5\%}$), while 20% EMR shows the slightly reduced thermal stability because of the presence of methyl ester chain of EMR. As a result, low-viscous EMR bioepoxy can be used as a potential diluent for the epoxy and can also replace the conventional diluents, which are already used in the coatings and structural applications (Figure 3.9).

Park et al. [63] prepared the biobased blends from ECO and DGEBA, which were polymerized via a latent thermal *N*-benzylpyrazinium hexafluoroantimonate (PBH) catalyst. The mechanical performance of cured blends systems was extensively enhanced with rising ECO bioepoxy. The pure epoxy resin exhibited 1.7 MPa $m^{1/2}$ fracture toughness (K_{IC}) value owing to its high cross-linked density and brittle nature. The K_{IC} value significantly enhanced with increasing content and demonstrated the highest value of 3.5 MPA $m^{1/2}$ at the 30 wt% ECO blend system. This is due to the incorporation of a large number of ECO soft segments into the epoxy matrix that enhances the flexible properties of blend systems and decreases the cross-linked density in the blend systems. Further, the flexural strength also enhances with the blend system from 87.6 to 117.1 MPa by the addition of 30% ECO content because of a bulky soft segment of ECO present in the epoxy.

The SEM micrograph of the fractured surface after K_{IC} toughness was also investigated. SEM micrographs show that the regular cracks in the fracture surface of the pure epoxy resin demonstrate its brittle nature. Conversely, the micrograph of blend systems reveals those tortuous cracks and various ridges with increasing ECO content. Finally, the epoxy/ECO blend system displays higher interfacial and mechanical properties as compared to the pure epoxy resin.

3.3.2 Soybean Oil-Based Bioepoxy Thermoset Polymer Blend

Currently, several researchers described that the preparation of ESO included biobased thermoset blend materials (Sahoo et al. [64] and Gupta et al. [65]). The ESO-based bioepoxy toughens the epoxy matrix extensively by performing as plasticizers or impact modifiers for the epoxy matrix. On the other hand, after the addition of ESO, the rigidity and strength of the amine-cured epoxy system decrease because of the lower cross-link density and phase separation. This is disagreeable for industry point of view. The performance of the biobased blend system was based on the functionality and characteristic of the epoxy resin and curing agents.

The synthesis of biobased ESO and epoxy resin blends examined the effect of ESO bioepoxy resin on petroleum-based epoxy resin at various ratios. The tensile strength (TS) of the blend systems was raised up to 20 wt% ESO [66]. Therefore, the epoxy/20%ESO blend systems exhibited a higher TS of 48.6 MPA as compared to the pure epoxy resin of 42.9 MPa. After the incorporation of the appropriate amount of ESO bioepoxy into the brittle epoxy resin, there is a significant decrease in the internal stress of the blend system. On the other hand, the ester linkage of flexible aliphatic ESO bioepoxy chain is capable of altering the bond angle and successfully decreasing the internal stress. Further, the character of the alternate side chain plays an extensive role to enhance the mechanical properties. As a result, it improves the TS owing to the better interaction among epoxy resin and ESO bioepoxy. Chen et al. [67] found similar results of epoxy/20%ESO bioepoxy blend systems.

Addition of higher content of ESO bioepoxy into the epoxy resin (17.2 J/m) enhanced the impact strength and achieved the maximum value 28.6 J/m at 20 wt% of bioepoxy owing to the plastic deformation of bioepoxy and epoxy as reported earlier by Ratna et al. [68]. Beyond 20%, the impact strength reduces, owing to the plasticization effect and the reactivity of the epoxy group in bioepoxy, which is lower than that of epoxy resin. This is mainly attributed to the steric hindrance of epoxy resin as depicted in Figure 3.10. The K_{IC} and G_{IC}

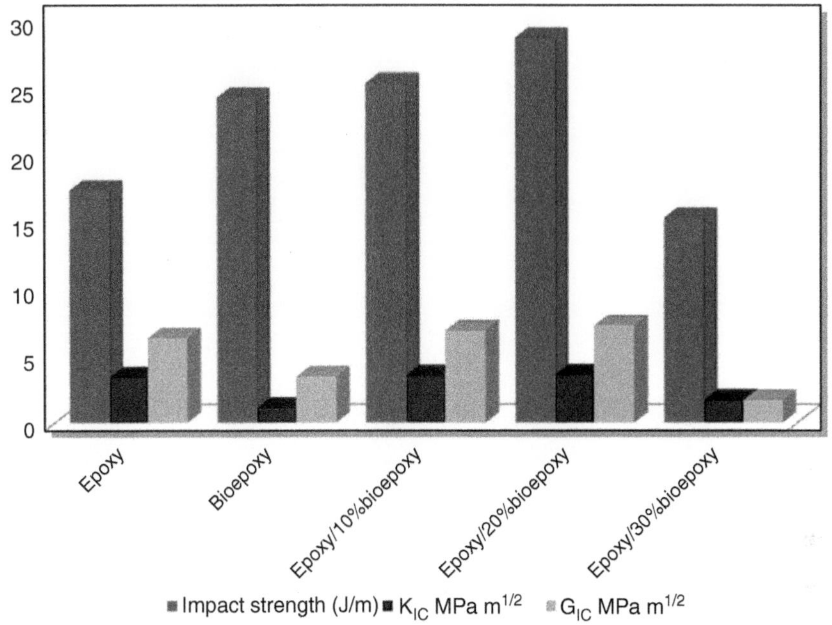

Figure 3.10 K_{IC}, G_{IC}, and notched impact strength of pure epoxy, bioepoxy, and its blends [66].

values also increased up to 20% bioepoxy (ESO) and effectively toughen and stiffen the epoxy matrix in an appropriate balance.

Further, SEM analysis of impact fractured surface sample of pure epoxy, bioepoxy (ESO), and its blends is displayed in Figure 3.11a–c. The SEM graph of pure epoxy resin demonstrated the extremely flat surface in fracture and confirms a brittle fracture owing to lower impact resistance, whereas the biobased epoxy/20%ESO blends show the rough surface with massive ridges and exhibited single-phased morphology. Although the ductility of ESO bioepoxy crack leads to a reduction in its brittle behavior, no phase separation indicates that bioepoxy is compatible with the epoxy matrix.

The developed epoxy/ESO blend systems with various wt% (5, 10, 20, and 30) were cured with triethylenetetramine (TETA) and investigated its mechanical and thermal characteristics to evaluate it to the commercial epoxy resin. As per the results, 20% ESO incorporated blends exhibited superior impact strength and toughness [68].

A bioepoxy resin synthesized from epoxidized soybean oil (ESO) via in situ polymerization is shown in Figure 3.12. The blends with epoxy at various ratios enhance the processability and toughness [69]. The 20 wt% ESO bioepoxy blend revealed higher impact strength as compared to pure epoxy. On the other hand, fracture toughness K_{IC} and G_{IC} values also got increased for epoxy resin after addition of ESO bioepoxy into the matrix, as illustrated in Figure 3.13.

Jin and Park [70] also reported the epoxy/ESO blend systems and had explored the effect of ESO bioepoxy on the mechanical performance of the epoxy/ESO blend system. The impact strength of the blends was extensively enhanced via

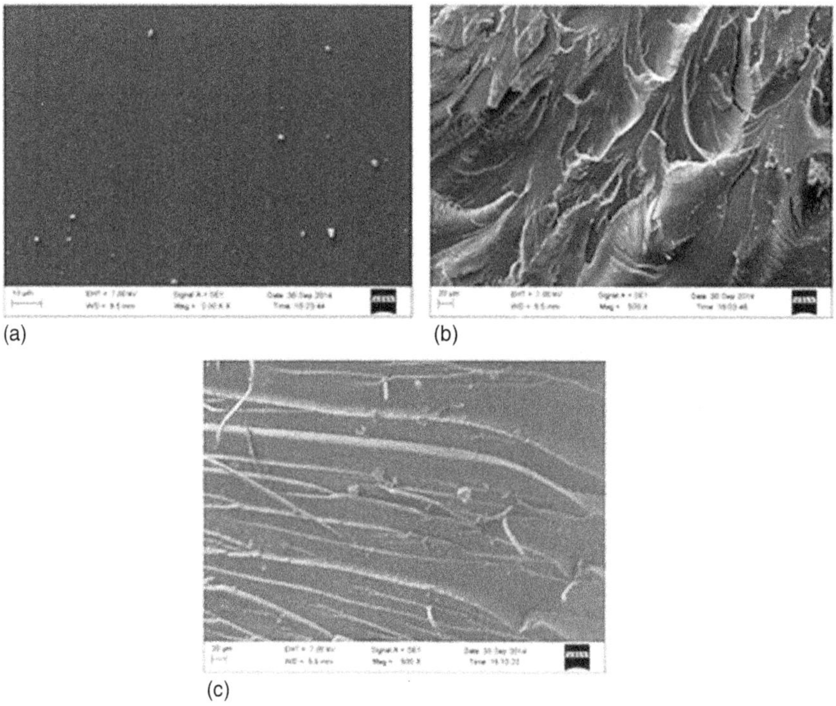

Figure 3.11 Morphology analysis of fracture surface of (a) pure epoxy, (b) bioepoxy, and (c) epoxy/20% bioepoxy. Source: Reproduced from Kumar et al. [66].

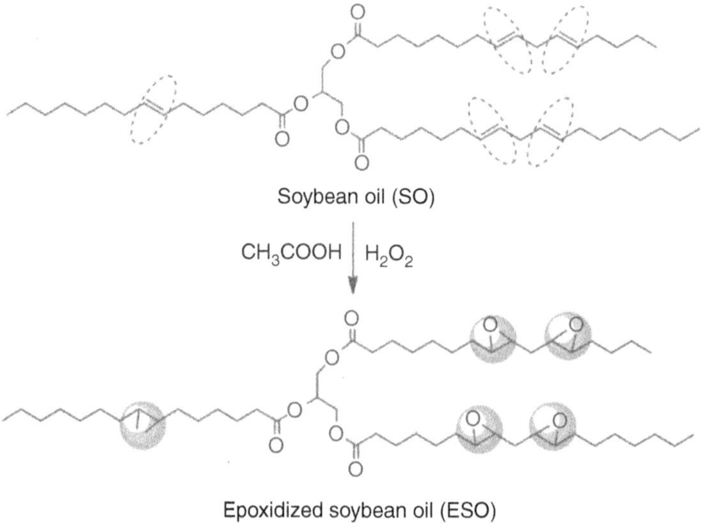

Figure 3.12 Synthetic route of bioepoxy resin from soybean oil.

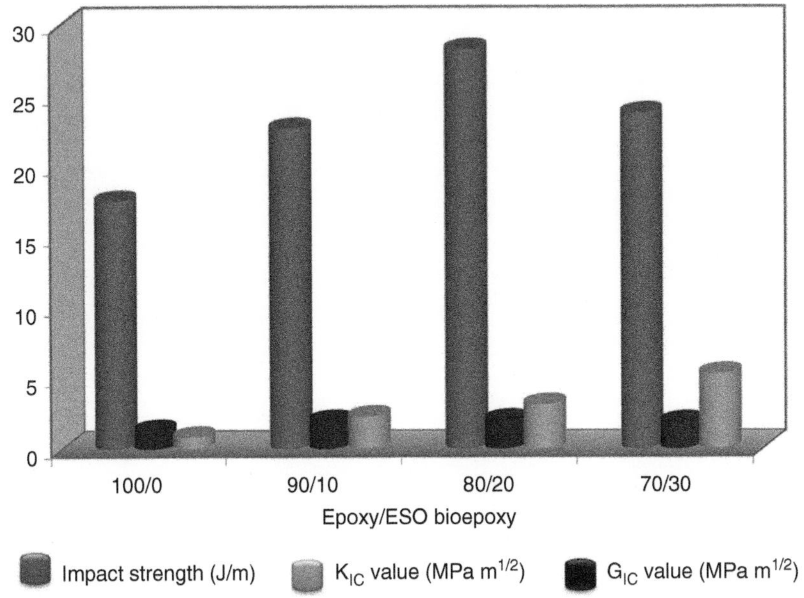

Figure 3.13 Fracture toughness and impact strength of epoxy and its bioepoxy blend systems [69].

the incorporation of ESO bioepoxy at 60 wt% because of the high intermolecular interaction in the blend system. Conversely, the lap shear strength of the blend system improved up to 40 wt% ESO. Further, the flexural strength and elastic modulus demonstrated similar values as compared to the pure epoxy as depicted in Figure 3.14. The results indicate that the ESO bioepoxy is a better applicant for employing it as a toughening agent.

3.3.3 Linseed Oil-Based Bioepoxy Thermoset Polymer Blend

Sahoo et al. [71] prepared a renewable resource-based toughened epoxy blend employing epoxidized linseed oil (ELO) and biobased curing agent. At various compositions of 10, 20, and 30 phr, the ELO bioepoxy was blended with a DGEBA epoxy as reactive diluents to decrease the viscosity for good processability and to reduce brittleness by curing with PKA. The TS and modulus of epoxy reduce steadily, other than elongation (%), which steadily improves with increasing ELO content. The toughened bioepoxy blend with 20–30 phr of ELO exhibits normal impact strength and improved elongation (%) from 7% to 13%. Addition of 20–30 phr of ELO enhances the enthalpy of curing without altering its curing peak temperature. Initially, ELO shows better miscibility with epoxy resin, and its blend confirms a single-phase morphology after curing. A decrease in viscosity of resin attributes to the highest possible insertion of bioepoxy, and addition of curing agent PKA results in an eco-friendly toughened epoxy, which is valuable for coatings and structural applications.

Miyagawa et al. [72] developed biobased epoxy materials by adding ELO and an anhydride cross-linker. A certain quantity of the DGEBF was substituted by ELO.

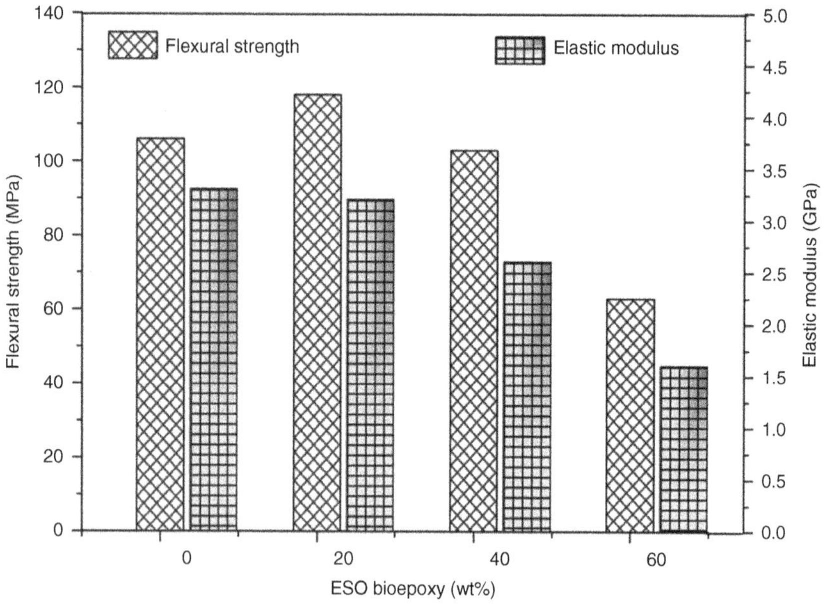

Figure 3.14 Mechanical properties of bioepoxy blends [70].

The addition of DGEBF, ELO, and an anhydride cross-linker proved to be a good blend to develop a novel biobased epoxy material exhibiting good elastic modulus, better T_g, and excellent heat distortion temperature (HDT) at higher content of ELO. With varying ELO content, the izod impact strength remains unaltered. The izod impact strength of the anhydride-cured rigid epoxy sample was comparatively less with an amine-cured network as they show a very high cross-link density. However, the izod impact strength reduces at more than 60 wt% ELO content. The structure of ELO is less rigid and branched than that of DGEBF. As a result, with the presence of more than 60 wt% ELO, the izod impact strength at room temperature becomes extremely less than T_g. This proves to be an ideal bioepoxy for various engineering industry applications in the future [73].

Sahoo et al. [74] reported a biobased epoxy resin from nonedible resources such as linseed oil (LO) and castor oil (CO). By an in-situ method, these oils were epoxidized and cross-linked with citric acid in the absence of an accelerator and their properties were evaluated with DGEBA. Relatively, ELO exhibits more TS and modulus than that of an ECO-based system. On the other hand, elongation (%) of ECO was extensively more as compared to both epoxy and ELO, which indicates its enhanced flexibility and toughness. The contact angle for bioepoxies is similar to that of epoxy and exhibits excellent chemical resistance. It was found that the TS and modulus of epoxy are 49.04 MPa and 1841.41 MPa, respectively, which are very close to that of commercially available cured TETA and thereby can substitute it with comparatively higher elongation (4.4%).

Asano et al. [75] reported that the TS and modulus of citric acid (CA)-cured DGEBA blend were 43 MPa and 2.1 GPa, correspondingly, with 3.6% highest strain. The higher cross-link density and aromatic structure in epoxy contribute to high TS and lower elongation (%) [76]. More cross-link density leads to good TS and modulus but with poor elongation (%), Conversely, lesser cross-link density results in poor TS and modulus but good elongation (%). On the other hand, the TS and modulus of ELO-based matrix are 5.25 and 53.58 MPa, respectively, that is relatively lower in comparison to epoxy. However, in contrast to epoxy, the elongation (%) of ELO is 300% more, which indicates its excellent toughness. Comparable behavior has also been observed for CA solution cured alkyd and epoxidized Jatropha oil (EJO) blends in the absence of an accelerator and exhibited relatively poor TS but better elongation (%) [77]. No hazardous catalysts and solvents were involved in this procedure, so it is eco-friendly.

Supanchaiyamat et al. [78] achieved related values of strength and modulus for bio-derived diacid-cured ELO specimens (Pripol 1009) in the presence of 1 wt% catalyst. In the same way, ECO shows poor TS (258 kPa) and modulus (1490 kPa) but exhibits excellent elongation (34.1%), which are extremely higher than those of cured epoxy and ELO systems. This attributes to the less epoxy functionality and more polyol content, which cause ductility in the matrix. The ESO exhibited a weak cross-link density because of the plasticizing effect of nonreactive sites of triglyceride that cause a remarkable effect on its mechanical performance. An excellent flexible soft rubbery material but with the notable toughening character was obtained because of the presence of low cross-linking in ECO. The lower modulus and higher elongation (%) corroborated with the elastomeric performance of bioepoxy.

3.3.4 Palm Oil-Based Bioepoxy Thermoset Polymer Blend

Tan and Chow [79] had prepared the blend of epoxidized palm oil (EPMO) with epoxy novolac, cycloaliphatic epoxide, and DGEBA. The results show much enhancement in the fracture toughness of the biobased epoxy blend system at high content of EPMO in the blend signifying that the addition of EPMO leads to toughen or flexibilize the brittle DGEBA epoxy blend. Because of the cross-linking, the toughness of the highly cross-linked epoxy decreases, owing to the absence of post-yield deformability and the comparatively low volume of plastic zone toward the notch tip [80]. Lu and Wool [81] had stated that plasticizing effect led to a considerable increase in its K_{IC} owing to the plasticizing effect of EPMO, which results in flexibility and improve the movement of molecular chains in the epoxy blend [82, 83]. Therefore, the ductility of the epoxy resin will be improved by inserting the flexible chains into the rigid bisphenol A backbone [84], which reduces the brittleness of the epoxy blend via incorporating EPMO into the epoxy matrix. Thus, at high EPMO content and lower cross-link densities, epoxy blends show higher K_{IC} value. Additionally, because of the cavitated holes dispersed all through the EB/EPMO epoxy matrix, there is an enhancement in the resistance to deform and obstruct the crack initiation in the epoxy blend system. Hence, EPMO-modified epoxy resins exhibit higher K_{IC} in contrast to pure epoxy resin.

Sarwono et al. [85] reported an epoxy resin synthesized by the reaction of DGEBA and *m*-xylylenediamine (*m*-XDA) and then 10 wt% of EPMO was added to modify it, as depicted in Figure 3.15. To the epoxy network, EPMO was

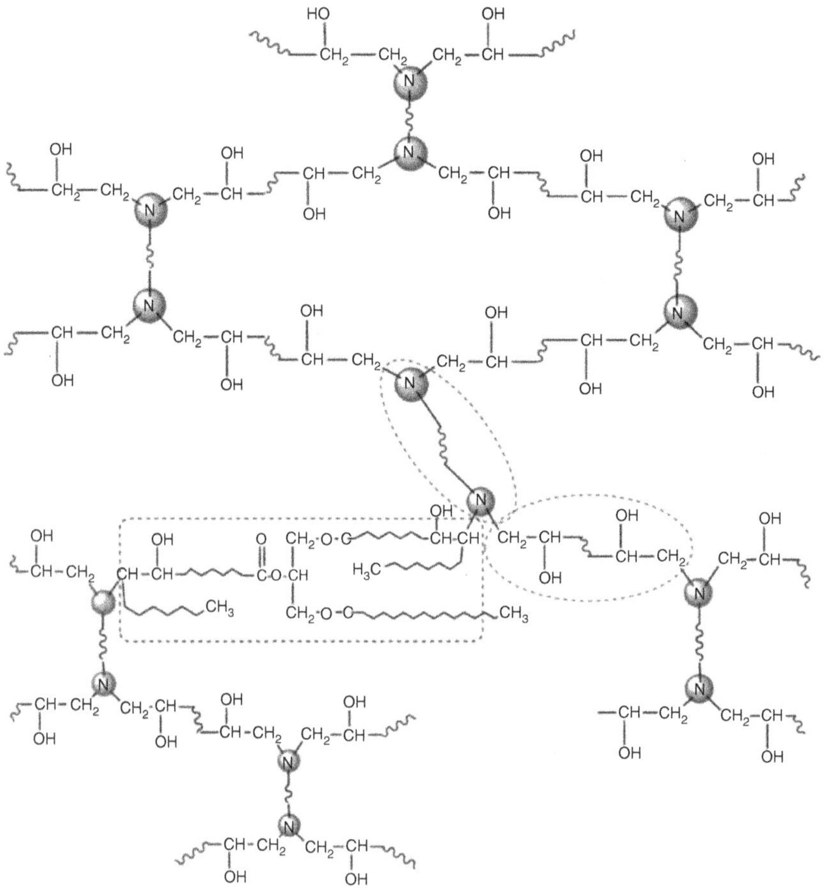

Figure 3.15 Schematic representation of the cross-linked network of EPMO/m-XDA/DGEBA epoxy systems.

grafted and a miscible blend was prepared by first reacting EPMO/m-XDA at 120 °C for at least two hours before the addition of the epoxy prepolymer. The morphological images indicate that at 10 wt% of EPMO sample preparation stage, the morphologies of the blend can change from heterogeneous to uniform phases depending on the type of inclusion method. Impact properties were highly influenced with the morphology of the blend indicated by the reduction in T_g, TS, and modulus as the phase varies from diverse to uniform. The decrease in cross-linking density and plasticizing effect and a rise in free volume result in lowering of T_g and mechanical properties of the modified epoxy resins. Although EPMO-modified epoxy resin exhibits better thermal stability in comparison to unmodified epoxy resin, the presence of EPMO perturbs the epoxy network, resulting in the decrease in storage modulus as we increase the mixing time.

Alsagayar et al. [86] reported that the TS and flexural property of epoxy/EPMO reduce when the content of EPMO raised. However, when EPMO content had attained 20%, impact strength, toughness, and strain at break were slightly improved. Results show that the impact strength enhances to around 21.91% and 22.59% from 48.28 J/m^2 for the pure epoxy to 58.86 and 59.19 J/m^2 for the biobased composites with 10 and 20 wt% EPMO, respectively. The toughness and impact strengths of the biocomposites improve by the incorporation of EPMO [87] because of the presence of cross-linked carbon–carbon double bonds in polymer networks, which inhibit the movement and facilitate in absorbing the impact energy. However, the impact strength decreases, as the bioepoxy content varies from 20 to 30 wt%. However, the blend exhibits more impact strength as compared to neat epoxy. Generally, at high EPMO content, the tensile strength and modulus along with flexural strength and modulus were decreased. As the EPMO content increases, the impact strength, toughness, and strain at break enhance, which attributes to the flexibility of the epoxy blend.

3.4 Application of Bioepoxy Polymer Blends

3.4.1 Paints and Coatings

The epoxy resin is extensively used in the anticorrosion coating application from the past few decades because of their excellent properties, such as higher mechanical and toughness performance, tremendous corrosion and chemical resistance, easy processability, low shrinkage on cure, and high safety. Further, also demonstrated the tremendous adhesion properties to numerous substrates and employs less energy compared to heat-cured powder coatings. On the other hand, the epoxy resin-coated product such as metal and containers are commonly utilized in packing applications to avoid rusting. Apart from this, the epoxy resin was also employed for decorative flooring applications [88–90]. Alam et al. [91] had also developed the plant oil-based eco-friendly coatings. Figure 3.16 demonstrated that the bioepoxy polymer blends are applied in various fields.

3.4.2 Adhesives

Further, the epoxy adhesive is also used in various industrial applications such as automobile, construction, boats, bicycle, golf, and aerospace because of their outstanding performances as well as good adhesion to the different substrate, high

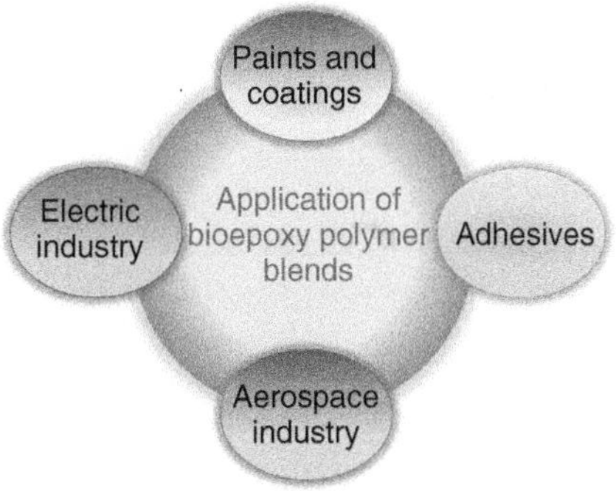

Figure 3.16 Applications of bioepoxy polymer in different fields.

temperature, good corrosion resistance, and low shrinkage [92]. Epoxy resin is used as a structural adhesive and high-performance applicant such as cryogenic engineering and also optimizes the shear strength at room and cryogenic temperatures [93–97].

3.4.3 Aerospace Industry

The epoxy resin is the main constituent in many adhesive formulation, and it is extensively utilized for structural adhesive purpose in the aerospace industry for the reason that it demonstrates high-adhesive strength and is inexpensive in nature. The epoxy resin reinforced with high strength glass, carbon, and Kevlar fibers are chemically compatible with the main substrate and favor to wet surfaces extremely without any difficulty and have the tremendous potential to be employed in the aerospace fields. On the other hand, the epoxy resin composites can fulfill the requirement for structure material applications in various areas such as civil, military, and aerospace to develop flooring panels, ducts, wings, and vertical and horizontal stabilizers [98–101].

3.4.4 Electric Industry

Conversely, the epoxy resin is also used in the electronic industries such as motors, switching, transformers, generators, and insulators because the epoxy resin is an excellent electrical insulator and protects the electrical component from moisture, dust, and short circuiting. Currently, metal-filled epoxy resins are broadly employed for electromagnetic interference shielding because they are less expensive and lighter than metal and can simply be molded into components. The performances of the epoxy-filled metal composites are based on the morphology of the metal particles [102]. The epoxy molding semiconductor composite materials are exactly employed as an encapsulation for semiconductor devices defending the integrated circuit devices from moisture mobile-ion impurities [103–105].

3.5 Conclusion

Typically, plant oil-based bioepoxy resins are extensively utilized in the preparation of biobased blends, which have various applications and also reduce the dependency toward petroleum resources, whereas the epoxidized plant oils are the potential green materials for the future prospects that are completely explored to develop an innovative opportunity for the bio-sector industry because of the limited fossil resources and dumping of huge quantity of waste materials by industrial sector on a daily basis. Further, the epoxidized plant oils are broadly synthesized and employed in the bio-sector industry in the last decade to replace partially and completely the toughened petroleum-based polymer owing to their sustainable and cost-effective nature. Epoxidized plant oil (EPO) possesses functional epoxide groups on their backbone chains, which cause its capability to create elastomeric network after curing and it is completely concluded that EPOs have the capability to entirely replace the existing petroleum-based polymeric products owing to the progress in oleo-chemistry technology, prolonged accessibility of POs and global market demand. On the other hand, epoxidized plant oils finally achieve popularity in nonfood applications, which also bestows the epoxidized plant oils to have a bright future.

References

1 Raquez, J.M., Deleglise, M., Lacrampe, M.F., and Krawczak, P. (2010). Thermosetting (bio) materials derived from renewable resources: a critical review. *Prog. Polym. Sci.* 35 (4): 487–509.
2 Auvergne, R., Caillol, S., David, G. et al. (2013). Biobased thermosetting epoxy: present and future. *Chem. Rev.* 114 (2): 1082–1115.
3 Kumar, S., Samal, S.K., Mohanty, S., and Nayak, S.K. (2018). Recent development of biobased epoxy resins: a review. *Polym. Plast. Technol. Eng.* 57 (3): 133–155.
4 Tan, S.G. and Chow, W.S. (2010). Biobased epoxidized vegetable oils and its greener epoxy blends: a review. *Polym. Plast. Technol. Eng.* 49 (15): 1581–1590.
5 Kumar, S., Krishnan, S., Samal, S.K. et al. (2018). Toughening of petroleum based (DGEBA) epoxy resins with various renewable resources based flexible chains for high performance applications: a review. *Ind. Eng. Chem. Res.* 57 (8): 2711–2726.
6 Suttie, E. (2012). *Bio-Resins in Construction: A Review of Current and Future Developments.* IHS BRE Press.
7 Kumar, S., Krishnan, S., Mohanty, S., and Nayak, S.K. (2018). Synthesis and characterization of petroleum and biobased epoxy resins: a review. *Polym. Int.* 67 (7): 815–839.
8 Gao, L.P., Wang, D.Y., Wang, Y.Z. et al. (2008). A flame-retardant epoxy resin based on a reactive phosphorus-containing monomer of DODPP and its thermal and flame-retardant properties. *Polym. Degrad. Stab.* 93 (7): 1308–1315.
9 Kumar, S.A. and Denchev, Z. (2009). Development and characterization of phosphorus-containing siliconized epoxy resin coatings. *Prog. Org. Coat.* 66 (1): 1–7.
10 Barcia, F.L., Amaral, T.P., and Soares, B.G. (2003). Synthesis and properties of epoxy resin modified with epoxy-terminated liquid polybutadiene. *Polymer* 44 (19): 5811–5819.
11 Zhang, X., Zhang, Z., Xia, X. et al. (2007). Synthesis and characterization of a novel cycloaliphatic epoxy resin starting from dicyclopentadiene. *Eur. Polym. J.* 43 (5): 2149–2154.
12 Gandini, A. (2008). Polymers from renewable resources: a challenge for the future of macromolecular materials. *Macromolecules* 41 (24): 9491–9504.

13 Holbery, J. and Houston, D. (2006). Natural-fiber-reinforced polymer composites in automotive applications. *JOM* 58 (11): 80–86.
14 Khot, S.N., Lascala, J.J., Can, E. et al. (2001). Development and application of triglyceride-based polymers and composites. *J. Appl. Polym. Sci.* 82 (3): 703–723.
15 Sharma, S. (2008). Fabrication and characterization of polymer blends and composites derived from biopolymers. Thesis (Ph.D.)–Clemson University, Publication Number: AAI3344251; ISBN: 9781109001648; vol. 70-01, Section: B, page: 0622; p. 247.
16 Meier, M.A., Metzger, J.O., and Schubert, U.S. (2007). Plant oil renewable resources as green alternatives in polymer science. *Chem. Soc. Rev.* 36 (11): 1788–1802.
17 Thulasiraman, V., Rakesh, S., and Sarojadevi, M. (2009). Synthesis and characterization of chlorinated soy oil based epoxy resin/glass fiber composites. *Polym. Compos.* 30 (1): 49–58.
18 Gupta, A.P., Ahmad, S., and Dev, A. (2010). Development of novel bio-based soybean oil epoxy resins as a function of hardener stoichiometry. *Polym. Plast. Technol. Eng.* 49 (7): 657–661.
19 Li, Y., Fu, L., Lai, S. et al. (2010). Synthesis and characterization of cast resin based on different saturation epoxidized soybean oil. *Eur. J. Lipid Sci. Technol.* 112 (4): 511–516.
20 Pin, J.M., Sbirrazzuoli, N., and Mija, A. (2015). From epoxidized linseed oil to bioresin: an overall approach of epoxy/anhydride cross-linking. *ChemSusChem* 8 (7): 1232–1243.
21 Wuzella, G., Mahendran, A.R., Muller, U. et al. (2012). Photocrosslinking of an acrylated epoxidized linseed oil: kinetics and its application for optimized wood coatings. *J. Polym. Environ.* 20 (4): 1063–1074.
22 Kong, X., Omonov, T.S., and Curtis, J.M. (2012). The development of canola oil based bio-resins. *Lipid Technol.* 24 (1): 7–10.
23 Paluvai, N.R., Mohanty, S., and Nayak, S.K. (2015). Epoxidized castor oil toughened Diglycidyl Ether of Bisphenol A epoxy nanocomposites: structure and property relationships. *Polym. Adv. Technol.* 26 (12): 1575–1586.
24 Kadam, A., Pawar, M., Yemul, O. et al. (2015). Biodegradable biobased epoxy resin from karanja oil. *Polymer* 72: 82–92.
25 Goud, V.V., Pradhan, N.C., and Patwardhan, A.V. (2006). Epoxidation of karanja (*Pongamia glabra*) oil by H_2O_2. *J. Am. Oil Chem. Soc.* 83 (7): 635–640.
26 Francucci, G., Cardona, F., and Manthey, N.W. (2013). Cure kinetics of an acrylated epoxidized hemp oil-based bioresin system. *J. Appl. Polym. Sci.* 128 (3): 2030–2037.
27 Warwel, S. (1995). Chemo-enzymatic epoxidation of unsaturated carboxylic acids. *J. Mol. Catal.* 1 (1): 29–35.
28 Dinda, S., Patwardhan, A.V., Goud, V.V., and Pradhan, N.C. (2008). Epoxidation of cottonseed oil by aqueous hydrogen peroxide catalysed by liquid inorganic acids. *Bioresour. Technol.* 99 (9): 3737–3744.

29 Kreye, O., Toth, T., and Meier, M.A. (2011). Copolymers derived from rapeseed derivatives via ADMET and thiol-ene addition. *Eur. Polym. J.* 47 (9): 1804–1816.

30 Arumugam, S. and Sriram, G. (2014). Synthesis and characterization of rapeseed oil bio-lubricant dispersed with nano copper oxide: its effect on wear and frictional behavior of piston ring–cylinder liner combination. *Proc. Inst. Mech. Eng., Part J: J. Eng. Tribol.* 228 (11): 1308–1318.

31 Clark, A.J. and Hoong, S.S. (2014). Copolymers of tetrahydrofuran and epoxidized vegetable oils: application to elastomeric polyurethanes. *Polym. Chem.* 5 (9): 3238–3244.

32 Tayde, S. and Thorat, P. (2014). Effect of epoxidized cottonseed oil on epoxy resin and its thermal behavior. *Int. J. Chem. Technol. Res.* 7: 260–268.

33 Guner, F.S., Yagcı, Y., and Erciyes, A.T. (2006). Polymers from triglyceride oils. *Prog. Polym. Sci.* 31 (7): 633–670.

34 Lu, Y. and Larock, R.C. (2009). Novel polymeric materials from vegetable oils and vinyl monomers: preparation, properties, and applications. *ChemSusChem* 2 (2): 136–147.

35 Schuchardt, U., Sercheli, R., and Vargas, R.M. (1998). Transesterification of vegetable oils: a review. *J. Braz. Chem. Soc.* 9 (3): 199–210.

36 Saurabh, T., Patnaik, M., Bhagt, S.L., and Renge, V.C. (2011). Epoxidation of vegetable oils: a review. International. *J. Adv. Eng. Technol.* 2 (4): 491–501.

37 Lu, J., Khot, S., and Wool, R.P. (2005). New sheet molding compound resins from soybean oil. I. Synthesis and characterization. *Polymer* 46 (1): 71–80.

38 Mercangoz, M., Kusefoglu, S., Akman, U., and Hortaçsu, O. (2004). Polymerization of soybean oil via permanganate oxidation with sub/supercritical CO_2. *Chem. Eng. Process.* 43 (8): 1015–1027.

39 Mallégol, J., Lemaire, J., and Gardette, J.L. (2000). Drier influence on the curing of linseed oil. *Prog. Org. Coat.* 39 (2–4): 107–113.

40 Keleş, E. and Hazer, B. (2008). Autooxidized polyunsaturated oils/oily acids: post-it applications and reactions with Fe (III) and adhesion properties. In: *Macromolecular symposia* (eds. E. Keles and B. Hazer), 154–160. Weinheim: Wiley-VCH.

41 Guler, O.K., Guner, F.S., and Erciyes, A.T. (2004). Some empirical equations for oxypolymerization of linseed oil. *Prog. Org. Coat.* 51 (4): 365–371.

42 Andjelkovic, D.D., Valverde, M., Henna, P. et al. (2005). Novel thermosets prepared by cationic copolymerization of various vegetable oils—synthesis and their structure–property relationships. *Polymer* 46 (23): 9674–9685.

43 Robertson, M.L., Chang, K., Gramlich, W.M., and Hillmyer, M.A. (2010). Toughening of polylactide with polymerized soybean oil. *Macromolecules* 43 (4): 1807–1814.

44 Knothe, G. and Derksen, J.T. (eds.) (1999). *Recent Developments in the Synthesis of Fatty Acid Derivatives.* The American Oil Chemists Society.

45 Patil, H. and Waghmare, J. (2013). Catalyst for epoxidation of oils: a review. *Discovery* 3 (7): 10–14.

46 Okieimen, F.E., Bakare, O.I., and Okieimen, C.O. (2002). Studies on the epoxidation of rubber seed oil. *Ind. Crops Prod.* 15 (2): 139–144.

47 Goud, V.V., Patwardhan, A.V., and Pradhan, N.C. (2006). Studies on the epoxidation of mahua oil (*Madhumica indica*) by hydrogen peroxide. *Bioresour. Technol.* 97 (12): 1365–1371.

48 Meyer, P.P., Techaphattana, N., Manundawee, S. et al. (2008). Epoxidation of soybean oil and Jatropha oil. *Thammasat Int. J. Sci. Technol.* 13: 1–5.

49 Petrovic, Z.S., Zlatanic, A., Lava, C.C., and Sinadinovic-Fiser, S. (2002). Epoxidation of soybean oil in toluene with peroxoacetic and peroxoformic acids—kinetics and side reactions. *Eur. J. Lipid Sci. Technol.* 104 (5): 293–299.

50 Cai, C., Dai, H., Chen, R. et al. (2008). Studies on the kinetics of in situ epoxidation of vegetable oils. *Eur. J. Lipid Sci. Technol.* 110 (4): 341–346.

51 Sinadinović-Fiser, S., Janković, M., and Petrović, Z.S. (2001). Kinetics of in situ epoxidation of soybean oil in bulk catalyzed by ion exchange resin. *J. Am. Oil Chem. Soc.* 78 (7): 725–731.

52 Hernandez, N.L.P., Bonon, A.J., Bahu, J.O. et al. (2017). Epoxy monomers obtained from castor oil using a toxicity-free catalytic system. *J. Mol. Catal. A: Chem.* 426: 550–556.

53 Dinda, S., Goud, V.V., Patwardhan, A.V., and Pradhan, N.C. (2011). Selective epoxidation of natural triglycerides using acidic ion exchange resin as catalyst. *Asia-Pac. J. Chem. Eng.* 6 (6): 870–878.

54 Karak, N. (2012). *Vegetable Oil-Based Polymers: Properties, Processing and Applications*. Elsevier.

55 Biermann, U., Bornscheuer, U., Meier, M.A. et al. (2011). Oils and fats as renewable raw materials in chemistry. *Angew. Chem. Int. Ed.* 50 (17): 3854–3871.

56 Biermann, U., Friedt, W., Lang, S. et al. (2000). New syntheses with oils and fats as renewable raw materials for the chemical industry. *Angew. Chem. Int. Ed.* 39 (13): 2206–2224.

57 Miao, S., Wang, P., Su, Z., and Zhang, S. (2014). Vegetable-oil-based polymers as future polymeric biomaterials. *Acta Biomater.* 10 (4): 1692–1704.

58 Gunstone, F.D. and Hamilton, R.J. (2004). *The Chemistry of Oils and Fats: Sources. Composition, Properties and Uses*, 141–143. Boca Raton, FL: CRC Press.

59 Niedermann, P., Szebényi, G., and Toldy, A. (2015). Characterization of high glass transition temperature sugar-based epoxy resin composites with jute and carbon fibre reinforcement. *Compos. Sci. Technol.* 117: 62–68.

60 Mubofu, E.B. (2016). Castor oil as a potential renewable resource for the production of functional materials. *Sustainable Chem. Processes* 4 (1): 11.

61 Sudha, G.S., Kalita, H., Mohanty, S., and Nayak, S.K. (2017). Biobased epoxy blends from epoxidized castor oil: effect on mechanical, thermal, and morphological properties. *Macromol. Res.* 25 (5): 420–430.

62 Sahoo, S.K., Khandelwal, V., and Manik, G. (2018). Renewable approach to synthesize highly toughened bioepoxy from castor oil derivative–epoxy methyl ricinoleate and cured with biorenewable phenalkamine. *Ind. Eng. Chem. Res.* 57 (33): 11323–11334.

63 Park, S.J., Jin, F.L., and Lee, J.R. (2004). Effect of biodegradable epoxidized castor oil on physicochemical and mechanical properties of epoxy resins. *Macromol. Chem. Phys.* 205 (15): 2048–2054.

64 Sahoo, S.K., Mohanty, S., and Nayak, S.K. (2015). Toughened bio-based epoxy blend network modified with transesterified epoxidized soybean oil: synthesis and characterization. *RSC Adv.* 5 (18): 13674–13691.

65 Gupta, A.P., Ahmad, S., and Dev, A. (2011). Modification of novel bio-based resin-epoxidized soybean oil by conventional epoxy resin. *Polym. Eng. Sci.* 51 (6): 1087–1091.

66 Kumar, S., Samal, S.K., Mohanty, S., and Nayak, S.K. (2017). Epoxidized soybean oil-based epoxy blend cured with anhydride-based cross-linker: thermal and mechanical characterization. *Ind. Eng. Chem. Res.* 56 (3): 687–698.

67 Chen, Y., Yang, L., Wu, J. et al. (2013). Thermal and mechanical properties of epoxy resin toughened with epoxidized soybean oil. *J. Therm. Anal. Calorim.* 113 (2): 939–945.

68 Ratna, D., Samui, A.B., and Chakraborty, B.C. (2004). Flexibility improvement of epoxy resin by chemical modification. *Polym. Int.* 53 (11): 1882–1887.

69 Sahoo, S.K., Mohanty, S., and Nayak, S.K. (2015). Synthesis and characterization of bio-based epoxy blends from renewable resource based epoxidized soybean oil as reactive diluent. *Chin. J. Polym. Sci.* 33 (1): 137–152.

70 Jin, F.L. and Park, S.J. (2008). Impact-strength improvement of epoxy resins reinforced with a biodegradable polymer. *Mater. Sci. Eng., A* 478 (1–2): 402–405.

71 Sahoo, S.K., Khandelwal, V., and Manik, G. (2018). Development of toughened bio-based epoxy with epoxidized linseed oil as reactive diluent and cured with bio-renewable crosslinker. *Polym. Adv. Technol.* 29 (1): 565–574.

72 Miyagawa, H., Mohanty, A.K., Misra, M., and Drzal, L.T. (2004). Thermo-physical and impact properties of epoxy containing epoxidized linseed oil, 1. *Macromol. Mater. Eng.* 289 (7): 629–635.

73 Miyagawa, H., Mohanty, A.K., Misra, M., and Drzal, L.T. (2004). Thermo-physical and impact properties of epoxy containing epoxidized linseed oil, 2. *Macromol. Mater. Eng.* 289 (7): 636–641.

74 Sahoo, S.K., Khandelwal, V., and Manik, G. (2018). Development of completely bio-based epoxy networks derived from epoxidized linseed and castor oil cured with citric acid. *Polym. Adv. Technol.* 29 (7): 2080–2090.

75 Asano, T., Kobayashi, M., Tomita, B., and Kajiyama, M. (2007). Syntheses and properties of liquefied products of ozone treated wood/epoxy resins having high wood contents. *Holzforschung* 61 (1): 14–18.

76 Shimbo, M., Ochi, M., and Konishi, Y. (1979). Fatigue behavior aliphatic of epoxide dicarboxylic by resin cured with acids. *Zairyo* 28: 319–325.

77 Gogoi, P., Boruah, M., Sharma, S., and Dolui, S.K. (2015). Blends of epoxidized alkyd resins based on Jatropha oil and the epoxidized oil cured with aqueous citric acid solution: a green technology approach. *ACS Sustainable Chem. Eng.* 3 (2): 261–268.

78 Supanchaiyamat, N., Shuttleworth, P.S., Hunt, A.J. et al. (2012). Thermosetting resin based on epoxidised linseed oil and bio-derived crosslinker. *Green Chem.* 14 (6): 1759–1765.
79 Tan, S.G. and Chow, W.S. (2010). Thermal properties, fracture toughness and water absorption of epoxy-palm oil blends. *Polym. Plast. Technol. Eng.* 49 (9): 900–907.
80 Levita, G., De Petris, S., Marchetti, A., and Lazzeri, A. (1991). Crosslink density and fracture toughness of epoxy resins. *J. Mater. Sci.* 26 (9): 2348–2352.
81 Lu, J. and Wool, R.P. (2008). Additive toughening effects on new bio-based thermosetting resins from plant oils. *Compos. Sci. Technol.* 68 (3–4): 1025–1033.
82 Clayton, A.M. (1988). *Epoxy Resins: Chemistry and Technology*, 1–4. Boca Raton, FL: CRC Press.
83 Gunstone, F.D. and Fred, B.P. (1997). *Lipid Technologies and Applications*, 765. Boca Raton, FL: CRC Press.
84 Ratna, D. and Banthia, A.K. (2004). Rubber toughened epoxy. *Macromol. Res.* 12 (1): 11–21.
85 Sarwono, A., Man, Z., and Bustam, M.A. (2012). Blending of epoxidised palm oil with epoxy resin: the effect on morphology, thermal and mechanical properties. *J. Polym. Environ.* 20 (2): 540–549.
86 Alsagayar, Z.S., Rahmat, A.R., Arsad, A., and Fakhari, A. (2015). Mechanical properties of epoxidized palm oil/epoxy resin blend. *Appl. Mech. Mater.* 695: 655–658.
87 Haq, M., Burgueño, R., Mohanty, A.K., and Misra, M. (2008). Hybrid bio-based composites from blends of unsaturated polyester and soybean oil reinforced with nanoclay and natural fibers. *Compos. Sci. Technol.* 68 (15–16): 3344–3351.
88 Gergely, A., Bertoti, I., Torok, T. et al. (2013). Corrosion protection with zinc-rich epoxy paint coatings embedded with various amounts of highly dispersed polypyrrole-deposited alumina monohydrate particles. *Prog. Org. Coat.* 76 (1): 17–32.
89 Hao, Y., Liu, F., and Han, E.H. (2013). Protection of epoxy coatings containing polyaniline modified ultra-short glass fibers. *Prog. Org. Coat.* 76 (4): 571–580.
90 Katariya, M.N., Jana, A.K., and Parikh, P.A. (2013). Corrosion inhibition effectiveness of zeolite ZSM-5 coating on mild steel against various organic acids and its antimicrobial activity. *J. Ind. Eng. Chem.* 19 (1): 286–291.
91 Alam, M., Akram, D., Sharmin, E. et al. (2014). Vegetable oil based eco-friendly coating materials: a review article. *Arabian J. Chem.* 7 (4): 469–479.
92 Kasemsiri, P., Neramittagapong, A., and Chindaprasirt, P. (2015). Curing kinetic, thermal and adhesive properties of epoxy resin cured with cashew nut shell liquid. *Thermochim. Acta* 600: 20–27.
93 Liu, Y., Yang, G., Xiao, H.M. et al. (2013). Mechanical properties of cryogenic epoxy adhesives: effects of mixed curing agent content. *Int. J. Adhes. Adhes.* 41: 113–118.

94 Jin, H., Miller, G.M., Pety, S.J. et al. (2013). Fracture behavior of a self-healing, toughened epoxy adhesive. *Int. J. Adhes. Adhes.* 44: 157–165.
95 Ullmann, F., Gerhartz, W., Yamamoto, Y.S. et al. (1985). *Ullmann's Encyclopedia of Industrial Chemistry*. VCH Publishers.
96 Bucknall, C.B. and Partridge, I.K. (1983). Phase separation in epoxy resins containing polyethersulphone. *Polymer* 24 (5): 639–644.
97 Kinloch, A.J. (2003). Toughening epoxy adhesives to meet today's challenges. *MRS Bull.* 28 (6): 445–448.
98 Guadagno, L., Raimondo, M., Vittoria, V. et al. (2014). Development of epoxy mixtures for application in aeronautics and aerospace. *RSC Adv.* 4 (30): 15474–15488.
99 Shamsuddoha, M., Islam, M.M., Aravinthan, T. et al. (2013). Characterisation of mechanical and thermal properties of epoxy grouts for composite repair of steel pipelines. *Mater. Des., (1980–2015)* 52: 315–327.
100 Ayad, M.M., El-Nasr, A.A., and Stejskal, J. (2012). Kinetics and isotherm studies of methylene blue adsorption onto polyaniline nanotubes base/silica composite. *J. Ind. Eng. Chem.* 18 (6): 1964–1969.
101 Azeez, A.A., Rhee, K.Y., Park, S.J., and Hui, D. (2013). Epoxy clay nanocomposites–processing, properties and applications: a review. *Compos. Part B: Eng.* 45 (1): 308–320.
102 Teh, P.L., Jaafar, M., Akil, H.M. et al. (2008). Thermal and mechanical properties of particulate fillers filled epoxy composites for electronic packaging application. *Polym. Adv. Technol.* 19 (4): 308–315.
103 Chand, N. and Nigrawal, A. (2008). Development and electrical conductivity behavior of copper-powder-filled-epoxy graded composites. *J. Appl. Polym. Sci.* 109 (4): 2384–2387.
104 Komori, S. and Sakamoto, Y. (2009). *Material for Advanced Packaging*, 339–363. Switzerland: Springer.
105 Ho, T.H. and Wang, C.S. (1996). Modification of epoxy resins with polysiloxane thermoplastic polyurethane for electronic encapsulation: 1. *Polymer* 37 (13): 2733–2742.

4

Cure Kinetics of Bio-epoxy Polymers, Their Blends, and Composites

P.A. Parvathy[1,2], Smitha Mohanty[3], and Sushanta K. Sahoo[1,2]

[1] *CSIR-National Institute for Interdisciplinary Science and Technology, Materials Science and Technology Division, Industrial Estate P O, Thiruvananthapuram 695019, India*
[2] *Academy of Scientific and Innovative Research (AcSIR), Ghaziabad- 201002, India*
[3] *CIPET, SARP-Laboratory for Advanced Research in Polymeric Materials, B-25, CNI Complex, Bhubaneswar 751024, India*

4.1 Introduction

Resin is a viscous or nanocrystalline material that gives a polymer after polymerization or curing. Thermoset resins can be natural or synthetic or may be a mixture of organic compounds. Epoxy, one of the engineering thermosets, is widely used in different fields such as coatings, adhesives, electronics, and structural parts because of outstanding mechanical properties, thermal stability, relatively less curing shrinkage, and high solvent resistance [1]. Nevertheless, extensive use of epoxy in industries is constrained by its poor crack resistance and high viscosity [2–4]. In order to solve this issue, a series of bio-soured diluents have been developed as alternatives or as secondary component modifiers to overcome the brittleness of the epoxy matrix. Further, rapid depletion of nonrenewable resources and environmental issues has motivated us to explore renewable resource-based polymers and also prepolymers to toughen the epoxy matrix in recent years [5]. The environmental concern from the manufacturer as well as consumers of petro-based epoxy polymers has pursued to look for bio-based epoxy resins as an alternative to substitute the commercially available thermosets [6–13]. Bio-based polymers derived from renewable feedstocks such as polysaccharides, glucose, plant oils, cashew nut shell liquid (CNSL), lignin, and soy proteins have received growing attention in both industrial use and academic purpose over petro-sourced polymers because of their low cost and eco-friendly nature. Lignin, the second most available biopolymer, is one of the most vital renewable resources [14, 15], and it has a rigid, macromolecular structure with diverse functionality, which can be chemically modified to synthesize lignin-based epoxy polymers in both solid and liquid phase. Similarly, CNSL-derived cardanol is an abundantly available industrial biomass and consists of functional compounds that are chemically modified to create high performance in epoxy resin [16, 17]. With a unique molecular structure, it is a

Bio-Based Epoxy Polymers, Blends, and Composites: Synthesis, Properties, Characterization, and Applications, First Edition. Edited by Jyotishkumar Parameswaranpillai, Sanjay Mavinkere Rangappa, Suchart Siengchin, and Seno Jose.

natural-sourced phenol compound with C_{15} unsaturated alkyl side chain and it can participate in chemical reactions and several polymerization processes. Out of all bio-resources, vegetable oils have received growing attention because of the benefits such as biodegradability, renewability, ease of processing, nonhazardous nature, and environmental friendliness [18]. Plant oils are triglycerides with unsaturated fatty acids such as linoleic (C18 : 3), linoleic (C18 : 2), and oleic (C18 : 1) acids. There are three routes to polymerize plant oils to polymers: (i) the direct polymerization of the triglyceride through the inherent double bonds; (ii) modifications by the introduction of readily polymerizable functional group including epoxidation, hydroformylation, and metathesis through the unsaturation of the triglycerides; and (iii) derive monomers from oil-based building blocks [19, 20].

4.2 Fundamentals of Curing Reaction Kinetics

The performance of thermosets and their blends depends on the type of the cross-linked polymeric network, curing agents, degree of cure, and curing conditions, which depend on the reaction mechanism and curing kinetics of the reaction [21–42]. Knowledge about curing reaction kinetics of modified resin executes a very significant role to control the curing reaction and consequently to achieve the optimized physical properties of the cured products [43–50] and also meaningful to manufacture high-performance epoxy thermosets. Generally, curing kinetic studies can be performed by differential scanning calorimetry (DSC) or Fourier transform infrared spectroscopy (FTIR) at isothermal or non-isothermal conditions [41, 42, 48]. The decrease in the intensity of epoxide groups during the reaction can be measured by quantitative analysis using FTIR spectroscopy. The degree of curing conversion is inversely proportional to the presence of epoxides and can be examined during the reaction time. Bands remaining unchanged in the curing process corresponding to the linkages that do not participate in the reaction can be taken as a reference for analysis. Among all known techniques, DSC is a commonly used technique to study the cross-linking kinetics of thermosets through various isoconversional and autocatalytic kinetic models [22, 23]. As per the previous reports, the isoconversional methods provide more consistent results because it discloses the curing kinetics as a function of conversion of the curing process [24–26]. The activation energy at each conversion is calculated by isoconversional methods, such as isoconversional models by Vyazovkin, Kissinger, Friedman, Kissinger–Akahira–Sunose (KAS), and Starink method [21–26].

4.2.1 Curing Kinetic Theories: Isothermal and Non-isothermal

The fractional curing conversion is given by the integration of exothermic DSC peak at any temperature T and it is normally represented as

$$\alpha = \frac{\Delta H_T}{\Delta H} \tag{4.1}$$

where ΔH_T is the enthalpy of curing reaction at temperature T and ΔH is the total enthalpy of reaction. Activation energy gets changed at each conversion because of the complexity during the curing process. In such case, the activation energy, E_a, is plotted as a function of curing conversion by the Kissinger method [19].

$$\ln\left(\frac{\beta}{T_{p^2}}\right) = C - \frac{E_a}{RT_p} \tag{4.2}$$

where β is the heating rate, T_p is the peak temperature for β, and E_a is the apparent activation energy.

E_a for each curing conversion can be determined from the slope of the linear plot of $\ln(\beta/T_p{}^2)$ against $1/T_p$, where C is a constant.

Usually, kinetic analysis of cross-linked epoxy resins by non-isothermal process is explained by Eq. (4.3).

$$\frac{d\alpha}{dt} = A\exp\left(\frac{-E_a}{RT}\right) f(\alpha) \tag{4.3}$$

where $d\alpha/dt$ is the rate of curing conversion, A is the pre-exponential factor, R is the gas constant, T is the absolute temperature, and $f(\alpha)$ is the function representing the kinetic model.

The mechanism of curing reaction of thermosets is usually of nth order or autocatalytic reactions.

The autocatalytic rate equation can be written as

$$\frac{d\alpha}{dt} = A\exp\left(\frac{-E_a}{RT}\right) \alpha^m (1-\alpha)^n \tag{4.4}$$

Kissinger, Friedman, and autocatalytic model can be employed to find the kinetic parameters. Considering International Confederation for Thermal Analysis and Calorimetry (ICTAC) recommendations, kinetic data and kinetic parameter computations are generally collected [22]. Malek [43] suggested another model that can be intended with the two functions $y(\alpha)$ and $z(\alpha)$ in line with the following equations.

$$y(\alpha) = \frac{d\alpha}{dt} \exp(x) \tag{4.5}$$

$$z(\alpha) = \pi(x) \left(\frac{d\alpha}{dt}\right) \frac{T}{\beta} \tag{4.6}$$

where $\pi(x) = \frac{x^3+18x^2+88x+96}{x^4+20x^3+120x^2+240x+120}$

$x = E_a/RT$ signifies the reduced activation energy and $\pi(x)$ as presented is the temperature integral. The normalized curves of $y(\alpha)$ and $z(\alpha)$ vs. the degree of conversion (α) are employed to find out the kinetic parameters. M and p are the maximum values of the $y(\alpha)$ and $z(\alpha)$ curves, respectively, and the average value of M is employed to determine the kinetic parameters. Replacing m/n by $\alpha_{Mm}/1-\alpha_{Mm}$, the equation can be converted into the following equation:

$$\ln\left[\left(\frac{d\alpha}{dt}\right)\right] \exp\left(\frac{E_a}{RT}\right) = \ln A + n\ln\left[\frac{\alpha_{Mm}}{\alpha^{1-\alpha_{Mm}}(1-\alpha)}\right] \tag{4.7}$$

where $p = m/n = \alpha_{Mm}/(1-\alpha_{Mm})$ α_{Mm} is the mean of α_M.

Similarly, the complexity of curing reaction of epoxy is explained by using a more complex model such as the Kamal model [20].

$$\frac{d\alpha}{dt} = [k_1(T) + k_2(T)\alpha^m](1 - \alpha)^n \tag{4.8}$$

where α is the value corresponding to the extension of conversion.

Unfortunately, this model does not satisfy the standard model-fitting techniques and hence unable to determine the complexity in the cross-linking process. Particularly, in a non-isothermal mode, this model yields more uncertain curing energy values. As per ICTAC, kinetic project isoconversional models or model-free kinetic methods are pre-eminent methods to counter the aforementioned challenges. The Vyazovkin et al. isoconversional methods [22–26] in this approach are more advanced and effective in determining certain values.

$$\left[\frac{d\ln\left(\frac{d\alpha}{dt}\right)}{dT^{-1}}\right]_\alpha = \frac{-E_\alpha}{R} \tag{4.9}$$

Vyazovkin method [24] proposed a set of n experiments performed at altered arbitrary heating programs, and the T_i is determined at any particular conversion α by finding the value of $E\alpha$

$$ø(E_\alpha) = \sum_{i=1}^{n}\sum_{j\neq 1}^{n} \frac{J[E_\alpha, T_i[t_\alpha]]}{J[E_\alpha, T_j[t_\alpha]]} \tag{4.10}$$

Sbirrazzuoli suggested a new model using the dependency of $E\alpha$ on α (or on T) to determine the curing parameters by integral isoconversional method [28]. The isoconversional principle, Eq. (4.9), is applied to the Kamal model (Eq. (4.8)) to obtain the following equation:

$$E_\alpha = E_\alpha(T) = \frac{\frac{A_1}{A_2}\exp\left(\frac{-E_1}{RT}\right)E_1 + \exp\left(\frac{-E_2}{RT}\right)E_2\alpha^m}{\frac{A_1}{A_2}\exp\left(\frac{-E_1}{RT}\right) + \exp\left(\frac{-E_2}{RT}\right)\alpha^m} \tag{4.11}$$

where $k_1 = A_1 \exp(-E1/RT)$ and $k_2 = A_2 \exp(-E_2/RT)$.

As the curing reactions proceed, the molecular motions get more confined and the rate-controlling stage switches over to diffusion mechanism from the chemical reaction. Thus, the effective activation energy E_{ef} becomes a function of the activation energies of both chemical and diffusion mechanisms.

$$E_{ef} = \frac{k(T)E_D + k_D(T,\alpha)E}{k(T) + k_D(T,\alpha)} \tag{4.12}$$

A new method was proposed by Sbirrazzuoli and coworkers [20, 28] to compute the kinetic parameters of Kamal's model using the dependency of activation energy on α (or on T).

$$E_\alpha = E_\alpha(T) = \frac{\frac{A_1}{D_0}\exp\left(\frac{-E}{RT}\right)E_D + \exp\left(\frac{-E_D}{RT} + k_\alpha\right)E}{\frac{A_1}{D_0}\exp\left(\frac{-E}{RT}\right) + \exp\left(\frac{-E_D}{RT} + k_\alpha\right)} \tag{4.13}$$

where K is the constant corresponding to the chemical reaction effect on the change in diffusivity, D_0 is the pre-exponential factor, and E_D is the activation energy of diffusion.

Vyazovkin proposed a new autocatalytic model based on Eqs. (4.9, 4.14) to forecast the cross-linking

$$\frac{d\alpha}{dt} = k_1\alpha^{m_1}(1-\alpha)^{n_1} + k_2\alpha^{m_2}(1-\alpha)^{n_2} \tag{4.14}$$

where $k_i = A_i \exp(-E_i/RT)$ for $i = 1, 2$ is the Arrhenius equation.

By means of isoconversional principles and by replacement of Eq. (4.14) in Eq. (4.9),

$$E_\alpha = \frac{E_1\alpha^{m_1-m_2}(1-\alpha)^{n_1-n_2}\exp\left(\frac{-E_1}{RT}\right) + \frac{A_2}{A_1}E_2\exp\left(\frac{-E_2}{RT}\right)}{\alpha^{m_1-m_2}(1-\alpha)^{n_1-n_2}\exp\left(\frac{-E_1}{RT}\right) + \frac{A_2}{A_1}\exp\left(\frac{-E_2}{RT}\right)} \tag{4.15}$$

4.3 Curing of Bio-thermosets

4.3.1 Curing Agents and Curing Reactions

The curing agent plays a vital role in the cross-linking of bio-resins through polyaddition or copolymerization reactions to form strong and dimensionally stable networks with required properties. The common cross-linkers used to cure bio-thermosets are listed in Table 4.1.

Although amine- and anhydride-based curatives provide excellent mechanical performance and improved thermal stability to epoxy network, they include the drawbacks such as toxicity, hazardous nature, difficulty in processing, and non-eco-friendliness. This concern leads to explore different renewable cross-linkers such as cardanol-based phenalkamine and bio-derived carboxylic acids (citric acid, sebacic acid, etc.) with and without the use of a catalyst. Currently, research works started concentrating on cross-linkers derived from chemically functionalized plant oils, bio-based acids, and anhydrides and amines, bio-sourced phenols, rosin acids, terpenes, and lignin in order to cure bio-thermosets.

The curing mechanism of epoxy-amine and epoxy blend-amine involves different steps such as initiation, propagation, and branching. In this curing reaction, a primary amine can react twice with two epoxy groups, whereas a secondary amine can react only once as displayed in Figure 4.1 in steps (1) and (2). A tertiary amine is not expected to react with the epoxy group because of the absence of active hydrogen. A secondary amine would react faster than a primary amine being more nucleophilic and generate more hydroxyl groups. In the last step (step (3) of Figure 4.1), the hydroxyl groups formed in the epoxy-amine reaction lead to form branched ether linkages through etherification reaction.

The acid anhydride curing reaction with epoxide proceeds through chain-wise polymerization compared to the stepwise addition reaction mechanism for epoxy/amine curing.

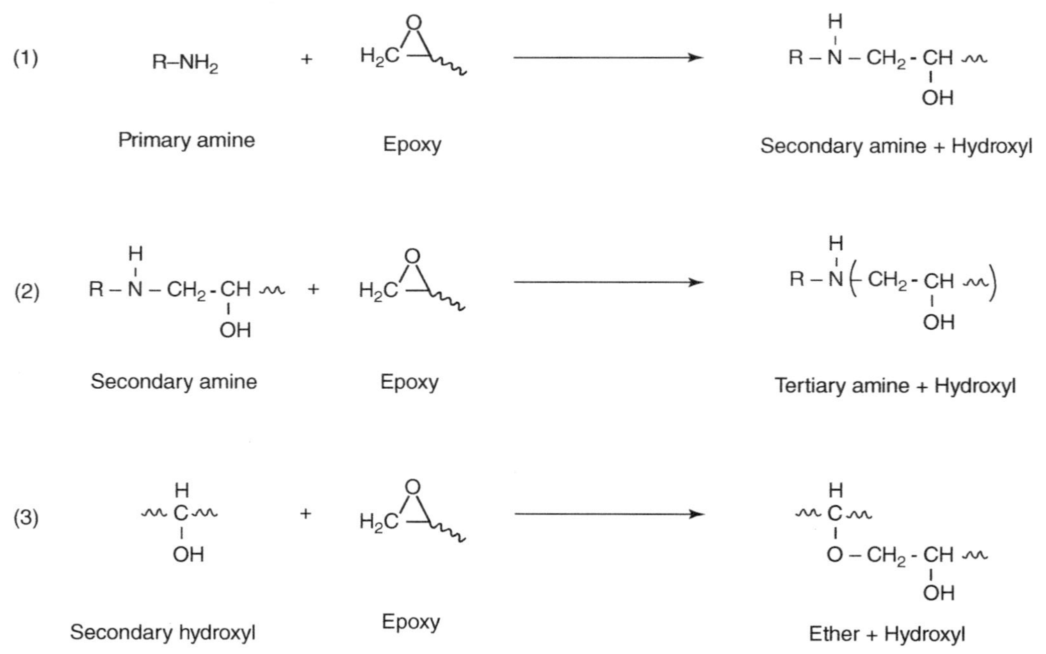

Figure 4.1 Curing reaction of epoxides with amine [33].

Table 4.1 List of cross-linkers used to cure bio-thermosets.

Amine-based curing agents	Anhydride-based curing agents	Acid-based curing agents
Aliphatic • Diethylenetriamine (DETA) • Triethylenetetramine (TETA) Cycloaliphatic • Isophoronediamine (IPDA) Aromatic • 4,4-Diaminodiphenyl methane (DDM) • Methylenedianiline (MDA)	• Hexahydrophthalic anhydride (HHPA) • Maleic anhydride (MA) • Methyl tetrahydrophthalic anhydride (MTHPA) • Tetrahydrophthalic anhydride (THPA) • Phthalic anhydride (PA) • Tetrachlorophthalic anhydride (TCPA) • Trimellitic anhydride (TMA)	• Citric acid • Adipic acid • Suberic acid • Sebacic acid • Succinic acid • Dodecanedioic acid • Decanedioic acid (tetra, hexa, octa)

At first, imidazole catalyst initiates the cross-linking by opening the oxirane group forming a zwitterion. The formed alkoxide anion may also attack an epoxide group forming an ether group that may subject to advance curing. Some side reactions are possible, such as homopolymerization and etherification, and also take place at high temperatures, when activated by imidazole initiator. The possible reaction scheme is depicted in Figure 4.2.

Similarly, dicarboxylic acid can also participate in the curing reaction with epoxides. The possible interactions of dicarboxylic acid and epoxy (as shown in Figure 4.3) in the presence of catalyst found to be more convoluted and they are etherification, esterification, hydrolysis, and condensation esterification [52].

4.4 Curing Kinetics of Bio-epoxies and Blends

Bio-based epoxide groups are cross-linked by curing agents or cross-linkers, such as, amines, acid anhydrides, or Lewis acids or bases as provided in (Table 4.2). Bio-epoxy resin features after curing are governed by the nature of epoxy and curing agent. Performing direct polymerization of bio-based epoxy is generally difficult. Usually, epoxidized oils have low reactivity because of internal epoxy groups and steric hindrance by pendant groups, which resulted in poor thermal stability and inferior mechanical properties. Reaction with amine with epoxides ends up with gel or cured products of poor quality. Despite the optimistic prospects of bio-epoxy resin, the lack of adequate cross-linking and lower mechanical properties restricts their wide exploitation. The direct polymerization of renewable resourced monomers or prepolymers is more thought-provoking and desired performance has not been achieved yet. Thus, curing optimization through kinetic models is essential to achieve required properties for specific applications. Numerous researchers investigated the curing kinetics of petro-based epoxy resin with anhydride cross-linkers because

Table 4.2 Literature summary on curing kinetics of bio-epoxy and blends.

Bio-epoxy	Curing agent and catalyst	Kinetic models	Summary	References
Epoxidized soybean oil and blends	Maleopimaric acid and 2-ethyl-4-methylimidazole (EMI)	Friedman, KAS, and Chen-Liu method; Kamal model	The predicted curing parameters agree well with the experimental data of ESO. The imidazole catalyzed curing process of ESO displayed autocatalytic behavior in the isothermal mode.	[51, 53]
	MHHPA and MI;	Kissinger method, Sestak–Berggren autocatalytic model, Malek method;	The curing reaction of epoxy/30 % ESO is more controlled through diffusion mechanism compared to other systems.	
	Sebacic acid	Kissinger–Akahira–Sunose (KAS); and Starink method		
Epoxidized linseed oil-based epoxy	Methyl-5-norbornene-2,3, dicarboxylic anhydride, imidazole MHHPA and MI	Friedman (FR), Kissinger–Akahira–Sunose (KAS), and Vyazovkin (VA) isoconversional methods	The predicted VA model is precisely matching with the experimental values. The diffusion-controlled reaction process was affected by the chemorheological influence	[54]
Epoxidized castor oil-based epoxy	Amine hardener	Kissinger method, Friedman's method, Flynn–Wall–Ozawa (FWO), Kissinger–Akahira–Sunose (KAS), and Starink methods	Šesták–Berggren model was the most suitable method to explain the curing process of bio-based epoxy resin	[55]
Epoxidized hemp oil based epoxy	Triethylenetetramine (TETA)	Kissinger method, Friedman's method, Flynn–Wall–Ozawa (FWO), and modified Kamal autocatalytic models	Negative activation energy is observed because of an unidentified competitive reaction. The proposed model agrees reasonably with experimental results.	[56]

(*continued overleaf*)

Table 4.2 (*Continued*)

Bio-epoxy	Curing agent and catalyst	Kinetic models	Summary	References
Acrylated oils and epoxy blends	MHHPA and triethyl benzylammonium chloride (TEBAC)	Flynn–Wall–Ozawa method, Sestak–Berggren autocatalytic model, and Malek method	The calculated kinetic parameters agree well with the experimental data for bio-based systems.	[57]
Tannic acid based epoxy		Kamal model, Sbirrazzuoli isoconversional method, and Vyazovkin method	Tannic acid-based epoxy exhibited a diverse activation energy dependence on curing.	[58]
Furan-based epoxy and blends	Methyl-5-norbornene-2,3, dicarboxylic anhydride, and imidazole	Friedman, KAS, FWO model, and Starink methods	Except the Friedman method, all other models matched with the experimental data. The curing reaction of neat epoxy and with poly(furfuryl alcohol) showed similar	[59, 60]
	Difurfuryl amine		kinetic reactions and autocatalytic models.	
	Diethylenetriamine (DETA)		Kinetic analysis showed that PFA has no effect on cure reaction mechanism in $\alpha < 0.75$ region and then decreased at $\alpha > 0.75$ because of curing of 15 wt% poly(furfuryl alcohol.	
Lignin-based epoxy	4,4-Diaminodiphenyl methane (DDM), diethylenetriamine (DETA), and MTHPA	Kamal model, Sbirrazzuoli method, and Vyazovkin method	In the early stage of conversion, the bio-based resins has raised E_a of epoxy slightly and E_a values of epoxy and lignin-epoxy blends follow a descending trend throughout the reaction.	[58, 61, 62]
Itaconic acid-based epoxy	MHHPA, methyl Imidazole	Kissinger method and Flynn–Wall–Ozawa method	The activation energy of epoxy raised on addition of itaconic acid-based bio-resin. Kissinger equation fitted well for the epoxy blend system	[63]
Cardanol-based epoxy blend	Isophorone diamine (IPD), Jeffamine	Friedman, Malek, and Sestak–Berggren method	Šesták–Berggren (SB) model was suitable for the kinetic reaction of lignin–cardanol blend system. All predicted parameters reach agreement with the obtained experimental results.	[47]
	MTHPA, TEBAC			

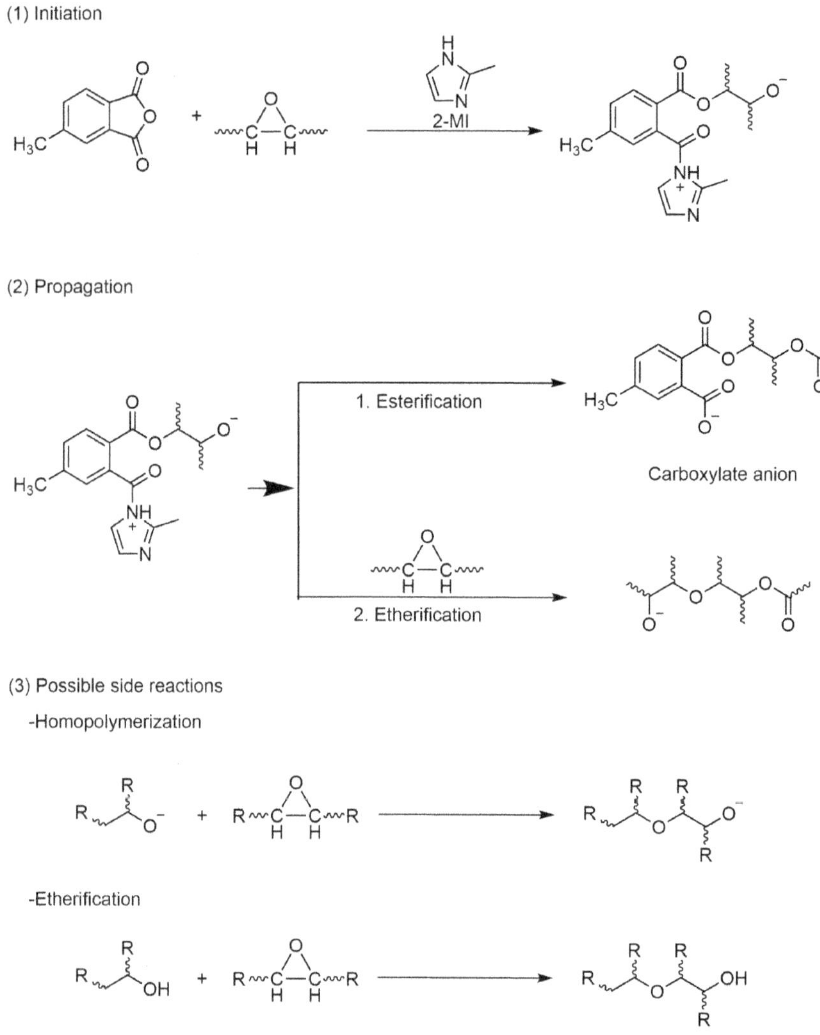

Figure 4.2 Curing reaction of epoxides with anhydride [51].

of the diverse advantages of anhydride curing agents [31–50]. Several researchers studied the curing ability of different curing agents and catalysts on performance of cross-linked epoxidized plant oils and other bio-sourced epoxies through different models [51–78].

Tan and Chow [74] investigated thermal curing of epoxidized soybean oil (ESO) resin at different temperatures using the methylhexahydrophthalic anhydride (MHHPA) cross-linker and 2-ethyl-4-methylimidazole (EMI) as a catalyst. Curing kinetics of the ESO/MHHPA systems was analyzed using FTIR, and it revealed that the EMI-catalyzed ESO/MHHPA system exhibited autocatalytic behavior in the isothermal curing reaction irrespective of curing temperatures and EMI content. Kinetic parameters such as reaction rate constants k_1 and k_2 value increased with curing temperature and EMI catalyst content. Activation energies reduced with increase in EMI content and reaction order $m+n$ value

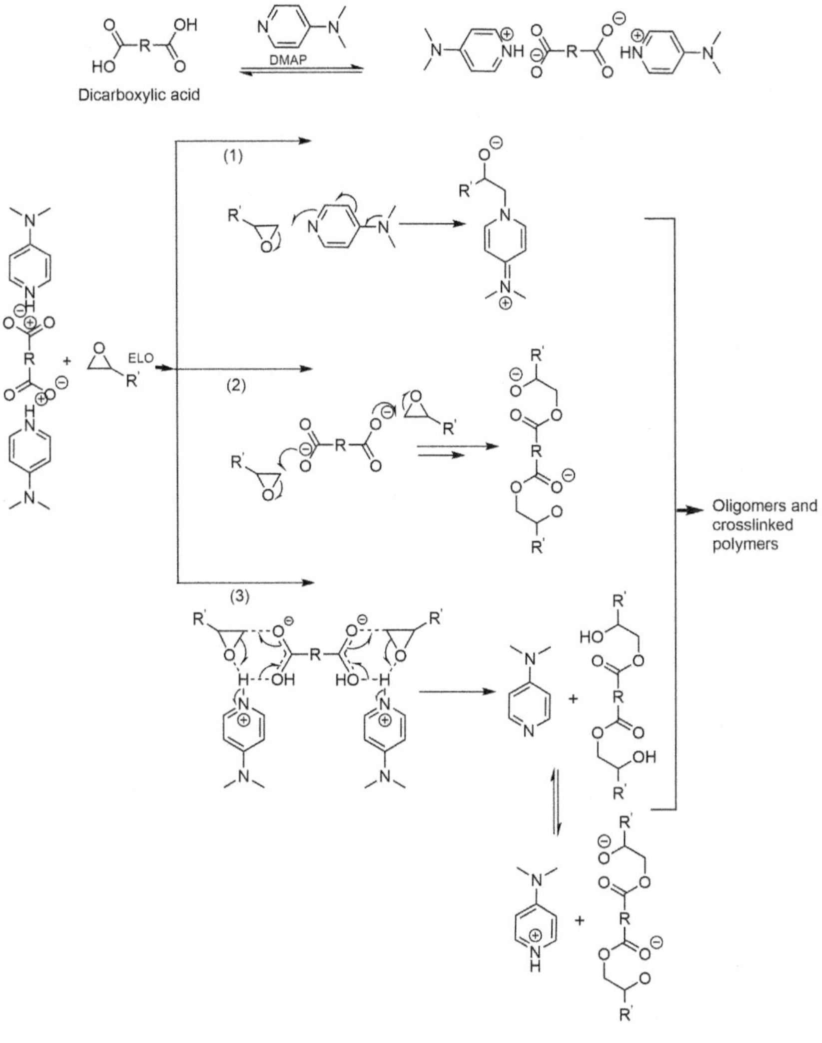

Figure 4.3 Curing reaction of epoxides with carboxylic acid [52].

was estimated to be within 1.5–3.0. The reported curing kinetic parameters help in developing processing technologies of ESO bio-epoxy thermoset to achieve maximum desirable properties.

Huo et al. [61] investigated the non-isothermal curing kinetics of the lignin and cardanol-based novolac epoxy resin (LCNE) and methyl tetrahydrophthalic anhydride (MTHPA) in the presence of dimethylbenzylamine (DMBA). The predicted values of kinetic reaction rates were found to be in good agreement with the obtained experimental values. Roudsari et al. [53] investigated the kinetics of the curing reaction of epoxy-ESO system cross-linked with a sebacic acid, a bio-sourced curing agent with non-isothermal DSC through KAS and Starink isoconversional methods. Epoxy with 10 and 20 wt% of ESO displayed a drop in activation energy in the initial stage of curing, followed by steady activation energy, while the epoxy/30 wt% of ESO showed an opposite trend. It was also observed that the activation energy of epoxy with sebacic acid is greater than amine-based cross-linking agent. However, this work paves the way for next generation of bio-epoxy network as novel green materials with acceptable properties.

Ferdosian et al. [77] examined the curing behavior of lignin-derived epoxy resins with two cross-linking agents: diethylenetriamine (DETA) and 4,4-diaminodiphenyl methane (DDM). The activation energies of the systems were computed and found that the system with DETA had higher activation energies than those of DDM, which indicates the more reactivity of aromatic amine than aliphatic amine.

Similarly, Pin et al. [76] stated that epoxidized linseed oil (ELO) cured with MHHPA and benzophenonetetracarboxylic dianhydride (BTDA) catalyzed by 2-MI and both networks displayed good thermal stability. For the ELO/MHHPA system, the DSC thermogram showed a single exothermic peak at about 150 °C, wherein ELO/BTDA showed an additional exothermic peak, at about 250 °C, which may be attributed to the occurrence of side reactions of homopolymerization and etherifications. The activation energy curve displayed a rapid increase in E_a at 220 °C, which could be liquefaction of unreacted BTDA. It was found that E_a is constant near 150 kJ/mol at a high temperature range beyond 200 °C, which is mostly associated with side reactions as homopolymerization and etherification. It was concluded after comparison that the cross-linking density improved in the case of BTDA.

Chen et al. [78] developed highly bio-sourced polymeric thermosets through ring-opening polymerization (ROP) of ESO cured with maleopimaric acid catalyzed by 2-EMI. It was noticed that the total heat of curing for the ROP of the system was only 31.7 kJ/mol epoxy group. Similarly, for anhydride/epoxy group (0.7 molar ratio), the E_a value reduced with the increase in catalyst amount from 0.5 to 1.5 phr. The predicted curing reaction parameters agreed well with the experimental data.

Curing kinetics of diglycidyl ether of bisphenol A (DGEBA)/20ESO, petro-sourced epoxy (DGEBA), and ESO was reported by Kumar et al. [51], in which MHHPA and 2-methylimidazole (2-MI) were taken as curing agent and catalyst, respectively. Enthalpy of curing reaction was found to be increased after integration of ESO bio-resin, which is due to improved reactivity of ESO bio-resin with anhydride. The calculated kinetic parameter of all bio-epoxy displayed reasonable agreement with the obtained values through experiment.

Epoxidized sucrose soyate (ESS)-based bio-epoxy resin was used for the study of curing kinetics, which is recently reported by Paramarta and Webster [62]. The curing was done with a MHHPA curing agent and a zinc-based catalyst. Anhydride and/or epoxy went through ring opening by metal coordination as a result of removal of carboxylate ligands. It accompanied change in the coordination of tetradentate to hexadentate. Considering the nth order of the reaction and the AutoFit model, it is found that curing takes place through a complex way and prevailed by a non-autocatalytic model (chain polymerization through alkoxide and carboxylate).

Esmaeili et al. [58] synthesized tannic acid (TA) based bio-based epoxy resin and proposed a new model to investigate the dynamic curing kinetics by DSC. Conventional kinetic models could not fit the curing profile; thus, a new model (Vyazovkin's advanced isoconversional method) was adopted, which forecasts the curing behavior of epoxidized tannic acid (ETA) resin successfully. The curing kinetic parameters showed that ETA had an unlike activation energy dependence on curing conversion compared with the petro-sourced epoxy.

Yamini et al. [59] modified epoxy with epoxidized lignosulfonate (ELS) and also with cyclocarbonated lignosulfonate (CLS) to develop a bio-epoxy blend system and then investigated the curing kinetics of both blends and compared with the unmodified system. CLS/DGEBA blend displayed higher value of curing heat of reaction than ELS/DGEBA system. It is noted that in the early stage of curing, the activation energy is slightly higher. Both the samples exhibited a similar trend in activation energy as attained by Vyazovkin advanced integral isoconversional method, which is ascribed to a similar reaction mechanism for all the systems.

Zolghadr et al. [60] integrated thermomechanical and the curing kinetics of epoxy with synthesized polyfurfuryl alcohol and Difurfuryl amine as additive and curative respectively. Friedman isoconversional method due to its high sensitivity displayed breakdown in fitting experimental results, but the other models (KAS, Flynn–Wall–Ozawa (FWO), Stranik) agree well with a high coefficient of determination. The cross-linking process of epoxy with phenol formaldehyde (PFA) displayed similar kinetic model as neat epoxy and both were found to be autocatalytic. The sample containing PFA displayed higher activation energy, which can be authorized to the higher viscosity, but still the values are similar to that of the petro-epoxy–amine system.

Hence, from the extensive literature survey, it is concluded that epoxidized plant oil is a better candidate for making bio-epoxy blends with improved toughened performance characteristics.

4.4.1 Curing Kinetics of Bio-epoxy Composites

In the past decades, implementation of nanotechnology to bio-based polymeric systems has explored new paths to improve the properties being advantageous and eco-friendly [27]. However, the enhancement in the properties of polymer nanocomposite majorly depends on the chemistry of polymer matrices, properties of nanofillers, matrix-filler compatibility, interface, and most importantly polymerization methods [27–33]. Inorganic particles such as layered silicates are the well-known fillers to increase the performance of polymers because of their unique structure and functional groups [34–36]. Other nanofillers such as polyhedral oligomeric silsesquioxane (POSS), nanocellulose, CNTs, graphene, and clays are used as reinforcing agents and bio-based epoxy resin as a base matrix to develop high-performance composites with higher stiffness and stability as well as other smart properties [79–85]. Especially, polymer clay nanocomposites have gained growing attention in industrial interest because of significant improvement in a large number of physical properties such as reduced gas permeability, lower expansion coefficient, excellent thermal stability, flame retardancy, and poor water uptake [34]. Nanoclays are the potential reinforcing agents for the bio-based epoxy blend to enhance the mechanical and thermal properties. Thus, this system has been opted as a case study for discussion in the Section 4.5. Tan et al [31] reported the co-catalytic effect of organo montmorillonite (OMMT) and reinforcing ability of ESO-based thermoset cured with MHHPA catalyzed by 2-EMI. The mechanical strength and stiffness were significantly improved, and fracture toughness was reduced with the increase in OMMT content. It was attributed to the higher gel content and cross-link density of the nanocomposite caused by better dispersion and catalytic activity of OMMT during the curing process.

4.5 Case Study: Non-isothermal Kinetics of Plant Oil–Epoxy–Clay Composite

More works have been reported on plant oil–clay-based thermoset nanocomposite in recent years [30–33]. As a case study, the effect of C30B nanoclay on the curing kinetic parameters of toughened epoxy/ESO blend has been discussed. After adding ESO into the epoxy, it was seen from DSC curve that the onset curing temperature and peak curing temperature raised slightly accompanied with an increase in heat of curing reaction (Figure 4.4). Minor increase in peak temperature and heat of curing indicates the poor inhibiting effect in curing because of long aliphatic chain of bio-resin and acceleration effect of low viscous resin system and better contact or reactivity with curing agent. For bio-epoxy/clay system, a decrease in peak cure temperature (T_{peak}) and an increase in the enthalpy of curing support the catalytic effect of C30B clay accelerating curing reaction.

Applying Kissinger equation, the activation energy (E_α) at various conversions (α) were determined from the slope of the linear plots of $\ln(\beta/T_\alpha{}^2)$ vs. $1/T_\alpha$ as depicted in Figure 4.5.

The disparity in the activation energy (E_α) at different α values was depicted in Figure 4.3. The change in the activation energy (E_α) confirmed complex curing process of epoxy blend systems [19, 33]. It was found that the activation energy of epoxy/ESO/C30B system was much lower as compared to the epoxy and epoxy/ESO blend across all curing conversions. The average activation energy was found to be 51.51 kJ/mol for virgin epoxy, 50.72 kJ/mol for epoxy/ESO blend, and 41.88 kJ/mol for epoxy/ESO/C30B system. Friedman method is widely used to find the kinetics of the curing process of the epoxy resin and its modified systems. Friedman proposed that the rate of reaction is only dependent on temperature at a constant degree of conversion and the equation as follows.

$$\ln\left(\frac{d\alpha}{dt}\right) = \ln[Af(\alpha)] - \frac{E_\alpha}{RT} \tag{4.16}$$

The order of the curing reaction depends on the curing agent and also the amount of fillers in the epoxy system. In order to explain the curing kinetic models for all epoxy systems (epoxy, epoxy/ESO blend, and epoxy/ESO/C30B

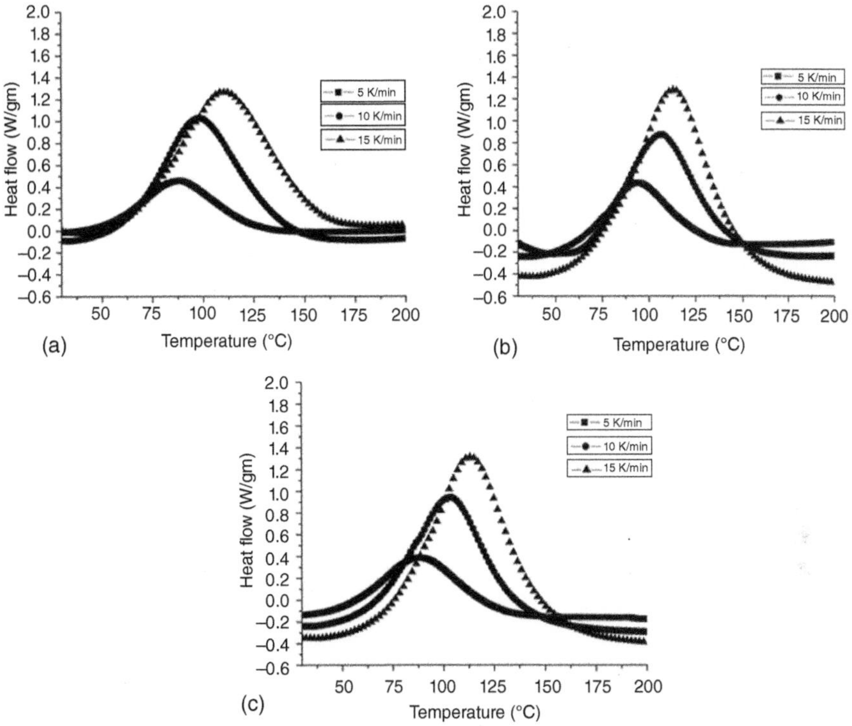

Figure 4.4 DSC curing thermogram of (a) epoxy, (b) epoxy/ESO blend, and (c) epoxy/ESO/C30B.

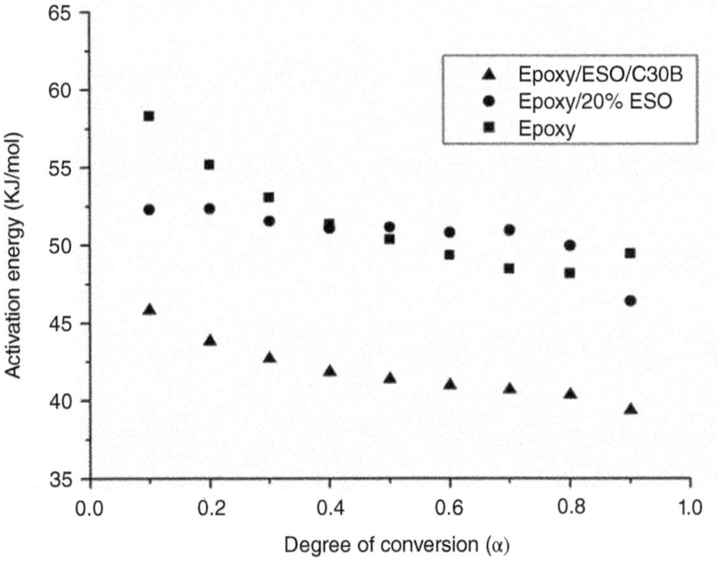

Figure 4.5 Activation energy vs. degree of conversion plot of epoxy systems.

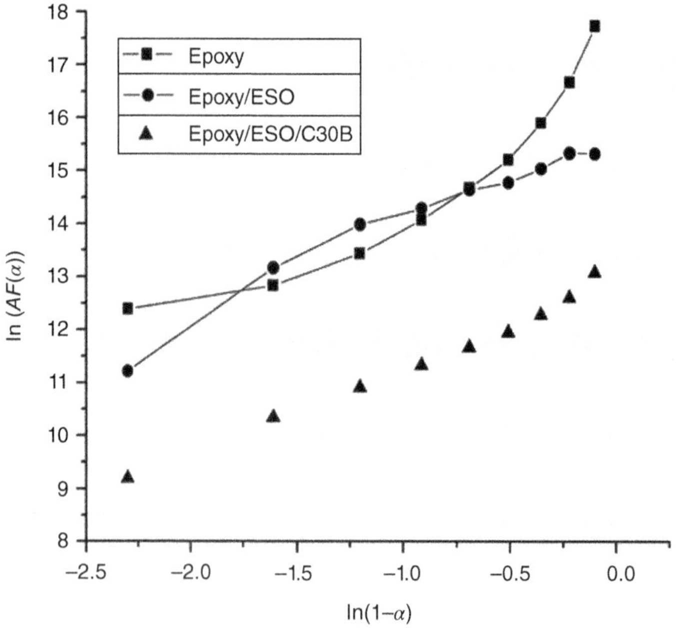

Figure 4.6 ln($A(f(\alpha))$) vs. ln(1−α) plot of virgin epoxy, epoxy/ESO, and epoxy/ESO/C30B at heating rate 10 °C/min.

nanocomposite), the activation energy derived from the Kissinger isoconversion equation within $\alpha = 0.1–0.9$ was used in Eqn. (4.16) and which used to determine the $\ln[Af(\alpha)]$ values. The mechanism of curing of epoxy generally occurs through two major kinetic reactions. The $\ln[Af(\alpha)]$ values for the nth order reaction may be written as

$$\ln[Af(\alpha)] = \ln A + n\ln(1-\alpha) \tag{4.17}$$

Friedman proposed that $\ln[Af(\alpha)]$ vs. ln(1−α) plot (Figure 4.6) should be a straight line whose slope corresponds to the order "n" of the reaction. Otherwise, the optimum value of ln(1−α) would be around −0.51 to −0.22, corresponding to α value within 0.2–0.4. Friedman plots of epoxy, epoxy/ESO, and epoxy/ESO/C30B system are depicted in Figure 4.5. It was clear from the figure that $\ln[Af(\alpha)]$ vs. ln(1−α) plot did not obey perfect linearity and exhibited proximity in the required conversion range. This signifies the fact that the curing reactions of all epoxy systems are autocatalytic in nature as reported earlier [19, 33]. Additionally, this fact was also confirmed from the plot of reaction rate ($d\alpha/dt$) against degree of conversion (α), in which it showed $\alpha = 0$ at the initial and final stage of curing, which indicated that curing processes are autocatalytic as described by Ryu et al. [35] in other similar reports.

The autocatalytic equation can be rewritten (by taking logarithm on both sides) as

$$\ln\left(\frac{d\alpha}{dt}\right) = \ln A - \frac{E_\alpha}{RT} + n(\ln(1-\alpha)) + m\ln\alpha \tag{4.18}$$

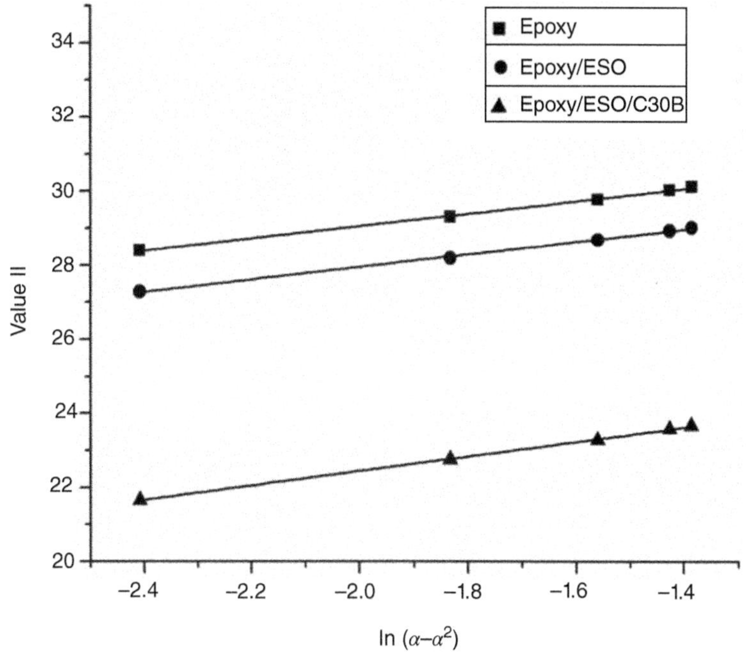

Figure 4.7 Linear plot of Value II vs. ln(α–α²) of epoxy, epoxy/ESO, and epoxy/ESO/C30B.

At (1–α) conversion, the rate equation can also be written as

$$\ln\left(\frac{\mathrm{d}(1-\alpha)}{\mathrm{d}t}\right) = \ln A - \frac{E_\alpha}{RT'} + m(\ln(1-\alpha)) + n\ln\alpha \tag{4.19}$$

Subtracting Eq. (4.18) from Eq. (4.19), we can get

$$\text{Value I} = (n-m)\ln\left(\frac{1-\alpha}{\alpha}\right) \tag{4.20}$$

where Value I $= \ln\left(\frac{\mathrm{d}\alpha}{\mathrm{d}t}\right) + \frac{E_a}{RT} - \ln\left(\frac{\mathrm{d}(1-\alpha)}{\mathrm{d}t}\right) - \frac{E_a}{RT'}$

Adding Eqs. (4.18, 4.19), we can obtain

$$\text{Value II} = 2\ln A + (m+n)\ln(\alpha-\alpha^2) \tag{4.21}$$

where Value II $= \ln\left(\frac{\mathrm{d}\alpha}{\mathrm{d}t}\right) + \frac{E_a}{RT} + \ln\left(\frac{\mathrm{d}(1-\alpha)}{\mathrm{d}t}\right) + \frac{E_a}{RT'}$

$n-m$ and $n+m$ values were obtained from the slope of the plot of Value I vs. ln[(1–α)/α] and Value II vs. $\ln(\alpha-\alpha^2)$ (Figure 4.7 and Figure 4.8), respectively. The values of $n-m$, $n+m$, and $2\ln A$ (determined from the intercept of plot in Figure 4.8) are presented in the Table 4.3.

It is clear that the overall order $(m+n)$ value of the reaction for all epoxy systems was more than 1, confirming the complex curing mechanism [19]. The pre-exponential factor of virgin epoxy is slightly decreased after addition of bio-resin and further reduced significantly on incorporation of nanoclay. Lower value of pre-exponential factor A and activation energy confirmed the accelerating effect of ESO and clay in the curing process.

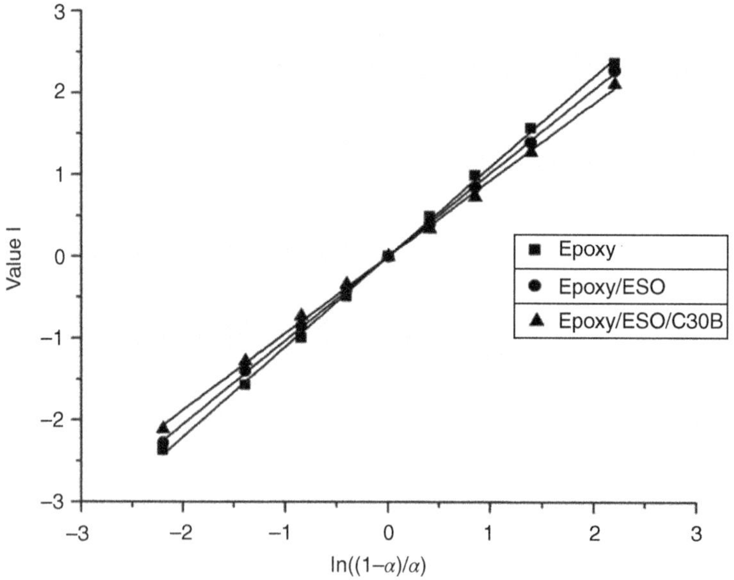

Figure 4.8 Linear plot of Value I vs. ln((1−α)/α) for virgin epoxy, epoxy/ESO, and epoxy/ESO/C30B.

Table 4.3 The kinetic parameters of the curing systems by non-isothermal method.

Sample	$n+m$	2 ln*A*	Correlation coefficient	$n-m$	Correlation coefficient	*M*	*n*	*A*
Epoxy	1.66	32.39	0.998	1.10	0.998	0.28	1.38	$e^{16.19}$
Epoxy/ESO	1.69	31.33	0.997	1.02	0.999	0.34	1.35	$e^{15.66}$
Epoxy/ESO/C30B	1.94	26.34	0.999	0.93	0.998	0.51	1.43	$e^{13.17}$

The reaction parameter m, n also indicates that ESO bio-resin favors the cross-linking process and catalytic effect of clay on curing reaction. The kinetic equations for all the three systems are expressed as follows:

Epoxy:

$$\frac{d\alpha}{dt} = e^{16.19}e^{-\frac{51.51}{RT}}(1-\alpha)^{1.38}\alpha^{0.28} \tag{4.22}$$

Epoxy/ESO:

$$\frac{d\alpha}{dt} = e^{15.66}e^{-\frac{50.72}{RT}}(1-\alpha)^{1.35}\alpha^{0.34} \tag{4.23}$$

Epoxy/ESO/C30B:

$$\frac{d\alpha}{dt} = e^{13.17}e^{-\frac{41.88}{RT}}(1-\alpha)^{1.43}\alpha^{0.51} \tag{4.24}$$

It is concluded that the addition of bio-resin decreased the activation energy of epoxy in beginning up to $\alpha = 0.7$ and slightly increased at later conversion. Total curing heat of reaction and overall order of reaction $m + n$ increased by addition of clay C30B in the blend. The peak cure temperature and activation energy are significantly reduced after incorporation of clay into the blend, confirming the catalytic effect of clay on reaction.

4.6 Conclusion and Future Prospective

The current chapter describes the fundamental of curing kinetic theories and bio-epoxy system and reviews the cross-linking of bio-epoxies and blends and composites through an extensive literature survey. Specifically, the presented case study on ESO-based epoxy composite show a great deal of promise regarding the use of bio-resins as a secondary component and organo nanoclay as a reinforcement. The projected bio-epoxy system ensured low activation energy of curing, easy processability, stiffness–toughening balance, and enhanced thermal stability. Future generation bio-epoxies has many challenges such as lower cross-link density, which resulted in inferior mechanical strength and modulus, lower thermal stability, and surplus ductility that are addressed through curing kinetic studies to make them suitable for applications. Recently, bio-based cross-linkers such as cardanol-derived phenalkamine, naturally occurring acids, and bio-based anhydrides have been used to cross-link renewable epoxy monomers to achieve desired properties with higher bio-based content. Similarly, nanofillers such as POSS, CNTs, and graphene are incorporated in bio-epoxy system to develop advance bio-thermoset nanocomposites. However, the curing kinetics of these bio-epoxy nanocomposites has remained as future research scenarios, and the influence of nanofillers on curing is studied in detail to optimize and design the formulation with excellent properties.

References

1 Fernández-Francos, X., Foix, D., Serra, À. et al. (2010). Novel thermosets based on DGEBA and hyperbranched polymers modified with vinyl and epoxy end groups. *React. Funct. Polym.* 70: 798–806.

2 Ratna, D. (2001). Mechanical properties and morphology of epoxidized soyabean-oil-modified epoxy resin. *Polym. Int.* 50: 179–184.

3 Muturi, P., Wang, D., and Dirlikov, S. (1994). Epoxidized vegetable oils as reactive diluents I. Comparison of vernonia, epoxidized soybean and epoxidized linseed oils. *Prog. Org. Coat.* 25: 85–94.

4 Yim, Y.J., Rhee, K.Y., and Park, S.J. (2017). Fracture toughness and ductile characteristics of diglycidyl ether of bisphenol-A resins modified with biodegradable epoxidized linseed oil. *Composites Part B* 131: 144–152.

5 Gandini, A. and Lacerda, T.M. (2015). From monomers to polymers from renewable resources: recent advances. *Prog. Polym. Sci.* 48: 1–39.

6 Lu, J. and Wool, R.P. (2008). Additive toughening effects on new bio-based thermosetting resins from plant oils. *Compos. Sci. Technol.* 68: 1025–1033.

7 Raghavachar, R., Letasi, R.J., Kola, P.V. et al. (1999). Rubber-toughening epoxy thermosets with epoxidized crambe oil. *J. Am. Oil Chem. Soc.* 76: 511–516.

8 Miyagawa, H., Misra, M., Drzal, L.T., and Mohanty, A.K. (2005). Fracture toughness and impact strength of anhydride-cured biobased epoxy. *Polym. Eng. Sci.* 45: 487–495.

9 Mustata, F., Tudorachi, N., and Rosu, D. (2011). Curing and thermal behavior of resin matrix for composites based on epoxidized soybean oil/diglycidyl ether of bisphenol A. *Composites Part B* 42: 1803–1812.

10 Jin, F.L. and Park, S.J. (2008). Thermomechanical behavior of epoxy resins modified with epoxidized vegetable oils. *Polym. Int.* 57: 577–583.

11 Sahoo, S.K., Mohanty, S., and Nayak, S.K. (2015). Toughened bio-based epoxy blend network modified with transesterified epoxidized soybean oil: synthesis and characterization. *RSC Adv.* 5: 13674–13691.

12 Park, S.J., Jin, F.L., and Lee, J.R. (2004). Thermal and mechanical properties of tetrafunctional epoxy resin toughened with epoxidized soybean oil. *Mater. Sci. Eng., A* 374: 109–114.

13 Altuna, F.I., Esposito, L.H., Ruseckaite, R.A., and Stefani, P.M. (2011). Thermal and mechanical properties of anhydride cured epoxy resins with different contents of biobased epoxidized soybean oil. *J. Appl. Polym. Sci.* 120: 789–798.

14 Naseem, A., Tabasum, S., Zia, K.M. et al. (2016). Lignin-derivatives based polymers, blends and composites: a review. *Int. J. Biol. Macromol.* 93: 296–313.

15 Upton, B.M. and Kasko, A.M. (2016). Strategies for the conversion of lignin to high-value polymeric materials: review and perspective. *Chem. Rev.* 116: 2275–2306.

16 Voirin, C., Caillol, S., Sadavarte, N.V. et al. (2014). Functionalization of cardanol: towards biobased polymers and additives. *Polym. Chem.* 5: 3142–3162.

17 Darroman, E., Durand, N., Boutevin, B., and Caillol, S. (2016). Improved cardanol derived epoxy coatings. *Prog. Org. Coat.* 91: 9–16.

18 Espinosa, L.M.D. and Meier, M.A.R. (2011). Plant oils: the perfect renewable resource for polymer science?! *Eur. Polym. J.* 47: 837–852.

19 Li, L., Zou, H., and Liang, M. (2014). Study on the effect of poly(oxypropylene)diamine modified organic montmorillonite on curing kinetics of epoxy nanocomposties. *Thermochim. Acta* 597: 93–100.

20 Kamal, M.R. (1974). Thermoset characterization for moldability analysis. *Polym. Eng. Sci.* 14: 231–239.

21 Vyazovkin, S. and Sbirrazzuoli, N. (1996). Mechanism and kinetics of epoxy–amine cure studied by differential scanning calorimetry. *Macromolecules* 29: 1867–1873.

22 Vyazovkin, S., Burnham, A.K., Criado, J.M. et al. (2011). ICTAC kinetics Committee recommendations for performing kinetic computations on thermal analysis data. *Thermochim. Acta* 520: 1–19.

23 Sbirrazzuoli, N., Mititelu-Mija, A., Vincent, L., and Alzina, C. (2006). Isoconversional kinetic analysis of stoichiometric and off-stoichiometric epoxy-amine cures. *Thermochim. Acta* 447: 167–177.

24 Vyazovkin, S. (2016). A time to search: finding the meaning of variable activation energy. *Phys. Chem. Chem. Phys.* 18: 18643–18656.

25 Vyazovkin, S. (2001). Modification of the integral isoconversional method to account for variation in the activation energy. *J. Comput. Chem.* 22: 178–183.

26 Sbirrazzuoli, N., Vincent, L., Mija, A., and Guigo, N. (2009). Integral, differential and advanced isoconversional methods Complex mechanisms and isothermal predicted conversion–time curves. *Chemom. Intell. Lab. Syst.* 96: 219–226.

27 Paul, D.R. and Robeson, L.M. (2008). Polymer nanotechnology: nanocomposites. *Polymer* 49: 3187–3204.

28 Alzina, C., Sbirrazzuoli, N., and Mija, A. (2010). Hybrid nanocomposites: advanced nonlinear method for calculating key kinetic parameters of complex cure kinetics. *J. Phys. Chem. B* 114: 12480–12487.

29 Alzina, C., Mija, A., Vincenta, L., and Sbirrazzuoli, N. (2012). Effects of incorporation of organically modified montmorillonite on the reaction mechanism of epoxy/amine cure. *J. Phys. Chem. B* 116: 5786–5794.

30 Liu, Z., Erhan, S.Z., and Xu, J. (2005). Preparation, characterization and mechanical properties of epoxidized soybean oil/clay nanocomposites. *Polymer* 46: 10119–10127.

31 Tan, S.G., Ahmad, Z., and Chow, W.S. (2014). Reinforcing ability and co-catalytic effects of organo-montmorillonite clay on the epoxidized soybean oil bio-thermoset. *Appl. Clay Sci.* 90: 11–17.

32 Sahoo, S.K., Mohanty, S., and Nayak, S.K. (2015). Study of thermal stability and thermo-mechanical behavior of functionalized soybean oil modified toughened epoxy/organo clay nanocomposite. *Prog. Org. Coat.* 88: 263–271.

33 Sahoo, S.K., Mohanty, S., and Nayak, S.K. (2015). A study on effect of organo modified clay on curing behavior and thermo-physical properties of epoxy methyl ester-based epoxy nanocomposite. *Thermochim. Acta* 614: 163–170.

34 Kotsilkova, R. (2007). *Thermoset Nanocomposites for Engineering Applications*, 325. Shrewsbury: Smithers Rapra Technology Limited.

35 Ryu, S.H., Sin, J.H., and Shanmugharaj, A.M. (2014). Study on the effect of hexamethylene diamine functionalized graphene oxide on the curing kinetics of epoxy nanocomposites. *Eur. Polym. J.* 152: 88–97.

36 Jana, S. and Zhong, W.H. (2009). Curing characteristics of an epoxy resin in the presence of ball-milled graphite particles. *J. Mater. Sci.* 44: 1987–1997.

37 Tao, Q., Su, L., Frost, R.L. et al. (2014). Effect of functionalized kaolinite on the curing kinetics of cycloaliphatic epoxy/anhydride system. *Appl. Clay Sci.* 95: 317–322.

38 Miyagawa, H., Mohanty, A.K., Burgueño, R. et al. (2006). Characterization and thermophysical properties of unsaturated polyester-layered silicate nanocomposites. *J. Nanosci. Nanotechnol.* 6: 464–471.

39 Boquillon, N. and Fringant, C. (2000). Polymer networks derived from curing of epoxidized linseed oil: influence of different catalysts and anhydride hardeners. *Polymer* 41: 8603–8613.

40 Dean, K. and Cook, W.D. (2002). Effect of curing sequence on the photopolymerization and thermal curing kinetics of dimetharylate/epoxy interpenetrating polymer networks. *Macromolecules* 35: 7942–7954.

41 Firouzmanesh, M.R. and Azar, A.A. (2005). Study of the effect of BDMA catalyst in epoxy novolac curing process by isothermal DSC. *J. Reinf. Plast. Compos.* 24: 345–353.
42 Fraga, F., Burgo, S., and Rodriguez, N.E. (2001). Curing kinetic of the epoxy system BADGE $n = 0/1,2$ DCH by Fourier transform infrared spectroscopy (FTIR). *J. Appl. Polym. Sci.* 82: 3366–3372.
43 Malek, J. (2000). Kinetic analysis of crystallization processes in amorphous materials. *Thermochim. Acta* 355: 239–253.
44 Heba-Laref, F., Mouzali, M., and Abadie, M.J.M. (1999). Effect of the crosslinking degree on curing kinetics of an epoxy-anhydride styrene copolymer system. *J. Appl. Polym. Sci.* 73: 2089–2094.
45 Hseih, K.H., Su, C.C., and Woo, E.M. (1998). Cure kinetics and inter-domain etherification in an amine-cured phenoxy/epoxy system. *Polymer* 39: 2175–2183.
46 Ivankovic, M., Incarnato, L., Kenny, J.M., and Nicolais, L. (2003). Curing kinetics and chemorheology of epoxy/anhydride system. *J. Appl. Polym. Sci.* 90: 3012–3019.
47 Natarajan, M. and Murugavel, S.C. (2016). Cure kinetics of bio-based epoxy resin developed from epoxidized cardanol–formaldehyde and diglycidyl ether of bisphenol–A networks. *J. Therm. Anal. Calorim.* 125: 387–396.
48 Omrani, A., Simon, L.C., Rostami, A.A., and Ghaemy, M. (2008). Cure kinetics FTIR study of epoxy/nickel-imidazole system. *Int. J. Chem. Kinet.* 40: 663–669.
49 Kim, M., Kim, W., and Choe, Y. (2002). Characterization of cure reactions of anhydride/epoxy/polyetherimide blends. *Polym. Int.* 51: 1353–1360.
50 Li, Y.F., Gao, J.G., Liu, G.D., and Zhang, R.Z. (2001). Cure kinetics and thermal property of bisphenol-S epoxy resin and phthalic anhydride. *Int. J. Polym. Mater.* 49: 441–455.
51 Kumar, S., Samal, S.K., Mohanty, S., and Nayak, S.K. (2017). Study of curing kinetics of anhydride cured petroleum-based (DGEBA) epoxy resin and renewable resource based epoxidized soybean oil (ESO) systems catalyzed by 2-methylimidazole. *Thermochim. Acta* 654: 112–120.
52 Ding, C., Shuttleworth, P.S., Makin, S. et al. (2015). New insights into the curing of epoxidized linseed oil with dicarboxylic acids. *Green Chem.* 17: 4000–4008.
53 Roudsari, G.M., Mohanty, A.K., and Misra, M. (2014). Study of the curing kinetics of epoxy resins with biobased hardener and epoxidized soybean oil. *ACS Sustainable Chem. Eng.* 2: 2111–2116.
54 Mahendran, A.R., Wuzella, G., Kandelbauer, A., and Aust, N. (2012). Thermal cure kinetics of epoxidized linseed oil with anhydride hardener. *J. Therm. Anal. Calorim.* 107: 989–998.
55 Lascano, D., Carrillo, L.Q., Balart, R. et al. (2019). Kinetic analysis of the curing of a partially biobased epoxy resin using dynamic differential scanning calorimetry. *Polymers* 11: 391.
56 Manthey, N.W., Cardona, F., Aravinthan, T., and Cooney, T. (2011). Cure kinetics of an epoxidized hemp oil based bioresin system. *J. Appl. Polym. Sci.* 122: 444–451.

57 Paluvai, N.R., Mohanty, S., and Nayak, S.K. (2015). Cure kinetics of exfoliated bio-based epoxy/clay nanocomposites developed from acrylated epoxidized castor oil and diglycidyl ether bisphenol A networks. *High Perform. Polym.* 27: 918–929.

58 Esmaeili, N., Vafayan, M., Salimi, A., and Zohuriaan-Mehr, M.J. (2017). Kinetics of curing and thermo-degradation, antioxidizing activity, and cell viability of a tannic acid based epoxy resin: from natural waste to value-added biomaterial. *Thermochim. Acta* 655: 21–33.

59 Yamini, G., Vafayanc, A.S.M., Zohuriaan-Mehr, M.J. et al. (2019). Cure kinetics of modified lignosulfonate/epoxy blends. *Thermochim. Acta* 675: 18–28.

60 Zolghadr, M., Zohuriaan-Mehr, M.J., Shakeri, A., and Salimi, A. (2019). Epoxy resin modification by reactive bio-based furan derivatives, curing kinetics and mechanical properties. *Thermochim. Acta* 673: 147–157.

61 Huo, S.P., Wu, G.M., Chen, J. et al. (2014). Curing kinetics of lignin and cardanol based novolac epoxy resin with methyl tetrahydrophthalic anhydride. *Thermochim. Acta* 587: 18–23.

62 Paramarta, A. and Webster, D.C. (2017). Curing kinetics of bio-based epoxy-anhydride thermosets with zinc catalyst. *J. Therm. Anal. Calorim.* 130: 2133–2144.

63 Kumar, S., Samal, S.K., Mohanty, S., and Nayak, S.K. (2019). Curing kinetics of bio-based epoxy resin-toughened DGEBA epoxy resin blend. *J. Therm. Anal. Calorim.* 137: 1567–1578.

64 Liang, G. and Chandrashekhara, K. (2006). Cure kinetics and rheology characterization of soy-based epoxy resin system. *J. Appl. Polym. Sci.* 102: 3168–3180.

65 Lin, B., Yang, L.T., Dai, H.H., and Yi, A.H. (2008). Kinetic studies on oxirane cleavage of epoxidized soybean oil by methanol and characterization of polyols. *J. Am. Oil Chem. Soc.* 85: 113–117.

66 Ma, Z.G. and Gao, J.G. (2006). Curing kinetics of o-cresol formaldehyde epoxy resin and succinic anhydride system catalyzed with tertiary amine. *J. Phys. Chem. B* 110: 12380.

67 Martini, D.D.S., Braga, B.A., and Samois, D. (2009). On the curing of linseed oil epoxidized methyl esters with different cyclic dicarboxylic anhydrides. *Polymer* 50: 2919–2925.

68 Jin, F.L. and Park, S.J. (2007). Thermal and rheological properties of vegetable oil-based epoxy resins cured with thermal latent initiator. *J. Ind. Eng. Chem.* 13: 808–814.

69 Ortiz, R.A., Lopez, D.P., Cisneros, M.L.G. et al. (2005). A kinetic study of the accelerated effect of substituted benzyl alcohols on the cationic photopolymerization rate of epoxidized natural oils. *Polymer* 46: 1536–1541.

70 Pielichowski, K., Czub, P., and Pielichowski, J. (2000). The kinetics of cure of epoxides and related sulphur compounds studied by dynamic DSC. *Polymer* 41: 4381–4388.

71 Rocks, J., Rintoul, L., Vohwinkel, F., and George, G. (2004). The kinetics and mechanism of cure of an amino-glycidyl epoxy resin by a co-anhydride as studied by FT-Raman spectroscopy. *Polymer* 45: 6799–6811.

72 Rosli, W.D., Kumar, R.N., Zah, S., and Hilmi, M.M. (2003). UV radiation curing of epoxidized palm oil–cycloaliphatic diepoxide system induced by cationic photoinitiator for surface coating. *Eur. Polym. J.* 39: 593–600.

73 Supanchaiyamat, N., Shuttleworth, P.S., Hunt, A.J. et al. (2012). Thermosetting resin based on epoxidized linseed oil and bio-derived crosslinker. *Green Chem.* 14: 1759–1765.

74 Tan, S.G. and Chow, W.S. (2011). Curing characteristics and thermal properties of epoxidized soybean oil based thermosetting resin. *J. Am. Oil Chem. Soc.* 88: 915–923.

75 Wang, H.H., Liu, B., Liu, X.Q. et al. (2008). Synthesis of biobased epoxy and curing agents using rosin and the study of cure reactions. *Green Chem.* 10: 1190–1196.

76 Pin, J.M., Sbirrazzuoli, N., and Mija, A. (2015). From epoxidized linseed oil to bioresin: an overall approach of epoxy/anhydride cross-linking. *ChemSusChem* 8: 1232–1243.

77 Ferdostian, F., Yuan, Z.S., Anderson, M., and Xu, C.C. (2015). Sustainable lignin-based epoxy resins cured with aromatic and aliphatic amine curing agents: curing kinetics and thermal properties. *Thermochim. Acta* 618: 48–55.

78 Chen, Y., Xi, Z., and Zhao, L. (2016). New bio-based polymeric thermosets synthesized by ring-opening polymerization of epoxidized soybean oil with a green curing agent. *Eur. Polym. J.* 84: 435–447.

79 Khandelwal, V., Sahoo, S.K., Kumar, A. et al. (2019). Bio-sourced electrically conductive epoxidized linseed oil-based composites filled with polyaniline and carbon nanotubes. *Composites Part B* 172: 76–82.

80 Sahoo, S.K., Khandelwal, V., and Manik, G. (2018). Development of toughened bio-based epoxy with epoxidized linseed oil as reactive diluent and cured with bio-renewable crosslinker. *Polym. Adv. Technol.* 29: 565–574.

81 Das, G., Deka, H., and Karak, N. (2013). Bio-based sulfonated epoxy/hyperbranched polyurea-modified MMT nanocomposites. *Int. J. Polym. Mater Po.* 62: 330–335.

82 Wang, R.P., Schuman, T., Vuppalapati, R.R., and Chandrashekhara, K. (2014). Fabrication of bio-based tannicepoxy-clay nanocomposites. *Green Chem.* 16: 1871–1882.

83 Tsujimoto, T., Uyama, H., Kobayashi, S. et al. (2015). Green nanocomposites from renewable plant oils and polyhedral oligomeric silsesquioxanes. *Metals* 5: 1136–1147.

84 He, M., Zhou, J.J., Zhang, H. et al. (2015). Microcrystalline cellulose as reactive reinforcing fillers for epoxidized soybean oil polymer composites. *J. Appl. Polym. Sci.* 132: 42488/1–42488/8.

85 Miyagawa, H., Mohanty, A.K., Drzal, L.T., and Misra, M. (2005). Nanocomposites from biobased epoxy and single-wall carbon nanotubes: synthesis, and mechanical and thermophysical properties evaluation. *Nanotechnology* 16: 118–124.

5

Rheology of Bioepoxy Polymers, Their Blends, and Composites

Appukuttan Saritha[1], *Battula D.S. Deeraj*[2], *Jitha S. Jayan*[1], *and Kuruvilla Joseph*[2]

[1] *Amrita Vishwa Vidyapeetham, Department of Chemistry, School of Arts and Sciences, Amritapuri, Kollam 690525, Kerala, India*
[2] *Indian Institute of Space Science and Technology Valiamala, Department of Chemistry, Thiruvananthapuram 695547, Kerala, India*

5.1 Introduction

The utility of epoxy resin in the global market is boosted as a result of its ever rising demand in various industries and its inevitable presence in high-end applications. Unfortunately, 75% of epoxy prepolymers that are produced around the world are synthesized by the condensation reaction of bisphenol A (BPA) and epichlorohydrin. BPA is an endocrine disruptor that causes a serious threat on human health. These health issues of BPA coupled with environmental issues forced the researchers and industries to find an appropriate alternative for BPA, and as a consequence, the bio-based epoxy resin started gaining attention. Bio-based thermosets can be defined as synthetic materials that are synthesized either by partial or by complete replacement of the conventional building blocks of a resin by matching building blocks based on renewable raw materials.

It is a trend in these recent years that researchers and manufacturers are focusing toward eco-friendly products as they have shown favorable results in terms of environmental impact and energy used to manufacture them [1]. The major issue concerned with BPA is its adverse impact on living beings together with its origin from fossil resources. Many advanced countries have banned the use of epoxy containing BPA, and hence, the alternative green pathway for the synthesis of epoxy resins became the immediate need of the hour.

Henceforth, several natural oils such as soya oil, cottonseed oil, algae oil, karanja oil, pine oil, corn oil, and palm oil have been epoxidized. The process of epoxidizing natural oil differs for each oil, and this in turn makes the bio-based resin production challenging [2]. Epoxy resins are one of the most widely used thermosets in different engineering fields because of their chemical resistance and thermomechanical properties. Bio-based thermoset resin systems are gaining importance because of the environmental factors arising from the effective utilization of bioresources, leading to substantial reduction in the use

Bio-Based Epoxy Polymers, Blends, and Composites: Synthesis, Properties, Characterization, and Applications, First Edition. Edited by Jyotishkumar Parameswaranpillai, Sanjay Mavinkere Rangappa, Suchart Siengchin, and Seno Jose.

of petroleum feedstocks. Bio-based epoxy materials have created a revolution that has helped in the creation of a sustainable environment with less global warming pursuits. Many of the bio-based epoxies cannot be effectively utilized in the fabrication of composites. This can be attributed to their toughness at 25 °C, inferior mechanical properties, or complex synthesis procedures [3, 4]. Because of this, diglycidyl ether of bisphenol A (DGEBA) is still considered as a viable choice for composite fabrication. However, Maiorana et al. [5] carried out the preparation of a family of bio-based epoxy thermoset resins called diglycidyl ether diphenolate *n*-alkyl esters (DGEDP). This novel bio-based epoxy can be tailored to be fit for various applications because its rheological and thermal properties are highly dependent on the *n*-alkyl side chain length [5].

Most prominent platforms for the production of bioepoxies are epoxidized vegetable oils that are already used as ingredients for protective coatings [6, 7]. Other precursors include epoxidized tree rosins, diglycidyl ethers and esters of furans, cashew nutshell liquid-based epoxy resins, multifunctional glycidyl flavonoids, diglycidyl ferulates, and diglycidyl diphenolates[8–14]. One of the factors that should be taken into consideration while preparing composites using bioepoxies is that their characteristic structures can cause a drop in final properties such as glass transition temperature and modulus [11, 15–20].

There are considerable works in the preparation and characterization of bioepoxy resins in the literature [21–24], but the number of rheological studies of bio-based resin systems [25, 26] is significantly less. Liquid composite molding (LCM) processes, such as vacuum-assisted resin transfer molding (VARTM), are often considered as cost effective as well as promising methods for the manufacture of composites [27, 28]. In this context, rheological properties and their change over time and temperature require a greater attention. Many studies on bio-based epoxy resins and their processing reports measure gel times [29] and/or viscosity measurements [30, 31] as a means to judge processability. Hence, it is highly mandatory that an authentic study of the rheology of these bio-based epoxies will definitely pave way toward their effective utilization in composite manufacturing process.

A detailed investigation of rheological properties and their variation with respect to time and temperature is a mandatory component during the processing of these bio-based epoxy composites. Hence, a thorough understanding of the rheology of these systems is highly essential. The present chapter focuses on the rheology of bioepoxy polymers, blends, and their composites.

5.2 Rheology of Bioepoxy-Based Polymers

Recent research in the field of bioepoxies point out that a lot of work is done toward the synthesis of bioepoxy resins, their characterization, and property analysis. Bio-based platforms are converted into bio-based epoxy polymers through various reaction routes. Among the various studies, the works that concentrate on the rheology of bio-based epoxy polymers prepared from various materials such as oils, lignocellulosic biomass, etc., are discussed.

5.2.1 Natural Oil-Based Epoxies

Surplus amounts of carbon–carbon double bonds in plant oils enable them to act as a common option for use in thermoset resins because these double bonds act as regions through which polymerizations can occur. Plant oils must be transformed either through direct polymerization or the conversion of double bonds to functional groups, followed by polymerization. They can also be used for the synthesis of bioepoxies after transforming them into fatty acids and then subjecting them to polymerization.

Omonov and Curtis [32] used epoxidized canola oil (ECO) as the precursor and phthalic anhydride (PA) as a curing agent for the production of a new bio-based epoxy thermoset. They found that the thermomechanical properties of the new epoxy thermoset could be varied by altering the ECO/PA ratio, which would make the system suitable for the fabrication of lignocellulosic fiber boards and particle boards. The rheology of the samples were analyzed at various curing temperatures such as 155, 170, 185, and 200 °C. It is seen that before curing, the samples exhibit a behavior similar to that of liquids with optimal low viscosity, which is suitable for various applications. Upon increase in curing temperature as well as the amount of curing agent, the gelation time of the mixtures decreased. T_g increases with an increase in the amount of curative but is significantly unaffected by curing temperature.

Stemmelen et al. [6] produced a polyamine from grape seed oil via the reaction of cysteamine chloride and UV-initiated thiolene chemistry. The polyamine thus obtained was used as a curing agent for a bio-based epoxy resin produced from linseed oil. The effect of diverse parameters such as the nature and amount of photoinitiator and the time of the reaction was regularly studied using various techniques. The thermal cross-linking of the bio-based polyamine from grapeseed oil and epoxidized linseed oil (ELO) was monitored by differential scanning calorimetry (DSC) and rheology. The kinetic parameters of the cross-linking reaction were measured by dynamic rheometry, and the thermomechanical behavior of the polymer was subsequently analyzed.

The rheological investigation was undertaken with a less value of angular frequency, and the temperature corresponding to the maximum of the G00 peak is taken as the reliable glass transition temperature of the cured material as depicted in Figure 5.1, and this value is in agreement with that obtained through DSC analysis.

Bio-based polyols with several functionalities were prepared from epoxidized soybean oil and castor oil-based fatty diol by Anthony et al. [33]. After a detailed study using nuclear magnetic resonance (NMR), FT-IR, gel permeation chromatography (GPC), and rheometry, it was found that by carefully controlling the reaction conditions, it is possible to alter the functionalities, molecule weight, and viscosity of these polyols [34]. The newly introduced hydroxyl group in epoxides would get involved in the ring-opening reaction after the consumption of highly reactive primary hydroxyl in castor oil diol, which results in oligomerization. The rheological behaviors of polyols with different reaction times are shown in Figure 5.2. It is found that, as the reaction time increases, the products become more viscous and shear thinning behavior emerges instead of

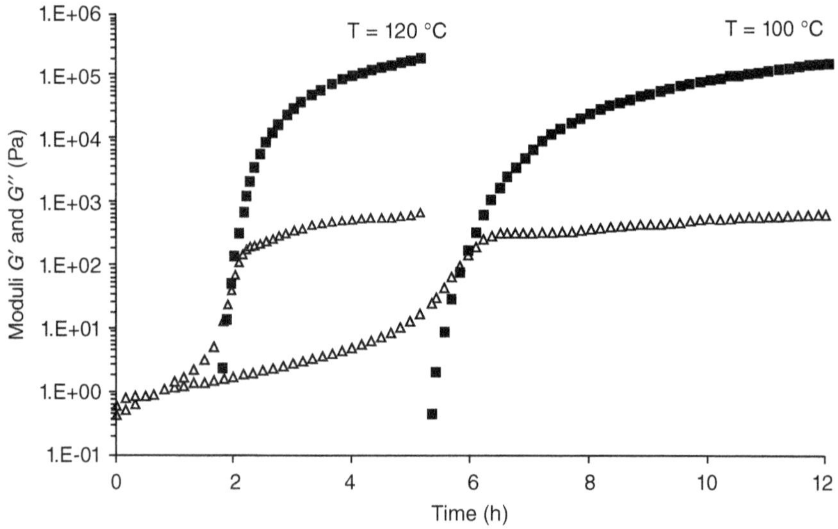

Figure 5.1 Rheological analysis of the isothermal curing of the polyamine grapeseed oil (AGSO)–ELO mixture for two temperatures and with G0 (*n*) and G00 (~). Modulus G′ are denoted as squares and G″ denoted as triangles. Source: Reproduced from Stemmelen et al. [6]. © 2011, John Wiley & Sons.

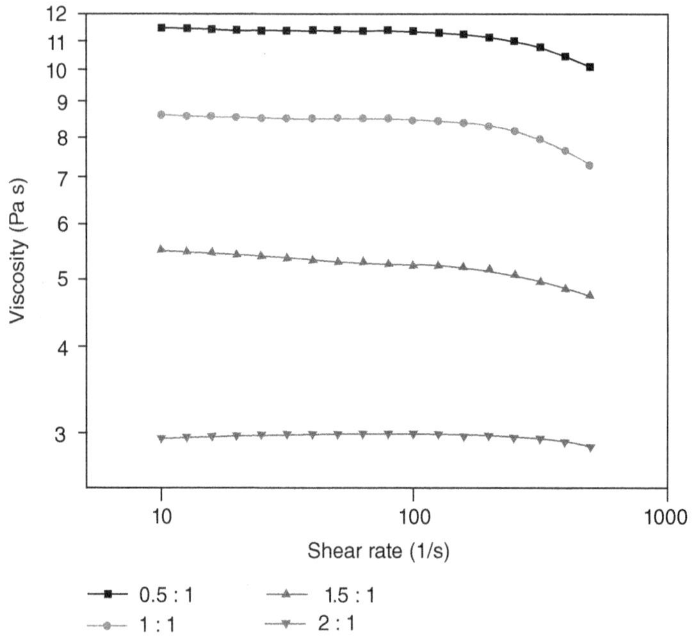

Figure 5.2 Rheological behaviors of polypols with different ratios between COD and ESBO. Source: Zhang et al. [34].

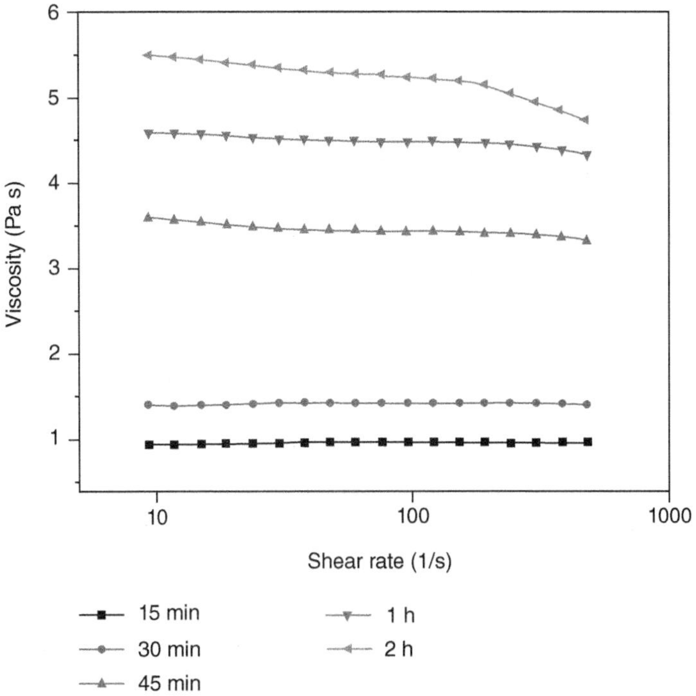

Figure 5.3 Rheological behaviors of polyols with different reactions. Source: Zhang et al. [34].

Newtonian behavior. The viscosity conforms to the molecular weight obtained by GPC.

Thus, a greater amount of castor oil diol (COD) provided more reactive primary hydroxyl groups to react with epoidized soybean oil (ESBO), and then, more epoxy groups were ring-opened by the primary hydroxyl groups in COD, instead of being ring-opened by the newly generated secondary hydroxyl groups. Therefore, the cross-linking densities decrease, resulting in relatively lower viscosities. The rheological behaviors of polyols with different reaction times are shown in Figure 5.3.

Jin and Park [35] noticed that, when the time of the reaction increases, the products become more viscous against the expected Newtonian behavior. The viscosity agrees with the molecular weight obtained by GPC. These bioepoxies were subsequently cured with a thermally latent initiator, *N*-benzylquinoxalinium hexafluoroantimonate (BQH), and it is observed that these systems differed in their glass transitions with ESO/BQH system, exhibiting a higher stability and lower glass transition temperature than ECO/BQH system. Moreover, the activation energy required for cross-linking of the former was higher than the latter due to steric hindrance.

Anthony et al.[33] carried out the synthesis of epoxy resins based on adipic acid by means of two different pathways. The former was a single step reaction, while the latter was through allylation of adipic acid, followed by its epoxidation. The viscosity at higher shear rates is reported and is 99% lower than that of DGEBA.

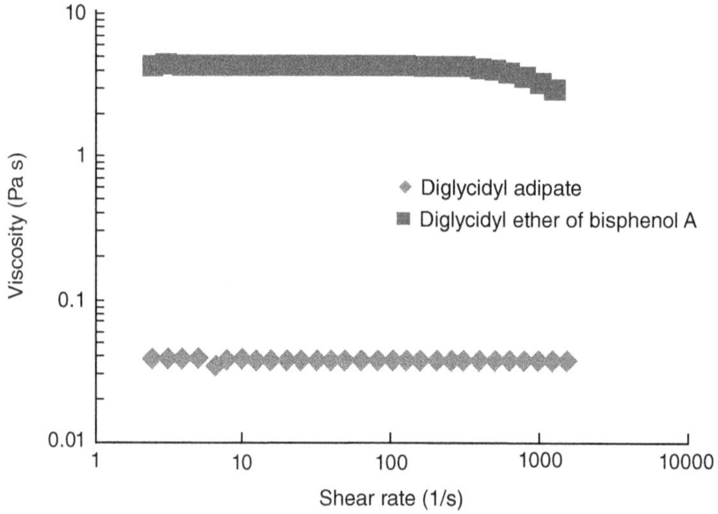

Figure 5.4 Comparing the diglycidyl ether of bisphenol A to diglycidyl adipate viscosities. Source: Reproduced from Maiorana et al. [33]. © 2016, John Wiley & Sons.

The low viscosity of diglycidyl adipate (Figure 5.4) along with the fact that it has two epoxy functionalities indicates that it has excellent potential for incorporation in epoxy resins used for vacuum infusion molding, high solid coatings, and as a reactive diluent for epoxy resins with higher viscosity.

Karger-Kocsis et al. [36] studied the curing, gelling, thermomechanical, and thermal degradation behavior of anhydride-cured EP-ESO compositions. The results showed that the dilution of EP with ESO was accompanied with marked changes in the curing, gelling behavior, and final properties. ECO-based thermoset epoxy resins were formulated with phthalic anhydride (PA) as the curing agent for different ratios of ECO to PA (1 : 1, 1 : 1.5, and 1 : 2 mol/mol) at curing temperatures of 155, 170, 185, and 200 °C. The gelation process of the epoxy resins and the viscoelastic properties of the systems during curing were studied by rheometry. It can be seen from Figure 5.5 that as the gelation progressed, the viscosity of the system increased dramatically. In an ideal case, after gelation, the viscosity of the reactive mixture should tend toward infinite values [32].

5.2.2 Isosorbide-Based Epoxy Resins

Isosorbide was initially used to improve the thermomechanical characteristics of polymeric systems. Recently, it has gained attention as a material that can be used as a precursor for the synthesis of unsaturated polyester resins and hence serve as excellent matrices for the production of composites in a cost-effective manner.

An easy way to replace bisphenol A in the preparation of epoxy resin is to substitute it with isosorbide-based derivatives. Marie Chrysanthos et al. [37] synthesized three different bio-based epoxy precursors from sorbitol and cardanol and a novel epoxy monomer that was successfully synthesized from

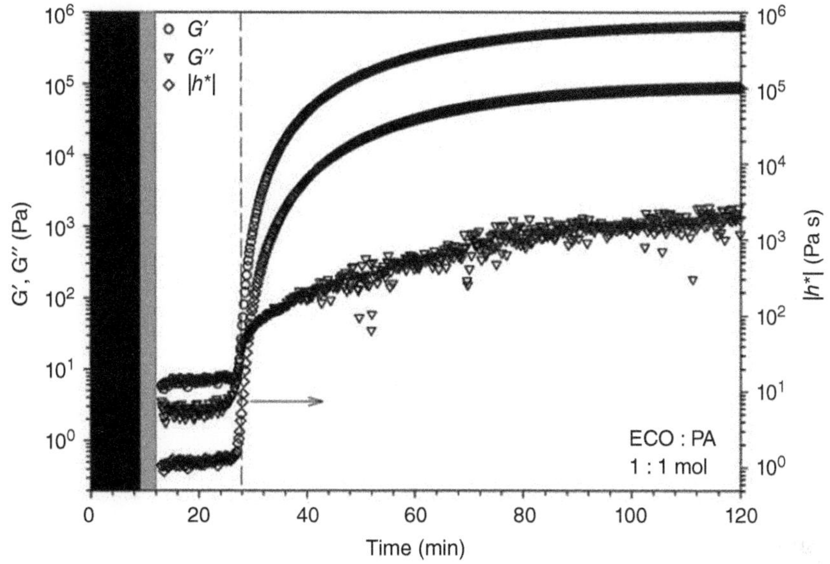

Figure 5.5 Viscoelastic properties of the ECO-based resin with an equimolar ratio of components cured at 155 °C [32].

isosorbide. Gelation and reactivity were analyzed through rheological measurements and were compared with a traditional epoxy resin. It was found that the cured sorbitol polyglycidyl ether-based epoxy resin exhibits the lowest gelation time at a temperature of 80 °C. They concluded that SPGE and diglycidyl ether of isosorbide could be a good candidate for the replacement of bisphenol-A-based prepolymers.

Oliver Goerz and Ritter [38] prepared an unsaturated polyester from isosorbide and itaconic acid and subjected it to polycondensation in the presence of acid. Because the reaction resulted in low molecular weight polymers, itaconic acid was replaced by succinic acid, and there was tremendous improvement in molecular weights. The synthesized copolyesters with higher contents of itaconic acid showed a gel point at an early stage of cross-linking with monomers and they exhibited shape memory effect.

Bio-based epoxy networks were derived from the isosorbide (1,4:3,6-dianhydro-D-sorbitol) (DAS) obtained by the double dehydration of sorbitol by epoxidizing it via two different reaction pathways. The isosorbide diglycidyl ether (or Diglycidyl Ether of DAS, DGEDAS) was subjected to curing with isophorone diamine (IPD) for the synthesis of the desired bio-based epoxy network. The structure of the bioepoxy thus prepared was thoroughly evaluated and comparative study was done with a conventional epoxy network based on DGEBA cured by IPD by Marie Chrysanthos et al. [39]. The gel time of the samples was evaluated using rheological measurements. It appears that the new bio-based reactive system $DGEDAS_n$ reacted with IPD and has a shorter gel time than the traditional one DGEBA-IPD, whereas the system $DGEDAS_0$IPD has almost the same gel time as DGEBA-IPD (Table 5.1).

Table 5.1 Gel time data for the reactive systems DGEBA/IPD, $DGEDAS_n$/IPD, and $DGEDAS_0$/IPD.

System	t_{gel} (min)	$\tan \delta_{gel}$	Δ
DGEBA-IPD	16	1.5	0.63
$DGEDAS_n$-IPD	7	1.0	0.50
$DGEBAS_0$-IPD	18	1.6	0.64

Source: Reproduced from Chrysanthos et al. [39]. © 2011, Elsevier.

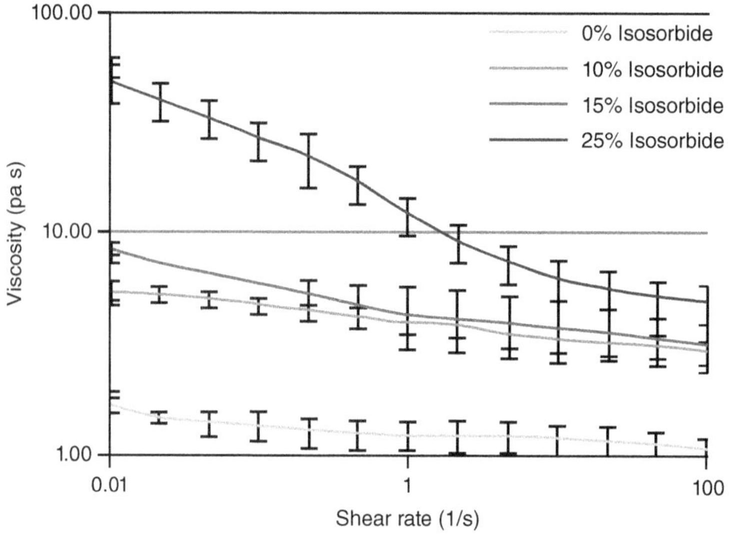

Figure 5.6 Isosorbide concentration effect on viscosity. Source: Reproduced from Omonov and Curti [40]. © 2015, John Wiley & Sons.

Joshua M. Sadler and coworkers [40] evaluated the conversion of isosorbide to unsaturated polyester resins and found that these materials have the potential to be developed as useful thermosetting materials in composite materials. They found that upon addition of higher quantities of this unique diol, the T_g and moduli of the cured polymers increase. Although viscosities show close proximity to commercial unsaturated polyester resin (UPE) resins, an increase in the concentration leads to an increase in the viscosity of the uncured resin, which could affect their processing in composite applications. The observation clearly depicts that the increase in isosorbide concentration causes an increase in shear thinning behavior (Figure 5.6). These results display the probability of tailoring the viscosity of resin system so that the properties can be tuned for discrete composite applications.

Water-soluble epoxy resins were prepared from diglycidyl ethers of isosorbide (DGEI) and isosorbide diamine (ISODA) by Hong et al. [41]. They investigated the effect of synthesis methods on the structure of DGEI and ISODA and their influence on the properties of the cured resins. They also compared the results with bisphenol-A-based commercial epoxy resins (DGEBA). DGEI and DGEBA were cured with diethylene triamine (DETA) and ISODA to give optically clear yellow to brown epoxy resins of different degrees of cross-linking. All systems containing isosorbide either on the epoxy or on the amine side exhibited faster curing rates than the standard petrochemical epoxy resin (DGEBA/DETA). The glass transition temperature of isosorbide-based resins was about 60 °C lower than those of DGEBA/DETA. The viscosity curves indicated a faster increase of viscosity and earlier gelation of isosorbide than bisphenol-A-based resins when cured with DETA at 25 °C. The gel time, taken to be arbitrarily at 500 Pa s, was reached within 5000 seconds for DGEI/DETA, while that for DGEBA/DETA was around 8000 seconds. Faster curing was aided by the higher content of hydroxyl groups in the former.

5.2.3 Phenolic and Polyphenolic Epoxies

Depending on the strength of the phenolic ring, phenols can be classified into various classes. Natural phenolic and polyphenols are effective in conferring superior properties to epoxy thermosets. Hence, a lot of research work is concentrated on the preparation and functionalization of epoxy resin based on these structures. Catechin[42], tannin [43], gallic acid[44–46], tannic acid[47], cardanol[48–50], etc., have been used as important platforms for the preparation of bioepoxies based on polyphenolics. In a novel preparation technique, tri- and tetra-derivatives of glycidyl ether of gallic acid (GEGA) were added with two primary amines, aliphatic and cycloaliphatic, and a tertiary amine. The former amines functioned as cross-linking agents, while the latter one acted as an ionic initiator for homopolymerization reaction. A complex three-step curing process was observed during the homopolymerization of GEGA in the presence of the tertiary amine initiator. The rheological isothermal experiments at room temperature showed Newtonian behavior for the GEGA precursor, with a measured viscosity value of 2 Pa s. The complex viscosity as a function of temperature for the three epoxy systems indicated the occurrence of the gelation event at temperatures 73, 92, and 114 °C for the three epoxy mixtures, respectively [51].

The inherent high viscosity possessed by commercial epoxy resins at room temperature makes its processability difficult, and hence, the addition of active epoxy-based diluents is of great interest in improving the processing parameters. Interestingly, bioepoxides derived from cardanol is found to function as effective diluents in the epoxy matrix. Polyepoxide cardanol glycidyl ether (PECGE) was synthesized from cardanol and used as a diluent in petroleum-based epoxy resin. The addition of the diluent resulted in a decrease in the viscosity of the epoxy matrix. It was seen that the viscosity values gradually decreased from 14 700 to 1892 mPa s, with the increase of the CGE content from 0% to 15%. A similar remarkable reduction of viscosity values was found in the sample with a CGE content of 15% and the sample with a PECGE content of 20% [52].

A novel bio-based organosilicon-grafted cardanol novolac epoxy resin (SCNER) was prepared and used as a reactive diluent and toughening agent to modify the anhydride-cured epoxy resin system. The results (Figure 5.7)

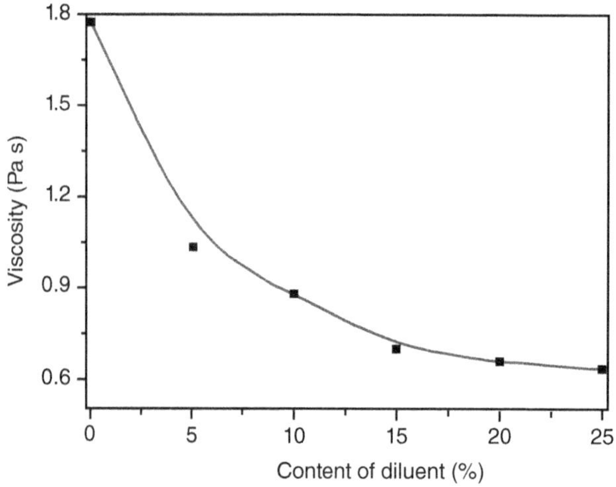

Figure 5.7 Effect of SCNER content on the viscosity of DGEBA at 25 °C. Symbol square denotes viscosity. Source: Reproduced from Chen et al. [53]. © 2015, American Chemical Society.

showed that SCNER had an outstanding dilution and toughening effect on the anhydride-cured epoxy resin system by Huo et al. [53].

5.2.4 Epoxidized Natural Rubber-Based Epoxies

Epoxidized natural rubber (ENR) is an important commercial product used for various applications. Al-Mansob et al. [54] ENR and analyzed the rheological properties of the ENR bitumen. Isochronal plots at 0.1 and 1 Hz of frequencies are plotted, and the increase in storage modulus shows the same trend except 12% of ENR modified one. It shows a tremendous increase in modulus at high temperature greater than 50 °C. Moreover, the viscosity of bitumen is increased by the addition of ENR and especially the increase at low temperature (Figure 5.8).

Nakason et al. [55] followed the method of performing epoxidation to prepare ENR. They have analyzed the rheological properties of ENR and cassava starch and compared it with the rheological properties of ENR, natural rubber, and cassava blends. Capillary rheometer was used to characterize the rheology; in all the cases, apparent shear stress is increased with the shear rate. ENR shows lower shear stress than the blends. Concentration of cassava starch is good enough to increase the shear stress, and blends showed shear thinning behavior. Nakason et al. [56] studied the rheological properties of ENR/poly(methyl methacrylate) blends. The rheological analysis confirms the pseudoplastic behavior of the blends, i.e. the shear thinning behavior, because of the disentanglement of entangled polymer chains. ENR blends containing more epoxide rings shows higher viscosity and it is shown in Figure 5.9.

Da Silva et al. [57] modified latex to get ENR and mixed it with hydrotalcite in order to analyze its rheology in detail. The rheological analysis helped to calculate the cross-link density, the physical cross-link density (X_{initial}) relative to entanglements, and the total cross-link density (X_{total}) was calculated by the help of the

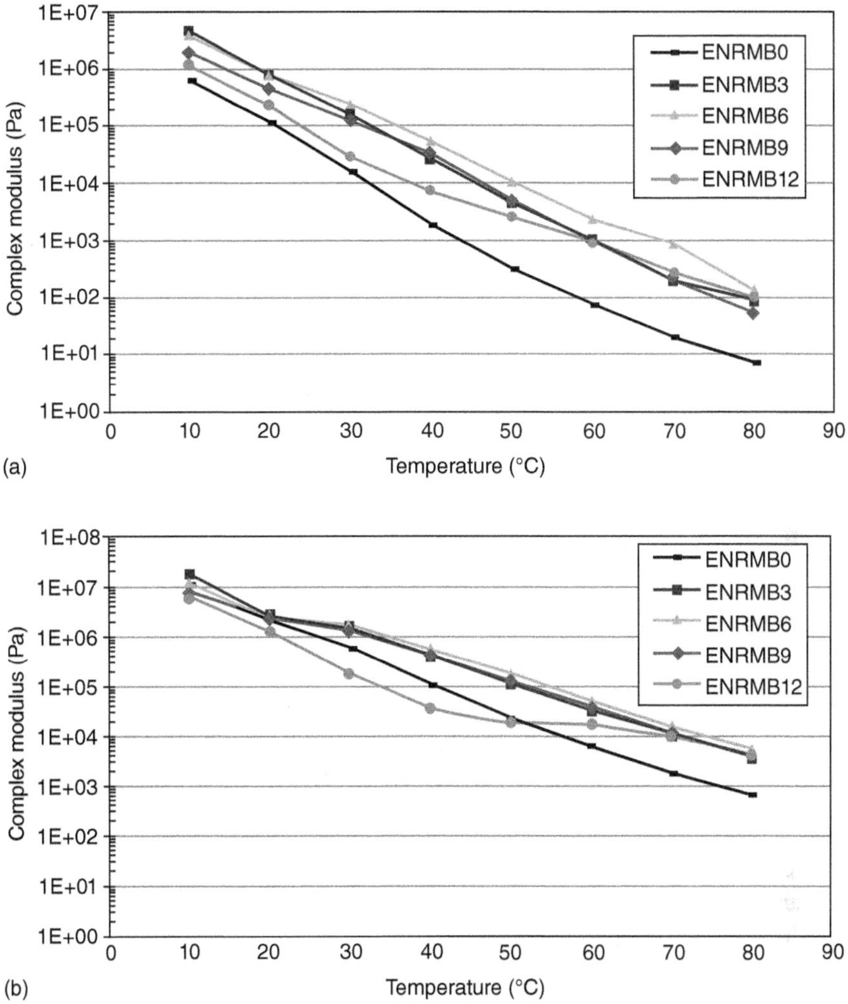

Figure 5.8 (a) Isochronal plots of complex modulus at 0.1 Hz for the epoxidized natural rubber modified bitumen (ENRMB) binders. (b) Isochronal plots of complex modulus at 10 Hz for the ENRMB binders. Source: (a,b) Reproduced from Al-Mansob et al. [54]. © 2014, Elsevier.

two equations.

$$X_{\text{total}} = \frac{G\prime_{\text{cured}}}{2RT}(0.5\ \text{Hz}) \tag{5.1}$$

$$X_{\text{initial}} = \frac{G\prime_{\text{uncured}}}{2RT}(5\ \text{Hz}) \tag{5.2}$$

where G' is the storage modulus.

Yunus et al. [58] made ENR-based magnetorheological elastomer (MRE) and mixed it with carbonyl iron particles (CIPs) by two roll mixing. The CIP affects the rheological behavior of ENR, and the storage modulus is decreased with

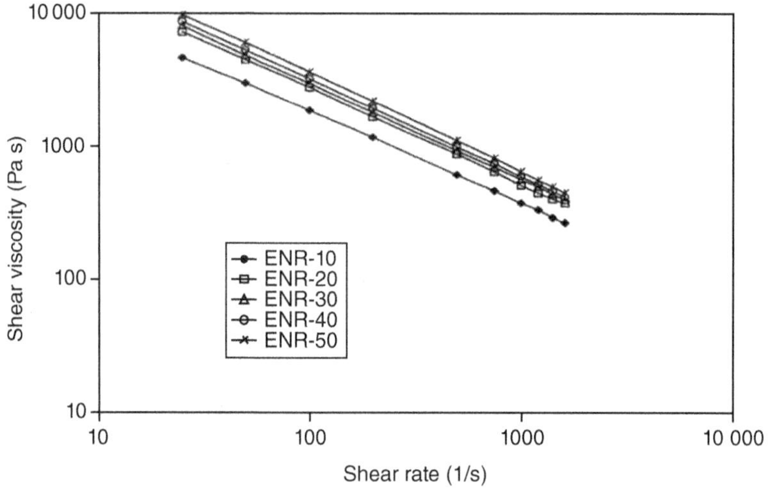

Figure 5.9 Effect of shear rate on the melt viscosity of various ENRs in 60/40 ENR/polymethyl methacrylate (PMMA). Source: Reproduced from Nakason et al. [56]. © 2004, John Wiley & Sons.

the increase in CIP loading. Hence, the ability of the material to store energy is decreasing with the loading and the loss factor is increased (Figure 5.10).

5.2.5 Epoxy Lignin Derivatives

Lignin is considered as a very important component that can be included into thermosetting resins because of the presence of alcoholic and phenolic groups in it. Hence, the use of lignin as a precursor for the synthesis of epoxy monomers is currently gaining attention. There are different methods employed to synthesize lignin-based epoxy systems such as blending the epoxy system with lignin or by functionalizing lignin derivatives such as Kraft lignin by direct glycidylation. The third method utilizes the technique of improving the reactivity of lignin derivatives and then subjecting it to glycidylation.

Junna Xin et al. [59] prepared a bio-based epoxy by the reaction between epichlorohydrin and partially depolymerized lignin. Effects of adding varying amounts of epoxy on the rheological properties of modified asphalt binder were carried out. In an attempt to make the lubricating industry green, the modification of castor oil with alkali lignin (AL)-based epoxy was carried out to prepare a gel-like lubricant by Esperanza Cortés-Triviño et al.[60]. The linear viscoelastic response of the synthesized gel-like dispersions in castor oil was decided by the index of epoxidation (EPI). Upon modeling of the rheological parameters, a power law model was used to explain the progression of G^* when it is plotted as a function of frequency. It can be seen that the power law parameters show a linear variation with EPI. Interestingly, it is seen that the linear viscoelastic functions show a hike upon aging up to a month as a result of the internal curing process, where the remnant epoxy groups still undergo reaction with the triglycerides in castor oil. The thermorheological characterization shows a softening temperature at around 50 °C.

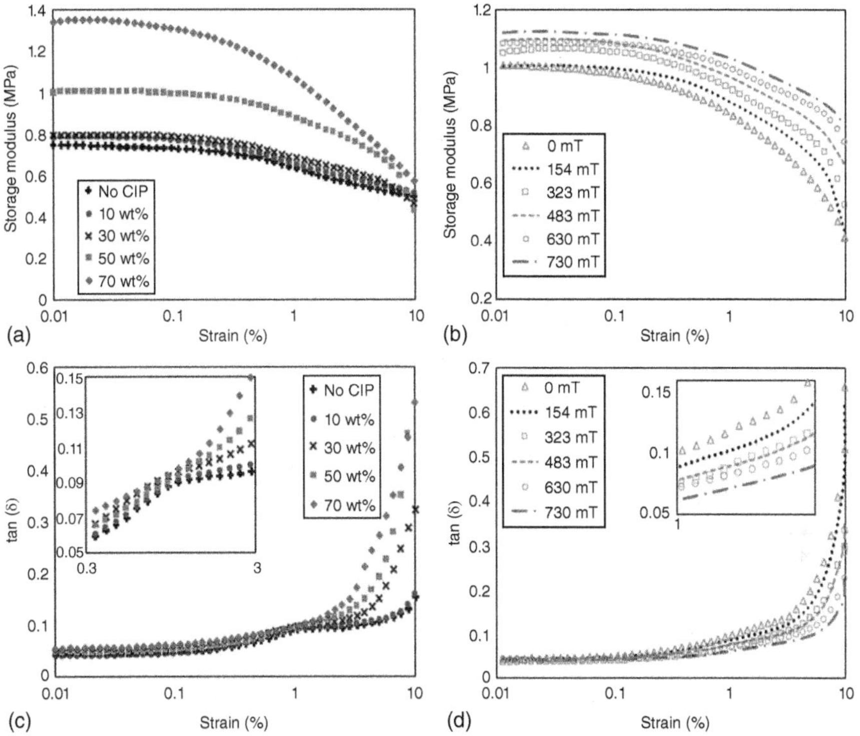

Figure 5.10 Results of (a, b) storage modulus and (c, d) loss factor of ENR-based MREs at different CIP contents and magnetic fields for various strain amplitudes. Source: From Yunus et al. [58].

In an attempt to find thickeners for castor oil, chemical modification of AL was carried out with poly(ethylene glycol) diglycidyl ether (PEGDE) for the development of eco-friendly lubricating greases [61]. By changing the PEGDE/AL ratio while carrying the epoxidation leads to the formation of lignin compounds with different epoxy indexes and residual NaOH amounts. This happens because of the competition between two reactions taking place in the process. The oleogels are essentially epoxidized lignin-based and are extreme structured systems that exhibit typical viscoelastic properties corresponding to solid-like gels yielding flow as observed in traditional lubricating greases. Furthermore, there exists a purely internal curing for a period of two months, which stimulates the progression of the rheological properties. Figure 5.11 shows viscous flow curves of stable epoxide-functionalized lignin-based oleogels, as a function of both epoxy index (Figure 5.11a) and thickener concentration (Figure 5.11b). A power law model satisfactorily describes the shear-thinning flow behavior observed within the shear rate range analyzed ($r_2 > 0.995$). As can be seen in Figure 5.12, the linear viscoelasticity response supports the idea that highly structured gels were obtained [38, 40], where the values of the storage modulus were always higher than those of the loss modulus, throughout the whole frequency range studied. The values of these viscoelastic moduli increased with the decrease of PEGDE/AL ratio, i.e.

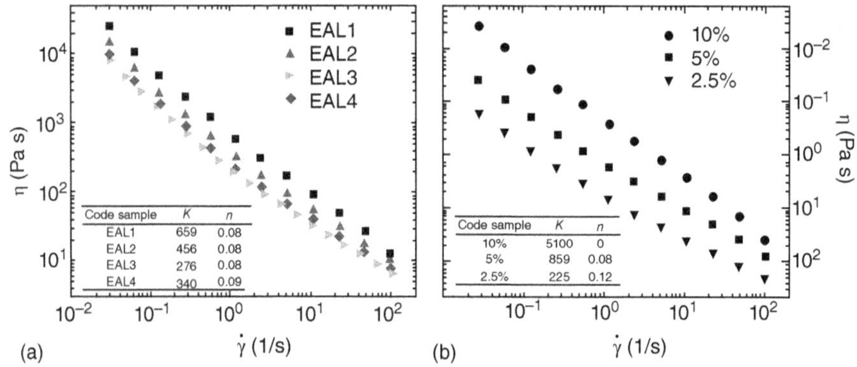

Figure 5.11 Viscous flow curves at 25 °C for epoxidized lignin-based oleogels as a function of (a) lignin epoxy index and (b) lignin concentration. Source: From Cortés-Triviño et al. [61].

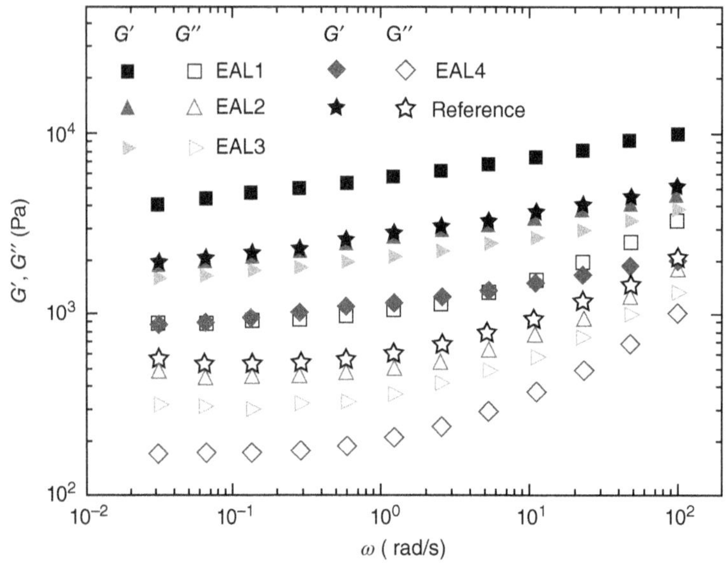

Figure 5.12 Evolution of the storage and loss moduli with frequency for epoxidized lignin based oleogels as a function of lignin epoxy index. Source: From Cortés-Triviño et al. [61].

with the increase of the epoxy index, as a consequence of the higher density of cross-linking points induced by the reaction of epoxy and hydroxyl groups. On the other hand, the high amount of NaOH present in the highly epoxidized lignin sample brings about an additional stabilizing effect because of partial saponification of castor oil, as can be inferred from the rheological response of the reference [61].

Owing to the fact that chemical modification of lignin can enhance the compatibility of lignin with polymers, there is a great deal of work that concentrates on lignin modification. Oana Petreus et al. [62] carried out a synthesis in which ammonium lignosulfonate was either directly subjected to epoxidation or first transformed into the acid form of the lignosulfonate, treated with phenol and then subjected to epoxidation. The liquid epoxy-modified lignin fractions and their mixtures were treated with a cyclic organic phosphorous compound and the structure of the resulted products was studied. Rheological data on liquid lignin–epoxy resins evidence a Newtonian behavior in the studied shear rate range, while the phosphorous derivatives present a Newtonian flow in another shear rate range. Viscosity data vs. shear rate for the (purified part of lignin-epoxy resin [RLE] + phenolated-lignin-epoxy resin [REO]) mixture indicate a good compatibility.

Lignin was extracted from Indulin AT (IND) and was used along with pyrogallol for the preparation of bio-based epoxy resins by Gunnar Engelmann and Ganster [63]. The beginning of curing of resin was examined rheologically with time sweeps at 80, 100, and 120 °C. These thermosets were employed for the synthesis of unidirectional composites with cellulosic fibers. It was found from viscosity measurements that the lignin content has to be restricted to 50% because of the increase in the viscosity of the resin upon increasing the lignin content.

5.2.6 Rosin-Based Resin

Deng et al. [64] made triglycidyl ester and glycidyl ether of different chain lengths from rosin and analyzed the viscosity and tabulated in Table 5.2. Atta et al. [65] synthesized epoxy from rosin condensed formaldehyde and cured with polyamides. The gel time of the curing process were analyzed by the help of a rheometer by measuring viscosity as a function of time. During the curing process, the viscosity is found to increase as the molecular weight as well as the length of the cross-links are increased. It was found that the enhancement in viscosity reduces the cure rate. In another study, Atta et al. [66] made tetra functional epoxy resins from rosin and characterized it with the help of rheological analysis. The cross-linking was characterized by plotting viscosity vs. time curve (Figure 5.13). They made rosin-acrylic acid (APA), rosin-maleic anhydride (MPA), and diabietyl ketone of resin acid (DAK). The activation energy for curing was also determined from the slope of logarithmic graph ln T_c (T_c – gel time) vs. $1/T$ (K) (T – temperature). The study showed that the increase in the temperature shows effects on the cross-link density and chain mobility.

5.3 Rheology of Bioepoxy-Based Composites

Di Landroand Janszen [67] fabricated biocomposites by incorporating natural fibers into the bioepoxy matrix. They have prepared hemp-reinforced bio-based epoxy laminates and analyzed the rheological and thermal properties of these laminates. They measured the viscosity evolution at different curing temperatures in order to analyze the processability of the bio-based epoxy resin. The laminates were prepared by following the resin transfer molding technique (RTM) after analyzing the viscosity using the rotational rheometer. The injection was carried out before the viscosity of resin increases appreciably, and hence, voids

Table 5.2 Epoxy resins produced from rosin.

Cured product	W(FPA) (%)	W(EGDE) (%)	Curing agent (phr)	N_1 [a)]	Prepolymer	N_2 [b)]	Viscosity @ 25 °C (Pa s)	Epoxy value (mol/100 g)
FPAE1	100	0	75.6	1	FPAE		>100	0.47 (0.51[c)])
FPAE2	100	0	151.2	2				
FPEG1-1	36	64	33.6	1				
FPEG1-2	36	64	67.2	2	FPEG1	10:5	>100	0.19 (0.23[c)])
FPEG1-J	36	64	84	2.5				
FPEG21	32	68	48.72	1				
FPEG2-2	32	68	97.44	2	FPEG2	12:5	43.5	0.26 (0.29[c)])
FPEG2-3	32	6S	121.8	2.5				
FPEG3-1	27	73	57.12	1				
FPEG3-2	27	73	114.24	2	FPEG3	15:5	7.8	0.32 (0.35[c)])
FPEG3-3	27	73	142.8	2.5				

Abbreviations: FPA, fumaropimaric acid; FPAE, triglycidyl ester of fumaropimaric acid.
The cure reactions were conducted at 110 °C for two hours and at 160 °C for three hours.
a) $N_1 = n(\text{MeHHPA}) : n(\text{epoxy group})$.
b) $N_2 = n(\text{epoxy group}) : n(\text{carboxy group})$.
c) Calculated value.
Source: Reproduced from Deng et al. [64]. © 2013, American Chemical Society.

or defects of the composites are avoided. Figure 5.14 shows the viscosity evolution at the time of curing process. The curves are good enough to understand the resin flow capability during the process of infusion. According to them, the laminates are to be prepared by the RTM technique and injection should be done before the increase in viscosity to synthesize defect-free composites.

Fombuena et al. [68] used calcium carbonate-based biofiller from sea shells and incorporated it into the bioepoxy matrix. They have monitored the thermal, mechanical, and rheological properties of these epoxy sea shell composites. The gel time was evaluated from the crossover point from storage and loss modulus curves. The obtained gel point of 150 seconds was good enough to avoid the particle precipitation. The cure time was also optimized from the rheological data, and it was around 20–25 minutes since the storage modulus attains a constant value as shown in Figure 5.15.

Meng et al. [69], in their study, prepared composites by compounding cellulose nanofibrils and epoxidized soybean oil with polylactic acid. The rheological analysis shows (Figure 5.16) that as the ESO content increases, the storage modulus decreases, indicating the plasticizing effect of ESO. In addition, as the cellulose nano fibrils (CNFs) content increases, it is capable of showing the dominant percolation in the ternary system. These studies confirm the plasticizing nature of bio-based epoxy.

Möller et al. [70] epoxidized linseed oil and carried out the rheological analysis of the same for the fabrication of fiber-reinforced composites by the method of resin infusion. The determination of gelation point is considered as a very important fact in figuring out the limit of the process of resin infusion. The gel time

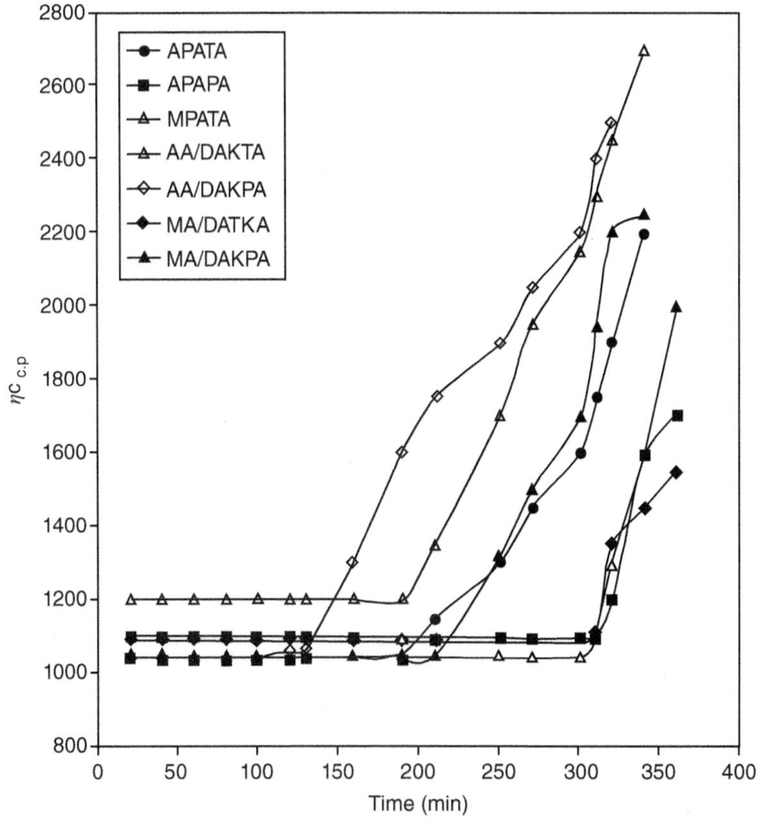

Figure 5.13 Variation of viscosity vs. time for acrylic acid (AA)/diabietyl ketone (DAK) epoxy resins at 293 K. Source: Reproduced from Atta et al. [66]. © 2005, Springer Nature.

of curing is determined using the two anhydride curatives such as Nadic methyl anhydride and methyltetrahydrophthalic anhydride (MTHPA and nadic methyl anhydride (NMA), and the curves are shown in Figure 5.17. The Arrhenius fitting shows that all the formulations are following Arrhenius behavior in the processed temperature range. In order to parameterize the infusion process, the inverse of viscosity, i.e. fluidity curves, was also analyzed in detail. The obtained results show that as the time increases, the fluidity decreases at a particular temperature.

Xin et al. [59] synthesized bioepoxy from lignin by treating it with epichlorohydrin and used it for the modification of asphalt. They studied the effect of the epoxy content on the rheological properties of modified asphalt powder. Using a parallel plate rheometer in dynamic shear condition, they have analyzed the rheological properties in detail. The bioepoxy-modified asphalt composites show better rutting resistance than the neat asphalt as shown in Figure 5.18. The addition of epoxy enhanced the maximum temperature.

Nissilä et al. [71] fabricated bio epoxy/cellulose composite for which the viscosity decreases with increasing temperature up to 80 °C. From the study, they have chosen 60 °C as the temperature for the preparation of cellulose based bioepoxy

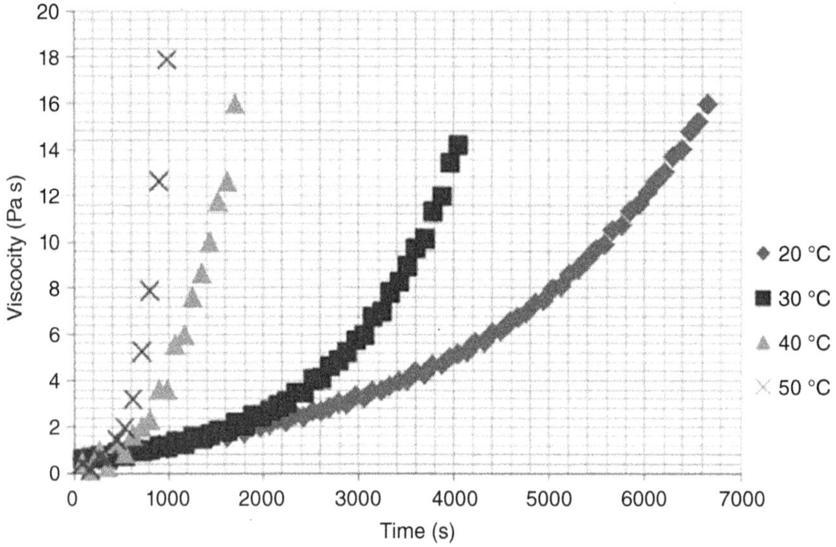

Figure 5.14 Curing curves at different temperatures. Source: Reproduced from Di Landro and Janszen [67]. © 2014, Elsevier.

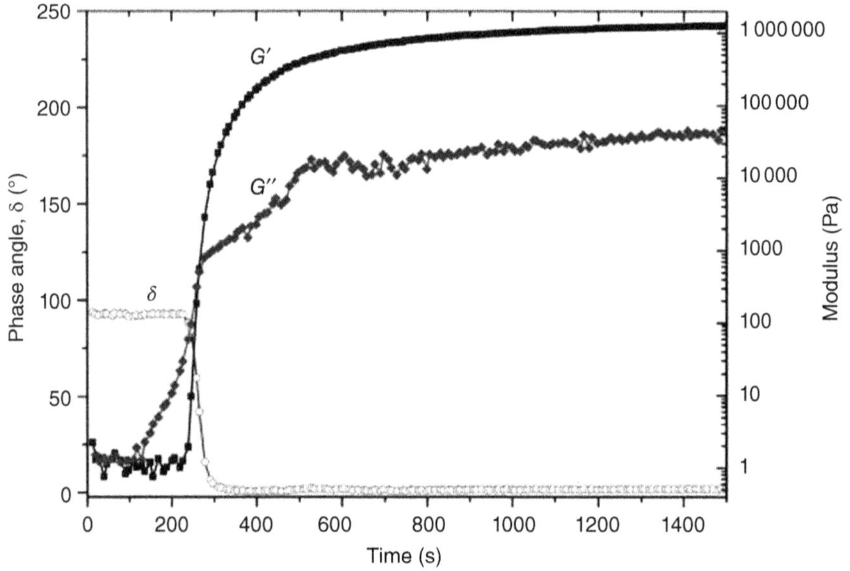

Figure 5.15 Curing profile of eco-friendly green epoxy resin at isothermal conditions (100 °C) by plate–plate rheometry. Source: Reproduced from Fombuena et al. [68]. © 2014, Elsevier.

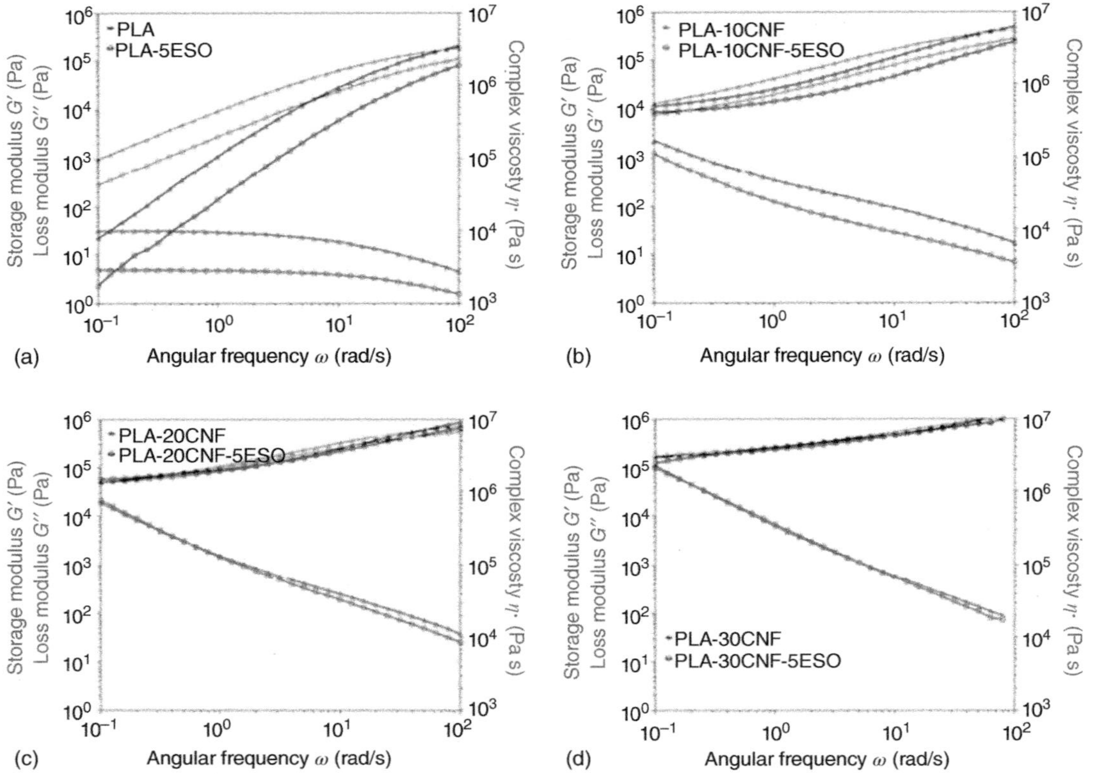

Figure 5.16 Comparison of storage moduli (G'), loss moduli (G''), and complex viscosity (η^*) of (a) PLA and PLA with 5 wt% ESO, (b) PLA-10 wt% CNF and PLA-10 wt% CNF with 5 wt% ESO, (c) PLA-20 wt% CNF and PLA-20 wt% CNF with 5 wt% ESO, and (d) PLA-30 wt% CNF and PLA-30 wt% CNF with 5 wt% ESO. Source: (a–d) Reproduced from Meng [69]. © 2018, Elsevier.

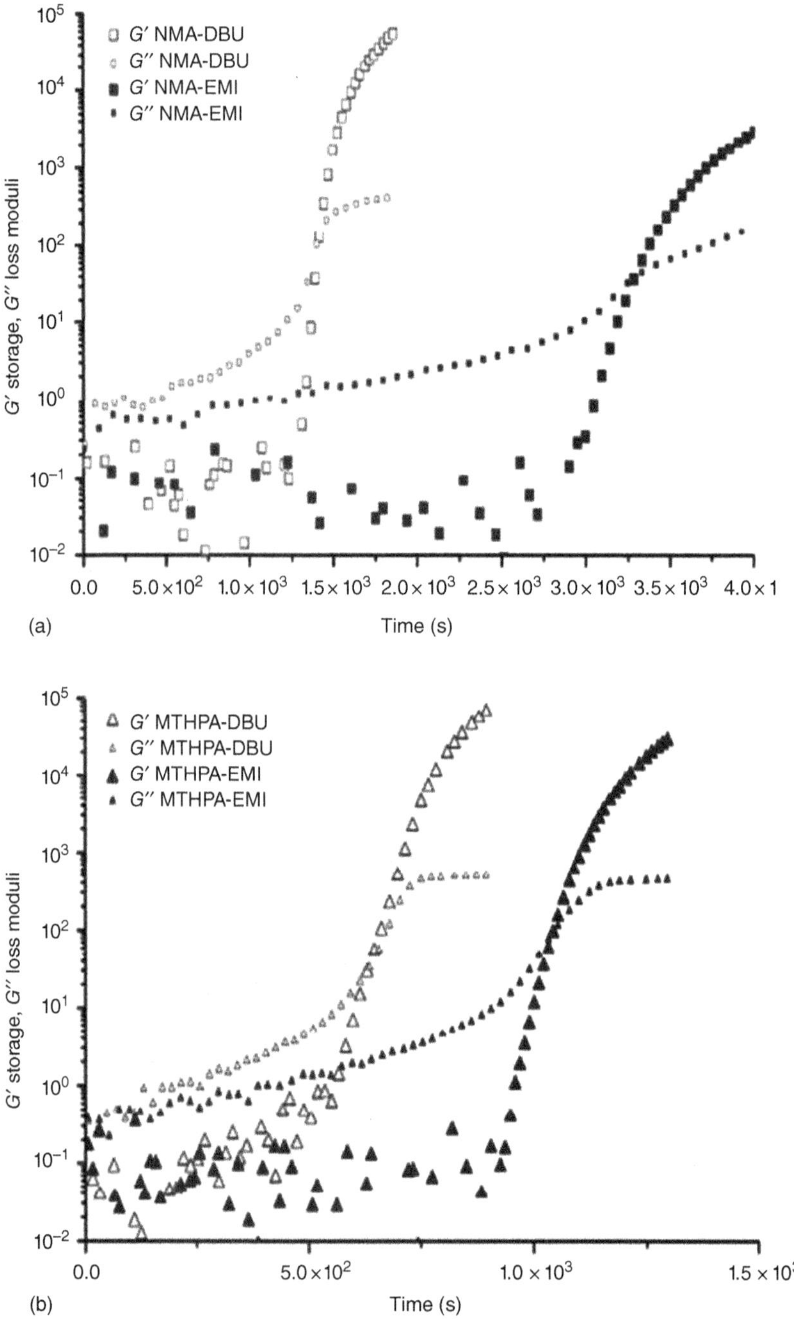

Figure 5.17 (a) Representative storage (G') and loss (G'') modulus curves for NMA-cured formulations at 120 °C Page and (b) representative storage (G') and loss (G'') modulus curves for MTHPA-cured formulations at 120 °C. Source: (a,b) Reproduced from Möller et al. [70]. © 2017, American Chemical Society.

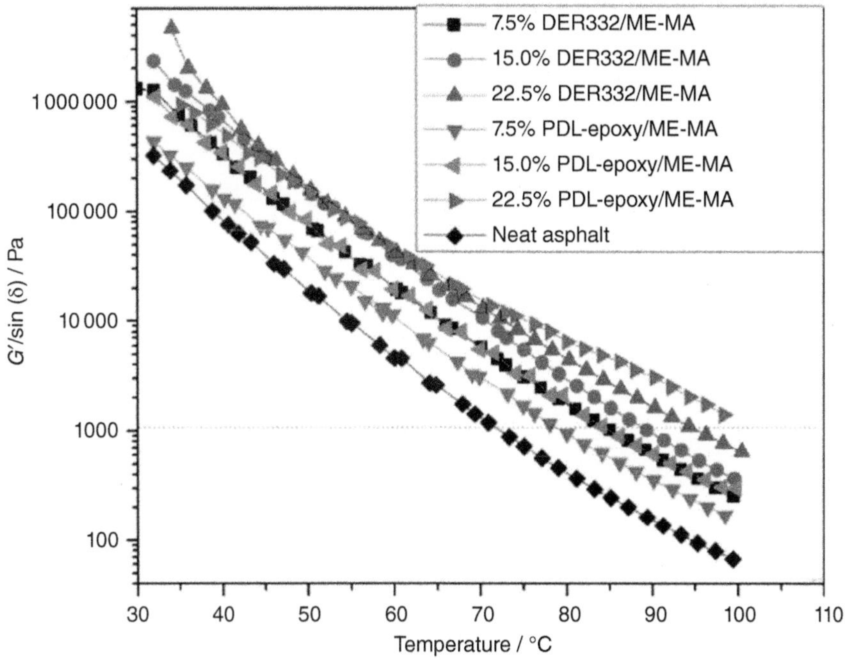

Figure 5.18 G^*/sin vs. temperature at 10 rad/s. At 60 °C, effects of frequency on the loss modulus. Source: Reproduced from Xin et al. [59]. © 2016, American Chemical Society.

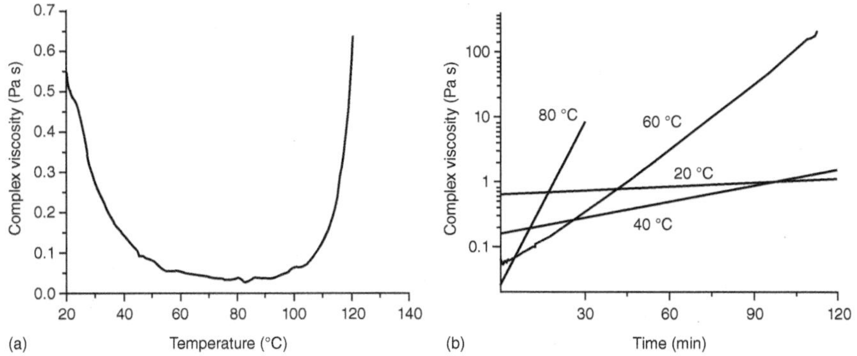

Figure 5.19 Vacuum infusion. Results of (a) a temperature ramp and (b) isothermal cure tests for the bio-epoxy resin. Source: (a,b) Reproduced from Nissilä et al. [71]. © 2019, Elsevier.

as the curing occurs before the impregnation at other temperatures. The rheological data are shown in Figure 5.19.

5.4 Rheology of Bioepoxy-Based Blends

Karger-Kocsis et al. [36] used epoxidized soybean oil in a standard epoxy resin to form composites and studied the curing, gelling, thermal, and mechanical

Table 5.3 Gel time data of the hybrid resins at various temperatures (mean data from three parallel tests) and activation energy (E_a) of gelling for the resins studied .

EP/ESO composition (wt%/wt%)	$t_{gel\,100\,°C}$ (s)	$t_{gel\,120\,°C}$ (s)	$t_{gel\,140\,°C}$ (s)	E_a (kJ/mol)
100/0	1024	290	199	52.9
75/25	1261	379	201	59.0
50/50	2050	618	240	66.4
25/75	2688	787	283	72.1
0/100	5746	1772	600	72.3

Source: Reproduced from Karger-Kocsis et al. [36]. © 2014, John Wiley & Sons.

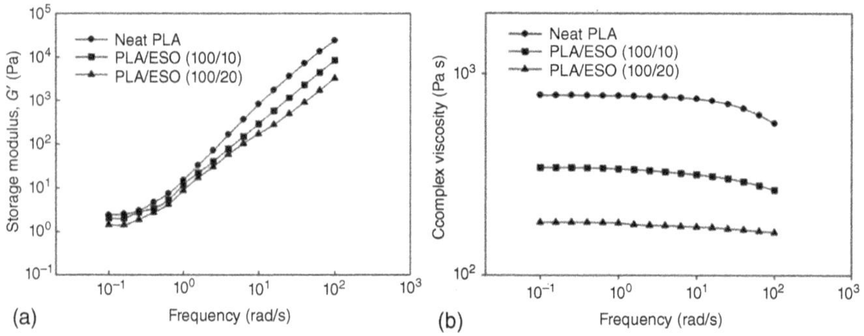

Figure 5.20 Frequency dependence of (a) dynamic storage modulus, G' and (b) complex viscosity, η^*, of neat PLA and PLA/ESO blends. Source: Reproduced from Ali [73]. © 2009, Springer Nature.

properties. They employed a dynamic stress rheometer for the chemorheological analysis of the various compositions. As the epoxidized contents in the blends decrease and the temperature increases, gelling takes place rapidly as shown in Table 5.3.

Xu et al.[72] analyzed the rheological properties of epoxidized soybean oil (ESO)-plasticized poly lactic acid (PLA). In the study, ESO act as a plasticizing agent for PLA. The rheological study reveals the relation between the ESO content and temperature on the rheology of PLA/ESO blends. The blends were showing shear thinning behavior, as with the raise in temperature, the viscosity got reduced. The viscosity of the blends were decreased with ESO content. Ali et al. [73] blended ESO with polylactic acid and carried out the rheological analysis in detail. The melt rheology shows that the storage modulus decreases with the ESO content in the blend and the linear viscoelastic limit of PLA is deformed as it is blended with ESO (Figure 5.20). The complex viscosity of the blends was less than that of the neat PLA, and it shows a non-Newtonian behavior at higher frequency range.

Sahoo et al. [74] blended ESO with petroleum-based epoxy resins and investigated the rheological study of uncured epoxy resin. The viscosity and shear rate of the uncured virgin epoxy was compared with that of the bioresins. For all

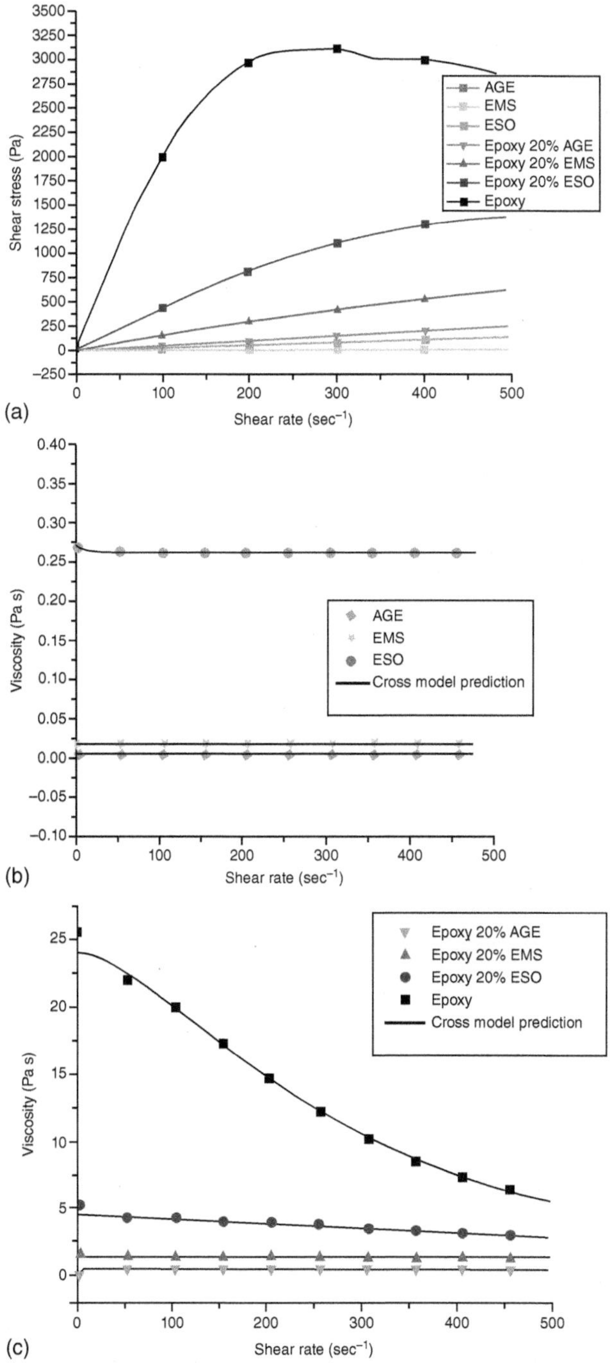

Figure 5.21 (a) Shear stress vs. shear rate plot, and (b) and (c) viscosity vs. shear rate. Source: (a–c) Reproduced with permission from Sahoo et al. [74]. © 2015, American Chemical Society.

the systems, the shear stress is increased with shear rate and the epoxies show pseudoplastic behavior, but the blend showed a transition from pseudoplastic to Newtonian as shown in Figure 5.21. Interestingly, the diluents in the study, epoxidized soybean oil (ESO), epoxy methyl soyate (EMS), and (alkyl(C12–C14) glycidyl ether (AGE), depicted Newtonian behavior at all shear rates.

5.5 Conclusions and Future Scope

Epoxy thermosets face the difficulty of getting recycled, and the industry is currently indulged in the search of feasible alternatives to reduce the carbon footprint generated during their production process. Concurrently, because of an increasing demand of green industries, a new natural feedstock has been used to develop new materials. In addition, the thermosetting polymer field is currently following the same line because of the low cost, sustainability, and lightweight that these materials can offer. Furthermore, a comparative and much detailed systematic study is inevitable to characterize the effect of various available precursors on the curing and rheological behavior, glass transition temperature, thermal and mechanical properties of various epoxy resin systems, and their composites. The lower glass transition temperature of most of the synthesized bioepoxies than the benchmark DGEBA epoxy resin limits its utility in nonstructural composite applications. Recent research has shown that there are a lot of possibilities in the development of bio-based binary and ternary composites where these bioepoxies could be employed to increase the segmental motion of the polymer chains, leading to the effective tailoring of the properties of the composites developed. A recent review by Eric Ramon et al. [51] shows a summary of the kinematic and dynamic viscosity of various bio-based systems. They have pointed out in the review that the viscosities of many bioepoxies are suitable for their utilization as precursors for the fabrication of composite components in aviation industry. The prospects of the utilization of these bioepoxies in various other industries through the replacement of traditional petroleum derivatives are yet to be deciphered. Although a lot of work is concentrated on the preparation and characterization of bioepoxy resins and their subsequent blends and composites, many works have not stressed on the rheological analysis. A systematic analysis of the rheology of various bioepoxy platforms is mandatory as it would give deep insights into deciding the processing parameters and would provide useful understandings in converting these bioepoxies into useful composite materials having high-end applications.

References

1 La Rosa, A.D., Recca, G., Summerscales, J. et al. (2014). Bio-based versus traditional polymer composites. A life cycle assessment perspective. *J. Cleaner Prod.* 74: 135–144.

2 Pawar, M., Kadam, A., Yemul, O. et al. (2016). Biodegradable bioepoxy resins based on epoxidized natural oil (cottonseed & algae) cured with citric and

tartaric acids through solution polymerization: a renewable approach. *Ind. Crops Prod.* 89: 434–447.

3 Auvergne, R., Caillol, S., David, G. et al. (2013). Biobased thermosetting epoxy: present and future. *Chem. Rev.* 114 (2): 1082–1115.

4 Maffini, M.V., Rubin, B.S., Sonnenschein, C., and Soto, A.M. (2006). Endocrine disruptors and reproductive health: the case of bisphenol-A. *Mol. Cell. Endocrinol.* 254: 179–186.

5 Maiorana, A., Spinella, S., and Gross, R.A. (2015). Bio-based alternative to the diglycidyl ether of bisphenol A with controlled materials properties. *Biomacromolecules* 16 (3): 1021–1031.

6 Stemmelen, M., Pessel, F., Lapinte, V. et al. (2011). A fully biobased epoxy resin from vegetable oils: from the synthesis of the precursors by thiol-ene reaction to the study of the final material. *J. Polym. Sci., Part A: Polym. Chem.* 49 (11): 2434–2444.

7 Patel, A., Maiorana, A., Yue, L. et al. (2016). Curing kinetics of bio-based epoxies for tailored applications. *Macromolecules* 49 (15): 5315–5324.

8 Brocas, A.-L., Llevot, A., Mantzaridis, C. et al. (2014). Epoxidized rosin acids as co-precursors for epoxy resins. *Des. Monomers Polym.* 17 (4): 301–310.

9 Sultania, M., Rai, J.S.P., and Srivastava, D. (2011). Process modeling, optimization and analysis of esterification reaction of cashew nut shell liquid (CNSL)-derived epoxy resin using response surface methodology. *J. Hazard. Mater.* 185 (2–3): 1198–1204.

10 Janvier, M., Hollande, L., Jaufurally, A.S. et al. (2017). Syringaresinol: a renewable and safer alternative to bisphenol A for epoxy-amine resins. *ChemSusChem* 10 (4): 738–746.

11 Hernandez, E.D., Bassett, A.W., Sadler, J.M. et al. (2016). Synthesis and characterization of bio-based epoxy resins derived from vanillyl alcohol. *ACS Sustainable Chem. Eng.* 4 (8): 4328–4339.

12 Llevot, A., Grau, E., Carlotti, S. et al. (2016). Selective laccase-catalyzed dimerization of phenolic compounds derived from lignin: towards original symmetrical bio-based (bis) aromatic monomers. *J. Mol. Catal. B: Enzym.* 125: 34–41.

13 Maiorana, A., Reano, A.F., Centore, R. et al. (2016). Structure property relationships of bio-based *n*-alkyl bisferulate epoxy resins. *Green Chem.* 18 (18): 4961–4973.

14 Ménard, R., Caillol, S., and Allais, F. (2017). Ferulic acid-based renewable esters and amides-containing epoxy thermosets from wheat bran and beetroot pulp: chemo-enzymatic synthesis and thermo-mechanical properties characterization. *Ind. Crops Prod.* 95: 83–95.

15 Maiorana, A., Yue, L., Manas-Zloczower, I., and Gross, R. (2016). Structure–property relationships of a bio-based reactive diluent in a bio-based epoxy resin. *J. Appl. Polym. Sci.* 133 (45): 1–8.

16 Kadam, A., Pawar, M., Yemul, O. et al. (2015). Biodegradable bio-based epoxy resin from karanja oil. *Polymer* 72: 82–92.

17 Liu, R., Zhang, X., Gao, S. et al. (2016). Bio-based epoxy-anhydride thermosets from six-armed linoleic acid-derived epoxy resin. *RSC Adv.* 6 (58): 52549–52555.

18 Garrison, M.D. and Harvey, B.G. (2016). Bio-based hydrophobic epoxy-amine networks derived from renewable terpenoids. *J. Appl. Polym. Sci.* 133 (45): 1–12.
19 Baroncini, E.A., Kumar Yadav, S., Palmese, G.R., and Stanzione, J.F. III, (2016). Recent advances in bio-based epoxy resins and bio-based epoxy curing agents. *J. Appl. Polym. Sci.* 133 (45): 1–19.
20 Yue, L., Pircheraghi, G., Monemian, S.A., and Manas-Zloczower, I. (2014). Epoxy composites with carbon nanotubes and graphene nanoplatelets–Dispersion and synergy effects. *Carbon* 78: 268–278.
21 Park, S.-J., Bae, K.-M., and Seo, M.-K. (2010). A study on rheological behavior of MWCNTs/epoxy composites. *J. Ind. Eng. Chem.* 16 (3): 337–339.
22 Apicella, A., Masi, P., and Nicolais, L. (1984). Rheological behaviour of a commercial TGDDM-DDS based epoxy matrix during the isothermal cure. *Rheol. Acta* 23 (3): 291–296.
23 O'Brien, D.J., Mather, P.T., and White, S.R. (2001). Viscoelastic properties of an epoxy resin during cure. *J. Compos. Mater.* 35 (10): 883–904.
24 Sun, L., Pang, S., Sterling, A.M. et al. (2002). Thermal analysis of curing process of epoxy prepreg. *J. Appl. Polym. Sci.* 83 (5): 1074–1083.
25 Amiri, A., Yu, A., Webster, D., and Ulven, C. (2016). Bio-based resin reinforced with flax fiber as thermorheologically complex materials. *Polymers* 8 (4): 153.
26 Abdelwahab, M.A., Misra, M., and Mohanty, A.K. (2015). Epoxidized pine oil-siloxane: crosslinking kinetic study and thermomechanical properties. *J. Appl. Polym. Sci.* 132 (37): 1–12.
27 Masoodi, R. and Pillai, K.M. (2011). Modeling the processing of natural fiber composites made using liquid composites molding. In: *Handbook of Bioplastics and Biocomposites Engineering Applications* (ed. S. Pilla), 43–74. Scrivener-Wiley.
28 Masoodi, R., Pillai, K.M., Grahl, N., and Tan, H. (2012). Numerical simulation of LCM mold-filling during the manufacture of natural fiber composites. *J. Reinf. Plast. Compos.* 31 (6): 363–378.
29 Liang, G. and Zhang, M. (2002). Enhancement of processability of cyanate ester resin via copolymerization with epoxy resin. *J. Appl. Polym. Sci.* 85 (11): 2377–2381.
30 Schuster, J., Govignon, Q., and Bickerton, S. (2014). Processability of biobased thermoset resins and flax fibres reinforcements using vacuum assisted resin transfer moulding. *Open J. Compos. Mater.* 4 (01): 1–11.
31 Laskoski, M., Dominguez, D.D., and Keller, T.M. (2005). Development of an oligomeric cyanate ester resin with enhanced processability. *J. Mater. Chem.* 15 (16): 1611–1613.
32 Omonov, T.S. and Curtis, J.M. (2014). Biobased epoxy resin from canola oil. *J. Appl. Polym. Sci.* 131 (8): –40142.
33 Maiorana, A., Subramaniam, B., Centore, R. et al. (2016). Synthesis and characterization of an adipic acid–derived epoxy resin. *J. Polym. Sci., Part A: Polym. Chem.* 54 (16): 2625–2631.

34 Zhang, J., Tang, J.J., and Zhang, J.X. (2015). Polyols prepared from ring-opening epoxidized soybean oil by a castor oil-based fatty diol. *Int. J. Polym. Sci.* 2015, 8 pages https://doi.org/10.1155/2015/529235.

35 Jin, F.-L. and Park, S.-J. (2007). Thermal and rheological properties of vegetable oil-based epoxy resins cured with thermally latent initiator. *J. Ind. Eng. Chem.* 13 (5): 808–814.

36 Karger-Kocsis, J., Grishchuk, S., Sorochynska, L., and Rong, M.Z. (2014). Curing, gelling, thermomechanical, and thermal decomposition behaviors of anhydride-cured epoxy (DGEBA)/epoxidized soybean oil compositions. *Polym. Eng. Sci.* 54 (4): 747–755.

37 Chrysanthos, M., Galy, J., and Pascault, J. (2013). Influence of the bio-based epoxy prepolymer structure on network properties. *Macromol. Mater. Eng.* 298 (11): 1209–1219.

38 Goerz, O. and Ritter, H. (2013). Polymers with shape memory effect from renewable resources: crosslinking of polyesters based on isosorbide, itaconic acid and succinic acid. *Polym. Int.* 62 (5): 709–712.

39 Chrysanthos, M., Galy, J., and Pascault, J.-P. (2011). Preparation and properties of bio-based epoxy networks derived from isosorbide diglycidyl ether. *Polymer* 52 (16): 3611–3620.

40 Sadler, J.M., Toulan, F.R., Palmese, G.R., and La Scala, J.J. (2015). Unsaturated polyester resins for thermoset applications using renewable isosorbide as a component for property improvement. *J. Appl. Polym. Sci.* 132 (30): 42315–42326.

41 Hong, J., Radojčić, D., Ionescu, M. et al. (2014). Advanced materials from corn: isosorbide-based epoxy resins. *Polym. Chem.* 5 (18): 5360–5368.

42 Nouailhas, H., Aouf, C., Le Guerneve, C. et al. (2011). Synthesis and properties of bio-based epoxy resins. Part 1. Glycidylation of flavonoids by epichlorohydrin. *J. Polym. Sci., Part A: Polym. Chem.* 49 (10): 2261–2270.

43 Benyahya, S., Aouf, C., Caillol, S. et al. (2014). Functionalized green tea tannins as phenolic prepolymers for bio-based epoxy resins. *Ind. Crops Prod.* 53: 296–307.

44 Tomita, H., and Yonezawa, K. (1985) Epoxy resin and process for preparing the same. US Patent 4 540 802.

45 Aouf, C., Lecomte, J., Villeneuve, P. et al. (2012). Chemo-enzymatic functionalization of gallic and vanillic acids: synthesis of bio-based epoxy resins prepolymers. *Green Chem.* 14 (8): 2328–2336.

46 Cao, L., Liu, X., Na, H. et al. (2013). How a bio-based epoxy monomer enhanced the properties of diglycidyl ether of bisphenol A (DGEBA)/graphene composites. *J. Mater. Chem. A* 1 (16): 5081–5088.

47 Unnikrishnan, K.P. and Thachil, E.T. (2008). Synthesis and characterization of cardanol-based epoxy systems. *Des. Monomers Polym.* 11 (6): 593–607.

48 Maffezzoli, A., Calo, E., Zurlo, S. et al. (2004). Cardanol based matrix biocomposites reinforced with natural fibres. *Compos. Sci. Technol.* 64 (6): 839–845.

49 Jaillet, F., Darroman, E., Ratsimihety, A. et al. (2014). New bio-based epoxy materials from cardanol. *Eur. J. Lipid Sci. Technol.* 116 (1): 63–73.

50 Darroman, E., Durand, N., Boutevin, B., and Caillol, S. (2015). New cardanol/sucrose epoxy blends for bio-based coatings. *Prog. Org. Coat.* 83: 47–54.

51 Ramon, E., Sguazzo, C., and Moreira, P.M.G.P. (2018). A review of recent research on bio-based epoxy systems for engineering applications and potentialities in the aviation sector. *Aerospace* 5 (4): 110.

52 Chen, J., Nie, X., Liu, Z. et al. (2015). Synthesis and application of polyepoxide cardanol glycidyl ether as bio-based polyepoxide reactive diluent for epoxy resin. *ACS Sustainable Chem. Eng.* 3 (6): 1164–1171.

53 Huo, S., Ma, H., Liu, G. et al. (2018). Synthesis and properties of organosilicon-grafted cardanol novolac epoxy resin as a novel bio-based reactive diluent and toughening agent. *ACS Omega* 3 (12): 16403–16408.

54 Al-Mansob, R.A., Ismail, A., Alduri, A.N. et al. (2014). Physical and rheological properties of epoxidized natural rubber modified bitumens. *Constr. Build. Mater.* 63: 242–248.

55 Nakason, C., Kaesaman, A., Sainamsai, W., and Kiatkamjonwong, S. (2004). Rheological behavior of reactive blending of epoxidized natural rubber with cassava starch and epoxidized natural rubber with natural rubber and cassava starch. *J. Appl. Polym. Sci.* 91 (3): 1752–1762.

56 Nakason, C., Panklieng, Y., and Kaesaman, A. (2004). Rheological and thermal properties of thermoplastic natural rubbers based on poly (methyl methacrylate)/epoxidized-natural-rubber blends. *J. Appl. Polym. Sci.* 92 (6): 3561–3572.

57 da Silva, V.M., Nunes, R.C.R., and de Sousa, A.M.F. (2017). Epoxidized natural rubber and hydrotalcite compounds: rheological and thermal characterization. *Polímeros* 27 (3): 208–212.

58 Yunus, N.A., Mazlan, S.A., Aziz, A. et al. (2019). Thermal stability and rheological properties of epoxidized natural rubber-based magnetorheological elastomer. *Int. J. Mol. Sci.* 20 (3): 746.

59 Xin, J., Li, M., Li, R. et al. (2016). Green epoxy resin system based on lignin and tung oil and its application in epoxy asphalt. *ACS Sustainable Chem. Eng.* 4 (5): 2754–2761.

60 Cortés-Triviño, E., Valencia, C., and Franco, J.M. (2017). Influence of epoxidation conditions on the rheological properties of gel-like dispersions of epoxidized kraft lignin in castor oil. *Holzforschung* 71 (10): 777–784.

61 Cortés-Triviño, E., Valencia, C., Delgado, M., and Franco, J. (2018). Modification of alkali lignin with poly(rthylene glycol) diglycidyl ether to be used as a thickener in bio-lubricant formulations. *Polymers* 10 (6): 670.

62 Petreus, O., Cazacu, G., Necula, A.M., and Ciolacu, D. (2008). Synthesis and characterization of phosphorus-containing lignin-epoxy resins. *Cellul. Chem. Technol.* 42 (9): 569.

63 Engelmann, G. and Ganster, J. (2014). Bio-based epoxy resins with low molecular weight kraft lignin and pyrogallol. *Holzforschung* 68 (4): 435–446.

64 Deng, L., Ha, C., Sun, C. et al. (2013). Properties of bio-based epoxy resins from rosin with different flexible chains. *Ind. Eng. Chem. Res.* 52 (37): 13233–13240.

65 Atta, A.M., Mansour, R., Abdou, M.I., and Sayed, A.M. (2004). Epoxy resins from rosin acids: synthesis and characterization. *Polym. Adv. Technol.* 15 (9): 514–522.

66 Atta, A.M., Mansour, R., Abdou, M.I., and El-Sayed, A.M. (2005). Synthesis and characterization of tetra-functional epoxy resins from rosin. *J. Polym. Res.* 12 (2): 127–138.

67 Di Landro, L. and Janszen, G. (2014). Composites with hemp reinforcement and bio-based epoxy matrix. *Compos. Part B Eng.* 67: 220–226.

68 Fombuena, V., Bernardi, L., Fenollar, O. et al. (2014). Characterization of green composites from bio-based epoxy matrices and bio-fillers derived from seashell wastes. *Mater. Des.* 57: 168–174.

69 Meng, X., Bocharova, V., Tekinalp, H. et al. (2018). Toughening of nanocellulose/PLA composites via bio-epoxy interaction: mechanistic study. *Mater. Des.* 139: 188–197.

70 Möller, J., Kuncho, C.N., Schmidt, D.F., and Reynaud, E. (2017). Rheological studies of high-performance bioepoxies for use in fiber-reinforced composite resin infusion. *Ind. Eng. Chem. Res.* 56 (10): 2673–2679.

71 Nissilä, T., Hietala, M., and Oksman, K. (2019). A method for preparing epoxy-cellulose nanofiber composites with an oriented structure. *Compos. Part A Appl. Sci. Manuf.* 125: 105515.

72 Xu, Y. and Qu, J. (2009). Mechanical and rheological properties of epoxidized soybean oil plasticized poly (lactic acid). *J. Appl. Polym. Sci.* 112 (6): 3185–3191.

73 Ali, F., Chang, Y.-W., Kang, S.C., and Yoon, J.Y. (2009). Thermal, mechanical and rheological properties of poly (lactic acid)/epoxidized soybean oil blends. *Polym. Bull.* 62 (1): 91–98.

74 Sahoo, S.K., Mohanty, S., and Nayak, S.K. (2015). Toughened bio-based epoxy blend network modified with transesterified epoxidized soybean oil: synthesis and characterization. *RSC Adv.* 5 (18): 13674–13691.

6

Dynamical Mechanical Thermal Analysis of Bioepoxy Polymers, Their Blends, and Composites

Angel Romo-Uribe

Research & Development, Advanced Science & Technology Division, Johnson & Johnson Vision Care Inc., 7500 Centurion Pkwy N., Jacksonville, FL 32256, USA

6.1 Focus

The ever-increasing pressure for reducing plastic waste is driving the development of biobased approaches where the plastic would be at least partially biodegradable. Epoxy polymers used as coatings, adhesives, or molded parts have rather broad applications because of the outstanding physical properties of these thermosets. However, traditional thermoset formulations are not degradable and in general not even recyclable. Hence, the justification for the development of bioepoxies, their blends, and composites [1–4]. The hope is to match the performance of traditional, synthetic thermosets, some of engineering grade suitable for electronics, and aerospace applications [5]. For this, bioepoxies are being blended with other polymers or reinforced with micro- and nano-sized particles and fibers of natural origin. However, phase separation and poor interfacial adhesion between the bioepoxy matrix and the fillers will be detrimental to the physical properties. Dynamic mechanical analysis (DMA) is an ideal analytical tool to investigate the viscoelastic response of materials and to characterize the extent of phase separation in polymeric blends [6]. Furthermore, modern DMA instrumentation now enables to determine the loss/improvement of mechanical properties under various conditions, namely, temperature, time, humidity, and so on. DMA is also playing a critical role in the recent development of smart polymer formulations, such as shape memory polymers, rubbers, and epoxies [7, 8]. This chapter will focus on the usefulness of DMA to characterize polymeric materials, providing examples on diverse polymeric systems. This approach is phenomenological and avoids unnecessary mathematical treatments that usually obscure key aspects of this technique to elucidate structure–property correlations. Finally, an up-to-date overview of the applications of DMA to bioepoxy systems is provided.

Bio-Based Epoxy Polymers, Blends, and Composites: Synthesis, Properties, Characterization, and Applications, First Edition. Edited by Jyotishkumar Parameswaranpillai, Sanjay Mavinkere Rangappa, Suchart Siengchin, and Seno Jose.

6.2 Bioepoxies and Reinforcers

The increasing stringent environmental regulations adopted across the world is fueling the research and development of natural alternatives to epoxies. This is reflected in the number of publications in the open literature. Bioepoxies have moderate mechanical and thermal properties; hence, they are being reinforced with diverse micrometer- and nanometer-sized fillers. Figure 6.1 provides an overview of the type of fillers utilized to reinforce the bioepoxies. The plot was produced with publications referenced at the end of this chapter.

A quick search of the open literature shows that natural fibers predominate overwhelmingly as the choice to reinforce bioepoxies. This is in order to maintain the biodegradability of the formulations and take advantage of the exceptional mechanical properties (and availability and low cost) of many natural fibers, as shown in Figure 6.2.

Among natural fibers, the derivatives of cellulose have been more utilized to produce bioepoxy composites, as shown in Figure 6.3.

In order to obtain a bioepoxy composite with decent thermomechanical properties, it is important to understand, at a molecular and bulk scale, the effect of adding a second component, either another polymer or a filler. The next step would be to optimize the physical properties by tuning the formulation. The applications of DMA to various polymeric systems, including composites and nanocomposites will be discussed in Section 6.3.

6.3 Dynamic Mechanical Analysis and Polymer Dynamics

The viscoelastic response of a polymeric material obtained for instance from a dynamic temperature ramp reflects molecular motions that translate into changes of the modulus; these are called molecular relaxations [9].

Figure 6.4 shows the viscoelastic response of a semicrystalline polymer over a broad temperature range. The dynamic tensile moduli, E', is over 30 GPa at $-100\,°C$ and slowly decayed as temperature was increased. However, note that at room temperature, $E' > 20$ GPa; this modulus is typical of a high-performance polymer.

The molecular relaxations activated by the changes in temperature are seen as step changes in E' or as peaks in the viscous modulus E''. These relaxations are traditionally termed α, β, and γ starting from the high-temperature range, as shown in Figure 6.4 [9, 10].

There is a wealth of information, at the molecular level, that can be extracted from the mechanical relaxations [9, 10]. That is:

a) The lower temperature relaxation γ is associated with valence bond vibrations. Note that valence bond vibrations are predominantly elastic, and therefore, there is negligible energy dissipation at the γ relaxation. Consequently, $E' \gg E''$, as exemplified in Figure 6.4.

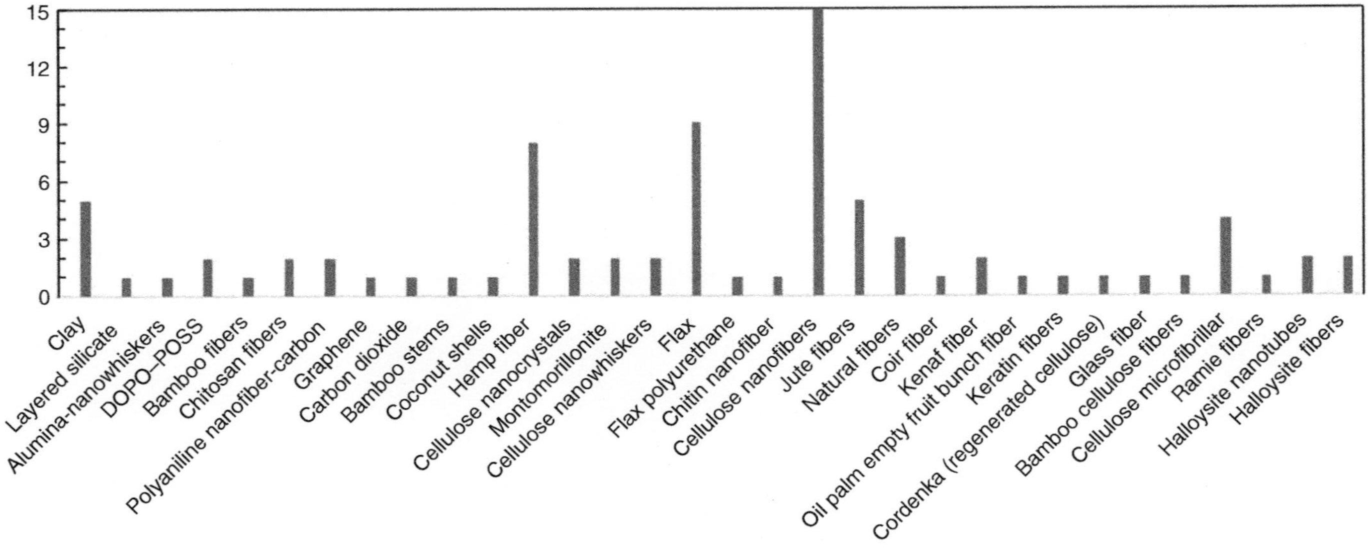

Figure 6.1 Type of fillers utilized to reinforce bioepoxies.

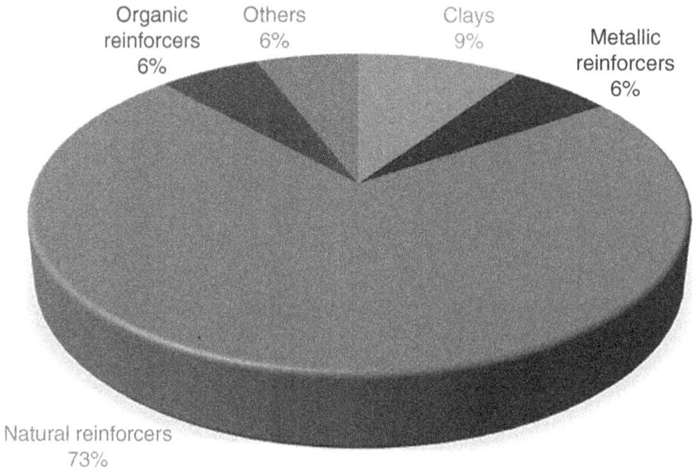

Figure 6.2 Distribution of materials utilized to reinforce bioepoxies.

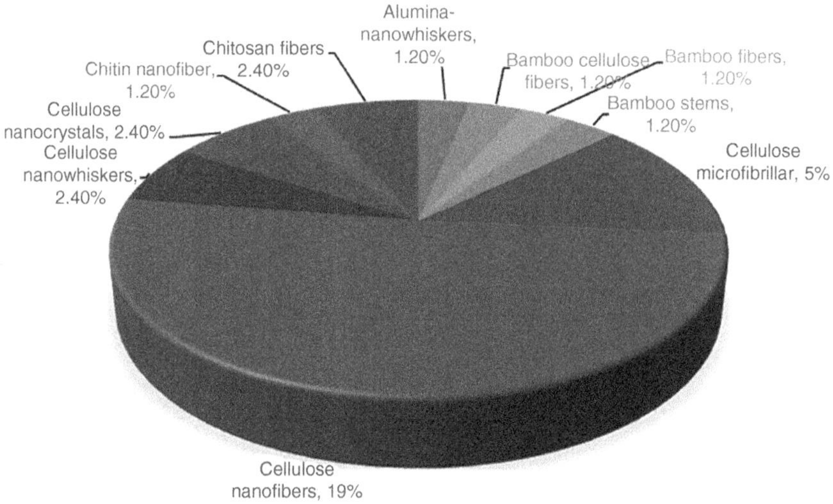

Figure 6.3 Distribution of natural fibers utilized to reinforce bioepoxies.

b) On increasing temperature, there is thermal dilation, increase of volume, and the energy barrier for molecular motions is lowered; hence, side group or crankshaft motions/rotations can be activated giving rise to the β relaxation.
c) The α relaxation is associated with segmental, cooperative motions typical of the glass transition where there is high dissipation of energy and the elastic modulus E' decays rapidly.

Back to the results shown in Figure 6.4, it can be seen that by increasing the temperature above the glass transition temperature (c. 100 °C), the elastic modulus rapidly decayed from c. 20 GPa to a plateau where $E' \sim 6$ GPa. The E' plateau after

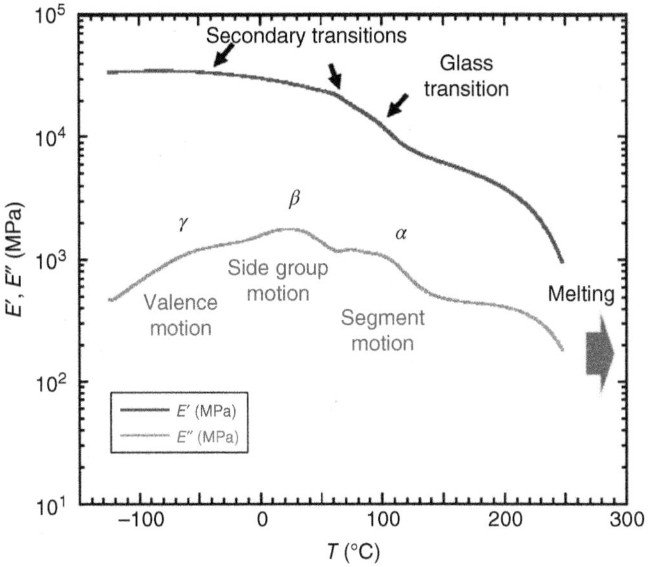

Figure 6.4 Dynamic temperature ramp of a semi-crystalline, high performance polymer. Dual cantilever mode and heating at 2°C/min, strain 0.1% and 1 Hz.

the glass transition is due to the crystalline fraction of the polymer, which acts as nanoreinforcers and prevents flow. Finally, increasing the temperature above 250 °C E' decayed again because of the melting transition (i.e. nanocrystalline domains gradually melt as temperature increases). Once the crystalline domains melt (in this case, about 300 °C), the polymer will be in the molten state and can be thermoformed.

Glassy polymers consist only of the amorphous phase; therefore, by heating above the glass transition temperature, the materials readily flow, as shown in Figure 6.5 for polyvinyl chloride (PVC). This is not an engineering polymer, and therefore, the tensile elastic modulus at −70 °C is only c. 5 GPa, nearly an order of magnitude smaller than the high-performance polymer of Figure 6.4.

The lack of nanoreinforcers in neat glassy polymers limits their useful temperature range to just about the glass transition temperature T_g, and in engineering, it is termed heat deflection temperature (HDT). However, utilizing fillers can mitigate the loss of mechanical properties, thus extending the useful temperature range of the polymer. This is exemplified in PVC reinforced with c. 4 wt% nanoclay bentonite, as shown in Figure 6.6. Increasing the temperature above the glass transition temperature shows that E' rapidly decayed but then held at c. 10 MPa up to 120 °C.

DMA is also very useful to understand the behavior of polymer blends, especially the response of the elastic modulus as this is critical when defining applications and upper working temperatures. Figure 6.7 shows the dynamic temperature ramp of a cross-linked copolymer of poly-ε-caprolactam (PCL) and polyhedral oligomeric silsesquioxane (POSS). The viscoelastic response shows

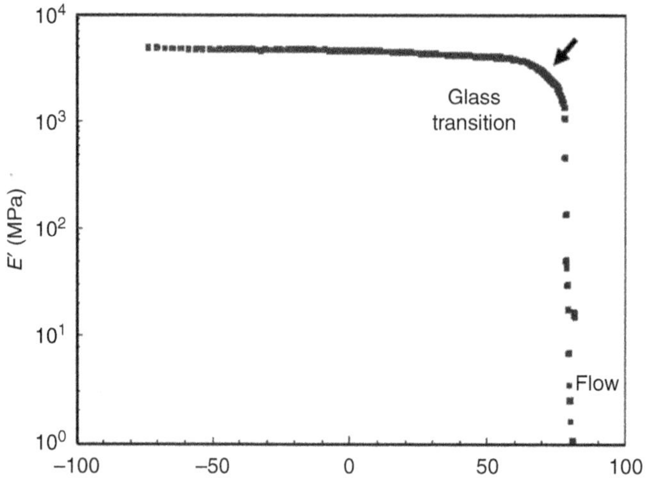

Figure 6.5 Dynamic temperature ramp of neat PVC, an amorphous polymer. Dualcantilever mode and heating at 2 °C/min and 1 Hz.

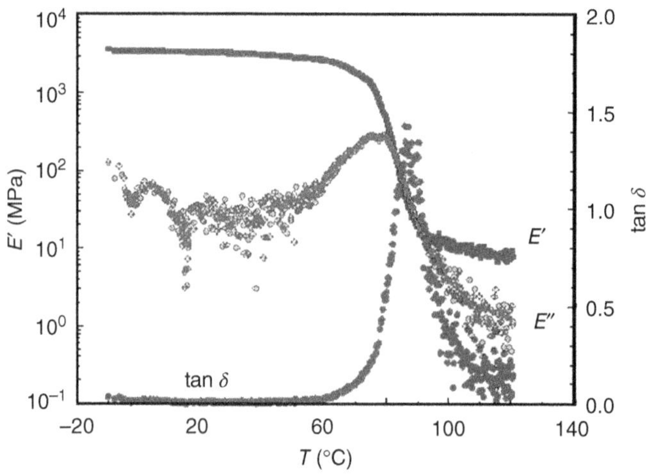

Figure 6.6 Dynamic temperature ramp of PVC reinforced with 4 wt% bentonite. Dualcantilever mode and heating at 2 °C/min and 1 Hz.

the glass transition temperature of PCL at c. −60 °C characterized by a decay in E' and a maximum in E'' (and the mechanical damping $\tan\delta = E''/E'$). PCL is semicrystalline; therefore, by increasing the temperature, there is a melting transition of PCL at c. 40 °C, where E' further decreases. The copolymer is cross-linked; hence, by further increasing the temperature, the material did not flow but exhibited a constant elastic modulus E', which is the rubber-like regime. POSS, which has dimensions of c. 1.5 nm, acts as a nanoreinforcer for the copolymer, thus increasing the rubbery modulus. The broad temperature interval between glass and melting transitions plus cross-linking enables this copolymer to display shape memory behavior [7, 11].

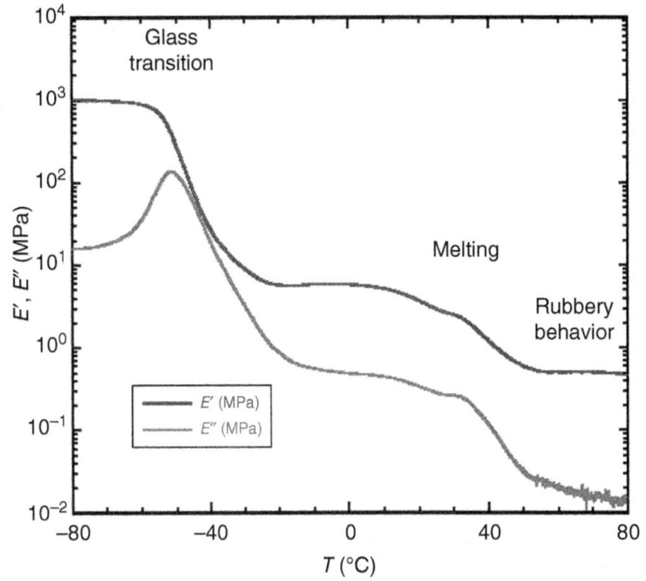

Figure 6.7 Dynamic temperature ramp of crosslinked PCL-POSS. Film tensile mode, heating at 2°C/min and 1 Hz. For more detail on PCL-POSS SMPs networks see Alvarado-Tenorio et al. [11].

DMA also enables understanding of the influence of fillers on the thermomechanical response of polymers. Figure 6.8 shows the influence of short glass fibers (SGFs) on the mechanical damping tan δ of a thermotropic polymer. Note that the intensity of the mechanical relaxations is greatly diminished after addition of 30 wt% SGF, as measured by tan δ. That is, the SGFs restrict molecular motions by confinement, thus reducing energy dissipation, and therefore, the mechanical relaxations are attenuated. Consequently, the elastic modulus E' increased, as shown in Figure 6.8c.

The behavior of polymeric materials as a function of frequency is an ideal tool to understand the dissipation of mechanical waves. This approach is ideal when targeting coatings for sound deadening applications. Figure 6.9 shows dynamic moduli as a function of frequency of an acrylic coating, testing over a range of temperatures from 10 to 90 °C. Note that the frequency range is quite restricted from 0.1 to 100 Hz.

However, using the TTS principle, the frequency range is greatly extended, fixing a reference temperature T_{ref} and shifting along the frequency axis all the moduli data.

The TTS principle developed by Williams, Landel, and Ferry (WLF) [13] enables to construct a master curve by shifting the dynamic moduli to a reference temperature using the empirical shifting factors a_T. The shifting factors are related to the coefficients c_1^{ref}, c_2^{ref} by the WLF equation:

$$\log\ a_T = \frac{-c_1^{\text{ref}}(T - T_{\text{ref}})}{c_2^{\text{ref}} + (T - T_{\text{ref}})} \tag{6.1}$$

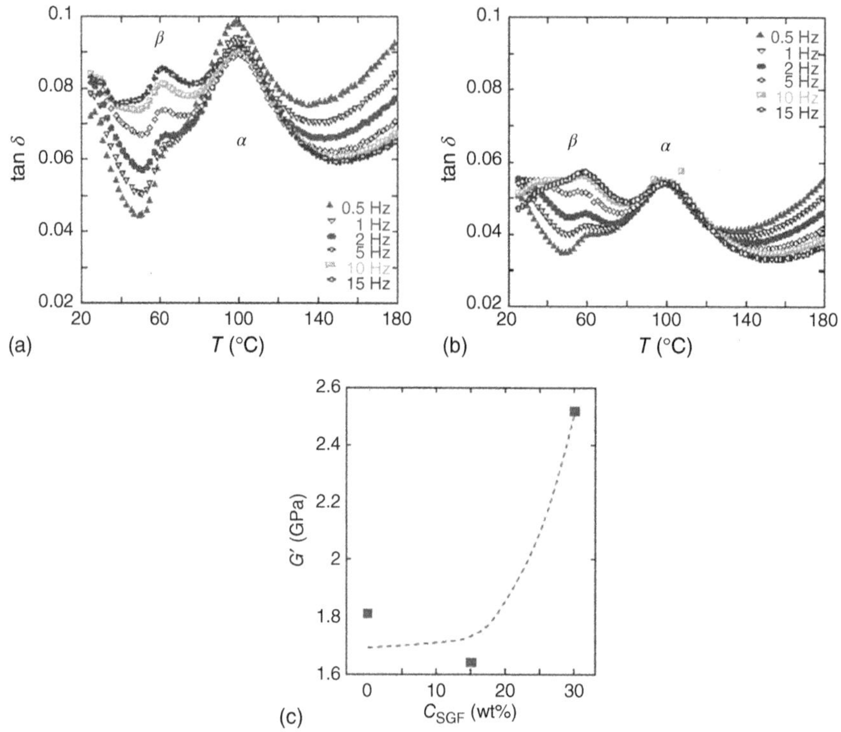

Figure 6.8 Dynamic temperature ramp of thermotropic liquid crystalline polymer (LCP): (a) neat and (b) reinforced with 30 wt% short glass fiber (SGF). Shear moduli as a function of temperature and frequency of oscillation. Torsion mode, heating at 2 °C/min and 1 Hz. Source: (a,b) Adapted from Romo-Uribe et al. [12].

where c_1^{ref} and c_2^{ref} are determined at the reference temperature T_{ref}. These parameters can be utilized to extract α_f, which is the temperature coefficient of fractional free volume, B is a constant found by Doolittle to be of the order of unity [13], and f_{ref} is the fractional free volume at the reference temperature T_{ref}.

$$c_1^{\text{ref}} = B/2.303 \cdot f_{\text{ref}} \tag{6.2}$$

$$c_2^{\text{ref}} = f_{\text{ref}}/a_f \tag{6.3}$$

Then, the WLF parameters at the glass transition temperature T_g were calculated using the following relationships [13, 14]:

$$c_1^g = \frac{c_1^{ref} c_2^{ref}}{c_2^{ref} + T_g - T_{ref}} \tag{6.4}$$

$$c_2^g = c_2^{\text{ref}} + T_g - T_{\text{ref}} \tag{6.5}$$

and the fractional free volume at the glass transition f_g can be determined from c_1^g by the following equation [13, 14]:

$$f_g/B = \frac{1}{\ln(10) \cdot c_1^g} \tag{6.6}$$

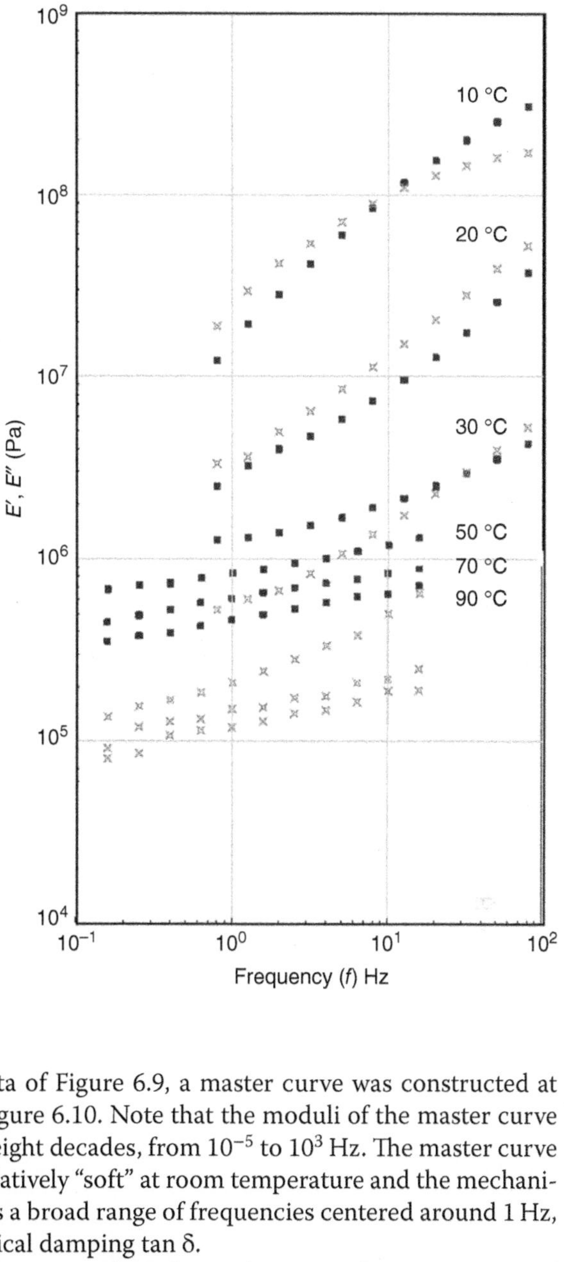

Figure 6.9 Dynamic frequency sweeps of an acrylic coating holding isothermally at the temperatures indicated in the plot. Film tensile mode.

Utilizing TTS and the data of Figure 6.9, a master curve was constructed at $T_{ref} = 20\,°C$, as shown in Figure 6.10. Note that the moduli of the master curve now extend over a range of eight decades, from 10^{-5} to 10^3 Hz. The master curve shows that this coating is relatively "soft" at room temperature and the mechanical energy dissipation covers a broad range of frequencies centered around 1 Hz, as measured by the mechanical damping tan δ.

The empirical shifting factors are plotted as a function of temperature and shown in Figure 6.11. The continuous line is the best fit to the WLF equation (Eq. (6.1)). It can be seen that the fit is quite good. The activation energy E_a of this coating can be extracted from the shifting factors using the Arrhenius equation. The results showed that $E_a = 153\,kJ/mol$.

The viscoelastic response as a function of frequency of a different coating is shown in Figure 6.12; this is the master curve constructed using $T_{ref} = 20\,°C$.

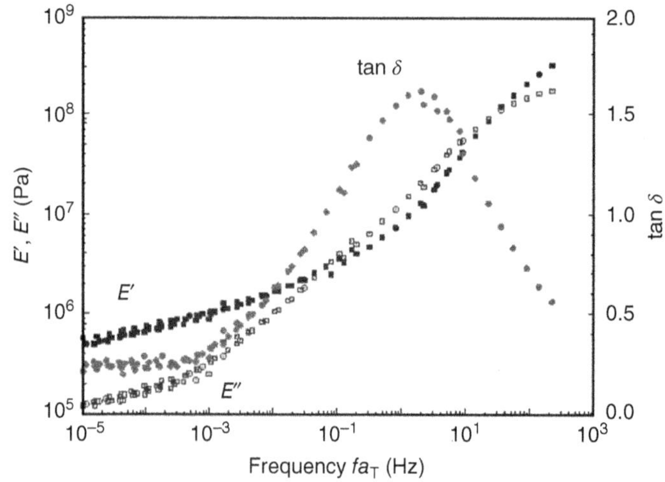

Figure 6.10 Master curve of an acrylic coating using TTS and the data of Figure 6.9, $T_{ref} = 20\,°C$.

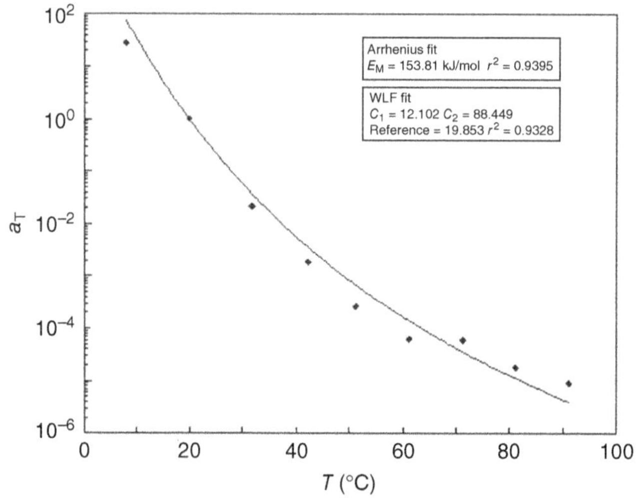

Figure 6.11 WLF shift factors a_T as a function of temperature T of an acrylic coating. The factors a_T were determined empirically by shifting the moduli data of Figure 6.9. The continuous line is the best fit to the WLF equation. $T_{ref} = 20\,°C$.

These results show that the coating is relatively "hard" at room temperature with elastic modulus varying from 0.1 to 1 GPa. Note that the moduli did not cross and therefore the intensity of the mechanical relaxation is greatly attenuated. Hence, the mechanical damping tan δ is small. Hence, this coating would have poor performance when damping mechanical waves. The activation energy E_a of this coating was determined using the Arrhenius equation, and the results showed that $E_a = 205$ kJ/mol. The greater activation energy of this coating illustrates the

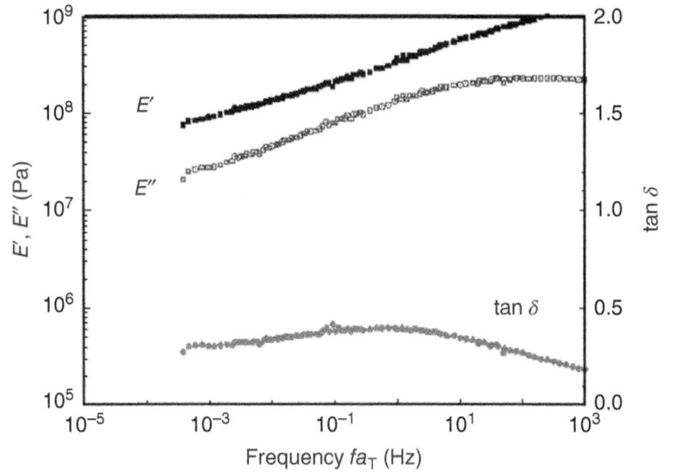

Figure 6.12 Master curve of a coating using TTS, $T_{ref} = 20\,°C$.

larger energy barrier that needs to be overcome to activate the mechanical relaxation of the material.

6.4 Applications

Bioepoxies have been reinforced with a variety of natural fibers comprising agave [15], flax [16–18], lignin [19, 20], keratin [21], jute [22], hemp [23, 24], tenax [25], coconut fiber [26–28], and sisal fiber [29]. Dynamic temperature ramps were utilized to determine the thermomechanical response of the neat epoxy and the influence of fiber content. The results reported by those research groups showed moderate increase of elastic modulus as fiber loading was increased, the greatest increase at room temperature. The fiber content was significant and did not influence the glass transition temperature. However, it is noted that keratin fibers loaded at 30 wt% significantly increased the elastic modulus and greatly reduced the mechanical damping [21].

Bioepoxy loaded with carbon fibers exhibited an increase of dynamic elastic modulus at room temperature when adding phosphorous intended to induce flame retardancy [30]. However, as temperature increased, the elastic modulus rapidly decayed, exhibiting a detrimental influence of the phosphorous.

Another approach as viable building blocks for renewable epoxies has been the addition of furanyl-based and phenyl-based diepoxy monomers to thermosets [31]. The addition of these compounds to the well-known bisphenol A diglycidyl ether (DGEBA) showed reduction of glass transition temperature T_g and a slight increase of elastic modulus at room temperature when using the furanyl-based monomer. The DMA results suggest some plasticization effect as temperature increases, but more viscoelastic characterization is needed to better understand the role of these monomers.

Another approach to reinforce bioepoxies has been the synthesis of hybrid reinforcers by forming SiO_2 particles on the surface of cellulose, named hybridized cellulose (mCNC@SiO_2). DMA temperature ramps showed increase of elastic modulus and increase of the onset to the glass transition temperature. The authors reported that mCNC@SiO_2 also reduced the internal stress, improved UV resistance, and also increased moisture resistance while maintaining high heat resistance of the composite. Hence, it was proposed that mCNC@SiO_2 is a green and multifunctional modifier of epoxy resin for high-performance LED encapsulation [32].

Bioepoxies based on vegetable oil (VO) have also been investigated [33–35]. However, it is challenging to produce a fully green biobased epoxy composite as they have relatively low strength, and this limits high-performance applications. Waste vegetable oils have been utilized as postuse cooking oil would be a promising source of triglycerides for manufacturing of epoxy composites. The results showed reduction of glass transition temperature and mechanical modulus relative to neat glass fiber and flex composites. That is, vegetable oil acted as a plasticizer in the composites [33]. More research is needed, and Mustapha et al. [35] has reviewed the recent research advances on VO-based epoxy resins and their composites with biobased hardeners incorporated with fibers or fillers.

Biobased fillers with micrometer- and nanometer-sized dimensions are attractive alternatives to reinforce biobased epoxy resins [36–42]. Nair et al. [36] showed that cellulose nanofibrils (CNFs) significantly increased the modulus, strength, and strain of the resulting composites when adding 18–23 wt% CNF. The addition of fibrils led to an overall increase in toughness, about 184 times for the composites compared to the neat epoxy. Moreover, the addition of CNFs did not affect the thermal stability of the epoxy, and the nanofibrils had a reinforcing effect in both glassy and glass transition region of the composites. DMA showed a significant decrease in intensity of the mechanical damping tan δ with the addition of CNFs, thus indicating a high interaction between fibrils and epoxy during the phase transition. On the other hand, Wu et al. [37] reported increase of elastic modulus above the glass transition temperature and decrease of glass transition temperature when adding cellulose nanowhiskers (CNWs) to a waterborne biobased epoxy (WTME) resin. The mechanical reinforcing effect and the attenuation of mechanical damping was attributed to the formation and increase of interfacial interaction by hydrogen bonds between the CNWs and the WTME matrix. Similar enhancement of mechanical properties was reported by Shibata and Nakai [38] when adding microfibrillated cellulose (MFC) to the biobased epoxy resin systems glycerol polyglycidyl ether (GPE) and sorbitol polyglycidyl ether (SPE) cured with tannic acid (TA).

Yue et al. [39] reported significant increase of mechanical modulus below and above the glass transition temperature of diglycidyl ether diphenolate ethyl ester (DGEDP-ethyl) when formulating with bacterial cellulose (BC) fiber mats. The BC mats were prepared by static cultivation of strain *Gluconacetobacter xylinus* ATCC 700178 in Hestrin–Schramm medium augmented with mannitol in sterile containers and then modifying the fibers with trimethylsilyl moieties (BCTMS). The composites were prepared by impregnation of BCTMS matrixes with the resin, followed by hot pressing and curing, and filler loading ranged from 5 to 30 vol%. The results suggest the use of prefabricated BC matrixes produced by microbial fermentation along with a biobased epoxy resin to produce high-performance biobased composites. Additionally, amine-functionalized cellulose nanocrystals provided excellent dispersion in epoxy resin systems and increased the mechanical tensile modulus and the glass transition temperature of a biobased epoxy resin derived from diphenolic acid. The results suggest that amine-treated cellulose nanocrystals are a viable route to the utilization of both biobased nanofillers and biobased epoxy resins [40].

Zhang et al. [41] reported that microcrystalline cellulose (MCC) grafted with *n*-octadecyl isocyanate (18C-g-MCC) thermomechanically reinforced the epoxidized soybean oil (ESO) polymer. That is, the glass transition temperature increased over 25 °C and the tensile elastic modulus E' increased over 400% above the glass transition temperature, thus resulting in a high-performance composite. Le Hoang et al. [42] reported enhancement of mechanical properties in BC/epoxy composites. The BC was prepared from Vietnamese nata-de-coco via an alkaline pretreatment, followed by a solvent exchange process, and the BC was modified with a silane coupling agent to induce the chemical affinity between BC and the epoxy resin. The authors report that the silane coupling agent had a vital role in improving the mechanical characteristics of the biobased composites.

Epoxy filled with magnetite (Fe_3O_4) clay (B), named (B-DPA-PANI@Fe_3O_4), was prepared at filler loading from 0.1 to 5 wt%, and the clay was surface modified with polyaniline (PANI). The epoxy nanocomposites exhibited significant increase of dynamic modulus E' and glass transition temperature when adding only 0.1 wt% of nanoreinforcer, and only 3 wt% of the filler increased the tensile strength of the composites by 256% and the glass transition temperature T_g by about 50 °C [43]. These results highlight a true "nano"-effect as loading of <1 wt% nanofiller already increased the tensile modulus E' by a factor of 2 and the T_g by c. 20 °C. Romo-Uribe has reported a similar "nanoreinforcement" effect but using only bear bentonite nanoclay to reinforce an acrylic coating [44]. The authors also reported significant improvement in corrosion inhibition; hence, the results suggest new routes for the preparation of efficient anticorrosion sensor coatings [43]. Another alternative to biobased thermosetting epoxy nanocomposite with efficient anticorrosive properties was based on PANI nanofiber–carbon dot nanohybrid (loading ranged from 0.50 to 1 wt%) and the epoxy of bisphenol A, sorbitol, and monoglyceride of castor oil (mole ratio of 16 : 3 : 1) [45].

The addition of nanoclays to bioepoxies has produced mixed results; in some cases, the elastic modulus E' and glass transition temperature T_g have exhibited some increase relative to the neat epoxy [46, 47]. In other cases, there has been reduction of elastic modulus and T_g, suggesting a plasticization effect, and the state of nanoclay dispersion did not play a role (i.e. intercalation vs. exfoliation) [48–50]. However, adding silanized alumina nanowhiskers to organo-montmorillonite (MMT) nanoclay increased the elastic modulus, the glass transition temperature T_g, and the fracture toughness [51]. On the other hand, nanoclay was incorporated in toughened epoxy modified with ESO [52]. DMA revealed an increase in rubber-like modulus and hence the cross-link density was reduced. The mechanical damping was reduced, and the glass transition temperature was increased upon addition of 5 wt% nanoclay [52].

Polyglycerol polyglycidyl ether (PGPE) cured with poly(l-lysine) (PL) and reinforced with MMT exhibited significant increase of elastic modulus E' and glass transition temperature T_g [53]. The epoxy–clay nanocomposites derived from renewable soybean oils and organo-modified MMT clay exhibited improved efficiency and performance through the application of low-viscosity glycidyl esters of epoxidized fatty acids (EGS) as the epoxy monomer and 4-methyl-1,2-cyclohexanedicarboxylic anhydride as the comonomer. Tensile testing showed that only 1 wt% of nanoclay improved the strength by 22% and the modulus by 13%, relative to the neat epoxy [54].

The eugenol-based epoxy monomer (TEU-EP) with a branched topology and a very rich biobased retention (80 wt%) was cured by 3,3′-diaminodiphenyl sulfone (33DDS), and the resultant TEU-EP/33DDS system exhibited adequate reactivity at high processing temperatures. Comparison with DGEBA/33DDS and TEU-EP/33DDS achieved a 33 °C, 39%, and 55% increment in the glass transition temperature, Young's modulus, and hardness, respectively. The authors also reported increase of creep resistance and dimensional stability and enhancement of electrical properties (i.e. reduced permittivity and dielectric loss factor) [55].

This chapter is intended as an overview of the state of the art in bioepoxy composites and the usefulness of DMA to further our understanding of these novel polymeric materials. The field is under development, and there is much room for innovation and technology development. There are other studies and excellent reviews on reinforced biobased epoxies, which have been published recently [56–59].

6.5 Conclusion

This overview has shown that biobased epoxy composites are a growing field and offer many opportunities for innovation. A rapid inspection of the literature has shown that nano-sized fillers have enormous impact on the thermomechanical properties of bioepoxies. In some cases, loadings within 3 wt% have been enough to induce significant changes of elastic modulus and glass transition temperature T_g. The required loadings when using natural fibers are as high as 30 wt%, and the changes in mechanical properties have been rather modest. DMA has proved a critical technique to assess the influence of fillers on the viscoelastic properties of the composites. It is noted that DMA offers much more possibilities to get insights into the molecular mechanisms behind the bulk properties of bioepoxy composites. This technique appears to be subutilized. The detailed analysis of filler content on mechanical relaxations and their activation energies will enhance our understanding of the influence of molecular structure, fillers, interfacial interactions, and/or confinement effects on the thermomechanical behavior of bioepoxy composites and blends.

References

1 Baroncini, E.A., Kumar Yadav, S., Palmese, G.R., and Stanzione, J.F. III, (2016). Recent advances in bio-based epoxy resins and bio-based epoxy curing agents. *J. Appl. Polym. Sci.* 133: 44103.

2 Mosiewicki, M.A. and Aranguren, M.I.A. (2013). A short review on novel biocomposites based on plant oil precursors. *Eur. Polym. J.* 49 (6): 1243–1256.

3 Kumar, S., Samal, S.K., Mohanty, S., and Nayak, S.K. (2016). Recent development of biobased epoxy resins: a review. *Polym. Plast. Technol. Eng.* 57 (3): 133–155.

4 Kumar, S., Krishnan, S., Mohanty, S., and Nayak, S.K. (2018). Synthesis and characterization of petroleum and biobased epoxy resins: a review. *Polym. Int.* 67 (7): 815–839.

5 Parameswaranpillai, J., Woo, E.M., Hameed, N., and Piontek, J. (eds.) (2016). *Handbook of Epoxy Blends.* Springer https://doi.org/10.1007/978-3-319-18158-5_23-1.

6 Nielsen, L.E. (1974). *Mechanical Properties of Polymers and Composites.* New York: Marcel Dekker Inc.

7 Romo-Uribe, A. (2016). Dynamic mechanical thermal analysis of epoxy/thermoplastic blends. In: *Handbook of Epoxy Blends* (eds. J. Parameswaranpillai, E.M. Woo, N. Hameed and J. Piontek), 1–31. Springer https://doi.org/10.1007/978-3-319-18158-5_23-1.

8 Romo-Uribe, A. (2020). Thermo mechanical properties of natural rubber-based composites and nanocomposites. In: *Natural Rubber based Composites and Nanocomposites* (ed. P.M. Visakh). Wiley-Scrivener.

9 Kaelble, D.H. (1969). Free volume and polymer rheology. In: *Rheology*, vol. 5 (ed. F.R. Eirich), 223–296. New York: Academic Press.

10 McCrum, N.G., Read, B.E., and Williams, G. (1991). *Anelastic and Dielectric Effects in Polymeric Solids.* Dover.

11 Alvarado-Tenorio, B., Romo-Uribe, A., and Mather, P.T. (2011). Microstructure and phase behavior of POSS/PCL shape memory nanocomposites. *Macromolecules* 44 (14): 5682–5692.

12 Romo-Uribe, A., Alvarado-Tenorio, B., Romero-Guzmán, M.E. et al. (2009). Dynamic mechanical analysis of thermotropic vopolyester – short glass fibers composites. *Polym. Adv. Technol.* 20: 759–767.

13 Ferry, J.D. (1980). *Viscoelastic Properties of Polymers*, 3e. New York: Wiley.

14 Dealy, J.M. and Larson, R.G. (2006). *Structure and Rheology of Molten Polymers. From Structure to Flow Behavior and Back Again.* Munich: Carl Hanser Verlag.

15 Mylsamy, K. and Rajendran, I. (2011). The mechanical properties, deformation and thermomechanical properties of alkali treated and untreated Agave continuous fibre reinforced epoxy composites. *Mater. Des.* 32: 3076–3084.

16 Adekunle, K., Cho, S., Patzelt, C. et al. (2011). Impact and flexural properties of flax fabrics and Lyocell fiber-reinforced bio-based thermoset. *J. Reinf. Plast. Compos.* 30: 685–697.

17 Cuinat-Guerraz, N., Dumont, M.-J., and Hubert, P. (2016). Environmental resistance of flax/bio-based epoxy and flax/polyurethane composites manufactured by resin transfer moulding. *Composites, Part A* 88: 140–147.
18 Duc, F., Bourban, P.E., and Månson, J.A.E. (2014). Dynamic mechanical properties of epoxy/flax fibre composites. *J. Reinf. Plast. Compos.* 33: 1625–1633.
19 Ferdosian, F., Zhang, Y., Yuan, Z. et al. (2016). Curing kinetics and mechanical properties of bio-based epoxy composites comprising lignin-based epoxy resins. *Eur. Polym. J.* 82: 153–165.
20 Zhao, S. and Abu-Omar, M.M. (2015). Biobased epoxy nanocomposites derived from lignin-based monomers. *Biomacromolecules* 16 (7): 2025–2031.
21 Hong, C.K. and Wool, R.P. (2005). Development of a bio-based composite material from soybean oil and keratin fibers. *J. Appl. Polym. Sci.* 95 (6): 1524–1538.
22 Jawaid, M., Abdul Khalil, H.P.S., Hassan, A. et al. (2013). Effect of jute fibre loading on tensile and dynamic mechanical properties of oil palm epoxy composites. *Composites, Part B* 45: 619–624.
23 Di Landro, L. and Janszen, G. (2014). Composites with hemp reinforcement and bio-based epoxy matrix. *Composites, Part B* 67: 220–226.
24 Palanivel, A., Veerabathiran, A., Duruvasalu, R. et al. (2017). Dynamic mechanical analysis and crystalline analysis of hemp fiber reinforced cellulose filled epoxy composite. *Polímeros* 27 (4): 309–319.
25 De Rosa, I.M., Santulli, C., and Sarasini, F. (2010). Mechanical and thermal characterization of epoxy composites reinforced with random and quasi-unidirectional untreated Phormium tenax leaf fibers. *Mater. Des.* 31: 2397–2405.
26 Suresh Kumar, S.M., Duraibabu, D., and Subramanian, K. (2014). Studies on mechanical, thermal and dynamic mechanical properties of untreated (raw) and treated coconut sheath fiber reinforced epoxy composites. *Mater. Des.* 59: 63–69.
27 Adekunle, K., Cho, S.-W., Ketzscher, R., and Skrifvars, M. (2012). Mechanical properties of natural fiber hybrid composites based on renewable thermoset resins derived from soybean oil, for use in technical applications. 124 (6): 4530–4541.
28 Saba, N., Jawaid, M., Alothman, O.Y., and Paridah, M.T. (2016). A review on dynamic mechanical properties of natural fibre reinforced polymer composites. *Constr. Build. Mater.* 106: 149–159.
29 Towo, A.N. and Ansell, M.P. (2008). Fatigue evaluation and dynamic mechanical thermal analysis of sisal fibre-thermosetting resin composites. *Compos. Sci. Technol.* 68: 925–932.
30 Toldy, A., Niedermann, P., Rapi, Z., and Szolnoki, B. (2017). Flame retardancy of glucofuranoside based bioepoxy and carbon fibre reinforced composites made thereof. *Polym. Degrad. Stab.* 142: 62–68.
31 Hu, F., La Scala, J.J., Sadler, J.M., and Palmese, G.R. (2014). Synthesis and characterization of thermosetting furan-based epoxy systems. *Macromolecules* 47 (10): 3332–3342.
32 Fan, X., Miao, J.-T., Yuan, L. et al. (2018). Preparation and origin of thermally resistant biobased epoxy resin with low internal stress and good UV

resistance based on SiO_2 hybridized cellulose for light emitting diode encapsulation. *Appl. Surf. Sci.* 447: 315–324.

33 Fernandes, F.C., Kirwan, K., Wilson, P.R., and Coles, S.R. (2019). Sustainable alternative composites using waste vegetable oil based resins. *J. Polym. Environ.* 27 (11): 2464–2477.

34 Fernandes, F.C., Kirwan, K., Lehane, D., and Coles, S.R. (2017). Epoxy resin blends and composites from waste vegetable oil. *Eur. Polym. J.* 89: 449–460.

35 Mustapha, R., Rahmat, A.R., Abdul Majid, R., and Mustapha, S.N.H. (2019). Vegetable oil-based epoxy resins and their composites with bio-based hardener: a short review. *Polym. Plast. Technol. Mater.* 58 (12): 1311–1326.

36 Nair, S., Dartiailh, C., Levin, D., and Yan, N. (2019). Highly toughened and transparent biobased epoxy composites reinforced with cellulose nanofibrils. *Polymers* 11 (4): 612.

37 Wu, G.-M., Liu, D., Liu, G.-F. et al. (2015). Thermoset nanocomposites from waterborne bio-based epoxy resin and cellulose nanowhiskers. *Carbohydr. Polym.* 127: 229–235.

38 Shibata, M. and Nakai, K. (2010). Preparation and properties of biocomposites composed of bio-based epoxy resin, tannic acid, and microfibrillated cellulose. *Polym. Phys.* 48 (4): 425–433.

39 Yue, L., Liu, F., Mekala, S. et al. (2019). High performance biobased epoxy nanocomposite reinforced with bacterial cellulose nanofiber network. *ACS Sustainable Chem. Eng.* 7 (6): 5986–5992.

40 Yue, L., Maiorana, A., Khelifa, F. et al. (2018). Surface-modified cellulose nanocrystals for biobased epoxy nanocomposites. *Polymer* 134: 155–162.

41 Zhang, H., Guo, Y., Yao, J., and He, M. (2016). Epoxidised soybean oil polymer composites reinforced with modified microcrystalline cellulose. *J. Exp. Nanosci.* 11 (15): 1213–1226.

42 Le Hoang, S., Vu, C., Pham, L.T., and Choi, H.J. (2017). Preparation and physical characteristics of epoxy resin/bacterial cellulose biocomposites. *Polym. Bull.* 75 (6): 2607–2625.

43 Jlassi, K., Radwan, A.B., Sadasivuni, K.K. et al. (2018). Anti-corrosive and oil sensitive coatings based on epoxy/polyaniline/magnetite-clay composites through diazonium interfacial chemistry. *Sci. Rep.* 8 (1): 13369.

44 Romo-Uribe, A. (2018). Polymers in 2D confinement. A nanoscale mechanism for thermo-mechanical reinforcement. *Polym. Adv. Technol.* 29: 507–516.

45 Saikia, A., Sarmah, D., Kumar, A., and Karak, N. (2019). Bio-based epoxy/polyaniline nanofiber-carbon dot nanocomposites as advanced anti-corrosive materials. *J. Appl. Polym. Sci.* 136: 47744.

46 Liu, Z., Erhan, S.Z., and Xu, J. (2005). Preparation, characterization and mechanical properties of epoxidized soybean oil/clay nanocomposites. *Polymer* 46: 10119–10127.

47 Lu, Y. and Larock, R.C. (2006). Novel biobased nanocomposites from soybean oil and functionalized organoclay. *Biomacromolecules* 7 (9): 2692–2700.

48 Miyagawa, H., Jurek, R.J., Mohanty, A.K. et al. (2006). Biobased epoxy/clay nanocomposites as a new matrix for CFRP. *Composites, Part A* 37 (1): 54–62.

49 Miyagawa, H., Misra, M., Drzal, L.T., and Mohanty, A.K. (2005). Biobased epoxy/layered silicate nanocomposites: thermophysical properties and fracture behavior evaluation. *J. Polym. Environ.* 13 (2): 87–96.
50 Miyagawa, H., Misra, M., Drzal, L.T., and Mohanty, A.K. (2005). Novel biobased nanocomposites from functionalized vegetable oil and organically-modified layered silicate clay. *Polymer* 46 (2): 445–453.
51 Miyagawa, H., Mohanty, A., Drzal, L.T., and Misra, M. (2004). Effect of clay and alumina-nanowhisker reinforcements on the mechanical properties of nanocomposites from biobased epoxy: a comparative study. *Ind. Eng. Chem. Res.* 43 (22): 7001–7009.
52 Sahoo, S.K., Mohanty, S., and Nayak, S.K. (2015). Study of thermal stability and thermo-mechanical behavior of functionalized soybean oil modified toughened epoxy/organo clay nanocomposite. *Prog. Org. Coat.* 88: 263–271.
53 Takada, Y., Shinbo, K., Someya, Y., and Shibata, M. (2009). Preparation and properties of bio-based epoxy montmorillonite nanocomposites derived from polyglycerol polyglycidyl ether and ε-polylysine. *J. Appl. Polym. Sci.* 113: 479–484.
54 Wang, R., Schuman, T., Vuppalapati, R.R., and Chandrashekhara, K. (2014). Fabrication of bio-based epoxy–clay nanocomposites. *Green Chem.* 16 (4): 1871–1882.
55 Wan, J., Zhao, J., Gan, B. et al. (2016). Ultrastiff biobased epoxy resin with high T_g and low permittivity: from synthesis to properties. *ACS Sustainable Chem. Eng.* 4 (5): 2869–2880.
56 Paramarta, A. and Webster, D.C. (2016). Bio-based high performance epoxy-anhydride thermosets for structural composites: the effect of composition variables. *React. Funct. Polym.* 105: 140–149.
57 Shibata, M. (2013). Bio-based epoxy resin/clay nanocomposites (Chapter 9. In: *Thermoset Nanocomposites* (ed. V. Mittal), 189–209. Wiley-VCH Verlag GmbH & Co.
58 Tang, Q., Chen, Y., Gao, H. et al. (2019). Chapter 8, Bio-based epoxy resin from epoxidized soybean oil. In: *Soybean-Biomass, Yield and Productivity* (ed. M. Kasai). IntechOpen https://doi.org/10.5772/intechopen.81544.
59 Shibata, M. (2011). Bio-nanocomposites using bio-based epoxy resins (Chapter 13. In: *Nanocomposites with Biodegradable Polymers: Synthesis, Properties, and Future Perspectives* (ed. V. Mittal). Oxford: Oxford Scholarship Online https://doi.org/10.1093/acprof:oso/9780199581924.003.0013.

7

Mechanical Properties of Bioepoxy Polymers, Their Blends, and Composites

Ahmad Y. Al-Maharma, Yousef Heider, Bernd Markert, and Marcus Stoffel

RWTH Aachen University, Institute of General Mechanics, Eilfschornsteinstraße 18, Aachen 52062, Germany

7.1 Introduction

The mechanical properties of bioepoxy-based materials of polymers, blends, and composites are characterized through using flexural, tensile, compression, and impact tests, which provide a good guidance in choosing the most appropriate material according to the structural application. Young's modulus reflects the stiffness of the characterized material. The biodegradable material characterized with higher elastic modulus experiences lower deformation under prescribed loading conditions. Conducting flexural test on bio-based materials is a necessary step before implementing them in structural applications, where the materials are subjected to the bending load. When the part is employed in bending applications, the interlaminar shear strength of the composite provides the necessary resistance to the bending loads. The energy absorption capacity of the biodegradable materials is a critical property determined if these materials can cope up with the energy transferred to the part by the applied impacting load [1]. The epoxy resin finds a wide application in coatings and hosting matrix for fiber-reinforced polymer (FRP) composites in different industrial fields including aerospace, automotive, electrical, and biomedical areas of applications, thanks to its outstanding mechanical properties (such as high tensile and flexural properties) and adhesive strength [2, 3]. The diglycidyl ether of bisphenol-A (DGEBA) is the most commonly type of synthetic epoxy resin implemented in the fabrication of conventional epoxy-based composites. This synthetic resin is characterized with effective adhesion capacity with a vast majority of materials, good resistance to moisture absorption from the surrounding ambient because of their long hydrophobic chains, and finally it has reasonable dimensional stability during the curing process [4, 5]. However, the synthetic epoxy resin has an adverse effect on the environment because it does not have biodegradation property. There is a wide range of bioepoxy-based materials that show weaker mechanical performance compared to their synthetic counterparts, but they have received an increasing interest because of their biodegradability and low cost. Consequently, the environmental impact of DGEBA epoxy resin can be alleviated through various techniques such

Bio-Based Epoxy Polymers, Blends, and Composites: Synthesis, Properties, Characterization, and Applications, First Edition. Edited by Jyotishkumar Parameswaranpillai, Sanjay Mavinkere Rangappa, Suchart Siengchin, and Seno Jose.

as (i) complete substitution with biomass-derived epoxy resins; (ii) partial replacement through developing blends of DGEBA and epoxidized vegetable oil (EVO)-derived resins; and (iii) using bio-based curing agents such as bio-based anhydride hardener to cure DGEBA [6]. In recent published literature, the synthetic epoxy resin is successfully substituted by bio-based polymers derived from lignin, sugar, furans, vegetable oils (VOs), and cardanol. Cardanol shows outstanding thermal and mechanical properties, which can be attributed to the aromatic and aliphatic structures in their chain. The sugar-based bioepoxy monomers are synthesized from maltitol, isosorbide, sucrose, and sorbitol. Bioepoxy monomers synthesized from isosorbide exhibit higher tensile properties compared to DGEBA when cured with hardeners of aliphatic amine [7–9].

The bioepoxy resin can be synthesized using one of the following methods: (i) traditional DGEBA cured by bio-based curing agents such as cardanol-based novolac; (ii) bio-based DGEBA (bisphenol-A) retrieved from bio-based glycerol-derived epichlorydrin; and (iii) blending bioepoxy resin with EVOs in the presence of an appropriate curing agent. Super-Sap© bioepoxy manufactured from the aforementioned third category is frequently used in preparing bioepoxy-based green composites reinforced with natural fibers [10]. The phenolic groups, which are derived from the lignin structure, have great potential to replace bisphenol-A during the preparing process of epoxy resin [11]. However, these bioepoxy-based composites have limited implementation in some important fields such as aviation industry because of their weak mechanical performance. The current research efforts are concentrated in improving the material properties of bioepoxy-based materials in order to broaden their implementation in structural applications [12]. Generally, the mechanical properties of bioepoxy-based materials are largely affected by the type of the epoxy polymer and curing agent in addition to the parameters of curing process such as temperature and pressure. The anhydride-based curing agents can be used in producing bioepoxy resins with superior chemical and physical properties [13, 14]. The mechanical properties of bioepoxy polymers can be improved through using the following methods: (i) reinforcing bioepoxy resins with fillers having outstanding mechanical properties to improve their strength and stiffness and (ii) blending bioepoxy resins with another polymer identified with improved properties in order to reduce their brittleness, which subsequently leads to improvement in the values of impact strength and fracture toughness [15, 16].

7.2 Mechanical Properties of Bioepoxy Polymers

VOs are the most important type of bioresources for synthesizing polymeric materials. The dominating ingredients of VOs are triglycerides – esters of glycerol with three fatty acids. VOs are appropriate for preparing hydrophobic polymers and complement ideally with other masses such as protein and carbohydrate, which are naturally hydrophilic [17]. Usually, the bioepoxy resins prepared from EVOs such as epoxidized soybean oil (ESO), epoxidized linseed oil (ELO), and self-cross-linked bioepoxy polymer derived from oil fatty acid have

Table 7.1 Mechanical properties of bioepoxy polymers.

References	Epoxy polymer	Tensile strength (MPa)	Tensile Young's modulus (GPa)	Elongation at break (%)
[18]	Epon	59	3.15	—
[19–22]	DGEBA	29.63–40	1.50–1.77	1.79
[23]	Commercial bioepoxy	62	3.0	6.0
[24]	Self-cross-linked bioepoxy	42.50	1.93	3.20
[19, 25]	Vanillin-based bioepoxy	14.86–30.58	0.96–2.01	1.31–3.77
[26]	Double methacrylated epoxidized sucrose soyate (DMESS)	30.30	1.09	4.40
[27]	Sorbitol polyglycidyl ether (SPE)/cysteine-2-ehtyl-4-methylimi-dazol (Cys-Imz)	41.30	2.27	15.50
[18]	ELO (adipic acid)	8.80	0.022	—
[28]	Karanja oil bioepoxy (citric acid)	10.60	0.0027	63
[29]	Bioepoxy (MPA)	6.0	0.043	61.50

lower mechanical properties relative to that of DGEBA epoxy because of their aliphatic structure as it can be inferred from the values listed in Table 7.1 [18, 24]. There are different methods that were suggested in improving the mechanical properties of bioepoxy resins prepared from VOs. Using styrenic monomers is one of these methods, which can be conducted through homopolymerizing these monomers with soybean oil [30]. Consequently, by modifying the internal structure of epoxy network, it is possible to produce modified bioepoxy-based resins and foams characterized with higher mechanical performance compared to the ones derived from the traditional epoxy polymer. For instance, the biodegradable epoxy network derived from phenolic acid, gallic acid, sugar, and rosin exhibits interesting properties and performance such as higher hardness, thermal stability, and viscoelastic properties relative to DGEBA-based epoxy [31, 32]. More specifically, the average tensile strength of bioepoxy resin is 55 MPa, which is comparably good with the strength of synthetic polymers such as polyester (22 MPa) and epoxy (31 MPa) resins. Based on this value of high tensile strength, it can be stated that the bio-based epoxy is considered as a promising hosting matrix for natural fibers to fabricate natural composites that can be implemented in different industrial applications [33]. Furthermore, the bioepoxy prepared from Nahor (*Mesua ferrea* Linn) oil shows an outstanding adhesive strength that can be utilized as an adhesive material for joining plastic materials [34].

VOs are used in manufacturing bioepoxy-based rigid foams with competitive mechanical performance and at lower price compared to their counterparts synthesized from synthetic epoxy resins. For instance, the mechanical properties of bioepoxy foam derived from acrylated epoxidized soybean oil (AESO) are promising with a compressive modulus of 20 MPa and a compressive strength of 1 MPa. These mechanical properties are comparable with those of semirigid industrial foam [35]. Additionally, it is possible to manufacture rigid and lightweight foam derived from plant oils such as fatty diamines and linseed oil, which adequately meet the requirements of automotive industry. Specifically speaking, these bioepoxy foams have the potential to be implemented as a core foam material in sandwich composite structures used in the interior parts of the vehicle [36, 37]. The mechanical properties of bioepoxy-based foams can be improved through reinforcing them with compatible materials such as reinforcing ESO-based epoxy foam with epoxidized mangosteen tannin (EMT). The incorporation of EMT serves dual function that makes the ESO-based foam stronger and enables it to restore more of its original shape. The capability of EMT on increasing the compressive strength of ESO-based foam is attributed to the high number of epoxy groups and rigid aromatic ring. However, the increasing loading of EMT increases the density of bioepoxy-based foam [38].

7.2.1 Effect of Modifying Bioepoxy Chemical Structure

In general, the mechanical properties of bioepoxy polymers can be improved through changing the chemical structure of the polymer because it has considerable effect on the cross-linking density and morphology of the epoxy network. For example, the thermomechanical properties of vanillin-based epoxy are enhanced through changing the length of the starting epoxy oligomer. The loose network structure decreases the flexural and tensile strengths of bioepoxy. It is noteworthy to mention that the changes conducted on the chemical structure of bioepoxy should be associated with precise controlling of reaction time and temperature because the increasing values of these two conditions play a critical role in improving the bonding strength of some bioepoxy polymers such as tannin-based epoxy acrylate resins [39–41]. However, it is a crucial matter to establish balance between mechanical and thermal properties of bioepoxy resin when modifying the formulation and chemical structure design of the polymer [42]. Different materials can be used as modifiers of bioepoxy network such as styrene and nitrate solution. It should be noted that the increasing quantity of styrene reduced both toughness and elongation at break properties of epoxy thermosets [26]. The nitrate solution is employed to accelerate the curing reaction of vanillin-based bioepoxy, leading to great reduction on curing time as well as enhancing considerably the mechanical properties such as Izod impact strength, tensile strength, and pull-off strength. This inorganic accelerator changes the architecture of bioepoxy, making its configuration nearly parallel to each other. Clearly, the polymer chains are retrieved in linear direction, resulting in improved elongation at break and tensile strength [19].

7.2.2 Effect of Curing Agents

The type of curing agents used in the preparation of bioepoxy polymers has significant effect on their physical and mechanical properties such as viscosity, strength, and stiffness. For example, the methacrylic anhydride is used to reduce the viscosity and improve the mechanical properties of methacrylate functional epoxidized sucrose soyate (MESS) through further functionalization of MESS with this type of curing agent [26]. Changing the epoxy to anhydride ratio has considerable impact on the mechanical properties of EVO-based thermosets such as epoxidized sucrose soyate (ESS) because it produces bioepoxy with a unique chemical network. In other words, the mechanical properties of bioepoxy thermoset rely on polymer backbone and cross-link density of the network. High elastic modulus and glass transition temperature can be attained with formulation including higher catalyst quantity and less than equimolar ratio of anhydride to epoxy. Nevertheless, when the VO-based bioepoxy is cured with over stoichiometry ratio of epoxy/anhydride, the impact strength is reduced because of the molecular segments, which are incapable to disentangle and respond against the rapid application of mechanical stress, leading to brittle behavior of this type of bioepoxy polymer upon impact [43, 44].

There are many hardeners including biodegradable and synthetic types that can be used in making the properties of bioepoxy polymers similar to their synthetic counterparts or even better than them. For example, the bioepoxy derived from eugenol and cured with commercial anhydride has similar storage modulus (2.8 GPa) to that of commercial bisphenol-A epoxy (2.9 GPa) cured with the same curing agent. In fact, these two epoxies have slightly higher storage modulus than that of eugenol-based bioepoxy (2.5 GPa) cured with rosin-derived bioanhydride [45]. The elastic modulus of epoxies cured with cashew nut shell liquid (CNSL)-based hardeners is equivalent to the ones prepared from synthetic hardeners. Therefore, the biodegradable loading can be increased in traditional epoxy resin through substituting the synthetic hardeners with CNSL-based hardeners [46]. Moreover, bioepoxy resin derived through glycidylation of 2,5-bis(4-hydroxy-3-methoxybenzylidene)cyclopentanone (DVCP), which is later cured with bio-based curing agents, shows comparable flexural strength and modulus to that of DGEBA cured with the same agents [25]. The ESO-based bioepoxy resin, which is synthesized through ring-opening reaction between ESO and bio-based maleopimaric acid (MPA) curing agent, has higher strength and modulus relative to the same resin cured by synthetic curing agent of hexahydro-4-methylphthalic anhydride (MHHPA) [29]. If the lignin-based raw material is appropriately used in a way that permits to utilize the full advantage of lignin's aromatic structure, it is supposed that the lignin-derived epoxy thermoset can exhibit the performance properties comparable to those of synthetic epoxy polymer [47]. For example, the bioepoxy resin derived from Japanese green tea has higher rigidity relative to DGEBA epoxy because of the combining effect of hard segment of bio-based curing agent (lignin) with average molecular weight and aromatic raw material of low molecular weight in its chemical structure [48]. The bioepoxy resin is prepared through curing isosorbide with amine curing agent of isophorone diamine. This bio-based resin shows higher rubbery modulus over DGEBA epoxy. It is important to highlight that the isosorbide-based bioepoxy has hygroscopic nature and the presence of moisture could deteriorate the aspects of epoxy networks, which are largely affected by the type of hardener used to cure the epoxy resin [49]. However, by using hydrophobic curing agent, the hydrophobicity and mechanical properties of isosorbide-based bioepoxy can be improved [50]. The properties of cross-linked soybean oil-based epoxy network can be changed through modifying the structure of soybean oil-based amine hardener. The soybean oil-based thermoset has very low moisture absorption tendency and less capacity to dissolve compared to the one cured with petrochemical hardeners [51].

7.3 Blends of Bioepoxy Resin

The most frequently used VOs for blending epoxy resins are linseed, soybean, hemp, and castor. During the synthesis of epoxy blend, the VOs are cross-linked into epoxy resins by adding a curing agent or a hardener such as anhydrides and aromatic and aliphatic amines. The properties of VO-derived epoxy resins depend on the ring-opening agent, composition of VO, type of epoxy resin, and degree of VO epoxidation. Partial substitution of viscous petrochemical based epoxy resin with functionalized EVOs creates materials characterized with reduced costs and moderate mechanical properties along with enhanced processability. The increasing quantity of EVO resin increases the capability of biodegradable epoxy blend on impact energy dissipation. This enhancement can be attributed to the flexible, long, and resilient network structure owned by EVO resins that increases the mobility of resulted epoxy blends, hence providing higher resistance to crack initiation and propagation. However, the incorporation of EVOs to epoxy resin reduces the flexural and tensile properties of bio-based epoxy blend. This mitigation is justified to poor interphase interaction between epoxy networks and EVOs because of less reactive oxirane rings located in triglyceride compared to pure synthetic epoxy resin. The progressive increase of EVO content inevitably contributes to higher stress concentration point, which reduces the strength aspects of the bioepoxy-based blend [8, 16, 52]. Further details about the toughening effect of EVO resins are coming through the discussion held in Section 7.3.1.

7.3.1 Toughening Effect of EVO-Based Resins

The flexible chains of bioepoxy-based monomers are blended with their synthetic epoxy counterparts in order to add to the cross-linked networks the required degree of flexibility. For example, the cardanol/benzoxazine epoxy blend system shows higher degree of flexibility and better storage modulus relative to homopolymers [53]. Actually, the EVOs serve as plasticizers when their loading in synthetic epoxy resin is higher than 10 wt% because of their effect on reducing the number of oxirane rings, which control the tensile strength and modulus of epoxy resin [20]. When EVOs are used to blend DGEBA epoxy resin at 30 wt%, they create second rubbery phase that imparts heterogeneous morphology to the structure of the blend system as indicated in Figure 7.1.

This rubbery matrix improves the impact strength of epoxy blend system. It can be anticipated from Figure 7.1a that the neat epoxy resin exhibits smooth glassy fractured surface with cracks because of its brittleness nature, which

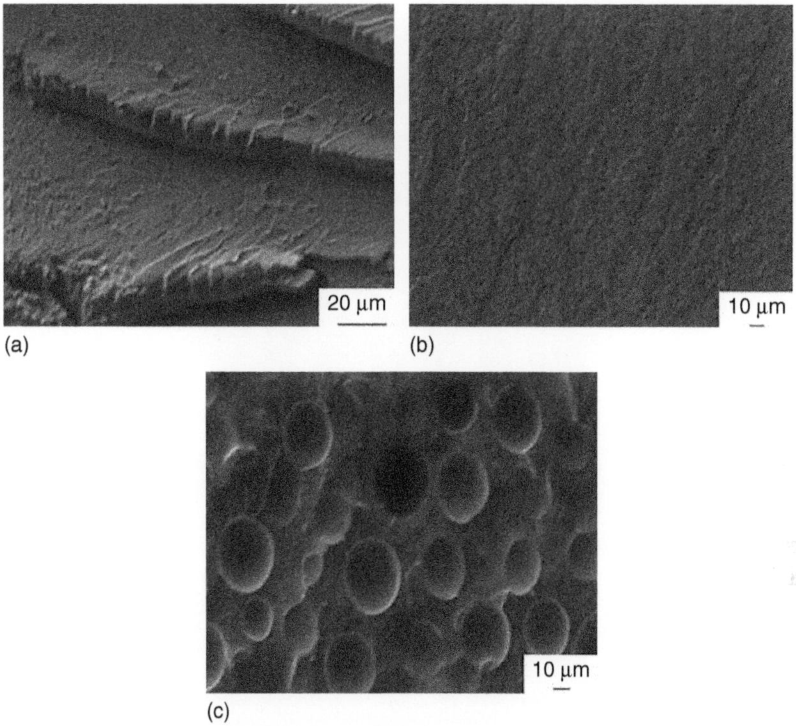

Figure 7.1 SEM micrographs of (a) pure synthetic epoxy, and blends of (b) epoxy/20% ESO, (c) epoxy/30% ESO. Source: (a–c) Reprinted from Sahoo et al. [52]. Copyright (2015), with permission from Springer Nature.

leads to weak impact strength property. In contrast, the 20 wt% ESO/80 wt% DGEBA blend system shown in Figure 7.1b exhibits enhanced impact strength with percentage of improvement of 60% relative to raw epoxy resin because of the plasticization effect of ESO resin. This plasticization happens because epoxy has good compatibility with ESO initially and phase separation is induced in the epoxy polymeric matrix during curing. Moreover, further blending of epoxy resin with 30 wt% of ESO creates spherical domains of oil, which can be easily noted in Figure 7.1c, having uniform dispersion in the structure of epoxy resin that enables the shear yielding process throughout the matrix and hence participates effectively in improving the toughness of the system. It can be concluded that the toughness improvement attained on the epoxy network is directly correlated with the particle size distribution of cavities in addition to the increasing content of ESO resin in epoxy blend system. This improvement is attributed to the existence of aliphatic long-chain structure in ESO resin [21, 52, 54]. Figure 7.2 highlights the effect of EVO-based resins on improving the impact strength and fracture toughness (G_{IC}) compared to various configurations of bioepoxy-based composites.

Beside VOs, the curing agent has a significant role in reducing the brittleness of DGEBA epoxy-based blend. For example, cardanol-based phenalkamine (PKA)

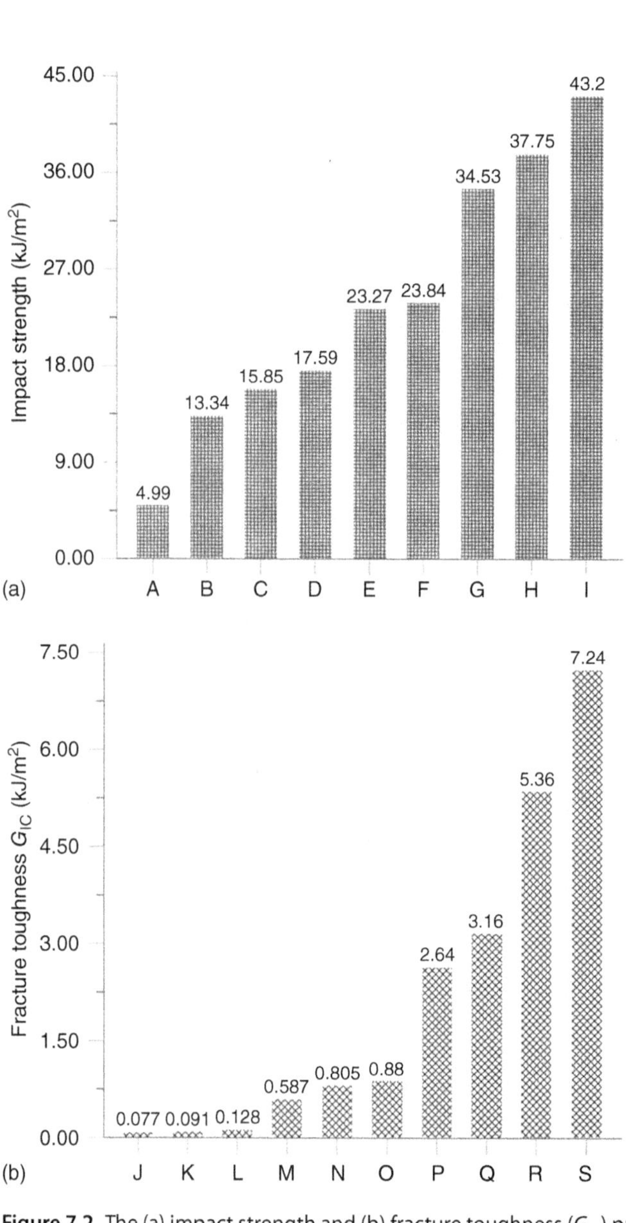

Figure 7.2 The (a) impact strength and (b) fracture toughness (G_{IC}) properties for bioepoxy-based materials: A: DGEBA/30 wt% ELO, B: bioepoxy/denim/Jute (40 wt%), C: AESO – 1,4-butanediol dimethacrylate (BDDMA)/bamboo fiber, D: DGEBA, E: Bioepoxy/denim (40 wt%), F: DGEBA/30 wt% ESO, G: AESO/denim (45 wt%), H: DGEBA/30 wt% EMS, I: AESO/30 wt% styrene/denim (60 wt%), J: DGEBF (diglycidyl ether of bisphenyl-F epoxy), K: DGEBF/50 wt% ELO, L: DGEBF/50 wt% ELO/5 wt% intercalated (OMMT) clay, M: 85 wt% DGEDP-ethyl + 15 wt% GE + glass mat, N: DGEBF/30 wt% ESO, O: DGEBA, P: DGEBA/30 wt% EMS, Q: DGEBA/20 wt% ESO, R: DGEBA/30 wt% ESO, S: DGEBA/30 wt% Itaconic acid (TEIA).

curing agent is used to increase the ductility of ELO/DGEBA blend system. The tensile strength of PKA-cured epoxy has equivalent value to that of epoxy cured using traditional aliphatic amine-based curing agents. The blend system contains 20 phr of ELO, showing high impact strength with moderate modulus and strength needed for structural applications. The improvement on impact strength can be explained to the long linear hydrocarbon chain existed in the curing agent (PKA) as well as ELO [55].

7.3.2 Effect of Chemical Interaction in Epoxy Blend

The polymeric blends that contained biodegradable derived resins utilize the synergy between biopolymers prepared from natural resources and synthetic polymers to attain desirable biodegradable, physical, and mechanical properties. The bioepoxy resins are used in improving the hydrophobic and mechanical properties of various polymeric blend systems, which are prepared based on synthetic polymers such as DGEBA, polyvinyl alcohol (PVA), and polylactic acid (PLA) [22, 56, 57]. Cardanol is one of the bioepoxy resins that is implemented in increasing the plasticization limit of cardanol and bio-based DGEBA blend system because of its hydrophobic nature [58]. There are different techniques that can be used in the preparation of bioepoxy blends such as traditional mechanical mixing and microwave-aided blending. The later blending method produces compatible blends of EVOs/synthetic resin in short time duration, which is characterized with improved ultimate tensile strength and elongation at break because of the increasing number of cross-links between biodegradable and synthetic ingredients via hydrogen bonding (OH of synthetic resin and oxirane of EVOs), providing strength to the blend system [59]. However, the conventional mechanical blending can produce bioepoxy blend systems with improved tensile strength properties such as the blend systems composed of two different bio-based resins of hyperbranched polyester (HPE) and hyperbranched epoxy (BHE). With increasing epoxy loading in the blend, the tensile strength of the blend system is enhanced because of the ideal cross-linking density and the existence of secondary interactions such as polar–polar interactions and hydrogen bonding that imposes a strict limitation on the free rotation of the ester linkages in the cured chemical structure of the blends [60]. In addition to mechanical properties, the chemical reaction between synthetic and biodegradable materials in the epoxy blend system critically controls the improvement attained on molding and drying properties of blend system [61].

7.3.3 Increasing Content Effect of EVOs in Bioepoxy Blend

The mechanical properties of EVO/epoxy blend system are mainly affected by the compatibility of EVO with epoxy resin, the content of EVO in the blend system, the type of epoxy resin, and the type of curing agent [62]. EVOs are copolymerized with conventional petroleum-derived epoxy (P-epoxy) to customize the aspects of the final resins. For instance, blend system containing around 30–40% of ESO shows the best performance in terms of impact strength and storage modulus [63]. Furthermore, the 41 wt% of synthetic epoxy matrix

Table 7.2 Tensile and flexural properties of bioepoxy blend systems.

References	Epoxy blend	Bioepoxy loading (wt%)	Tensile strength (MPa)	Tensile Young's modulus (GPa)	Flexural strength (MPa)	Flexural modulus (GPa)
[18]	ESO/Epon	30	60	3.19	99	2.91
	EAS (epoxidized allyl soyate)/Epon	10	53	2.97	127	3.50
	ESO/DGEBA	60	—	—	106	3.30
	EMS/Epon	30	59	3.15	98	2.84
[66]	E-epoxy/P-epoxy	50	63.60	1.36	—	—
[67]	DOL/bisphenol-A epoxy	25	240.50	19.95	—	—
	DKL/bisphenol-A epoxy		248.80	19.47	—	—
[68]	Polyfurfuryl alcohol/P-epoxy	5	16	1.08	—	—
[69]	Rosin/P-epoxy	25	—	4.02	—	—
[22]	Itaconic acid (TEIA)/DGEBA	30	47.59	2.52	67.32	3.56
[54]	DGEBA/soy epoxy	30	77.80	7.56	97.40	1.43
	DGEBA/ soy epoxy (20% bismaleimides)		89.8	8.99	127.9	1.82
[55]	DGEBA/ELO	10	42.29	1.27	—	—
[52]	DGEBA/ESO	20	30.27	1.35	81.91	2.36
[70]	Epoxy/epoxidized methyl soyate	20	42.9	1.76	71.10	2.42

reinforced with either glass or carbon fibers can be substituted with ELO resin, leading to considerable improvement in the impact strength [64, 65]. Additionally, the incorporation of some types of bioepoxy resins introduces considerable improvements to the values of tensile and flexural properties of epoxy blend system as it can be concluded from the data listed in Table 7.2. The flexural strength and fracture toughness of DGEBA epoxy resin blended with epoxidized castor oil (ECO) are significantly improved with increasing content of ECO in the blend system without affecting its Young's modulus [62]. The incorporation of rosin- and lignin-based bioepoxies such as those prepared from DEO-polymerized Kraft (DKL)/organosolv lignin (DOL) in P-epoxy resin up to 50% by weight considerably increases the tensile, flexural, and adhesion strengths of the epoxy blend system [67, 69, 71, 72]. Biodegradable resins prepared from cardol/cardanol with episulfide group (CCES) and epoxy group (CCEO) are prepared and further blended with conventional DGEBA epoxy polymer. Based on the outcomes of lap sheer strength test, the CCES–DGEBA blend epoxy system shows comparable adhesion strength relative to CCEO–DGEBA and DGEBA polymers. Additionally, the CCES–DGEBA blend system including 20 wt% of CCES exhibits improved corrosion resistance behavior [73]. Furan-based bioepoxy resin, which is blended with DGEBA at minor quantities, increases the strength and modulus of the epoxy blend system because of the interfacial adhesion in the compatible blend system originating from polar to polar interaction. However, the overloading of furan-based bioepoxy resin in the synthetic epoxy blend reduces its mechanical properties because if decreasing cross-linking density. Even though there is a reduction occurred in the mechanical properties at higher furan-based bioepoxy loadings, they are still equal to those of a neat epoxy network [68].

The bark extractive-derived epoxy resin (E-epoxy) has similar tensile strength to P-epoxy. E-epoxy has higher tensile strength (63 MPa) relative to ESO derived epoxy (29 MPa) and liquefied wood-based epoxy (58 MPa). Relative to the long-chain structure of ESO, the liquefied wood-derived epoxy or E-epoxy has higher opportunity to be implemented in the aerospace and automotive industries. The lower tensile modulus of E-epoxy/P-epoxy blend system relative to pure P-epoxy resin could be explained to the bulky ring structures in the extractives that reduce the cross-linking density of the blend [74]. Although the E-epoxy resin can enhance the toughness properties of the epoxy blend system, the over loadings of this bioresin has negative impact on the mechanical performance of the epoxy blend. Hence, the highest mechanical strength is achieved by replacing 10% of P-epoxy with E-epoxy bioresin. After incorporating 10% of E-epoxy, the strain of epoxy blend is enhanced from 6.6% to 12.3%, leading to toughness enhancement of 84% while the tensile strength is reduced slightly from 76 to 70 MPa [66]. Using bismaleimides to modify the soy based matrices at suitable concentrations can considerably improve their mechanical properties. The addition of bismaleimide-modified soy epoxy resin to DGEBA up to 30 wt% improves the values of tensile modulus and strength, while it reduces the impact strength to the value directly related to the content of bismaleimide in the blend system. The enhancement in the tensile strength can be justified to the rigidity imparted by the heteroaromatic ring and homopolymerization of bismaleimides in addition to the increased cross-linked network between aromatic and aliphatic epoxy resin. The reduction on the impact strength can be attributed to the limited chain mobility because of the creation of heteroaromatic bismaleimide rings. It should be noted that the tensile strength of DGEBA epoxy-based blend containing more than 30 wt% of soy-derived epoxy is noticeably reduced because of the incomplete curing of the aliphatic epoxy resin [54]. It is possible to maintain proportionally high Young's modulus value of unsaturated polyester resin (UPE) while having improved impact strength and fracture toughness properties when it is blended with 20% and 25 wt% of epoxidized palm oil (EPO) and epoxidized methyl soyate (EMS), respectively. The increasing loading of these epoxidized oils decreases the cross-linking density of blend system, which reduces the values of storage modulus. In order to increase the fracture toughness of UPE/EMS blend, it is recommended to minimize the number of rubber particles that have a size of 1.0 μm. The morphology of second rubbery phase can be controlled through choosing various functionalized initiators such as benzoyl peroxide and VOs [75–77].

7.4 Bioepoxy-Based Composites

A wide range of realistic combinations of VOs, polymerization routes, chemical modification, nature of fibers, and fillers, which are implemented as reinforcements, enable flexible customization of the composite aspects to conform to the requirements of functional or structural materials. Hence, various types of nano-, micro-, and macrofillers have been suggested as reinforcements for bioepoxy-based composites, including natural or synthetic, inorganic, and organic types, in order to provide sufficient reinforcing solutions for particular structural requirements. The compatibility between filler and matrix critically dominates the overall properties of composite such as the interfacial strength and moisture absorption. This compatibility can be improved through modifying the surface of filler or using compatibilizing agents [23, 63, 78]. Plasma treatment is used to improve the compatibility between bioepoxy resin and natural fibers retrieved from the leafs of the giant reed *Arundo donax* L. without affecting their tensile properties [79]. When the bioepoxy resins are used to substitute their synthetic counterparts, the environmental harm of FRP composites is reduced by more than 85% [80]. In addition to their role as a hosting matrix for natural and synthetic fibers, the bioepoxy resins can be implemented to strengthen the reinforced concrete beams, which emphasize their implementation in construction applications [81, 82]. Both AESO and commercial bioepoxy resins are appropriate for structural applications. The strength and stiffness properties of biocomposites fabricated from commercial bioepoxy resins are better than those synthesized from AESO resin. However, the AESO-based composites possess higher impact strength compared to bioepoxy composites [83–85].

7.4.1 Undesirable Effect of Moisture Absorption

The bioepoxy resins absorb higher amount of moisture compared to traditional epoxy resin because the hydroxyl group and cellulose are used to synthesize bioepoxies [86]. The chemical structure, resin composition, and cross-linking density of bioepoxy thermoset resins have critical influence on the mechanical and aging resistance performances of biocomposites fabricated from them. The aspects of natural fibers have higher impact on the moisture absorption tendency of biocomposites compared to resin composition [87]. The moisture absorbed by biocomposites has a plasticizing effect on the composites because it alters their failure mode; in addition, it reduces the stiffness of biocomposites through degrading the interfacial adhesion at a fiber/matrix interface [88]. By improving the interfacial adhesion strength at the fiber/matrix interface through using chemical and physical treatments, the amount of moisture absorbed by the biocomposite can be reduced [89–91]. Acetic anhydride (AA)-treated unidirectional (UD) flax fibers are used to reinforce bioepoxy resin. This type of treatment decreases the amount of moisture absorbed by bioepoxy composite because of the enhanced adhesion at the fiber/matrix interface. The low concentration (2%) of AA enhances the tensile strength, modulus, and bond shear strength by 55%, 58%, and 7%, respectively [92].

7.4.2 Fiber-Reinforced Bioepoxy Composite

The mechanical properties of bioepoxy composites depend on the type of VO, type of epoxy resin, type of biodegradable hardener, type of filler used, and physical and chemical modifications and treatments conducted on VOs and reinforcements [16]. The most frequently used fillers in the fabrication of bioepoxy-based composites along with their mechanical properties are summarized in Table 7.3. Manufacturing parts from biocomposites fortified with natural fibers reduce their weights and introduces significant improvements on their specific aspects such as strength, stiffness, and impact strength [95, 105–107]. The vast majority of bioepoxy resins can be used as reactive diluents, which mitigate viscosity, enhance toughness, and provide better impregnation and wetting of reinforcing fibers [66]. EMS prepared from soybean oil is used to toughen and decrease the viscosity of the epoxy resin system. The ESO resin is considered as an effective green toughening agent to reduce the brittleness of epoxy. Nevertheless, the over plasticization and decreasing cross-linking density lower the strength and stiffness considerably, making the epoxy blend inappropriate for wide implementation in industrial fields. In order to improve the physical properties of this toughened blend, epoxy resin can be reinforced with natural fibers and inorganic fillers [70].

The natural fibers have better compatibility with bio-based resins compared to synthetic fibers. Therefore, biocomposites reinforced with natural fibers show higher mechanical performance and have lower cost relative to the ones reinforced with synthetic fibers. By combining two types of natural fibers, additional strength can be added to the natural composite [108]. Currently, biocomposites are regarded as a valid replacement of fiberglass-reinforced polymer composites in various applications because they are less expensive and biodegradable. However, the biocomposites, specifically those fabricated from bioepoxy resins, still suffer from poor mechanical performance, limiting their wide implementation in structural applications because of lower mechanical properties of bioepoxy and natural fibers compared to their synthetic counterparts [8, 109].

7.4.2.1 Natural Fiber-Reinforced Bioepoxy Composites

The lignocellulosic fillers are widely used to reinforce VO-based epoxy composites because of their low cost, lightweight, and competitive biodegradable properties. The effect of lignocellulose fillers on improving the mechanical properties of their biocomposites is directly controlled by the structure and properties of the reinforcement such as the length, diameter, loading, architecture (UD, woven), and orientation/dispersion of the reinforcing fiber [63]. Hemp, flax, and bamboo fibers (BFs) are the most common natural fibers that are used to improve the tensile and bending stiffness properties of bioepoxy resins [110]. The biocomposites composed of natural fiber-reinforced VO-based

Table 7.3 Tensile strength and modulus properties of bioepoxy-based composites.

References	Biocomposite	Tensile strength (MPa)	Tensile Young's modulus (GPa)
[20]	DGEBA/10 wt% EVO/milled recycled carbon (MCF)	52.91	3.19
[64]	P-epoxy/40 wt% ELO/carbon fiber	365	84.40
	P-epoxy/10 wt% ELO/glass fiber	395	25.10
[84]	AESO – 1,4-butanediol dimethacrylate (BDDMA)/bamboo fiber (styrene free)	40.13	—
[93]	1.0 epoxidized sucrose soyate (ESS)/0.75 methylhexahydrophthalic anhydride (MHHPA)/50 vol% functionalized flax	222.90	25.30
[94]	ELO/amino silane-treated basalt	362.44	18.75
[95]	Glucofuranoside bioepoxy (GFTE)/carbon fiber	498.46	14.54
	DGEBA/carbon fiber	471.36	13.66
[72]	25 wt% depolymerized hydrolysis lignin (DHL)/75 wt% DGEBA/fiberglass	187	23.20
	DGEBA/fiberglass	214	17.50
[96]	Bioepoxy/hemp fiber	63	5.87
[97, 98]	Bioepoxy/bark cloth	33	3.0
[99]	ELO/polyaniline filler (15 wt%)	15.91	0.25
[90]	Bioepoxy/silanized flax (40 vol%)	230	28.80
	Bioepoxy/alkalized flax (40 vol%)	256	28.50
[100]	Bioepoxy/UD raw flax fiber	222.94	22.30
	Bioepoxy/nonwoven thick mat flax	76.28	8.04
[101]	20 wt% ESO/epoxy/15 wt% short sisal	34.83	2.48
[70]	20 wt% EMS/epoxy/UD sisal	103.14	3.24
[102]	Bioepoxy/epoxy-coated bamboo fibers	807.43	50.48
	Bioepoxy/silanized spherical silica fume (SiO_2)-coated bamboo fibers	732.24	47.91
[103, 104]	Bioepoxy/core = UD flax/skin = nonwoven mat flax (50 wt% fiber)	185	16.67
[83]	Bioepoxy/denim/jute (40 wt% fiber)	43.31	7.45
	Bioepoxy/denim (40 wt% fiber)	30.06	5.17
	AESO/denim (45 wt% fiber)	41.26	7.10
	AESO + 30 wt% styrene/denim (60 wt% fiber)	7.38	1.23

resins have two main drawbacks of weak bonding at the fiber/matrix interface and poor resin properties. In order to resolve these structural issues, the surface treatments (such as using alkali solution and/or silane agents) of natural fibers and choosing appropriate curing agents for cross-linking VO-based resins are recommended [93, 94, 111]. These surface treatments eliminate many organic materials with poor mechanical properties, inherent to the structure of natural fiber. This elimination increases the amount of hydroxyl groups located at the surface of natural fibers, which effectively contribute in increasing the fiber's interaction with the bioepoxy matrix [90, 112]. By improving the interfacial adhesion at the fiber/matrix interface, the mechanical and aging properties of the treated composite are considerably improved [113]. Moreover, the alkalized bark cloth-reinforced bioepoxy composite has a tensile strength of 33 MPa, which exceeds a threshold strength of 25 MPa needed for composites used for manufacturing dashboard panels or car instrument [97, 98]. The strong interfacial adhesion at hemp/bioepoxy interface justifies the higher damage resistance and postimpact damage tolerance relative to that of hemp/synthetic epoxy composite [114]. The increasing loading of natural fibers in bioepoxy-based composites considerably increases the strength and modulus of the composite, while it reduces the elongation at break values [99, 115]. The biocomposites reinforced with adequate loading of natural fibers such as sisal and flax fibers need more energy to break because there are more interfaces existing at the crack path relative to composites fortified with low loadings of fibers [101].

The packing arrangements of natural fibers inside the structure of bioepoxy composites play critical role in determining the degree of improvement introduced by these fillers in the values of tensile strength and modulus. For instance, the mechanical performance of UD sisal fiber-based composite is considerably higher than that of discontinuous and short fiber-based biocomposites [116, 117]. Additionally, the UD flax/bioepoxy composite has a tensile strength of 222.94 MPa, which is higher than the 185 MPa threshold strength needed for structural applications [100]. Therefore, this biocomposite can be used to substitute traditional short glass fiber/polyester and glass/bioepoxy in manufacturing composite surfboard in addition to load-bearing parts such as the panels of vehicle body, body chassis, crash elements, and wind turbine blade [118–122]. When natural fibers are used in twill-woven fabric form, the contact area between the fiber and resin is higher, bonding between matrix and fiber is stronger than other arrangements such as plain weave, leading to higher Young's modulus value [123]. In order to produce flax fibers/bioepoxy composite at a feasible cost and improved mechanical performance, a hybrid bioepoxy-based composite reinforced with two types of flax fibers' fabric arrangements are needed to meet this goal. These arrangements are UD flax fibers used at the hybrid composite core, while the external surfaces of the composite are fabricated from a nonwoven mat of flax fibers. This mat provides protection for internal UD flax fiber-based layers in the internal core from the surrounding moisture that may reduce their mechanical performance [103, 104]. The stacking sequence of bioepoxy/natural fiber's layers considerably affected the delamination area in the laminate after exerting dynamic loads on biocomposite. Hence, it is recommended to decrease the angle difference of fiber orientation in two neighboring layers [124].

7.4.2.2 Synthetic Fiber-Reinforced Bioepoxy Composites

Carbon and E-glass fibers are the most frequently used synthetic fibers in the fabrication of bioepoxy composites. These fibers improve the strength and modulus of epoxy biocomposites in addition to their tribological properties [125]. Glass fiber mats are used to reinforce bioepoxy resins synthesized from monoglycidyl ether of eugenol (GE) and bio-based diglycidyl ethers of diphenolate (DGEDP) with ethyl (DGEDP-ethyl) and pentyl (DGEDP-pentyl) esters. The resulting bioepoxy-based composites have good potential to substitute traditional DGEBA/glass fiber composite. The GE is implemented as a reactive diluent because of its capability to reduce the viscosity of the DGEDP–ester epoxy resins [126].

7.4.2.3 Hybrid Fiber-Reinforced Bioepoxy Composites

The hybridization of natural fibers/bioepoxy composites with other synthetic or mineral fibers characterized with outstanding aging and mechanical properties can significantly improve the overall properties of the resulting hybrid composite, making them more suitable to be used in industrial applications [127, 128]. By using basalt fiber to hybridize the jute/epoxy composite, the resulting hybrid composite exhibits higher flexural and aging properties in addition to higher ability to absorb impact energy relative to laminates fabricated from jute/epoxy composite. The best arrangement for hybrid composites in terms of optimum aging resistance is sandwich-like structure with natural fibers having higher moisture barrier properties should be used to fabricate the external layers [129, 130]. Hemp fabric-reinforced bioepoxy composite is characterized with mechanical properties that qualify it as an ideal core for sandwich-like composite structures [131]. The implementation of multiple natural fibers in the same biocomposite structure such as sisal, banana, and coconut provides the bioepoxy-based hybrid composite with sufficient strength, making it an ideal biomaterial for manufacturing helmet shell structure and mechanical seal in piping industry [132, 133]. Carbon fiber is one of synthetic fibers that is commonly used to hybridize vegetable fiber-based composites, especially flax fibers. The hybridization of biocomposites with carbon reinforcements decreases the cost and brittleness of hybrid composite and improves its capability on energy dissipation [134].

7.4.3 Bioepoxy-Based Nanocomposites

Nanocomposites fabricated from bioepoxies are characterized with superior properties such as fracture toughness, modulus, strength, durability, and flame retardancy in addition to their ease of processability and low density [8, 135]. The mechanical properties of synthetic epoxy resin such as fracture toughness and impact strength can be improved through blending epoxy resin with EVOs (such as ESO and ELO) to form blend systems cured with appropriate curing agents. However, blending synthetic epoxy with EVOs reduces the tensile strength and modulus because of bulky ring structure presented in the extractives, which reduces the cross-link density of the blended system. These blends can be further improved through reinforcing them with suitable nanofillers such as those listed in Table 7.4 in order to produce bioepoxy nanocomposites characterized with good tensile strength and modulus properties at moderate plasticity [9, 66]. These nanofillers are capable of tremendously improving various aspects of

Table 7.4 Mechanical properties of bioepoxy-based nanocomposites.

References	Epoxy biocomposite	Tensile strength (MPa)	Tensile Young's modulus (GPa)	Elongation at break (%)
[66, 136]	10 wt% E-epoxy/ P-epoxy/25 wt% acetone-treated CNFs	132	4.94	—
[137]	Sorbitol glycidyl ether (SGE)/triethylenetetramine (TETA)/5 wt% OMMT clay	40.50	1.7	—
[27]	SPE-Cys/1.0 wt% chitosan nanofiber (CsNF)	32.50	1.15	11.10
[138]	Hyperbranched epoxy/0.5 wt% GO	76.41	—	38.37
[139]	Waterborne terpene-maleic ester epoxy (WTME)/8 wt% cellulose nanowhiskers (CNWs)	—	800.10	—
[140]	Sulfone epoxy/30 wt% hyperbranched polyurethane/ 5 wt% nanoclay	19.78	—	50.14
[141]	20 wt% ESO/epoxy/1 wt% nanoclay	30.4	1.62	2.58
[142]	Sulfone epoxy/5 wt% nanoclays treated with hyperbranched polyuria	27.47	—	27.10
[143]	Bioepoxy/walnut shell and coconut shell	69.50	—	21.82

bioepoxy-based nanocomposites as it can be inferred from the improvements summarized in Table 7.5.

The major challenge for VO-based nanocomposite is to achieve ideal dispersion of nanofillers with minimum agglomeration, so perfect interaction including load transfer with hosting matrix can be attained [14, 152]. The agglomeration of nanofillers is undesirable during the preparation of nanocomposites because the agglomerates decrease the reinforcing efficiency of nanofillers and reduce the degree of matrix cross-linking. For example, at 4 wt% of CNTs, the nanocomposite shows higher value of impact strength relative to pure epoxy resin because the CNTs are uniformly distributed in the matrix [153].

7.4.3.1 Nanoclay-Reinforced Bioepoxy Composites

Nanoclays are commonly used in enhancing different aspects of nanocomposites fabricated from either pure bioepoxy resin or bioepoxy blends such as strength, modulus, impact strength, and fracture toughness. The intercalated clay particles are more effective in improving the fracture toughness of bioepoxy-based nanocomposites. For intercalated clay reinforced nanocomposites, the strength of clay aggregation and strong interfacial adhesion at bioepoxy/clay interface hinder cracks from propagating [9, 154]. By incorporating nanoclay into epoxy resin

Table 7.5 The improvements attained on the properties of bioepoxy-based nanocomposites.

References	Bioepoxy nanocomposite	Filler loading (wt%)	Characterization
[63, 144]	Bioepoxy/graphene oxide (GO)	0.3	Graphene filler improves the tensile strength and modulus by 23% and 35%, respectively.
[138]	Bio-based hyperbranched epoxy (HBE)/GO	0.5	The toughness, adhesive strength, tensile strength, and elongation at break are improved by 263%, 189%, 161%, and 159%, respectively. The oxygenated functionalities located at GO sheets contribute in improving the impact strength and adhesive strength.
[145]	Bioepoxy/lignin nanoparticles	<50	The functional groups existed at the surface of lignin particles improve the adhesion at fiber/matrix interface. A remarkable enhancement in the value of impact toughness is recorded.
[146]	Bioepoxy/triethylammonium hydrogen sulfate ionic liquid (IL)-treated lignin particles	2.0	The flexural modulus, flexural strength, toughness, and tensile strength are increased by 57%, 80%, 23%, and 52%, respectively.
[147]	Acrylated and epoxidized soybean oil (AESO)/plasma-enhanced chemical vapor deposition (PECVD)-treated grinded coconut waste particles	40	The elongation at break, tensile strength, and elastic modulus are increased by 20.43%, 167.92%, and 50%, respectively.
[148]	Bioepoxy/calcium carbonate biofillers attained from seashells	30	The calcium carbonate biofillers increase the flexural modulus with over 50%.
[149]	Bioepoxy/oil palm ash (OPA)	3.0	The OPA fillers increase the density of nanocomposite while they introduce considerable improvement in the values of flexural and tensile strengths.
[150]	Bioepoxy/carbonized cattle bone particles	15	The increasing content of the carbonized bone particles decreases the elongation at break values. The tensile and flexural strengths are increased by 88.02% and 103.95%, respectively.
[143]	Bioepoxy/coconut and walnut shells powders	50	At this high loading of powders, the tensile and flexural strengths are slightly affected with moisture absorption.
[27]	Sorbitol polyglycidyl ether (SPE)-amino acid hardeners of lysine (Lys)/ chitin nanofiber (ChNF)	3.0	The tensile modulus, tensile strength, and elongation at break are increased by 47.02%, 70.79%, and 85.71%, respectively.
	SPE-cysteine (Cys)/chitosan nanofiber (CsNF)	1.0	The tensile modulus and tensile strength are increased by 96.92% and 225%, respectively, while the elongation at break is reduced by 43.08%.
[151]	Isosorbide-based epoxy/SiO_2 nanoparticles	0.15	The resulting biocomposite shows hydrophobic behavior with effective mechanical robustness against sand erosion.

The percentages of enhancement attained on bio-composite is evaluated based on neat epoxy resin values.

modified with ESO, the damping behavior, storage modulus, and cross-linking density are improved. With increasing load of nanoclay in nanocomposite, the tensile strength values are unaffected while the tensile modulus has been increased. The epoxy system filled with clay nanoparticles shows higher impact strength values relative to pure epoxy resin. By adding 5 wt% of C30B nanoclay to epoxy resin, the dimensional stability and resistance to degradation of the resulting nanocomposite are improved [141]. The DGEBA epoxy is toughened with acrylated ECO (AECO) and reinforced with organically modified montmorillonite nanoclays (OMMTs). The bioepoxy-based nanocomposites composed of DGEBA/10 wt% AECO/1 wt% OMMT and DGEBA/20 wt% AECO/1 wt% OMMT show ideal material properties for various industrial uses [155]. The tensile strength and scratch hardness are improved by two and fivefolds, respectively, through modifying epoxy resin with *M. ferrea* L. seed oil-based polyurethane and then reinforcing the resulting blend system with clay nanoparticles [156].

Epoxy resins containing polyester chains are synthesized from lignin, glycerol, and saccharides. The thermal and mechanical properties of the aforementioned resin can be improved by reinforcing it with clay nanoparticles [157].

The chemical modification of clay nanoparticles is ideally needed in order to attain adequately the highest possible dispersion of clay in bioepoxy resin and to enhance the interfacial strength between clay and surrounding epoxy resin. Typical treatments of filler surface can be conducted through using silane agents, amines, isocyanate, and alkali solution [63]. The OMMT nanofillers significantly improve the mechanical properties of sorbitol glycidyl ether (SGE)/triethylenetetramine (TETA) polymeric system. This improvement is a result of both good compatibility between clay/clay modifier and epoxy network in addition to high cross-linking density [137]. In order to increase the biodegradable content in synthetic epoxy resin, this resin is cured and reinforced with bio-based curing agent (phenalkamines cardolite) and montmorillonite nanoclay (MMT), respectively. At 0.5 phr of MMT, the flexural and tensile moduli of MMT/epoxy are improved with 9.38% and 10.2%, respectively, relative to raw epoxy resin. However, the elongation at break and Izod impact strength are slightly decreased upon the incorporation of MMT nanoclays to epoxy matrix [158]. Nanoclays treated with hyperbranched polyurea are used to reinforce VO-modified sulfone epoxy resin at various loadings (1–5 wt%). Epoxy nanocomposite containing 5 wt% of nanoclay shows higher than 300% of enhancement in tensile strength, even though the elongation at break is reduced with increasing content of nanoclays [140, 142].

7.4.3.2 Cellulose Nanofiller-Reinforced Bioepoxy Composites

Cellulose nano fillers (CNFs) provide comprehensive improvements to the tribological and mechanical properties of bioepoxy-based nanocomposites such as tensile strength and modulus in addition to the storage modulus, fracture toughness, impact strength, and elongation at break. CNFs treated with silane agent have better wettability and improved interfacial bonding strength with bioepoxy matrix because the nanocomposite reinforced with silylated CNFs contains much less voids or bubbles relative to raw CNFs/bioepoxy nanocomposites. The surface silanization of CNFs enhances the curing of bioepoxy resin through decreasing the activation energy. Additionally, the silylated CNFs improves the overall mechanical performance of bioepoxy nanocomposite such as fracture toughness and reduces the brittleness of bioepoxy resin [159]. It should be noted that the CNFs can improve the tensile strength and modulus of bioepoxy composites by 32% and 12%, respectively, at filler loading of 3 wt% because beyond this limit, they contribute effectively in reducing the cross-linking density of epoxy network. However, the increasing content of CNFs increases the elongation at break and subsequent impact strength property [160].

CNWs show excellent reinforcing effect on emphasizing the values of tensile strength and elastic modulus of waterborne terpene-maleic ester epoxy (WTME)-based nanocomposites because of the improved interfacial interactions created via hydrogen bonds between WTME matrix and CNWs nanofillers [139]. Bioepoxy resin prepared from bark extractives is considered as an appropriate option to partially replace the traditional synthetic epoxy thermoset. Blending the aforementioned bioepoxy with P-epoxy at 50 wt% produces a blend system that exhibits tensile strength comparable to that of pure P-epoxy resin. In order to additionally enhance the strength and modulus properties of this blend, uniformly dispersed CNFs are used as reinforcements. By mixing 10% of bioepoxy with neat epoxy resin, the toughness can be improved up to 84% relative to neat P-epoxy, while the tensile strength and modulus of CNFs/bioepoxy nanocomposites prepared from this blend are improved to around two to four times, respectively [66]. In addition to epoxy resins, other types of polymers can be blended and reinforced with EVOs and CNFs, respectively, in order to improve their strength, modulus, and fracture toughness properties such as polyester and PLA. For example, ESO and CNFs are combined with PLA resin in order to improve its mechanical properties. By adding ESO to PLA/CNF nanocomposite reinforced with 10 wt% of CNFs, the ductility of the resulting hybrid composite is improved by 5–10 times with only slight reduction in modulus and strength. The outstanding mechanical properties of CNFs/PLA/ESO composite can be explained to balance between percolation of CNF and plasticization effect of ESO. Consequently, to produce material with desired strength and toughness, it is vital to manage the quantities of CNF and ESO in tertiary composite system [161].

Even though there is great effort paid in employing epoxy nanocomposites with CNF reinforcements in various industrial disciplines in order to achieve higher sustainability, the mechanical performance of these biocomposites is low because of poor compatibility at fiber/matrix interface. Consequently, the synthetic epoxy resin is blended with bioepoxy monomers, which play a main role in improving both the distribution of nanofillers and resin penetration through them [136]. With increasing content of cellulose nanofibers, the amount of porosity is increased in CNFs/epoxy nanocomposite, which reduces the mechanical properties to the values lower to the expected ones. Hence, composite preparation methods with higher consolidation pressures are needed to enhance the consolidation processes. Managing the relationship between porosity and cellulose fillers leads to considerable enhancements in stiffness, fracture behavior, and strength [162]. An average improvement of 15% in the values of Young's modulus is registered for composites prepared through hot press method with cellulosic filler loading levels of 5–7.5% [163].

7.4.4 Multiscale Bioepoxy Composites

The bio-based nanofillers improve various properties of bioepoxy composites other than the mechanical ones such as reducing the weight of these composites and improving their tribological properties. These nanofillers enhance the load transfer among the reinforcements and bioepoxy matrix in addition to their role in increasing the values of strength, stiffness, and impact strength of multiscale composite [164]. The mechanical performance of this type of composite reinforced with particles such as lemon peels are affected by various factors such as the strength of natural fiber and hosting resin, loading of particles, and interfacial bonding strength between epoxy matrix and reinforcing fillers and particles [165, 166]. 3 wt% of bagasse ash fillers can considerably improve the tensile, flexural, and impact strength of bio-based hybrid epoxy composite reinforced with combined fibers of banana and flax. Maximum values of tensile and flexural strength are achieved when 5 wt% of bagasse ash fillers are used to reinforce bio-based hybrid epoxy composite emphasized with a combination of banana and kenaf fibers [167]. Furthermore, the titanium oxide nanoparticles play a dominant role in improving the interfacial bonding strength between epoxy and flax fibers, enhancing the stress transfer between them when they are homogenously dispersed in epoxy matrix. This bonding improvement leads to noticeable increase in the tensile strength property of flax/epoxy composite [168]. Coating the surface of BFs with bioepoxy coating contains silanized spherical silica fume (SiO_2), makes the fibers more hydrophobic, and improves the flexural strength and modulus of BFs/bioepoxy composite up to 25% and 20%, respectively, indicating a better matrix/fiber interface. The scanning electron microscopy (SEM) micrographs shown in Figure 7.3 clearly highlight the function of SiO_2 in improving the bonding strength at the interface between BFs and bioepoxy [102].

Basalt nanoparticles can effectively enhance the tensile and fracture toughness properties of basalt/epoxy and basalt/jute/epoxy composites when they are used to reinforce these composites at 3 wt%. This improvement takes place because of the increasing frictional force and interfacial adhesion at epoxy/basalt nanoparticles and reinforcing fiber/epoxy interface, respectively, which subsequently strengthens the mechanical interlocking in the composite system, leading to enhanced load transfer from hosting resin toward nanoparticles and reinforcing fibers in addition to delaying in small cracks' initiation and propagation in multiscale composite. Hence, this type of composites has higher capability to bear loads relative to traditional fiber/matrix composite system and participate in increasing the tensile modulus and fracture toughness of the multiphase composite [169].

7.5 Conclusion

The biodegradable content can be increased in epoxy-based materials through synthesizing pure bioepoxy from green resources such as VOs, using green curing agents, and blending bioepoxy with its synthetic counterpart up to 50 wt%. The curing agents have considerable impact in improving the aging and mechanical properties of bioepoxy-based materials. These agents enhance

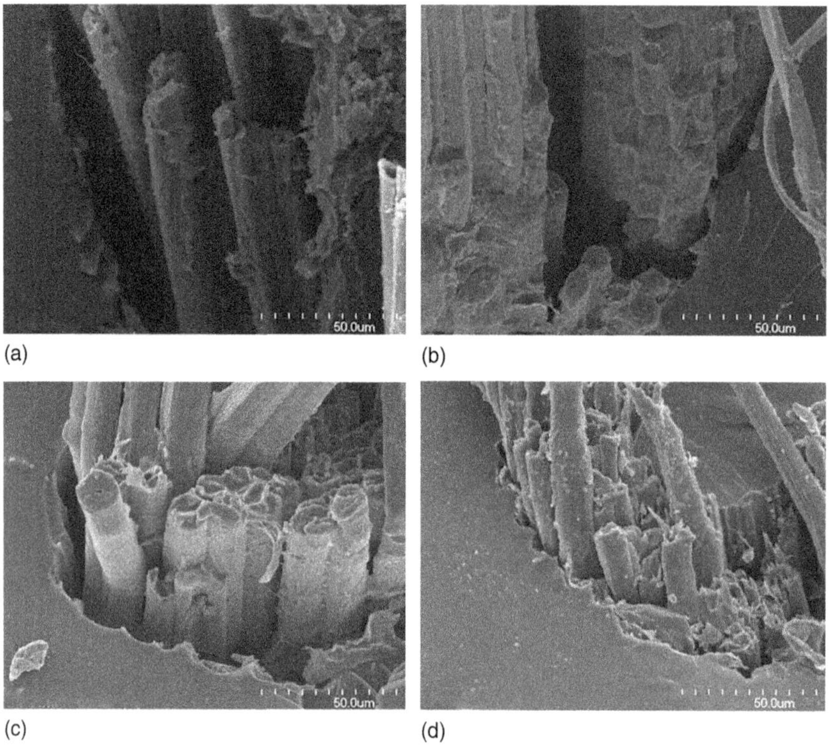

Figure 7.3 Scanning electron microscopy (SEM) micrographs of bamboo/bioepoxy failure area: (a) EP/raw-BF, (b) EP/BF-Ep, (c) BF–starch nanocrystals (StN), and (d) BF–SiO_2. Source: (a–d) Reprinted from Gauvin et al. [102]. Copyright (2016), with permission from John Wiley and Sons.

the aspects of bioepoxy polymers through modifying their chemical structures by changing the cross-linking density and morphology of the epoxy network. The bioepoxy resins prepared from VOs exhibit poor mechanical performance relative to their petrochemical counterparts. In order to improve the properties of these biodegradable resins, they are blended with synthetic epoxy resins such as DGEBA and Epon. This mixing significantly improves the ductility behavior of traditional epoxy resin because of the capability of VO-based resins on creating a rubbery phase inside the epoxy structure; hence, the values of impact strength and fracture toughness are increased for epoxy blend system. Furthermore, the bioepoxy resins are used as reactive diluents to reduce the viscosity of conventional epoxy resin and improve the impregnation of reinforcing fiber. However, the increasing loading of VO-derived bioepoxy resin in epoxy blend system decreases the strength and modulus properties of this blend because of the effect of bioepoxy on reducing the cross-linking density of synthetic epoxy network upon blending. In order to increase these properties and keeping at the same time good toughness properties, the epoxy blend can be reinforced with natural/synthetic fillers at specific contents. The most frequently used fillers for fabrication of bioepoxy composite are carbon, glass, bamboo, sisal, flax, and hemp fibers, nanoclays, and CNFs. The last two fillers are used as nanoreinforcements for bioepoxy-based nano- and multiphase composites. These nanofillers play multiple roles in multiscale bioepoxy composites because of their function on both strengthening the bioepoxy-based matrix and enhancing the stress transfer efficiency between reinforcement and hosting matrix. However, the bioepoxy-based composites suffer from two structural flaws of poor interfacial adhesion at the fiber/matrix interface and weak mechanical performance of hosting resin. The interfacial adhesion strength can be improved through conducting physical and/or chemical treatments on the surface of reinforcing fillers.

7.6 Future Perspectives and Recommendations

Additional studies on modifying the chemical structure of EVO polymers are needed through using new types of curing agents and modifiers to cure these polymers. Moreover, the studies that address the effect of man-made fibers on improving the mechanical properties of bioepoxy composites are quite limited. Hence, it is recommended to develop new classes of bioepoxy composites prepared from epoxy blends, which are reinforced with carbon-based synthetic fillers such as carbon, CNTs, and graphene oxide (GO). These composites have higher strength and modulus relative to the ones reinforced with natural fillers.

Acknowledgment

The first author would like to appreciate the financial support provided by German Academic Exchange Service (DAAD) for the purpose of accomplishing this research.

References

1 Bajpai, P.K., Ahmad, F., and Chaudhary, V. (2017). Processing and characterization of bio-composites. In: *Handbook of Ecomaterials* (eds. L.M.T. Martínez, O.V. Kharissova and B.I. Kharisov), 1–18. Cham: Springer https://doi.org/10.1007/978-3-319-48281-1.
2 Jahanshahi, S., Abdulkhani, A., Doosthoseini, K. et al. (2016). Study of physical and mechanical properties of produced strawboard using bio epoxy tannins resin. *Iran. J. Wood Pap. Ind.* 7 (2): 271–282.
3 Declet-Vega, A., Sepúlveda-Ramos, N., Crespo-Montoya, S. et al. (2019). Bio-composites reinforced with strontium titanate nanoparticles: mechanical behavior and degradability. *J. Compos. Sci.* 3 (1): 7.
4 Pradhan, S., Pandey, P., Mohanty, S. et al. (2015). Insight on the chemistry of epoxy and its curing for coating applications: a detailed investigation and future perspectives. *Polym.-Plast. Technol. Eng.* 55 (8): 862–877.

5 Kumar, S., Krishnan, S., Mohanty, S. et al. (2018). Synthesis and characterization of petroleum and biobased epoxy resins: a review. *Polym. Int.* 67 (7): 815–839.
6 Bledzki, A.K., Urbaniak, M., Boettcher, A. et al. (2013). Bio-based epoxies and composites for technical applications. *Key Eng. Mater.* 559: 1–6.
7 Ogueri, K.S., Farhana, Z., Aris, Z.F.M. et al. (2015). Thermosetting resin compositions based on bio-derived phenols and sugars. *Proceedings of 73rd Annual Technical Conference and Exhibition of the Society of Plastics Engineers*, Orlando, USA (23 March 2015). Washinton DC, USA: The American Chemical Society.
8 Paluvai, N.R., Mohanty, S., and Nayak, S.K. (2014). Synthesis and modifications of epoxy resins and their composites: a review. *Polym.-Plast. Technol. Eng.* 53 (16): 1723–1758.
9 Drzal, L., Misra, M., Miyagawa, H. et al. (2005). Bio-based epoxy, their nanocomposites and methods for making those. US Patents 10/966,624, filed 2 June 2005.
10 Corona, A. (2013). Quantitative sustainability assessment of bio-based materials for wind turbine rotor blades. Master thesis. Technical University of Denmark.
11 Ferdosian, F., Yuan, Z., Anderson, M. et al. (2014). Synthesis of lignin-based epoxy resins: optimization of reaction parameters using response surface methodology. *RSC Adv.* 4 (60): 31745–31753.
12 Bachmann, J., Hidalgo, C., and Bricout, S. (2017). Environmental analysis of innovative sustainable composites with potential use in aviation sector—a life cycle assessment review. *Sci. China Technol. Sci.* 60 (9): 1301–1317.
13 Jin, F.-L., Li, X., and Park, S.-J. (2015). Synthesis and application of epoxy resins: a review. *J. Ind. Eng. Chem.* 29: 1–11.
14 Bobade, S.K., Paluvai, N.R., Mohanty, S. et al. (2016). Bio-based thermosetting resins for future generation: a review. *Polym.-Plast. Technol. Eng.* 55 (17): 1863–1896.
15 Jha, K., Kataria, R., Verma, J. et al. (2019). Potential biodegradable matrices and fiber treatment for green composites: a review. *AIMS Mater. Sci.* 6 (1): 119–138.
16 Mustapha, R., Rahmat, A.R., Abdul, M.R. et al. (2019). Vegetable oil-based epoxy resins and their composites with bio-based hardener: a short review. *Polym.-Plast. Technol. Mater.* 58 (12): 1311–1326.
17 Miao, S., Wang, P., Su, Z. et al. (2014). Vegetable-oil-based polymers as future polymeric biomaterials. *Acta Biomater.* 10 (4): 1692–1704.
18 Ramon, E., Sguazzo, C., and Moreira, P. (2018). A review of recent research on bio-based epoxy systems for engineering applications and potentialities in the aviation sector. *Aerospace* 5 (4): 110.
19 Nikafshar, S., Zabihi, O., Hamidi, S. et al. (2017). A renewable bio-based epoxy resin with improved mechanical performance that can compete with DGEBA. *RSC Adv.* 7 (14): 8694–8701.
20 Fernandes, F.C., Kirwan, K., Lehane, D. et al. (2017). Epoxy resin blends and composites from waste vegetable oil. *Eur. Polym. J.* 89: 449–460.

21 Kumar, S., Krishnan, S., Samal, S.K. et al. (2018). Toughening of petroleum based (DGEBA) epoxy resins with various renewable resources based flexible chains for high performance applications: a review. *Ind. Eng. Chem. Res.* 57 (8): 2711–2726.

22 Kumar, S., Samal, S.K., Mohanty, S. et al. (2018). Bio-based tri-functional epoxy resin (TEIA) blend cured with anhydride (MHHPA) based cross-linker: thermal, mechanical and morphological characterization. *J. Macromol. Sci., Part A* 55 (6): 496–506.

23 Budd, R. and Cree, D. (2019). Effect of fire retardants on mechanical properties of a green bio-epoxy composite. *J. Appl. Polym. Sci.* 136 (16): 47398.

24 Huang, K., Liu, Z., Zhang, J. et al. (2015). A self-crosslinking thermosetting monomer with both epoxy and anhydride groups derived from tung oil fatty acids: synthesis and properties. *Eur. Polym. J.* 70: 45–54.

25 Shibata, M. and Ohkita, T. (2017). Fully biobased epoxy resin systems composed of a vanillin-derived epoxy resin and renewable phenolic hardeners. *Eur. Polym. J.* 92: 165–173.

26 Yu, A.Z., Rahimi, A., and Webster, D.C. (2018). High performance bio-based thermosets from dimethacrylated epoxidized sucrose soyate (DMESS). *Eur. Polym. J.* 99: 202–211.

27 Shibata, M., Fujigasaki, J., Enjoji, M. et al. (2018). Amino acid-cured bio-based epoxy resins and their biocomposites with chitin- and chitosan-nanofibers. *Eur. Polym. J.* 98: 216–225.

28 Kadam, A., Pawar, M., Yemul, O. et al. (2015). Biodegradable biobased epoxy resin from karanja oil. *Polymer* 72: 82–92.

29 Chen, Y., Xi, Z., and Zhao, L. (2016). New bio-based polymeric thermosets synthesized by ring-opening polymerization of epoxidized soybean oil with a green curing agent. *Eur. Polym. J.* 84: 435–447.

30 Ronda, J.C., Lligadas, G., Galià, M. et al. (2013). A renewable approach to thermosetting resins. *React. Funct. Polym.* 73 (2): 381–395.

31 Aouf, C., Nouailhas, H., Fache, M. et al. (2013). Multi-functionalization of gallic acid. Synthesis of a novel bio-based epoxy resin. *Eur. Polym. J.* 49 (6): 1185–1195.

32 Rapi, Z., Szolnoki, B., Bakó, P. et al. (2015). Synthesis and characterization of biobased epoxy monomers derived from d-glucose. *Eur. Polym. J.* 67: 375–382.

33 Johnson, R.D.J., Arumugaprabu, V., Rajasekar, E. et al. (2018). Mechanical property studies on environmental friendly bio epoxy resin. *Mater. Today: Proc.* 5 (2): 6815–6820.

34 Dinda, S., Ravisankar, V., and Puri, P. (2016). Development of bio-epoxide from nahor (*Mesua ferrea Linn*) oil. *J. Taiwan Inst. Chem. Eng.* 65: 399–404.

35 Bonnaillie, L.M. and Wool, R.P. (2007). Thermosetting foam with a high bio-based content from acrylated epoxidized soybean oil and carbon dioxide. *J. Appl. Polym. Sci.* 105 (3): 1042–1052.

36 Mazzon, E., Habas-Ulloa, A., and Habas, J.-P. (2015). Lightweight rigid foams from highly reactive epoxy resins derived from vegetable oil for automotive applications. *Eur. Polym. J.* 68: 546–557.

37 Dworakowska, S., Cornille, A., Bogdał, D. et al. (2015). Formulation of bio-based epoxy foams from epoxidized cardanol and vegetable oil amine. *Eur. J. Lipid Sci. Technol.* 117 (11): 1893–1902.
38 Khundamri, N., Aouf, C., Fulcrand, H. et al. (2019). Bio-based flexible epoxy foam synthesized from epoxidized soybean oil and epoxidized mangosteen tannin. *Ind. Crops Prod.* 128: 556–565.
39 Jahanshahi, S., Pizzi, A., Abdulkhani, A. et al. (2016). Analysis and testing of bisphenol A-free bio-based tannin epoxy-acrylic adhesives. *Polymers (Basel)* 8 (4): 143.
40 Kalita, H. and Karak, N. (2012). Epoxy modified bio-based hyperbranched polyurethane thermosets. *Des. Monomers Polym.* 16 (5): 447–455.
41 Fache, M., Viola, A., Auvergne, R. et al. (2015). Biobased epoxy thermosets from vanillin-derived oligomers. *Eur. Polym. J.* 68: 526–535.
42 Li, C., Liu, X., Zhu, J. et al. (2013). Synthesis, characterization of a rosin-based epoxy monomer and its comparison with a petroleum-based counterpart. *J. Macromol. Sci., Part A* 50 (3): 321–329.
43 Paramarta, A. and Webster, D.C. (2016). Bio-based high performance epoxy-anhydride thermosets for structural composites: the effect of composition variables. *React. Funct. Polym.* 105: 140–149.
44 Gupta, A.P., Ahmad, S., and Dev, A. (2010). Development of novel bio-based soybean oil epoxy resins as a function of hardener stoichiometry. *Polym.-Plast. Technol. Eng.* 49 (7): 657–661.
45 Qin, J., Liu, H., Zhang, P. et al. (2014). Use of eugenol and rosin as feedstocks for biobased epoxy resins and study of curing and performance properties. *Polym. Int.* 63 (4): 760–765.
46 Benega, M.A., Raja, R., and Blake, J.I. (2017). A preliminary evaluation of bio-based epoxy resin hardeners for maritime application. *Procedia Eng.* 200: 186–192.
47 Koike, T. (2012). Progress in development of epoxy resin systems based on wood biomass in Japan. *Polym. Eng. Sci.* 52 (4): 701–717.
48 Basnet, S., Otsuka, M., Sasaki, C. et al. (2015). Functionalization of the active ingredients of Japanese green tea (*Camellia sinensis*) for the synthesis of bio-based epoxy resin. *Ind. Crops Prod.* 73: 63–72.
49 Chrysanthos, M., Galy, J., and Pascault, J.-P. (2011). Preparation and properties of bio-based epoxy networks derived from isosorbide diglycidyl ether. *Polymer* 52 (16): 3611–3620.
50 Bellido-Aguilar, D.A., Zheng, S., Zhan, X. et al. (2019). Effect of a fluoroalkyl-functional curing agent on the wettability, thermal and mechanical properties of hydrophobic biobased epoxy coatings. *Surf. Coat. Technol.* 362: 274–281.
51 Frias, C.F., Serra, A.C., Ramalho, A. et al. (2017). Preparation of fully biobased epoxy resins from soybean oil based amine hardeners. *Ind. Crops Prod.* 109: 434–444.
52 Sahoo, S.K., Mohanty, S., and Nayak, S.K. (2014). Synthesis and characterization of bio-based epoxy blends from renewable resource based epoxidized soybean oil as reactive diluent. *Chin. J. Polym. Sci.* 33 (1): 137–152.

53 Rao, B.S. and Palanisamy, A. (2013). Synthesis of bio based low temperature curable liquid epoxy, benzoxazine monomer system from cardanol: thermal and viscoelastic properties. *Eur. Polym. J.* 49 (8): 2365–2376.

54 Sithique, M.A., Ramesh, S., and Alagar, M. (2008). Mechanical and morphological behavior of bismaleimide-modified soy-based epoxy matrices. *Int. J. Polymer. Mater. Polymer. Biomater.* 57 (5): 480–493.

55 Sahoo, S.K., Khandelwal, V., and Manik, G. (2018). Development of toughened bio-based epoxy with epoxidized linseed oil as reactive diluent and cured with bio-renewable crosslinker. *Polym. Adv. Technol.* 29 (1): 565–574.

56 Xiong, Z., Ma, S., Fan, L. et al. (2014). Surface hydrophobic modification of starch with bio-based epoxy resins to fabricate high-performance polylactide composite materials. *Compos. Sci. Technol.* 94: 16–22.

57 Li, R., Zhang, Y., Xiong, Z. et al. (2018). Liquefied banana pseudo stem-based epoxy composites: incorporation of phenol derivative and its characterization. *Composites Part B* 143: 28–35.

58 Verge, P., Toniazzo, V., Ruch, D. et al. (2014). Unconventional plasticization threshold for a biobased bisphenol-A epoxy substitution candidate displaying improved adhesion and water-resistance. *Ind. Crops Prod.* 55: 180–186.

59 Riaz, U., Ashraf, S.M., and Sharma, H.O. (2011). Mechanical, morphological and biodegradation studies of microwave processed nanostructured blends of some bio-based oil epoxies with poly (vinyl alcohol). *Polym. Degrad. Stab.* 96 (1): 33–42.

60 Saikia, A., Hazarika, D., and Karak, N. (2019). Tough and biodegradable thermosets derived by blending of renewable resource based hyperbranched epoxy and hyperbranched polyester. *Polym. Degrad. Stab.* 159: 15–22.

61 Kanehashi, S., Oyagi, H., Lu, R. et al. (2014). Development of bio-based hybrid resin, from natural lacquer. *Prog. Org. Coat.* 77 (1): 24–29.

62 Tan, S.G. and Chow, W.S. (2010). Biobased epoxidized vegetable oils and its greener epoxy blends: a review. *Polym.-Plast. Technol. Eng.* 49 (15): 1581–1590.

63 Zhang, C., Garrison, T.F., Madbouly, S.A. et al. (2017). Recent advances in vegetable oil-based polymers and their composites. *Prog. Polym. Sci.* 71: 91–143.

64 McIsaac, A. and Fam, A. (2018). The effect of bio-based content in resin blends on tensile properties of FRP wet layup systems. *Constr. Build. Mater.* 168: 328–337.

65 Yim, Y.-J., Rhee, K.Y., and Park, S.-J. (2017). Fracture toughness and ductile characteristics of diglycidyl ether of bisphenol-A resins modified with biodegradable epoxidized linseed oil. *Composites Part B* 131: 144–152.

66 Kuo, P.-Y. (2016). Development and characterization of an extractive-based bio-epoxy resin from beetle-infested lodgepole pine (*Pinus contorta* var. *latifolia*) bark. Doctoral thesis. University of Toronto.

67 Ferdosian, F., Zhang, Y., Yuan, Z. et al. (2016). Curing kinetics and mechanical properties of bio-based epoxy composites comprising lignin-based epoxy resins. *Eur. Polym. J.* 82: 153–165.

68 Zolghadr, M., Zohuriaan-Mehr, M.J., Shakeri, A. et al. (2019). Epoxy resin modification by reactive bio-based furan derivatives: curing kinetics and mechanical properties. *Thermochim. Acta* 673: 147–157.

69 Demircan, G., Kisa, M., and Özen, M. (2017). Mechanical properties of rosin-based bio-epoxy Resin. *El-Cezerî J. Sci. Eng.* 5 (2): 387–393.

70 Sahoo, S.K., Mohanty, S., and Nayak, S.K. (2015). Effect of lignocellulosic fibers on mechanical, thermomechanical and hydrophilic studies of epoxy modified with novel bioresin epoxy methyl ester derived from soybean oil. *Polym. Adv. Technol.* 26 (12): 1619–1626.

71 Jung, J.Y., Park, C.-H., and Lee, E.Y. (2017). Epoxidation of methanol-soluble kraft lignin for lignin-derived epoxy resin and its usage in the preparation of biopolyester. *J. Wood Chem. Technol.* 37 (6): 433–442.

72 Ferdosian, F., Yuan, Z., Anderson, M. et al. (2016). Synthesis and characterization of hydrolysis lignin-based epoxy resins. *Ind. Crops Prod.* 91: 295–301.

73 Lv, J., Liu, Z., Zhang, J. et al. (2017). Bio-based episulfide composed of cardanol/cardol for anti-corrosion coating applications. *Polymer* 121: 286–296.

74 Kuo, P.-Y., Sain, M., and Yan, N. (2014). Synthesis and characterization of an extractive-based bio-epoxy resin from beetle infested Pinus contorta bark. *Green Chem.* 16 (7): 3483–3493.

75 Miyagawa, H., Mohanty, A.K., Burgueño, R. et al. (2007). Novel biobased resins from blends of functionalized soybean oil and unsaturated polyester resin. *J. Polym. Sci., Part B: Polym. Phys.* 45 (6): 698–704.

76 Mustapha, S.N.H., Rahmat, A.R., Arsad, A. et al. (2014). Novel bio-based resins from blends of functionalised palm oil and unsaturated polyester resin. *Mater. Res. Innovat.* 18 (Suppl. 6): S6-326–S326-330.

77 Tan, S.G. and Chow, W.S. (2010). Thermal properties, fracture toughness and water absorption of epoxy-palm oil blends. *Polym.-Plast. Technol. Eng.* 49 (9): 900–907.

78 Mosiewicki, M.A. and Aranguren, M.I. (2013). A short review on novel biocomposites based on plant oil precursors. *Eur. Polym. J.* 49 (6): 1243–1256.

79 Scalici, T., Fiore, V., and Valenza, A. (2016). Effect of plasma treatment on the properties of *Arundo donax* L. leaf fibres and its bio-based epoxy composites: a preliminary study. *Composites Part B* 94: 167–175.

80 La Rosa, A.D., Recca, G., Summerscales, J. et al. (2014). Bio-based versus traditional polymer composites. A life cycle assessment perspective. *J. Clean. Prod.* 74: 135–144.

81 McSwiggan, C. and Fam, A. (2017). Bio-based resins for flexural strengthening of reinforced concrete beams with FRP sheets. *Constr. Build. Mater.* 131: 618–629.

82 McIsaac, A. and Fam, A. (2018). Durability under freeze–thaw cycles of concrete beams retrofitted with externally bonded FRPs using bio-based resins. *Constr. Build. Mater.* 168: 244–256.

83 Temmink, R., Baghaei, B., and Skrifvars, M. (2018). Development of biocomposites from denim waste and thermoset bio-resins for structural applications. *Composites Part A* 106: 59–69.

84 Fei, M.-e., Liu, W., Jia, A. et al. (2018). Bamboo fibers composites based on styrene-free soybean-oil thermosets using methacrylates as reactive diluents. *Composites Part A* 114: 40–48.

85 Zhan, M. and Wool, R.P. (2013). Design and evaluation of bio-based composites for printed circuit board application. *Composites Part A* 47: 22–30.

86 Masoodi, R. and Pillai, K.M. (2012). A study on moisture absorption and swelling in bio-based jute-epoxy composites. *J. Reinf. Plast. Compos.* 31 (5): 285–294.

87 Liu, W., Chen, T., Fei, M.-e. et al. (2019). Properties of natural fiber-reinforced biobased thermoset biocomposites: effects of fiber type and resin composition. *Composites Part B* 171: 87–95.

88 Alexander, M.G., Benzarti, K., Chlela, R. et al. (2018). Durability of flax/bio-based epoxy composites intended for structural strengthening. *MATEC Web Conf.* 199: 07014.

89 Cuinat-Guerraz, N., Dumont, M.-J., and Hubert, P. (2016). Environmental resistance of flax/bio-based epoxy and flax/polyurethane composites manufactured by resin transfer moulding. *Composites Part A* 88: 140–147.

90 Perremans, D., Guo, Y., Baets, J. et al. (2014) Improvement of the interphase strength and the moisture sensitivity of flax fibre reinforced bio-epoxies: effect of various fibre treatments. *Proceedings of the ECCM-16*, Seville, Spain (22–26 June 2014).

91 Zhu, J., Zhu, H., Njuguna, J. et al. (2013). Recent development of flax fibres and their reinforced composites based on different polymeric matrices. *Materials (Basel)* 6 (11): 5171–5198.

92 Loong, M.L. and Cree, D. (2017). Enhancement of mechanical properties of bio-resin epoxy/flax fiber composites using acetic anhydride. *J. Polym. Environ.* 26 (1): 224–234.

93 Taylor, C., Amiri, A., Paramarta, A. et al. (2017). Development and weatherability of bio-based composites of structural quality using flax fiber and epoxidized sucrose soyate. *Mater. Des.* 113: 17–26.

94 Samper, M., Petrucci, R., Sanchez-Nacher, L. et al. (2015). Properties of composite laminates based on basalt fibers with epoxidized vegetable oils. *Mater. Des.* 72: 9–15.

95 Niedermann, P., Szebényi, G., and Toldy, A. (2015). Characterization of high glass transition temperature sugar-based epoxy resin composites with jute and carbon fibre reinforcement. *Compos. Sci. Technol.* 117: 62–68.

96 Di Landro, L. and Janszen, G. (2014). Composites with hemp reinforcement and bio-based epoxy matrix. *Composites Part B* 67: 220–226.

97 Rwawiire, S., Tomkova, B., Militky, J. et al. (2015). Development of a biocomposite based on green epoxy polymer and natural cellulose fabric (bark cloth) for automotive instrument panel applications. *Composites Part B* 81: 149–157.

98 Johnson, R., Prabu, V.A., Amuthakkannan, P. et al. (2017). A review on biocomposites and bioresin based composites for potential industrial applications. *Rev. Adv. Mater. Sci.* 48 (1): 112–121.

99 Khandelwal, V., Sahoo, S.K., Kumar, A. et al. (2018). Electrically conductive green composites based on epoxidized linseed oil and polyaniline: an insight

into electrical, thermal and mechanical properties. *Composites Part B* 136: 149–157.

100 Avril, C., Bailly, P., Njuguna, J. et al. (eds.) (2012). Development of flax-reinforced bio-composites for high-load bearing automotive parts. *Proceeding of ECCM*, Venice.

101 Sahoo, S.K., Mohanty, S., and Nayak, S.K. (2016). Mechanical, thermal, and interfacial characterization of randomly oriented short sisal fibers reinforced epoxy composite modified with epoxidized soybean oil. *J. Nat. Fibers* 14 (3): 357–367.

102 Gauvin, F., Richard, C., and Robert, M. (2018). Modification of bamboo fibers/bio-based epoxy interface by nano-reinforced coatings. *Polym. Compos.* 39 (5): 1534–1542.

103 Zhu, J. (2015). Development of novel flax bio-matrix composites for non-structural and structural vehicle applications. Doctoral thesis. Cranfield University.

104 Schuster, J., Govignon, Q., and Bickerton, S. (2014). Processability of biobased thermoset resins and flax fibres reinforcements using vacuum assisted resin transfer moulding. *Open J. Compos. Mater.* 04 (01): 1–11.

105 Chandramohan, D. and Marimuthu, K. (2011). A review on natural fibers. *Int. J. Res. Rev. Appl. Sci.* 8 (2): 194–206.

106 Barragán, C.A.H., Valencia, B.A.R., and Valencia, F.R. (2014). Development of new materials based on bio-epoxy resins with natural fibers for automotive industry. *Revista Ambiental Agua, Aire Y Suelo* 5 (2): 3228. ISSN: 1900-9178.

107 Dash, A. and Tripathy, S. (2018). Mechanical characteristics of chicken feather teak wood dust epoxy filled composite. *IOP Conf. Ser.: Mater. Sci. Eng.* 377 (1): 012111.

108 Sankar, N. and Chandramohan, V. (2015). A study on mechanical behaviour of natural fiber reinforced composite. *Int. J. Sci. Eng. Res.* 3 (5): 3221.

109 Maino, A., Janszen, G., and Di Landro, L. (2019). Glass/Epoxy and Hemp/Bio based epoxy composites: manufacturing and structural performances. *Polym. Compos.* 40 (S1): E723–E731.

110 Pil, L., Bensadoun, F., Pariset, J. et al. (2016). Why are designers fascinated by flax and hemp fibre composites? *Composites Part A* 83: 193–205.

111 Fathi, B., Foruzanmehr, M., Elkoun, S. et al. (2019). Novel approach for silane treatment of flax fiber to improve the interfacial adhesion in flax/bio epoxy composites. *J. Compos. Mater.* 53 (16): 2229–2238.

112 Michelena, A.H., Graham-Jones, J., Summerscales, J. et al. (2015). Eco-friendly epoxy resin/flax fibre composite system interface improvement with the chemical treatments of the flax surface and resin formulation. *2nd ICNF2015*, São Miguel.

113 Liotier, P.-J., Pucci, M.F., Le Duigou, A. et al. (2019). Role of interface formation versus fibres properties in the mechanical behaviour of bio-based composites manufactured by Liquid Composite Molding processes. *Composites Part B* 163: 86–95.

114 Scarponi, C., Sarasini, F., Tirillò, J. et al. (2016). Low-velocity impact behaviour of hemp fibre reinforced bio-based epoxy laminates. *Composites Part B* 91: 162–168.

115 Ahmad, F. and Bajpai, P.K. (2018). Evaluation of stiffness in a cellulose fiber reinforced epoxy laminates for structural applications: experimental and finite element analysis. *Defence Technology* 14 (4): 278–286.

116 Mancinoa, A., Marannano, G., and Zuccarello, B. (2018). Implementation of eco-sustainable biocomposite materials reinforced by optimized agave fibers. *Procedia Structural Integrity* 8: 526–538.

117 Giuliani, P., Giannini, O., and Panciroli, R. (2018). Viscoelastic experimental characterization of flax/epoxy composites. *Procedia Structural Integrity* 12: 296–303.

118 Michelena, A.H., Graham-Jones, J., Summerscales, J. et al. (2015). Eco-friendly epoxy resin/flax fibre composite system as a material for surf boards production. *2nd ICNF2015*, São Miguel.

119 Zhu, J., Immonen, K., Avril, C. et al. (2015). Novel hybrid flax reinforced supersap composites in automotive applications. *Fibers* 3 (4): 76–89.

120 Graceraj, P.P., Venkatachalam, G., Pandi, C. et al. (2016). Analysis of mechanical behavior of hybrid fibre (Jute-Gongura) reinforced hybird polymer matrix composites. *U.P.B. Sci. Bull.* 78 (3). ISSN 1454-2358.

121 Fangueiro, R. and Rana, S. (2016). *Natural Fibres: Advances in Science and Technology Towards Industrial Applications*. Dordrecht: Springer.

122 Corona, A., Markussen, C.M., Birkved, M. et al. (2015). Comparative environmental sustainability assessment of bio-based fibre reinforcement materials for wind turbine blades. *Wind Eng.* 39 (1): 53–63.

123 Jamshaid, H., Mishra, R., Militky, J. et al. (2016). Mechanical, thermal and interfacial properties of green composites from basalt and hybrid woven fabrics. *Fibers Polym.* 17 (10): 1675–1686.

124 Moeini, M., Salehi, M., and Yarmohammadi, M. (2016). Dynamic behavior of composite laminated plate with eco-friendly matrix and natural fibers and bio-inspired stacking. *40th Solid Mechanics Conference*, Warsaw.

125 Omrani, E., Barari, B., Moghadam, A.D. et al. (2015). Mechanical and tribological properties of self-lubricating bio-based carbon-fabric epoxy composites made using liquid composite molding. *Tribol. Int.* 92: 222–232.

126 Yue, L., Maiorana, A., Patel, A. et al. (2017). A sustainable alternative to current epoxy resin matrices for vacuum infusion molding. *Composites Part A* 100: 269–274.

127 Chandramohan, D. and Bharanichandar, J. (2013). Impact test on natural fiber reinforced polymer composite materials. *Carb Sci Technol* 5: 314–320.

128 Ragunath, S., Velmurugan, C., Kannan, T. et al. (2018). Evaluation of tensile, flexural and impact properties on sisal/glass fiber reinforced polymer hybrid composites. *NISCAIR-CSIR* 25 (5): 425–431.

129 Fiore, V., Scalici, T., Badagliacco, D. et al. (2017). Aging resistance of bio-epoxy jute-basalt hybrid composites as novel multilayer structures for cladding. *Compos. Struct.* 160: 1319–1328.

130 Fiore, V., Scalici, T., Badagliacco, D. et al. (2016). Durability behavior of bio-epoxy/jute-basalt hybrid composites for cladding. *Abstracts from*

National Young Researchers' Forum on Materials Science and Technology, XIII AIMAT National Congress, Ischia, (July 2016), Milano. Italy: Wichtig Publishing.

131 Boccarusso, L., Durante, M., and Langella, A. (2018). Lightweight hemp/bio-epoxy grid structure manufactured by a new continuous process. *Composites Part B* 146: 165–175.

132 Bharath, B., Kumar, G.C., Shivanna, G. et al. (2018). Fabrication and mechanical characterization of bio-composite helmet. *Materials Today: Proceedings* 5 (1): 2716–2720.

133 Senthilathiban, A., Ravi Kumar, L., and Chandramohan, D. (2016). Characterization of mechanical seal using hybrid of natural materials. *Indian Journal of Science and Technology* 9 (48): 1–5.

134 Santulli, C. (2019). Mechanical and impact damage analysis on carbon/natural fibers hybrid composites: a review. *Materials (Basel)* 12 (3): 517.

135 Mistretta, M.C., Botta, L., Morreale, M. et al. (2018). Injection molding and mechanical properties of bio-based polymer nanocomposites. *Materials (Basel)* 11 (4): 613.

136 Kuo, P.Y., Barros, L.A., Yan, N. et al. (2017). Nanocellulose composites with enhanced interfacial compatibility and mechanical properties using a hybrid-toughened epoxy matrix. *Carbohydr. Polym.* 177: 249–257.

137 Rigail-Cedeño, A. and Schmidt, D.F. (2017). Bio-based epoxy clay nanocomposites. *AIP Conference Proceedings* 1914 (1): 030023.

138 Baruah, P. and Karak, N. (2016). Bio-based tough hyperbranched epoxy/graphene oxide nanocomposite with enhanced biodegradability attribute. *Polym. Degrad. Stab.* 129: 26–33.

139 Wu, G.-m., Liu, D., Liu, G.-f. et al. (2015). Thermoset nanocomposites from waterborne bio-based epoxy resin and cellulose nanowhiskers. *Carbohydr. Polym.* 127: 229–235.

140 Das, G., Kalita, R.D., Deka, H. et al. (2013). Biodegradation, cytocompatability and performance studies of vegetable oil based hyperbranched polyurethane modified biocompatible sulfonated epoxy resin/clay nanocomposites. *Prog. Org. Coat.* 76 (7–8): 1103–1111.

141 Sahoo, S.K., Mohanty, S., and Nayak, S.K. (2015). Study of thermal stability and thermo-mechanical behavior of functionalized soybean oil modified toughened epoxy/organo clay nanocomposite. *Prog. Org. Coat.* 88: 263–271.

142 Das, G., Deka, H., and Karak, N. (2013). Bio-based sulfonated epoxy/hyperbranched polyurea-modified MMT nanocomposites. *Int. J. Polym. Mater.* 62 (6): 330–335.

143 Chandramohan, D. and Presin Kumar, A.J. (2017). Experimental data on the properties of natural fiber particle reinforced polymer composite material. *Data Brief* 13: 460–468.

144 Loeffen, A., Cree, D., Sabzevari, M. et al. (2018). Graphene oxide reinforced bio-epoxy polymers. *Proceedings of The CSME International Congress*, Toronto, Canda, (27–30 May 2018). Canada: York University.

145 Collins, M.N., Nechifor, M., Tanasa, F. et al. (2019). Valorization of lignin in polymer and composite systems for advanced engineering applications – a review. *Int. J. Biol. Macromol.* 131: 828–849.

146 Nisha, S.S., Nikzad, M., Al Kobaisi, M. et al. (2019). The role of ionic-liquid extracted lignin micro/nanoparticles for functionalisation of an epoxy-based composite matrix. *Compos. Sci. Technol.* 174: 11–19.

147 Kocaman, S., Karaman, M., Gursoy, M. et al. (2017). Chemical and plasma surface modification of lignocellulose coconut waste for the preparation of advanced biobased composite materials. *Carbohydr. Polym.* 159: 48–57.

148 Fombuena, V., Bernardi, L., Fenollar, O. et al. (2014). Characterization of green composites from biobased epoxy matrices and bio-fillers derived from seashell wastes. *Mater. Des.* 57: 168–174.

149 Abdul Khalil, H.P.S., Fizree, H.M., Bhat, A.H. et al. (2013). Development and characterization of epoxy nanocomposites based on nano-structured oil palm ash. *Composites Part B* 53: 324–333.

150 Harish, S., Sanjeeva, M., and Venkatesh Guptha, N.S. (2018). The study of tensile and flexural strength of cattle bone particulate reinforced epoxy. *Materials Today: Proceedings* 5 (10): 20927–20931.

151 Zheng, S., Bellido-Aguilar, D.A., Huang, Y. et al. (2019). Mechanically robust hydrophobic bio-based epoxy coatings for anti-corrosion application. *Surf. Coat. Technol.* 363: 43–50.

152 Soundhar, A., Rajesh, M., Jayakrishna, K. et al. (2019). Investigation on mechanical properties of polyurethane hybrid nanocomposite foams reinforced with roselle fibers and silica nanoparticles. *Nanocomposites* 5 (1): 1–12.

153 Huo, L., Wang, D., Liu, H. et al. (2016). Cytotoxicity, dynamic and thermal properties of bio-based rosin-epoxy resin/ castor oil polyurethane/ carbon nanotubes bio-nanocomposites. *J. Biomater. Sci. Polym. Ed.* 27 (11): 1100–1114.

154 Rajaei, M., Kim, N.K., Bickerton, S. et al. (2019). A comparative study on effects of natural and synthesised nano-clays on the fire and mechanical properties of epoxy composites. *Composites Part B* 165: 65–74.

155 Paluvai, N.R., Mohanty, S., and Nayak, S.K. (2015). Cure kinetics of exfoliated bio-based epoxy/clay nanocomposites developed from acrylated epoxidized castor oil and diglycidyl ether bisphenol A networks. *High Perform. Polym.* 27 (8): 918–929.

156 Dutta, S., Karak, N., Saikia, J.P. et al. (2009). Biocompatible epoxy modified bio-based polyurethane nanocomposites: mechanical property, cytotoxicity and biodegradation. *Bioresour. Technol.* 100 (24): 6391–6397.

157 Hirose, S., Hatakeyama, T., and Hatakeyama, H. (2012). Novel epoxy resins derived from biomass components. *Procedia Chem.* 4: 26–33.

158 Mustapha, R., Rahmat, A.R., Majid, R.A., and Mustapha, S.N.H. (2018). Mechanical and thermal properties of montmorrillonite nanoclay reinforced epoxy resin with bio-based hardener. *Materials Today: Proceedings* 5 (10): 21964–21972.

159 Barari, B., Omrani, E., Dorri Moghadam, A. et al. (2016). Mechanical, physical and tribological characterization of nano-cellulose fibers reinforced bio-epoxy composites: an attempt to fabricate and scale the 'Green' composite. *Carbohydr. Polym.* 147: 282–293.

160 Wongjaiyen, T., Brostow, W., and Chonkaew, W. (2018). Tensile properties and wear resistance of epoxy nanocomposites reinforced with cellulose nanofibers. *Polym. Bull.* 75 (5): 2039–2051.

161 Meng, X., Bocharova, V., Tekinalp, H. et al. (2018). Toughening of nanocellulose/PLA composites via bio-epoxy interaction: mechanistic study. *Mater. Des.* 139: 188–197.

162 Masoodi, R., El-Hajjar, R.F., Pillai, K.M. et al. (2012). Mechanical characterization of cellulose nanofiber and bio-based epoxy composite. *Mater. Des. (1980–2015)* 36: 570–576.

163 University of Wisconsin System Solid Waste Research Program (2013). Characterization of Regenerated Cellulose for Bio-based Epoxy Fibrous Composites. Student Project Report.

164 Gopal, P., Raja, V.B., Chandrasekaran, M. et al. (2017). Wear study on hybrid natural fiber epoxy composite materials used as automotive body shell. *ARPN J. Eng. Appl. Sci.* 12 (8).

165 Patil, A.Y., Hrishikesh, N.U., Basavaraj, G.D. et al. (2018). Influence of bio-degradable natural fiber embedded in polymer matrix. *Mater. Today: Proc.* 5 (2): 7532–7540.

166 Gope, P.C. (2018). Maximum tangential stress coupled with probabilistic aspect of fracture toughness of hybrid bio-composite. *Eng. Sci. Technol., Int. J.* 21 (2): 201–214.

167 Vivek, S. and Kanthavel, K. (2019). Effect of bagasse ash filled epoxy composites reinforced with hybrid plant fibres for mechanical and thermal properties. *Composites Part B* 160: 170–176.

168 Prasad, V., Suresh, D., Joseph, M.A. et al. (2018). Development of flax fibre reinforced epoxy composite with nano TiO_2 addition into matrix to enhance mechanical properties. *Mater. Today: Proc.* 5 (5): 11569–11575.

169 Mishra, R., Jamshaid, H., and Militky, J. (2017). Basalt nanoparticle reinforced hybrid woven composites: mechanical and thermo-mechanical performance. *Fibers Polym.* 18 (12): 2433–2442.

8

Bio-epoxy Polymer, Blends and Composites Derived Utilitarian Electrical, Magnetic and Optical Properties

RaviPrakash Magisetty and Naga Srilatha CH

Defence Institute of Advanced Technology, Materials Engineering Division, Pune 411025, India

8.1 Introduction

Polyepoxides are a class of reactive prepolymers consisting of epoxide groups reacted with polyfunctional hardeners or with themselves to form a thermosetting polymer [1–4]. This kind of epoxide thermosets often exhibit electrical, magnetic, and optoelectronic properties and are enabled by neat resins, blends, and composites facilitating for functional applications including high-tension electrical insulators, structural adhesives, bar-reinforced plastic materials, and metal coatings for electronic and electrical components [1–4]. According to the recent research report, production intensity of bio-based polymers, their derivative blends, and composites will reach from 3.5 million tonnes in 2011 to nearly 12 million tonnes by 2020 [5]. In recent years, applications derived by bio-based polymers and their composites are rapidly expanding because of the growing awareness of harmful chemical manufacturing processes and related environmental concerns, as well as rapid depletion of petroleum resources [5]. Among them, thermosetting bio-based resin materials are more popular because of their availability of states such as liquid state, viscous state, or soft solid; even these states of matter can be reversible with the help of curing agents or solvents [6]. Moreover, these thermoset materials exhibit high strength, high modulus, and excellent chemical and thermal resistances; are widely used in the form of coatings, adhesives, and composites; and are utilized in the field ranging from large electrical systems such as electrical transformers, generators, and motors and their components to miniaturization of electronic circuit boards. Bioepoxies consists of a minimum of two epoxide groups, also known as glycidyls [5–7]. Cationic homopolymerization of these bioepoxies can be done with the presence of coreactants such as alcohols, acids, thiols, anhydrides, phenols, and polyfunctional amines, which are also referred as hardeners. This wide range of coreactants enabled homopolymerized bioepoxies, their blends, and composites that found applications in various manufacturing solicitations (such as electrical/electronic laminates, adhesives, flooring and paving applications, and coatings as well as high-performance composites for electrical applications)

Bio-Based Epoxy Polymers, Blends, and Composites: Synthesis, Properties, Characterization, and Applications, First Edition. Edited by Jyotishkumar Parameswaranpillai, Sanjay Mavinkere Rangappa, Suchart Siengchin, and Seno Jose.

requiring excellent adhesion, superior strength, excellent performance, good chemical resistance even at elevated temperatures and stable electrical and electronic properties [5–7]. Also, strategic reconfiguration of bioepoxies has been enlightly adapted as adhesives and sealants for high-performance electronic applications such as foldable printed circuit boards, highly bendable transparent thin film transistors, and flexible microfluidic devices. Importantly, superior adhesion caused by hydroxyl groups of these bioepoxy materials have attracted considerable attention for electrical interconnections between semiconductor chips and substrates such as flip chip bondings [5–8]. Although the magnificent progress is achieved in understanding and obtaining the functional properties of bio-based epoxy resin-based materials, there remains a challenge to properties other than mechanical and thermal properties, such as electrical, magnetic, and optical properties that are an essential prerequisite for many functional applications [5–9]. These properties are probably attributable to the cured stoichiometric or near-stoichiometric quantities of curatives, as well as their blend configurations and a wide range of filler compatibility within a composite system. In addition to the other class of thermosets, different epoxy-grade blends, including plasticizers, additives, or fillers with sophisticated processing methods, are predominant to achieve final functional properties with reduced cost [5–9].

In this context, current advancements in bioepoxies, their blends, and composite materials, in addition to the significance of bioepoxy-based materials, have been discussed. Bioresins and curing agents realized via various synthetic routes from bio-based resources are highlighted, and their electrical, magnetic, and optical properties are reviewed. Further, wide functionalities of bioepoxy-based materials facilitated that efficient and low-cost electrical and electronic devices are expounded.

8.2 Significance of Bioepoxy-Based Materials

Epoxy materials are a class of high-performance thermosetting precursors, wherein epoxy resin contains epoxide groups that can be cured by a diversity of curing agents, such as amines and anhydrides, to form a three-dimensional network [2–4]. This kind of cured 3D network system, its derivatives, and composites hold a dominant position in the commercial application of polymer market, especially automotive, construction, electrical and electronics, aerospace, and military applications because of their superior chemical resistance, high strength, minimum shrinkage, and volatility of by-products while maintaining the electrical and electronic properties [2–4]. Thus, recent studies report that the global market demand of epoxies, their blends, and composites was estimated at about US$ 18.6 billion in 2013 and was predicted to reach US$ 25.8 billion by 2018. This reflects an increment in the value of epoxy material in the global market over the coming years. However, the experimental studies reveal that the major content of bisphenol in epoxy resin induces toxicity [1]. Therefore, in order to avoid the usage of this toxic material, researchers have experimentally investigated and synthesized bioepoxies tremendously with the help of natural renewable compatible biomaterials, which proved biocompatibility and degradability to determine their feasibilities as alternatives [1]. Vegetable oil, liquefied wood, and lignin are considered as the most promising materials to realize bio-based epoxies and their blends [8, 10]. Epoxidized vegetable oils are already commercialized and commonly used as plasticizers, and this material chemical structure is based on aliphatic chains and aromatic groups [8, 10]. Moreover, bark extractives are used as a new type of bioepoxy resins [11–13]. The benefits include its abundance, renewability, and richness in phenolic compounds. Also, many research investigations and techniques are being studied and applied to find a suitable solution to convert the waste bark into useful products. However, the most common approach to synthesize epoxy resin monomers is to etherify the phenolic hydroxyl groups using epichlorohydrin [11–13]. The mechanism involves using the hydroxyl groups as nucleophiles to attack carbons on epichlorohydrin (ECH). Compared to the other methods such as double bond oxidation, the ECH method is more cost-effective and industrially preferred and has a simple purification process [11–13]. Although bioepoxy resins have been produced through the ECH procedure, bark extractives contain complex and diverse compounds, resulting in irregular reactivity [11–13]. A potential bioepoxy not only requires high yield and reactivity of monomer but also demands good mechanical, thermal, and electrical properties of cured resin, which could be further realized via bioepoxy blends and composites [5–9]. Although the dimensional composition and structures of bioepoxies induce electronic properties to facilitate functional electrical, electronic, and optoelectronic applications, the processability, reactivity, moisture, and oxygen stability could be the major problems that make them incompatible for functional applications [5–9]. To overcome this, bioepoxies, their blends, and composites were extensively explored as alternative epoxy materials by introducing a wide range of derivatives and filler-incorporated materials that can enable desirable functional advantages [5–9]. This intense interest of low-dimensional systems with high potential electrical, electronic, and optoelectronic properties could lead to an investigation of the number of components and devices. Moreover, interesting architecture rendered a wide range of macromolecules based on polymer systems and their blend that can be prepared via synthetic routes, thereby extended the scope to various molecular-level electrical, electronic, and optoelectronic applications [5–9]. Importantly, researchers have extensively explored the properties of long-chain oriented epoxy systems; this long-chain base material possesses a number of conjugated double bonds to lead to intensified electronic conductivity and optical characteristics [5–9]. Further, the chemical and structural anisotropy of this conjugated conducting thermosetting polymer probably possesses anisotropic electronic characteristics. Conjugated soluble polymers possess significant interest because of their anisotropic functional advantages such as electrical and optical characteristics while maintaining the solubility [14]. However, in electronic devices especially designing of large voltages enabled molecular design field effect transistors (FETs) induce extensive power consumption, though high dielectric materials can facilitate for this purpose, negatively affected charge carrier mobility leads to energetic disorder [14, 15]. To improve this, low k dielectrics are feasible, which compensate to reduce capacitance effect [14, 15]. Bioepoxies and their blends or composites could possess low dielectric characteristics; this suggests that it could be a potential candidate for electronic as well as moletronic applications. Also, these conjugated systems exhibit magnetic and optoelectronic properties, which are probably attributable to the conjugation in the main backbone chain in addition to the solitons or antisolitons effect that may yield magnetic, optoelectronic, and charge transport properties [14, 15]. However, all the above research delineations suggest that many bioepoxies, blends, and composites have been synthesized successfully, but their performance is often insufficient to high-performance material industry. Although magnificent progress is achieved in understanding and obtaining the functional properties of bio-based epoxy resin-based materials, there remains a challenge to properties other than mechanical and thermal properties, such as electrical, magnetic, and optical properties, which are an important prerequisite for many industrial and commercial applications.

8.3 Bioepoxy-Derived Utilitarian Electrical, Magnetic, and Optical Properties

8.3.1 Bioepoxy-Based Material: Electrical and Electronic Properties

Current advancement in electrical and electronic applications demanding the thermoset epoxy resins in the form of sealants, coatings, impregnants, and adhesives, which is probably because of its insulating characteristics. Also, potting and molding compounds with the help of an epoxy-based material were used to produce void-free insulation around components [2–4]. However, for a desirable application, thermoset epoxy material selection demands understanding of insulating characteristics, which requires not only electrical and electronic properties but also other functional characteristics such as chemical resistance, mechanical strength, thermal cycling, operating temperature range, dimensional stability, and resistance to vibration and shock [2–4]. Researchers have categorized epoxies into two kinds for electrical and electronic systems such as epoxies with amine groups and those coupled with anhydride curing agents [2–4]. Among them, the amine group-enabled epoxies are most frequently used materials such as sealants, adhesives, coatings, and impregnants. Also, anhydride-based epoxies can be utilized as potting and encapsulation materials, particularly the components that require high temperature demands or broad volume usage [2–4]. Ultraviolet light energy curable epoxies have been increased in recent days because of their high-performance sealing, bonding, and coating to applications [2–4]. Moreover, insulating characteristics of these epoxy resins such as loss tangent and dielectric constant vary with the type of resin. Aromatic cured epoxy resins are better because of their low loss tangent, high dielectric constant, and moderate conductivity as compared to the aliphatic amine-cured resins [2–4]. Higher conductivities are probably attained with the aid of amine salts or adducts with water, which highly dissociated into separate ions than either amines or organic carboxylic acids themselves. However, recently, researchers have proved that in the presence of organic acids, aliphatic amines produce conducting

Table 8.1 Electrical and electronic properties, method of epoxy casting technique, and ASTM test measurement methods [2–4].

Electrical properties	Method of epoxy casting systems	Test methods
Volume resistivity	Vacuum pressure impregnation (VPI)	ASTM D257
Surface resistivity	Pressure gelation and vacuum casting	ASTM D257
Dialectic constant	Vacuum casting and pressure gelation	ASTM D150
Arc resistance	Vacuum casting	ASTM D495
Dielectric strength	Vacuum casting	ASTM D149
Dissipation factor	State of loss of energy by oscillation in a dissipative system	ASTM D150

properties [2–4]. Similarly, aromatic anhydride-cured resins produce the best dielectric properties. The high softening and hardening are usually associated with the lower electrical conductivity, loss factor, and dielectric constant because of its reduced internal dipolarity and ionic mobility [2–4, 16–23].

Conversely, higher dielectric losses are more likely to exist in flexible resins. Recently, researchers have reported that the epoxy resin temperature depends on the dielectric loss tangent characteristics at 60-Hz frequency [2–4, 16–23]. The effect of temperature and frequency, also other conditions, also influences the electrical characteristics; the shift of peak provides evidence for the factors influencing the electrical properties [2–4, 16–23]. In all epoxy materials, these significant characteristics are always not prominent and are attributable to the 3D molecular network-instigated small polar groups. The common electrical and electronic properties, method of epoxy casting technique, and ASTM test measurement methods are listed in Table 8.1. Also, electrical systems, method of epoxy casting technique, and methodological advantages are provided in Table 8.2 [2–4].

Khandelwal et al. have developed electrically conductive panianiline (PANI) and carbon nanotube-filled biosourced epoxidized linseed oil composites [1]. The experimental results of these composites suggest that the composites' conductivity is increased with the addition of 0.1 wt% of carbon nanotubes into the epoxidized linseed oil (ELO)/PANI system, which further elucidates the synergetic effect taking place among the two different filler types. Further improvement in conductivity was obtained by the addition of 0.4 wt% of carbon nanotubes (CNTs), which was 4.5×10^{-5} S/cm. This electrical conductivity is attributable to the PANI dispersion in ELO matrix system. Moreover, they have suggested that the conductivity might be attributable to the Newtonian behavior, since, the addition of CNT to ELO/PANI15 mixture leads to enhancement in the viscosity and displayed near thinning behavior rendered systematic network leads to systematic mobility of charged electrons, thus further attributable to the high conductivity. Also, Young's modulus, tensile strength, increment in T_g, and unaffected thermal degradation in spite of the incorporation of fillers make them suitable candidates for specific applications

Table 8.2 Electrical systems, method of epoxy casting technique, and methodological advantages.

Electrical systems	Method of epoxy casting systems	Advantages
Energy generators and drives	Vacuum pressure impregnation (VPI)	Excellent viscosity and stability
Indore current and voltage transformers	Pressure gelation and vacuum casting	Outstanding toughness and low shrinkage
Indoor switchgear equipment	Vacuum casting and pressure gelation	High T_g and high toughness
Power transformer	Vacuum casting	Excellent impregnation and high toughness
HV-RIP bushings	Vacuum impregnation	Long pot life, long bushing, and excellent impregnation
Outdoor resistance	Numerous processes are available	High toughness, flexibility, hydrolysis stability, and hydrophobicity
Busbar system	Potting, casting, and vacuum impregnation	High filler load, ambient cure, and long storage
Ignition coil system	Vacuum casting	High thermal shock resistance
Potting	Potting and vacuum casting	Flame-retardant, bromine-free, ambient cure, and high thermal conductivity

such as antistatic applications [1]. Further, the research reports suggest that the effective treatment/processing methods of nanofiller-incorporated composites only improve electrical properties [5]. Li et al. [5] have reported the fabrication of multifunctional environmental friendly epoxy nanocomposites via rich functional group-enabled abundant protein utilized as biosurfactant-treated multiwalled carbon nanotubes. However, experimental investigation results elucidate that, for improvement in functional properties, gelatin-CNTs are more efficacious than the chemically treated CNTs [5]. Gelatin treatment induces improved wettability, CNT dispersion, conductivity, and other physical properties correlated with the CNT/epoxy mixture and amino-functionalized systems. Thus, the experimental evidence of improved electrical conductivity characteristics of 0.5% of gelatin-CNTs epoxy nanocomposites is 2 orders of magnitude higher than the NH2-CNT/epoxy system [5]. A significant green synthetic mechanism is responsible for improved multifunctional properties of nanomodified epoxy composite system [5]. Moreover, recently, researchers have proposed 2D materials with high electron intrinsic mobility, large surface area, and good electrical and thermal conductivity. These interesting properties are attributable to the structure and its configuration [6, 24]. Functional groups with oxygen moieties such as carboxyl, hydrozyl, carbonyl, epoxy, lactone, and phenol, among them, epoxy and hydroxyl groups are present on the basal plane are most important to improve biological activities with wide range of functionalities [7, 25, 26]. These oxygenated groups induce electrochemical, electronic, and mechanical properties. The adoption of 2D material with high dispensability and dispersity makes them unique for electrical and electronic applications even for optoelectronic components [7]. However, polymers are gaining wider use as dielectric materials. This is due to the easier processing, flexibility able to tailer made for specific uses and better resistance to the chemical attack. Among them, thermoset bio-based diglycidyl ether of diphenolate esters (DGEDP) exhibit comparable electrical properties as petroleum-derived diglycidyl ether of bisphenol-A, whereas DGEDP derived from levulinic acid is safe and readily renewable. To determine the potential requirement of dielectric properties of diglycidyl ether of bisphenol A (DGEBA), McMaster et al. have investigated and experimentally studied on synthesized series of DGEDP esters (methyl, ethyl, propyl, and butyl esters). The spectroscopic results elucidated that the DGEDP propyl induces highest dielectric constant among the series, which could be attributable to the side chain motions and segmental small local, free volume, and steric hindrances [8]. Recently, De and Karak have reported the reaction of bisphenol-A and epichlorohydrin, followed by reacting with pentaerythritol to develop low-viscous hyperbranched epoxy resin. The developed bioepoxy dielectric constant, dielectric loss, and moisture absorption were 1.8%, 0.009%, and 0.09%. Also, this resin exhibits an adhesive strength of 3429 MPa and a toughness of about 1432 MPa, especially the toughness is more than 800%; therefore, low brittleness leads to a potential material for fabrication of flexible electronic screens and devices. Further, the moisture phobility provides material stability in hazardous environmental conditions while maintaining efficient performances [27]. Moreover, researchers have suggested that the epoxies do not contain epichlorohydrin and ultraviolet chromosphere groups; hence they exhibit high resistance to ultraviolet and are durable for outdoor applications [28] such as electrical insulators and heavy electrical applications such as motors, generators, transformers, and their components [28]. Trifunctional cycloaliphatic epoxides generally have low viscosity and good adhesion strength, which are free from chlorides [28]. Therefore, in order to protect against corrosion in microelectronic industry, these resins are used as packaging materials because of their unlikely nature of chloride groups [28]. Furthermore, Gao et al. have developed transparent cycloaliphatic epoxy resin; these resins could be used as a packaging material in optoelectronic device packaging. It was prepared by addition of 1,3,5,7-tetramethylcyclotetrasiloxane and 1,2-epoxy-4-vinyl-cyclohexane through hydrosilylation. This resin exhibits excellent hydrophobicity, thermal stability, and high thermal/ultraviolet resistance [9]. However, electrical, electronic, and optical device performance depends on environmental factors, among which corrosive surfaces drastically degrade the component and device performance. In order to overcome these drawbacks, application of anticorrosive coating on metallic surfaces needs to be employed to enhance the durability of components and devices because those devices or components are enclosed by metallic surfaces. Conductive filler nanomaterials incorporated polyaniline composites, and their hybrids are also used as a coating material against anticorrosion. Recently, biobased anticorrosive coatings have been tremendously developed because of their wide functionality and application compatibility. Thus, bio-based epoxy nanocomposites are fabricated by an *in situ* method using polyaniline nanofiber carbon dot nanohybrid with respect to the epoxy as a anticorrosive material. The epoxy was obtained by polycondensation of bisphenol-A sorbitol, and monoglyceride of castor oil is used as a hybrid compound with epichlorohydrin. The anticorrosion study of nanocomposites showed excellent corrosion protection efficiency (corrosion rate: 5.68×10^{-3} miles per year) in 3.5 wt% of NaCl compared to the pristine system. Therefore, this suggests that the bio-based epoxies could be prominent candidates to protect electronic and electrical components and devices from corrosion [29]. Nobrega et al. have developed bio-inspired ultra high frequency (UHF) sensor for partial discharge detection in power transformers, where a microchip antenna was designed with a radiating element shape based on the leaf of *Jatropha mollissima* Baill plant. This antenna was protected by epoxy coating against corrosion and provides mechanical support, external noise immunity, and lifetime compatibility with power transformers [30].

The researchers explore the epoxy exposure-induced electronic properties for packaging of grapheme-based biosensors. The effects of epoxy and its curing agents on electrical characteristics of grapheme-based field effect transistor are also investigated by Uddin et al. recently [31]. The experimental results suggest that the Dirac point shift of 36 V was observed after two hours of epoxy exposure, while after 15 minutes of exposure to epoxy curing agent, the Dirac point changed more than 130 V from 63 to −68 V, completely changing the doping type of grapheme from strong p-type to strong n-type because of electron-donating nature of the adsorbed molecules. This significant change in the electronic properties of graphene is probably due to the strong electron-donating nature of the amine group present in the epoxy curing agent [31]. During the fabrication of electrical and electronic sensors, interconnection is a predominant strategy to match the two systems reliably without altering its sensing efficiency. Bruck et al. have developed nanosized exfoliated graphite (EG) dispersed in latex network sheets as strain and thermal measurement sensors. However, such sensors are always enabled commonly with the sensing element or substrate. In order to provide external connection, conductive silver epoxies or paints are feasible. Bioepoxies or blend or conductive filler-incorporated nanocomposite materials could be a reliable solution used as an impedance matching adhesive in between a mechanoelectrical interface [32]. Recently, researchers have developed a dielectric sensor based on distributed architecture topology. The measured sensor delivers dielectric information as a notch frequency between 21 and 25 GHz. This proposed sensor performs as a dielectric sensor to identify biomaterials but can be used for other applications as well. It is fabricated in a slandered micrometer size via the SiGe BiCMOS process with HBTs and a consumable power of about 13.4 mW DC power. The sensor was used to conduct measurements on air, epoxy resin, and honey. Owing to the distributed concepts and instructive interference, the sensitivity of 0.32 GHz per unit permittivity was achieved. Based on these measurements, the presented results illustrate that this technique could be a promising solution for biomaterial sensing and characterization [33]. Advanced air vehicles are a class of bio-inspired flying robots that require a wide range of power electronic components for efficient operation. Recently, Krishnan and Harursampath proposed a novel material, and the said material is a kind of piezoelectric material based on piezoelectric fibers incorporated in epoxy-based polymeric material. Microfiber composites are made by embedding piezoelectric fibers in epoxy resin, and it has a multilayer thin plate-like structure consisting of a core piezoelectric layer supplied with a dynamic electric field through outer layers by interweaving comb teeth pair of electrodes referred to as interdigitate electrodes [34]. Moreover, greener materials for the printed circuit boards (PCBs) are explored recently. The bio-based composites from soybean oil resins, chicken feathers, and E-glass fibers could potentially replace the traditional E-glass fiber-reinforced epoxy composites used in PCBs. The flame retardancy of the PCB was achieved by halogen-free melamine polyphosphate and diethylphosphinic salt. The essential properties of biocomposite Bio-PCBs were described to be compatible and reliable, including mechanical, electrical, and thermal; flammable; and peel strength. These functional properties, along with the electrical and electronic characteristics, are promising for PCB applications [10]. Moreover, Kosbar and Gelorme have reported bioepoxy resins for components and the protections in computers [35]. Recently, paper manufacturing industry produced paper by-products which are in demand because of its eco-friendly and biocompatibility properties; with these paper by-products, researchers are developing newly biocompatible and biodegradable resins and applying to develop printed circuit boards for electronic industry. Thus, this kind of biocompatible and degradable epoxy materials could be futuristic materials to develop eco-friendly sustainable components. Epoxy-based waste with biobased materials could minimize environmental pollution as a replacement of current sustainable petroleum. However, PCBs required functional properties, such as thermal stability, high moisture resistance, dielectricity, high glass transition temperature, and flame retardancy. Biopolymer such as lignin-based material is a commonly used phenolic resin naturally exhibiting hydrophobicity and shows good thermal stability. Lignin/epoxy copolymers produce electrical/electronic characteristics applicable to industrial needs [35].

8.3.2 Bioepoxy-Based Material: Magnetic and Optoelectronic Properties

Recently, bioepoxies and their blends and composites are in demand, which enabled novel applications such as theranostics, spintronics, and memory chips. Excellent cellular uptake, biocompatibility, bioconjugation possibilities, chemical modification flexibilities, its enabled characteristics, and broad wave length absorbance have attracted much interest for imaging modalities. However, magnetic dynamic spin capabilities of the epoxies are very low; introducing defects or manipulating oxygen functionalities probably improves its magnetic spin response, rendering an excellent candidate for magnetic resonance imaging contrast agents. Enayati et al. suggested that this probability of dynamic magnetic imaging contrast could be attributable to the introduction of hydroxy and epoxide groups [36]. Recently, Fraga et al. have developed paramagnetic epoxy resin, and the experimental results suggest that the loss of chlorine atom ligand during the cure of the resin indicates that Fe(III) can act as a Lewis acid catalyst for the cross-linking reactions. Consequently, at high bisphenol

A diglycidyl ether (BADGE), $n = 0$/hemin rations, the formation of ether and ester bonds occurs simultaneously during the process. On the other hand, the Fe(III) ion remains in the high spin state during the epoxy/hemin cross-linking reaction. As a result, the material exhibits the magnetic properties of hemin [37]. It has been proved that the ligand affects the electronic state of the five d-electrons in Fe(III) complexes, and depending on the ligand, energy ordering of the various electronic states of Fe(III) complexes can change. Strong field of Fe(III) complexes can change because the strong field ligands' d-electrons are in a low spin configuration ($S = 1/2$), while weak field ligands, e.g. chlorides, fluoride, or water, cause the d-electrons remain unpaired in high spin state ($S = 1/2$). This further reveals that the dissolved hemin is in thermal equilibrium between low and high spin states, while solid hemin is mostly in the high spin state, which leads to magnetic properties. Thus, magnetic susceptibility studies suggest that Fe(III) remains in the high spin states during hemin epoxy cross-linking reaction. At low temperature, the susceptibility increases, which can be associated with the strengthening of magnetic short-range interactions [37]. Although the difference between low and high spin states is important for biological activity, state spin influences the binding affinity and the connecting structure between Fe(III) and porphyrin ring leading to the probability of obtaining a wide range of bioepoxy derivatives, and their blends' composition consequently enabled magnetic properties. Such kind of composite material enabled by magnetic properties was developed for purification of ϒδT cells in order to protect from cancer. The method for cell separation by using purified human ϒδ2 antibody-immobilized mPDGs is used as a useful tool in various biological applications. The greater advantage of magnetic particles is rapidity separation upon applying magnetic fields. Thus, when coincubation with PBMCs or in vitro Vϒ9Vϒδ2 T cell cultures, the interaction between monoclonal antibody-immobilized particles and Vδ2 T cell receptor (TCR) of Vϒ9Vδ2 cells allows excellent separation. Because of this easy processing, performance, and short incubation time, the use of antibody-immobilized epoxy-functionalized mPDGs appear attractive for the preparation of Vϒ9Vδ2 T cells. Moreover, researchers have suggested that bioepoxy materials such as grafted functional epoxide groups over the magnetic nanoparticles induce wide functionalities and have been attracted in diverse applications, for example, in biosensing, drug delivery, and cell labeling. Shen et al. [38] have developed and reported magnetic nanocomposites, where poly(glycidyl methacrylate) (PGMA) with epoxy functionality were chosen as versatile materials because they are reactive, inexpensive, biocompatible, and nontoxic [39]. Also, epoxide group-instigated functional polymer was used because of its good stability during radicle polymerization and high activity in various chemical reactions, while 7-amino-4-methylcoumarin was adopted as the florescent die because of its high quantum efficiency, photostability, and presence of aminofunctional groups for easy reaction with epoxy groups [40]. The developed composites are based on numerous epoxide reactive groups grafted over magnetic–particle surface; therefore, the targeting ligands are attracted by applying magnetic force fields. Further, the resulted fluorescent and superparamagnetic properties of these nanocomposites are extensively explored for bioimaging and more. The PGMA brushes obtained via atom transfer radicle polymerization could facilitate a large number of epoxide groups onto magnetic nanoparticles enabling multilevel functionalities, which could be used as a florescent dye, where biomolecules act as targeting moieties for a specific recognition and epoxy groups act as anchors for attaching onto the nanoparticle surface. These newly developed magnetic nanocomposite-grafted epoxide groups with enabled wide functional activities lead to a novel versatile platform to develop other possibilities in nanomedicine and biological chemistry [38]. Biologically modified magnetic beads are most promising materials that offer new attractive possibilities in biomedicine and bioanalysis because they can attach with the biological molecules and they can also be manipulated by an external magnetic field. Wide range of magnetic beads categories even surface functionalized nanosized particulate compatibilized beads are commercially available. Electrochemical biosensing strategies enabled by the integrated magnetic beads provide improved analytical performances [41]. Zhu et al. have reported surface-functionalized magnetic microspheres for the separation of vascular endothelial growth factor nucleic acid and lactase enzyme immobilization [42]. Results suggest that for the separation of vascular endothelial growth factor (VEGF) nucleic acid by coupling with the corresponding primer, carboxyl-modified magnetic microspheres with its proportional enzyme composition lead to wide application prospects in the field of biology and medicine [42]. Also, researchers have suggested that epoxy can be utilized as a protection system for bioelectronics, which was proposed by Yang et al.; they have developed a biosensor system, where bonding of a chip to a PCB platform is necessary to provide power supply to the system. The bonding wires used are thin conductive wires made of aluminum and are exposed to the environment. The biological fluid based epoxy was utilized on a chip surface as a protection layer in the experiment. The probability of wires could make a contact as a result of shortening of system. Epoxy may be used as an insulating material to encase the bonding system. This process is soft enough to shape into a well and will immediately dry up and harden upon exposure to UV light [43].

Recent attention in the field of polymer science has focused on optical and magnetic characteristics of polymers, which are pre-eminently depending on their chemical composition-instigated backbone structure [44]. Moreover, experimental research investigations suggest that the polymers and their blends/their various combinational derivatives could exhibit magnetic and optoelectronic properties. These special properties are probably attributable to the backbone structure and molecular chain conjugation. Also, some experimental research reports reveal that solitons/antisolitons may yield magnetic, transport, and optical properties; this phenomenon was propounded for the first time for polyacetylenes [45]. Later on, the possibility of doping through the generation of nonlinear solitons in polymers has been attracted considerable attention. Experimental results from measurements of magnetic and optical properties of lightly doped samples are in agreement with the soliton mechanism [46]. However, they suggest that in accord to the Curie law, soliton contribution decreases upon doping, but the Pauli law of susceptibility reveal that the soliton concentration is small and apparently zero in the limit of completely homogeneous doping.

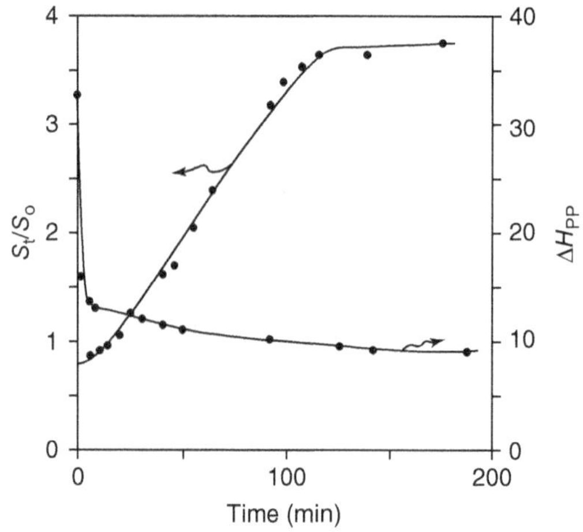

Figure 8.1 Poly(DPDPM) line width (♦) and free spin population (⧫) vs. exposure time to iodine at 25 °C in vacuum; S_0, initial spin population and S_t, the spin population at each time "*t*." Source: Reproduced with permission from Choi et al. [48]. © 2000, American Chemical Society.

Thus, localized states generated by dilute doping are nonmagnetic spins. Further, this significantly suggests in the view of dramatically increasing electrical conductivity that the charge carriers are spinless [46]. Although they would suggest that the higher doping level leads to electrical conductivity rendered spinless nonmagnetic moments, the magnetic properties advocated probably due to the spin dynamics, in which further improvements in spin dynamics could be attained because of the doping phenomena. This was experimentally investigated by Jang et al., whose results suggest that the undoped poly(DPDPM) spin density was 1018 spin/g and the line width (H_{pp}) was 33 G at $g = 2.0011$ [47]. According to Figure 8.1, poly(DPDPM), changes in spin population and line width would be attained with the aid of *in situ* doping [47]. Thus, doping of substituent-instigated radicals and nonequivalent double bonds in the backbone chain induce spin density [49, 50]. Also, electron spin resonance (ESR) of a conducting polymer has been shown in Figure 8.2. This illustration suggested that the ESR is straight forwardly connected to the spin density, i.e. with the ESR spectra of poly(dihexyldipropargyl ammonium bromide) (PDHDPAB), spin density could be quantified, and the quantified spin density is 1020 spin/g, upon doping, and its time decreases. This is further related to the spin–spin and spin–lattice relaxations, which were performed experimentally at 90 MHz; proton NMR spin–spin (τ_1) and spin–lattice (τ_2) relaxation measurements were conducted in the temperature range of 140–440 K. The results suggested that spin–spin relaxation animalities existed within the relaxation $\tau_1 = 10$ ms at 140 K to $\tau_2 = 14$ ms at 440 K. The spin–lattice relaxation (Figure 8.3) was present in the absence of τ_2 transition, τ_1 is minimum, which was due to spin diffusion to spin sites, for example, the motion of spin side or end groups to induce relaxation. With a slight shift of τ_1 toward longer values, the behavior revealed and appeared to be similar to the molecular process, which was further responsible for ESR change line width [45]. This suggests that bioepoxy or its derivative systems could exhibit magnetic properties, which would further depend on the doping level and element as a result of conjugation.

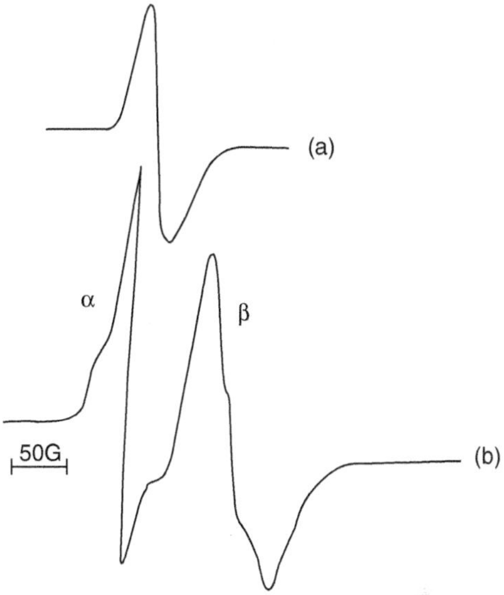

Figure 8.2 Poly(DPDPM) electron spin resonance (ESR) spectra vs. heating time at 230 °C under nitrogen gas: (a) zero time and (b) after 15 minutes. Source: Reproduced with permission from Choi et al. [48]. © 2000, American Chemical Society.

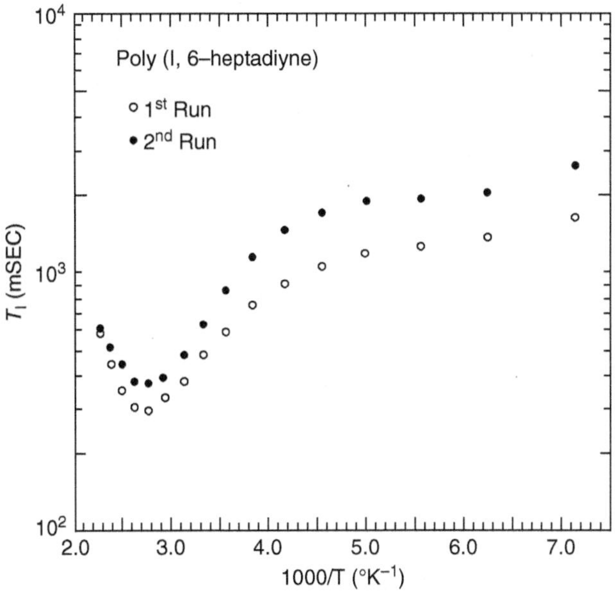

Figure 8.3 Temperature dependence (140–440 K) of spin–lattice interaction time in poly(1,6-heptadiynes) measured by the 180-T-90 sequence at 90 MHz at a Bruker CXP spectrometer. Source: Gibson et al. [45]. © 1983, American Chemical Society.

Organic polymers with π-conjugation such as PA, polydiacetylene, polythiophene, and phenylenevinylene and their derivatives and blends are third-order nonlinear optical χ^3 materials; optical nonlinearity was ranging from 10^{-12} to 10^{-9} esu, [51–53] which could be suggested for applications such as

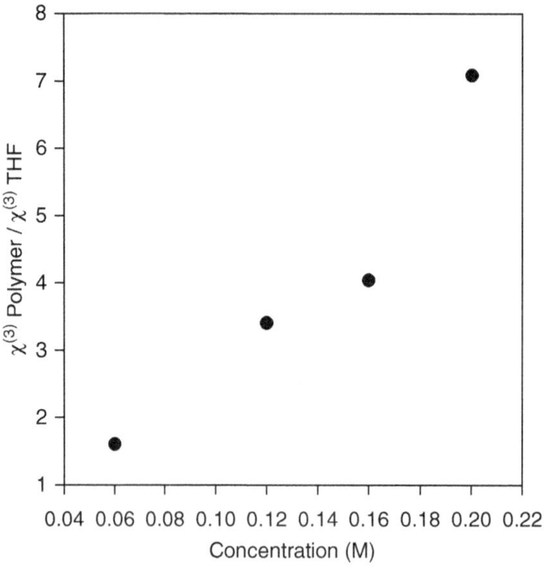

Figure 8.4 Concentration dependence of susceptibility χ^3 of poly(1,6-heptadiynes) dissolved in THF. Source: Reproduced with permission from Choi et al. [48]. © 2000, American Chemical Society.

optoelectronic and moletronic switches and ultrafast-optical devices [54, 55]. Optical nonlinearity could be effectively monitored, analyzed, and optimized with the aid of current molecular design and optimization tools. However, because of the limitations of suggested polymers and its derivatives and blends, practical applications are less explored. Bioepoxies and their derivatives and blends could be promising potential candidates as third-order optical materials because of their functional advantages over the above suggested organic materials. The nonlinear optical properties were analyzed via degenerate four-wave mixing method (DFWM), and this method was implemented for conjugated polymers during the measurement of optical susceptibility; such experimental investigations can be utilized as an optical property prediction technique in order to analyze the reliability of optoelectronic materials. Moreover, the results suggest that the optical properties are dependent on the change in the concentration of polymer solution (Figure 8.4).

8.4 Conclusion

In this context, current advancement in bioepoxy materials and their significance have been described to realize functional application probabilities. A wide range of bio-based resources enabled green synthesis of curing agents and bioepoxy resins, and their electrical, electronic, magnetic, and optical properties are reviewed. Further, this report suggests that the bioepoxy-based materials with extensive functionalities can lead to excellent functional materials to deliver efficient and low-cost electrical and electronic devices.

References

1 Khandelwal, V., Sahoo, S.K., Kumar, A. et al. (2019). Bio-sourced electrically conductive epoxidized linseed oil based composites filled with polyaniline and carbon nanotubes. *Compos. Part B Eng.* 172: 76–82.

2 Iyer, G., Gorur, R., Richert, R. et al. (2011). Dielectric properties of epoxy based nanocomposites for high voltage insulation. *IEEE Trans. Dielectr. Electr. Insul.* 18 (3): 659–666.

3 Mazzanti, G., Montanari, G.C., Peruzzotti, F., and Zaopo, A. (1996). Some remarks regarding the test cells used for electric strength measurements. In: *Conference Record of the 1996 IEEE International Symposium on Electrical Insulation*, vol. 2, 474–477. IEEE.

4 Floyd, D.E., Peerman, D.E., and Wittcoff, H. (2007). Characteristics of the polyamide-epoxy resin system. *J. Appl. Chem.* 7 (5): 250–260.

5 Li, Y., Li, R., Fu, X. et al. (2018). A bio-surfactant for defect control: Multifunctional gelatin coated MWCNTs for conductive epoxy nanocomposites. *Compos. Sci. Technol.* 159: 216–224.

6 Szabó, T., Berkesi, O., Forgó, P. et al. (2006). Evolution of surface functional groups in a series of progressively oxidized graphite oxides. *Chem. Mater.* 18 (11): 2740–2749.

7 Kim, F., Cote, L.J., and Huang, J. (2010). Graphene oxide: surface activity and two-dimensional assembly. *Adv. Mater.* 22 (17): 1954–1958.

8 McMaster, M.S., Yilmaz, T.E., Patel, A. et al. (2018). Dielectric properties of bio-based diphenolate ester epoxies. *ACS Appl. Mater. Interfaces* 10 (16): 13924–13930.

9 Gao, N., Liu, W., Yan, Z., and Wang, Z. (2013). Synthesis and properties of transparent cycloaliphatic epoxy–silicone resins for opto-electronic devices packaging. *Opt. Mater. (Amst).* 35 (3): 567–575.

10 Zhan, M. and Wool, R.P. (2013). Design and evaluation of bio-based composites for printed circuit board application. *Compos. Part A Appl. Sci. Manuf.* 47: 22–30.

11 Kuo, P.-Y. (2016) Development and characterization of an extractive-based bio-epoxy resin from beetle-infested lodgepole pine (*Pinus contorta* var. latifolia) bark. PhD Thesis. University of Toronto.

12 Kuo, P.-Y., Sain, M., and Yan, N. (2014). Synthesis and characterization of an extractive-based bio-epoxy resin from beetle infested *Pinus contorta* bark. *Green Chem.* 16 (7): 3483–3493.

13 Kuo, P.-Y., de Assis Barros, L., Sain, M. et al. (2016). Effects of reaction parameters on the glycidyl etherification of bark extractives during bioepoxy resin synthesis. *ACS Sustain. Chem. Eng.* 4 (3): 1016–1024.

14 Jung, H., Jung, K., Hong, M. et al. (2018). Understanding the origin of the regioselectivity in cyclopolymerizations of diynes and how to completely switch it. *J. Am. Chem. Soc.* 140 (2): 834–841.

15 Veres, J., Ogier, S., Lloyd, G., and de Leeuw, D. (2004). Gate insulators in organic field-effect transistors. *Chem. Mater.* 16 (23): 4543–4555.

16 Magisetty, R., Shukla, A., and Kandasubramanian, B. (2018). Dielectric, hydrophobic investigation of ABS/$NiFe_2O_4$ nanocomposites fabricated by

atomized spray assisted and solution casted techniques for miniaturized electronic applications. *J. Electron. Mater.* 47 (9): 5640–5656.

17 Magisetty, R.P., Shukla, A., and Kandasubramanian, B. (2018). Magnetodielectric microwave radiation absorbent materials and their polymer composites. *J. Electron. Mater.* 47 (11): 6335–6365.

18 Malik, A., Magisetty, R., Kumar, V. et al. (2020). Dielectric and conductivity investigation of polycarbonate-copper phthalocyanine electrospun nonwoven fibres for electrical and electronic application. *Polym. Technol. Mater.* 59: 154–168.

19 Magisetty, R., Shukla, A., and Kandasubramanian, B. (2019). Terpolymer (ABS) cermet (Ni-$NiFe_2O_4$) hybrid nanocomposite engineered 3D-carbon fabric mat as a X-band electromagnetic interference shielding material. *Mater. Lett.* 238: 214–217.

20 Magisetty, R., Prajapati, D., Ambekar, R. et al. (2019). β-Phase Cu-phthalocyanine/acrylonitrile butadiene styrene terpolymer nanocomposite film technology for organoelectronic applications. *J. Phys. Chem. C* 123 (46): 28081–28092.

21 Magisetty, R. and Cheekuramelli, N.S. (2019). Additive manufacturing technology empowered complex electromechanical energy conversion devices and transformers. *Appl. Mater. Today* 14: 35–50.

22 Magisetty, R., Kumar, P., Kumar, V. et al. (2018). $NiFe_2O_4$/Poly(1,6-heptadiyne) nanocomposite energy-storage device for electrical and electronic applications. *ACS Omega* 3 (11): 15256–15266.

23 Magisetty, R., Kumar, P., Gore, P.M. et al. (2019). Electronic properties of Poly(1,6-heptadiynes) electrospun fibrous non-woven mat. *Mater. Chem. Phys.* 223: 343–352.

24 Gao, W., Alemany, L.B., Ci, L., and Ajayan, P.M. (2009). New insights into the structure and reduction of graphite oxide. *Nat. Chem.* 1 (5): 403–408.

25 Eda, G., Mattevi, C., Yamaguchi, H. et al. (2009). Insulator to semimetal transition in graphene oxide. *J. Phys. Chem. C* 113 (35): 15768–15771.

26 Eda, G. and Chhowalla, M. (2010). Chemically derived graphene oxide: towards large-area thin-film electronics and optoelectronics. *Adv. Mater.* 22 (22): 2392–2415.

27 De, B. and Karak, N. (2013). Novel high performance tough hyperbranched epoxy by an $A_2 + B_3$ polycondensation reaction. *J. Mater. Chem. A* 1 (2): 348–353.

28 Kumar, S., Krishnan, S., Mohanty, S., and Nayak, S.K. (2018). Synthesis and characterization of petroleum and biobased epoxy resins: a review. *Polym. Int.* 67 (7): 815–839.

29 Saikia, A., Sarmah, D., Kumar, A., and Karak, N. (2019). Bio-based epoxy/polyaniline nanofiber-carbon dot nanocomposites as advanced anticorrosive materials. *J. Appl. Polym. Sci.* 136 (27): 47744.

30 Nobrega, L., Xavier, G., Aquino, M. et al. (2019). Design and development of a bio-inspired UHF sensor for partial discharge detection in power transformers. *Sensors* 19 (3): 653.

31 Uddin, M.A., Bayram, F., Koley, G. et al. (2016). Epoxy exposure induced electronic properties change of graphene. *IEEE Sensors* 2016: 1–3.

32 Bruck, H.A., Smela, E., Yu, M. et al. (2017). *A new multiscale bioinspired compliant sensor*, vol. 7, 163–169. Springer.

33 Jamal, F.I., Guha, S., Eissa, M.H. et al. (2016). A 24 GHz dielectric sensor based on distributed architecture. In: *2016 German Microwave Conference (GeMiC)*, 173–176. IEEE.

34 Krishnan, V.J. and Harursampath, D. (2015). High voltage power electronic drive circuit for flapping wing MAV. In: *2015 International Conference on Computation of Power, Energy, Information and Communication (ICCPEIC)*, 0524–0527. IEEE.

35 Kosbar, L.L. and Gelorme, J. (1997). Biobased epoxy resins for computer components and printed wiring boards. In: *Proceedings of the 1997 IEEE International Symposium on Electronics and the Environment. ISEE-1997*; San Fransisco, CA (5–7 May 1997; Code 46508), 28–32. IEEE.

36 Enayati, M., Nemati, A., Zarrabi, A., and Shokrgozar, M.A. (2019). *The Role of Oxygen Defects in Magnetic Properties of Gamma-Irradiated Reduced Graphene Oxide*. Elsevier.

37 Fraga, F., Jover, A., and Meijide, F. (2017) Paramagnetic epoxy resin. (September 2016).

38 Shen, Y., Zhao, L., Qi, L. et al. (2012). Reactive polymer as a versatile toolbox for construction of multifunctional superparamagnetic nanocomposites. *Chem. – A Eur. J.* 18 (43): 13755–13761.

39 Dong, X., Zheng, Y., Huang, Y. et al. (2010). Synthesis and characterization of multifunctional poly(glycidyl methacrylate) microspheres and their use in cell separation. *Anal. Biochem.* 405 (2): 207–212.

40 Kramer, T., Röder, T., Huber, K., and Kitzerow, H.-S. (2005). Surface modification of epoxy-functionalized acrylate colloids. *Polym. Adv. Technol.* 16 (1): 38–41.

41 Pividori, M.I. and Alegret, S. (2010). Micro and nanoparticles in biosensing systems for food safety and environmental monitoring. An example of converging technologies. *Microchim. Acta* 170 (3–4): 227–242.

42 Zhu, Y.H., Wang, Q.B., Gu, H.C. et al. (2002). Preparation of surface functional magnetic microspheres and their application in nucleic acid separation and enzyme immobilization. *Acta Acad. Med. Sin.* 24 (2): 118–123.

43 Yang, H.C., Uy, K.J., Ye, Y.Y., and Chung, W.Y. (2013). Magnetic biosensing system based on TSMC 0.35 um BioMEMS process for sensing magnetic nanobeads. *Micro & Nano Lett.* 8 (6): 288–290.

44 Masuda, T., Abdul Karim, S., and Nomura, R. (2000). Synthesis of acetylene-based widely conjugated polymers by metathesis polymerization and polymer properties. *J. Mol. Catal. A Chem.* 160 (1): 125–131.

45 Gibson, H.W., Bailey, F.C., Epstein, A.J. et al. (1983). Poly(1,6-heptadiyne), a free-standing polymer film dopable to high electrical conductivity. *J. Am. Chem. Soc.* 105 (13): 4417–4431.

46 Ikehata, S., Kaufer, J., Woerner, T. et al. (1980). Solitons in polyacetylene: magnetic susceptibility. *Phys. Rev. Lett.* 45 (13): 1123–1126.

47 Jang, M.S., Kwon, S.K., and Choi, S.K. (1990). Cyclopolymerization of diphenyldipropargylmethane by transition metal catalysts. *Macromolecules* 23 (18): 4135–4140.

48 Choi, S.K., Gal, Y.S., Jin, S.H., and Kim, H.K. (2000). Poly (1, 6-heptadiyne)-based materials by metathesis polymerization. *Chem. Rev.* 100 (4): 1645–1682.
49 Nechtschein, M., Devreux, F., Genoud, F. et al. (1983). Magnetic-resonance studies in undoped *trans*-polyacetylene $(CH)_x$. II. *Phys. Rev. B* 27 (1): 61–78.
50 Heeger, A.J., Kivelson, S., Schrieffer, J.R., and Su, W.-P. (1988). Solitons in conducting polymers. *Rev. Mod. Phys.* 60 (3): 781–850.
51 Torruellas, W.E., Neher, D., Zanoni, R. et al. (1990). Dispersion measurements of the third-order nonlinear susceptibility of polythiophene thin films. *Chem. Phys. Lett.* 175 (1–2): 11–16.
52 Halliday, D.A., Burn, P.L., Bradley, D.D.C. et al. (1993). Large changes in optical response through chemical pre-ordering of poly(p-phenylenevinylene). *Adv. Mater.* 5 (1): 40–43.
53 Bredas, J.L., Adant, C., Tackx, P. et al. (1994). Third-order nonlinear optical response in organic materials: theoretical and experimental aspects. *Chem. Rev.* 94 (1): 243–278.
54 Kajzar, F. and Messier, J. (1985). Third-harmonic generation in liquids. *Phys. Rev. A* 32 (4): 2352–2363.
55 Kajzar, F., Etemad, S., Baker, G.L., and Messier, J. (1987). X(3) of *trans*-$(CH)_x$: Experimental observation of 2Ag excited state. *Synth. Met.* 17 (1–3): 563–567.

9

Spectroscopy and Other Miscellaneous Techniques for the Characterization of Bio-epoxy Polymers, Their Blends, and Composites

Mohammad Khajouei[1], Peyman Pouresmaeel-Selakjani[1], and Mohammad Latifi[2]

[1] *Babol (Noshirvani) University of Technology, School of Chemical Engineering, Nanotechnology Research Institute, Department of Chemical Engineering, Shariati Av., P.O. Box 484, Babol, Iran*
[2] *Polytechnique Montréal, Department of Chemical Engineering, Process Engineering Advanced Research Lab (PEARL), Montréal, Quebec, Canada*

9.1 Introduction

Development of sustainable and eco-friendly materials is crucial for the environmental protection [1]. As such, synthetic epoxy polymers are castoff in a diversity of applications demanding low weight with suitable tensile characteristics [2, 3]. Nowadays, bio-based polymers resulting from renewable resources [4, 5] have turned out to be progressively important as sustainable and eco-proficient products, which can substitute the petrochemical-derived products of stocks. From all polymers manufactured, the leading part is thermoplastics, whereas the insignificant part is thermosets, 82% and 18%, respectively [6].

Monomers of bio-epoxy polymers can be formed from bio-oils, i.e. natural renewable oils extracted from plants, such as linseed, soybean, karanja, and canola [7]. The most considered, studied, and applied group of bio-epoxy polymers and blends/mixtures is the bio-epoxy resins.

The lack of sufficient thermal stability causes a limit in the widespread usage of polymer materials in fiber-covered polymer mixture structures, i.e. inapplicability at high temperature and poor fire performance.

Research studies intend to evaluate additives combined into bio-epoxy compositions and to assess their mechanical properties and thermal stability. Various methods are employed to characterize the different polymer groups such as bio-epoxy polymers and their mixtures. Tensile and Charpy durability of fractured surfaces can be studied via SEM (scanning electron microscopy). All outcomes of these investigations are a first step in sympathetic mechanical/thermal performance of additives in bio-epoxy polymers. Different types of vibrational spectroscopies (i.e. infrared, near-infrared, and Raman) are broadly employed for the understanding and characterization of epoxy polymers. In the research and development (R&D) steps, infrared and Raman spectra test results of polymers give vision into the epoxy structures at molecular level (due to the polymer

Bio-Based Epoxy Polymers, Blends, and Composites: Synthesis, Properties, Characterization, and Applications, First Edition. Edited by Jyotishkumar Parameswaranpillai, Sanjay Mavinkere Rangappa, Suchart Siengchin, and Seno Jose.

chains conformations and orientations). DSC or differential scanning calorimetry would also be utilized to calculate the thermal behavior of various bio-epoxy polymers.

The chemical configurations of such bio-based epoxy polymers have been evaluated by SEC, ESI-TOF MS, FTIR, ^{1}H NMR, MALDI-TOF mass spectrometry, and ^{13}C NMR analysis [8]. Other methods of polymer evaluation are dynamic mechanical analysis (DMA), rheological measurements, and thermogravimetric analysis (TGA). In addition, size exclusion chromatography (SEC) can be employed for the measurement of molar mass distribution of the epoxy polymers. The density of the various epoxy mixtures can be obtained via utilization of classical Archimedes procedures by immersing the polymer in water.

In the following sections (section 9.2 and its subsections), the most important techniques for the characterization of bio-epoxy polymers are described in detail.

9.2 Various Methods for Epoxy Polymer Characterization

9.2.1 FTIR Spectroscopy

For *in situ* monitoring of various processes such as phase separation, curing process, or similarity of aging, the study of spectral clarification and the band assignments are critical. There are numerous epoxy polymers with diverse characters, different polymerization degrees, etc. The IR spectroscopy method would be employed to typify the epoxy nature.

Mid-infrared spectroscopy is generally employed for organic compound characterization, which is adequate of reliable data, and libraries of spectra may simply be established. Quantitative and qualitative data would be attained via this procedure, although its functions in epoxy systems are quite limited because of the intensity and location of the oxirane ring absorptions.

Two oxirane ring absorption characteristics are detected in the ranges of 4000 and 400 cm^{-1}. The former is found at 915 cm^{-1} and credited to oxirane group C–O distortion, whereas in the research done by Dannenberg[9]. It was suggested that the mentioned band does not correspond completely to the deformation of C–O but also correspond to other unidentified progressions.

The second one was found at around 3050 cm^{-1} and credited to epoxy ring C–H methylene group pressure. This band is not completely valuable; meanwhile, the band intensity was small and is thus near to O–H strong absorptions. However, in low polymerization degrees, epoxy monomers would be employed as the qualitative revealing factor for the presence of epoxy group.

Near-infrared spectra showed extremely higher suitability for epoxies. The n-IR spectrum can cover a strong implication of vibrations in m-IR and grouping or mixed bands. In the mentioned sort, scarcer bands were detected, so it has been employed via numerous researchers [10–12] to monitor the reactions of curing process. The band intensity in the section is far lower than that in the m-IR range, thereby permitting the utilization of thicker and undiluted models to obtain reasonable data quality. In this region, two different absorptions correlated can be explained regarding the oxirane group:

1- In near 4530 cm^{-1}: This region matches to the second stretching of epoxy ring overtone combined with primary stretching of C–H at around 2725 cm^{-1}. Nonetheless, the band is appropriately divided from others, and it is proper for quantitative analysis [10–13]
2- In near 6070 cm^{-1}: The first terminal of CH_2 overtone in stretching type [14]. The band affects with the aromatic stretching overtone of C–H at near 5969 cm^{-1} [15]; therefore, there were some aromatic rings in the configuration (such as DGEBA), which was not proper for quantification.

It is noteworthy that the oxirane mixture band (combination of bending + stretching), which is frequently positioned at around 4530 cm^{-1} for typical epoxies, cannot be detected in the n-IR results of 3,4-epoxycyclohexylmethyl-3′, 4′-epoxycyclohexane carboxylate or ECC possibly because of its overlap by combination bands of C–H. Moreover, the second overtone of C=O that is typically sited at the range of from 5100 to 5200 cm^{-1} is obviously discovered. The key structures detected in this spectrum are stretching band overtones of CH_2 and C–H. While n-IR spectroscopy may not offer much valuable information for these types of polymers, it can still be employed to investigate on the curing processes over the development of bands allocated to the curing agents.

The amino group displays clear absorptions in the both of m- and n-IR ranges. The most important absorptions in m-IR are N—H bond deformation and stretch. The bands moreover replicate various variances between the primary and secondary groups of amines as described below:

- While the N–H stretching can be found in the range of 3500 cm^{-1} besides 3300 cm^{-1}, a primary group of amines demonstrate a doublet (stretching modes of symmetric–antisymmetric), whereas the secondary group of amines display a sole band.
- The deformation of N–H band is set at the range of 1650–1500 cm^{-1} in primary amines, whereas it is moved to lower wavenumbers of range 1580–1490 cm^{-1} in secondary amines, which also is frequently frail.

The measureable usage of these bands is restricted because of its spectral position: the stretching of N–H is actual near the strong absorption band of O–H (insignificant water quantity disturbs its zone), whereas the deformation band is positioned in an area where numerous signals corresponding to organic bonds capture.

In the n-IR range, the amine bands are intense and well defined. Moreover, some differences are between the primary/secondary groups of amine absorptions. The first N–H overtone stretching of primary amines contains two symmetric and antisymmetric bands that are located in the range of 6897–6452 cm^{-1}, and the symmetric one is more strong. There is a single band in the secondary amines. When both kinds exist, the secondary band cannot be utilized because of the overlap of the two. N–H stretching and bending combination would be detected at the ranges around 4900–5000 cm^{-1}, and it can be employed for quantitative drives regarding Weyer and Lo [31].

9.2.1.1 How Phase Separation Process Can Affect the IR Spectrum

The blend-phase separation includes the change of a second phase that regularly shows a dissimilar refractive index and would be identified via the presence of turbidity. Most of the investigations study the supposed as "cloud point" (could point is instant once the model is no longer obvious) evaluating visible transmission. Scatter light of particles when the size is same as to the incident radiation wavelength; when ranges of IR radiation are from 780 nm to 15 μm for normal analysis (or from 780 nm to 1.1 μm for nearby ranges and also in the mid-range of 1.1–15 μm), the phase separation onset is detected via employing n-IR or m-IR although with lower accurateness in comparison with employing visible light. Consequently, infrared experiments obtained delayed data for the cloud point values. Even more modern methods such as SAXS might give data of the incipient process of phase separation. However, IR has an advantage in this regard: It offers extra chemical information throughout the process of phase separation. The m-IR, with regard to its longer wavelength, is infrequently employed for characterizing phase separation, while it is employed to characterize additional structures including bigger size particles or to evade interventions of the system color.

Turbidity can also be detected in IR as a baseline upsurge. This factor would be utilized to track the process of phase separation in a band-free region, like 6300 cm^{-1}. Otherwise, this technique has been employed to monitor initially immiscible classification compatibilization.

Characteristic cured epoxy thermosets revealed a diversity of hydrogen bonds such as OH···OH, OH···NH, and OH···N. In the presence of PMMA, intramolecular relations convert to redistribute as the PMMA carbonyl groups interrelate with the initially existed and also lately shaped OH groups as it can be detected via the presence of bonding of carbonyl-OH hydrogen that is placed at 3500 cm^{-1}. The shifts in IR results can be correlated with the miscibility in compound systems.

In some bio-based epoxy analysis, FTIR epoxidized lignin extractives, called L-epoxy, and epoxidized cellulose extractives, known as C-epoxy, can also be discovered. L-epoxy revealed some durable absorption bands of aromatic stretch at around 3052 cm^{-1} (that is corresponds to C-H stretch), range of 2263–2033 cm^{-1} that show the ring substitution configuration, and also the range of 1632–1501 cm^{-1} of ring stretch. In the contrary, C-epoxy has a band absorption of long chain at 710 cm^{-1} and a strong vibration of alcohol at around 1113 cm^{-1}, which are also detected on the results of E-epoxy IR spectrum. In the following, there are some typical bands that are assigned to the different factors that can be found in epoxy polymers (Table 9.1).

9.2.2 Nuclear Magnetic Resonance (NMR) Spectroscopy

Various groups of atomic nuclei perform (as though spin on their axis) same as to the Earth. As they are charged positively, they produce an electro-magnetic field again same as the Earth does. Therefore, in the effect view, they can act as tiny bar magnetics. Not all of them perform in this manner, but luckily, both ^{1}H and ^{13}C show nuclear spins and respond to this method.

Table 9.1 Typical bands in the IR spectra.

Band (cm^{-1})	Assignment
3300	–OH stretching
2960	Methyl-oxirane (epoxy) of C–H stretching of $-CH_2$
1702	acrylic acid CO stretching
1616	C=C stretching in aromatic ring
1507	aromatic ring C–C stretching
1359	Aromatic –OH
1295	cyclic ether (epoxy)Stretching
1044	C–O–C ethers stretching
905	C–O stretching of Oxirane (epoxy) group
835	C–O–C in oxirane group

Once an exterior magnetic field is absent, the spin direction of nuclei will be erratically oriented. However, when this nuclei model is employed in an external magnetic field, the nuclear spins would borrow exact orientations similar to a compass needle that responses to the Earth's magnetic field and bring into line. Two probable orientations can be assumed with the external field, parallel and with the same direction of external field or against the field such as antiparallel to field.

The energy quantity and therefore, the exact EM radiation frequency that are essential for resonance to be happen (reliant on both of magnetic field strength that applied and the nuclei type) are being investigated. Because the strength of magnetic field increases, the energy variance of two spin states upsurges and a higher EM radiation frequency (that means more energy) desires to be applied to attain a spin-flip.

Superconducting magnets are employed to generate very resilient magnetic field at the demand of 21 tesla (T). Lower strengths may also be utilized in the range between 4 and 7 T. At the mentioned levels, the necessary energy to pass the nuclei into resonance is in MHz and resembles the radio wavelength energies (for instance, at 4.7 T + 200 MHz ^{1}H, nuclei come into resonance and in 50 MHz, ^{13}C comes into resonance). This energy is significantly lower compared to that vital for IR spectroscopy method, which is around 10 to 4 kJ/mol vs. near 5–50 kJ/mol.

^{1}H and ^{13}C are not sole in the capability of undergoing NMR. All odd number nuclei in their protons such as ^{1}H, ^{2}H, ^{14}N, ^{19}F, ^{31}P, etc., or odd neutron number nuclei (like ^{13}C) display the magnetic possessions requisite for NMR. Only even proton and neutron number nuclei such as ^{12}C and ^{16}O do not have the obligatory magnetic characteristics.

Many important pieces of data are obtainable from ^{1}H-NMR. The first information is the peak chemical shift. This is useful for identifying the H atom type that formed the signal. The second information is the peaks' integration ratios. The area below a ^{1}H-NMR peak is right relational to the number of H atoms

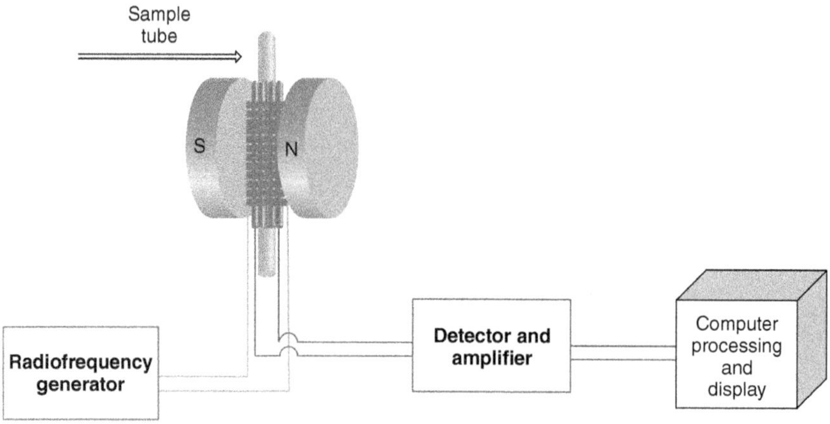

Figure 9.1 The schematic of NMR analysis processor. Source: Lipton and Purdue [16].

that generated the peak. The area is computing via the area integrating, which is perform routinely by the software.

The primary NMR spectrometer arrangement is shown below (Figure 9.1). The sample is placed in a small-sized glass tube and then fixed between a strong magnetic pole. A generator of radio frequency pulses the tube and stimulates the nuclei to cause a spin-flip. The spin-flip is captured via detectors and then the signal will be sent to a computer to analyze and study.

The spectra of NMR are shown as an applied radio frequency plot vs. absorption. The functional frequency rises from left to right, so the left side is the lower field side (also known as the downfield or deshielded side) and the plot right side is the high field (also named as up-field or shielded side) as it can be seen in Figure 9.2.

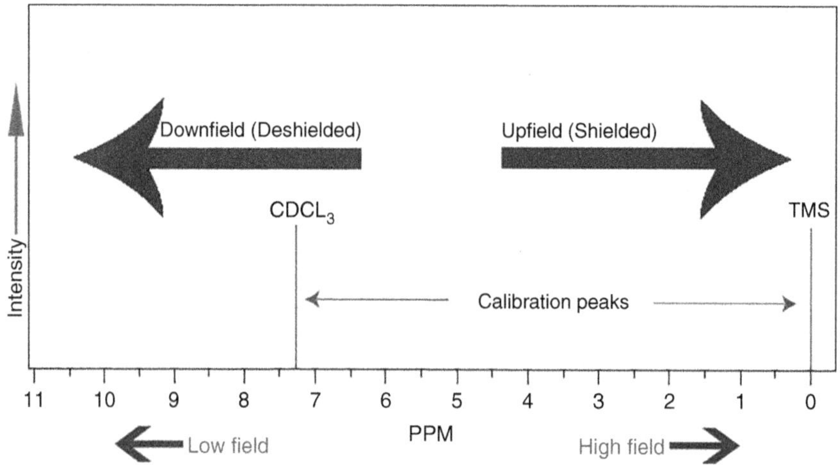

Figure 9.2 Sample of NMR plot as a result of analysis.Source: Lipton and Purdue [16].

Table 9.2 Sample of electronegativity effects and chemical shift.

Compound, CH_3X	CH_3F	CH_3OH	CH_3Cl	CH_3Br	CH_3I	CH_4	$(CH_3)_4Si$
Electronegativity of X	4.0	3.3	2.9	2.6	2.3	2.0	1.6
Chemical shift δ (ppm)	4.31	3.42	3.00	2.7	2.14	0.21	0

The plot position that the nuclei absorbs is named as chemical shift. As it contains an arbitrary value, a standard reference point should be employed. Two standards that are the most communal are TMS (tetramethylsilane with chemical formula of $Si(CH_3)_4$), which is allocated a zero chemical shift, and $CDCl_3$ or deuterochloroform, which is the 7.26 chemical shift for ^1H and 77 for ^{13}C NMRs.

Molecule structural features have an influence on the magnetic field exact magnitude that experienced by a specific nucleus. It indicates that the atoms of hydrogen that have dissimilar chemical environments may experience diverse chemical shifts. This makes NMR so beneficial to determine the structure of organic chemistry. There are three chief sorts that can affect the nucleus shielding: electronegativity, hydrogen bonding, and magnetic anisotropy (Π) of systems.

The surrounded electrons of the nucleus are not motionless, and therefore, they made their specific electromagnetic field. The exact magnetic field may be in conflict with the applied field and thus diminishes the field, which was experienced by the nucleus. The electrons that shield the nucleus are named accordingly. Electron-withdrawing sets may reduce the nucleus electron density, deshield the nucleus, and lead to a greater chemical shift. Now, we compare the data of Table 9.2.

As can be inferred from Table 9.2, when the electronegativity of X increases, the chemical shift of δ rises. This is due to the halide atom effect that pulls the density of electron away from the methyl group. This disclosures the C and H atom nuclei deshielding the nuclei and shifting the resulted peak downfield.

Protons that are captive in H bonding such as –OH or –NH are regularly detected over an extensive chemical shift range. This phenomenon is about the deshielding process that happens in the hydrogen bond. Because H bonds are motioned, regularly generating, breaking, and generating again, there would be a widespread range of H bond strengths and therefore a wide range of deshielding. This effect as well as solvation, acidity, temperature, and concentration made it very challenging to predict the chemical shifts of the atoms.

^1H-NMR has an exclusive signal for each dissimilar H atom type that exists in the compound. Because the shielding amount is reliant on the resident chemical environment, the particular chemical shift for hydrogen atoms may diverge extensively. There are three primary approaches that may employ to control if H atoms are indistinguishable:

1- *Substitution approach*: The modest although slowest procedure. The approach is to substitute each H atom, an atom at a time with another one. For instance, a Cl atom to see if you create a dissimilar complex. Each dissimilar creation specifies an altered type of H atom.

2- *Verbal explanation*: This method obliges to define each H atom verbally. If we have a dissimilar depiction, then the hydrogen atoms are unlike. For instance, the –NH bond is not similar to a –CH bond with regard to the atom which the H is committed to. Also, a $-CH_3$ bond is dissimilar from a $-CH_2-$ bond because of the number of committed H atoms. Moreover, based on this route, sp^3 C–H is dissimilar from a sp^2 C—H bond that is also dissimilar to a sp C—H bond.

9.2.3 Differential Scanning Calorimetry (DSC)

The DSC is a system that can be used to investigate what results when polymers are heated. It can be employed to study the polymer thermal transitions. Thermal transitions are the variations that take place in a polymer when it is heated. The crystalline polymer melting and glass transition are examples of thermal transitions. Practically, the alteration of heat flux to a pan that contained the sample and an empty one is observed. DSC is commercially accessible into forms of power-compensating or heat flux DSC. It calculates the heat amount required to increase the sample temperature and then it controls the heat amount essential to increase reference material temperature. This permits the transition detection such as melts, curing, phase changes, and glass transitions. The DSC analysis result is a heat flux curve vs. temperature or vs. time. Two different conventions can be estimated: The ample exothermic reactions revealed a positive–negative peak, which depends on the equipment type that is employed in the research. The obtained curve may be utilized to compute transition enthalpies. This is performed via the peak integration corresponding to a specified transition.

DSC is employed in polymer investigations for principally three types of experiments:

1) Glass–rubber transition temperature known as T_g value calculations
2) Melting or recrystallization temperature and heat calculations known as T_m/T_c value and H_f/H_c value, respectively
3) Reacting system measurements or cure measurements

DSC is utilized generally to examine the polymeric materials and to control their thermal transitions and structural compounds. The structure of unknown materials can be determined via employing complementary methods such as IR spectroscopy. Both temperatures of glass transition and melting points for most of the polymers existed in standard collations, and the technique shows degradation of polymer via expected melting temperature lowering method. T_m is influenced by the polymer molecular weight and also by its thermal characterizations.

The percentage of polymer crystalline content may be calculated from the DSC melting or crystallization peaks in the resulted graph employing fusion reference heats, which are found in the literature. DSC technique is used to investigate the polymers thermal degradation using approaches such as OOT factor (oxidative onset temperature/time).

Moreover, analysis of slight events in the first data of heat thermal studies may be valuable as these deceptively "anomalous peaks" may actually also be illustrative of procedure or storage thermal history of materials or physical aging of epoxy polymer. Data evaluations of first and second heat, which are poised at steady heating rates, allow the specialist to study both polymer treating history and material characteristics.

Thermogravimetric examination known as TGA may be more practical to determine the decomposition behavior of the polymers. The polymer impurities are detected via examining thermograms for irregular anomalous peaks, and plasticizers are identified at their specific boiling points.

9.2.4 Thermogravimetric Analysis (TGA)

Thermogravimetric analysis or TGA is an analytical method that can be employed for the determination of material thermal stability and their volatile component fraction via supervising the changes of weight, which will occur when the sample is heated in constant rate. Thermogravimetry has been established around 1900 and fundamentally includes the mass designation of an undercontrolled sample in the isothermal mode or linearly changing temperature in a specified atmosphere. The thermogravimetric test result is a mass plot that is a function of time/temperature [17].

The TG analysis, similar to DSC, is common in polymer research, especially to investigate the thermal stability of polymeric structures under operational conditions. The TGA thermograms reveal the degradation temperature of epoxy polymer or polymer matrix and also the bio- or nanocomposites near each other and lower than those of each element.

TGA depends on a high grade of accuracy in three different dimensions: weight, temperature, and the variation of weight based on temperature. The analysis feature is to control temperature degradation, inorganic/organic component level of the various materials, temperature decomposition peaks, and also residues. A derivative curve of weight loss may be employed to detect the point at which weight loss is most deceptive.

The thermogravimetric analysis is also used to control the α-cellulose and hemicellulose amount with improved correctness compared with frequently utilized "wet chemical routes." This analysis has been used in the biochemical composition investigation such as lignin, cellulose, and hemicellulose of different compounds such as beechwood, Avicel, alkaline lignin, corn stover, and switch grass after pretreatment processes. TGA is so valuable because of its (at-line) high-throughput nature and is perfect to assess the bio-epoxy polymers such as the lignin-based one, although the technique is sometimes destructive such as temperature ramping that can lead to the change in polymer analyses.

9.3 Various Bio-Based Epoxy Polymers, Theirs Uses, and Methods of Characterization in Review

This part primarily draws consideration toward the worldwide-performed characterization of renewable asset-based epoxy polymers determined from furan, itaconic corrosive, lignocellulosic biomass, bio-oil, tannins, rosin corrosive, etc. Various renewable feeds, such as isosorbide, epoxidized plant oils, lignin derivatives, cardanol, itaconic acid, and vanillin, are utilized to produce bio-based epoxy thermosets. Previously, we had discussed about the different strategies to characterize the diverse bio-based epoxy blends from renewable feedstock, and here, we will discuss about the industrial and pilot-scale projects to generate and characterize various bio-based epoxy polymers.

9.3.1 Fire-Retardant-Based Epoxy

Wang and co-workers [18] created an exceedingly fire retardant, steady (in thermal points) and pressure-sensitive adhesives (PSAs) with a great peel quality that is practically equivalent to the marketable pressure-sensitive cement PSAs. PSAs have been produced from epoxidized soybean oils and carboxylic corrosive-ended polyesters. In this work, TGA examination uncovers that in nearness of N_2 and O_2 environments, the onset deterioration temperatures upgrades by the addition of two flame-retardant monomers of DDP[1] and DOPOHQ-HE[2]. By assessing the fire retardancy of PSAs utilizing the combustion microscale calorimetry, restricting oxygen record analysis, UL-94, and the experiment strategy, it was established that the fire retardancy of two PSAs containing phosphorus improves via expanding the sum of flame-retardant monomers.

The PSAs are totally bio-based ingredients with great warm steadiness and fire retardancy beside the absence of all sorts of unstable natural composites, which gives numerous variation applications.

Recently, it has been confirmed that vanillin, a monoaromatic subordinate from lignin, is the foremost reasonable mechanically created fabric for tall execution polymers, while combination of diepoxies is troublesome. In any case, accomplishing the bio-based epoxy tars of tall execution and great fire retardancy is a most extreme challenging assignment.

Because of the excellent intumescent and strong char arrangement capacity of cured vanillin epoxy polymers, they show UL-94 V0 rating and tall LOI (restricting oxygen list) of around 32.8%. Furthermore, these tars also illustrate tall T_g at around 214 °C, pliable quality, and pliable modulus of 80.3 and 2709 MPa individually, which is greatly higher than the cured classical DGEBA that was demonstrated by Wang et al. [19].

It might be concluded that the starting corruption temperature was due to the chemical configuration of arrange and the crosslink thickness. The constantly growing starting degradation temperature of EADI-modified epoxy resins featured to the great crosslink density of the DGEBA epoxy system.

Deng et al. [20] portrayed an arrangement of unique rosin corrosive siloxane epoxy gums called as AESE copolymers via the response of ethylene–glycol–diglycidyl-ether-altered AP-EGDE (acrylpimaric corrosive) with poly(methyl-phenyl-siloxane). In this investigation, ductile tests uncover that epoxy/anhydride identical proportion and the substance of silicone influence the cross-link thickness of epoxy and thereafter impact the mechanical properties. The adjusted epoxy tar AESE possesses the pliable qualities of almost between 20 and 40 MPa that are somewhat fewer than that of APEGDE. However, their prolongation at break was higher than that of AP-EGDE. Warm gravimetric investigation (TGA) detailed that the warm solidness of AESE epoxy tars is nice compared to AP-EGDE because of the advancement of a protecting buildup and avoiding gas advancement. The results described that the copolymers may be employed as bio-based epoxy and its usage was wide by the siloxane consolidation.

9.3.2 (Lignocellulosic Biomass)-Based Epoxy Polymers

Lignin illustrated an appealing asset for fragrant-renewable feedstock, existing within the huge sum within the Earth [21]. After an auxiliary adjustment, lignin can be utilized within the union of bio-based epoxy gum. Either different lignin-based epoxy systems have been created form on lignin determined phenols (LDPs) for illustrations such as vanillin, eugenol, and dihydroeugenol or coordinate utilized of bulk lignin have been reported.

Development of lignin epoxy systems or LBEN from organosolv lignin and also lignin-inferred phenol, which is known as dihydroeugenol, is delivered by different chemical alterations such as phenolation, demethylation, and phenol-formaldehyde response detailed by Zhao and Abu-Omar [22]. In difference to a regular synthesis method where lignin was epoxidized earlier than mixing with co-monomers (LBEN), epoxy derivatives (LINEN) indications enhanced crosslink density (ρ), α-relaxation temperature ($T\alpha$) and storing modulus in glassy state (Eg') as demonstrated from dynamic mechanical analysis (DMA), and enhanced thermal stability was confirmed via thermos-gravimetric analysis (TGA).

Nikafshar et al. [23] synthesized vanillin-based epoxy and asserted that the tensile strength and Izod impact strength of this epoxy structure develops in the presence of inorganic accelerators in comparison with the DGEBA system. Meanwhile, it is noteworthy that the inorganic accelerators effect on the durability of the synthesized vanillin-based epoxy resin was evidenced via the scanning electron microscopy (SEM) study of the Izod impact test fractured sample surfaces.

An innovative triglycidyl eugenol subordinate with the commercial name of 3EPO-EU has been arranged, which performs as an introductory monomer for blend of unused bio-based thiol-epoxy thermosets prior detailed by Guzman et al [24]. Through diverse thiols as the fundamental catalysts such as tetrathiol determined from PETMP (pentaerythritol), a trithiol obtained from eugenol commercially known as 3SH-EU and a hexathiol arranged from 6SH-SQ (squalene) have been employed in the nearness of 4-(*N*,*N*-dimethylamino)pyridine. In addition to the change in details comprising squalene, adaptable diglycidyl ether obtained from 2EPO-HEX (hexanediol) was also included. he ingredients arranged are inflexible at absolute temperature (room, C) and shows T_g up to 103 °C. The warm soundness and thermomechanical and mechanical features have been explored that vary with changing basic properties of the arranged ingredients.

9.3.3 Furan-Based Epoxy Resin

The square chemical configuration of polymeric materials plays a critical part in deciding the features of the manufactured compounds. Recently, the bio-based epoxy tar inferred from furan subsidiaries acts as a possible supernumerary to petroleum-based compounds with regard to the nearness of backbone fragrant ring arrangement.

Shen et al. [25] manufactured a BOF (2,5-bis((2-oxiranylmethoxy)methyl)furan) bio-based epoxy from 2,5-furan-dimethanol and strengthened by its Diels–Alder or DA response additive BOF/DA that comprises of an adaptable pendant chain. From that point, different weight percentages of BOF/DA (0%, 25%, 50%, and 75%) consolidated to the BOF epoxy gum and then treated with isophorone-diamine (IPDA). The warm BOF/DA reversibility was affirmed by differential filtering calorimetry and ultraviolet–visible (UV–vis) spectroscopy. The retro-Diels–Alder degree (r-DA degree) response from UV–vis was almost 25–27% in different cured frameworks. At 50% BOF/DA, an affect quality increments to almost 85% without showing much impact on its pliable quality and modulus.

9.3.4 Rosin Corrosive-Based Epoxy

Rosin could be a characteristic widely available epoxy within the conifers and pines. It is additionally extricated via refining of tall oil or as a by-product of Kraft mash preparation. Rosin is the blend of neutral (10%) and acidic (90%) composites. The acidic parts recognized as a rosin acid and it may be a blend of abietic type acid (40–60%) and pimaric type (9–27%) acid based on the total rosin weight. Rosin corrosive displayed the expansive ring structure of hydrogenated phenanthrene and comparable in inflexibility to petroleum-based cyclic aliphatic or fragrant complexes. Therefore, throughout the reactions of the carboxylic acid (–COOH) group and C=C double bond, the resultant rosin derivatives can serve as alternates for petroleum-based cyclic aliphatic and aromatic composites employed in polymer synthesis. Hence, long time later, rosin has found to gain increasing attention as a renewable feed in bio-based epoxy tar production.

9.3.5 Itaconic Corrosive-Based Epoxy

Because of the developing mindfulness of the lessening of nonrenewable assets and outflow of greenhouse gasses, there are expanding temptations in epoxy tar blend from bio-based components such as itaconic corrosive and successor to tars containing BPA because it is harmful to human well-being. Itaconic corrosive (IA) has been delivered via aging of carbohydrate, such as glucose, within the *Aspergillus terreus* nearness. It could be an incredible potential to BPA supplant in chemical engineering because of the presence of two bunches of –COOH, one bond of C=C, molecular short chain, and no adaptable bond or unbending ring, and it was chosen as one of the best 12 possible bio-based stage chemicals by the American Office of Energy [26]. These days, it has been affirmed to be a huge comonomer for bio-based polymer amalgamation and broadly utilized as paper coatings, emulsion paints, and latexes.

9.3.6 Self-mending Epoxy Resin

Self-healing characteristics of epoxy tar was credited with regard to the disengagement and recombination of the linkages over DA and retro-DA responses. Amid warm cure of the harmed test, the test completely patches at a temperature of 125 °C in a period of 20 minutes and at a temperature of 80 °C for 72 hours within the work of Tian et al. [27]. Besides, the self-healing epoxy tar shows much better mechanical features that are analogous to the characteristics of the bisphenol A epoxy gum and the MHHPA. Billiet and coworkers [28] arranged a self-healing framework for epoxy compounds over Michael expansion response of maleimide materials. Exceptionally tall recuperation proficiency of around 121% has been reported.

Khoee and Kachoei [29] created an amine nano-container healants established on poly(glycidylmethacrylate-comethylmethacrylate) commercially known as GMA : MMA. The adhesives utilized to joint copper with exhibited cohesive fracture, which is necessary for self-healing examination based on nano-containers by the lap-shear strength study. The outcomes exposed that after combination of 7.5 wt% GMA : MMA-50 nano-containers into epoxy matrix, the healing effectiveness was accomplished 85% high. Moreover, reported the self-healing ability of the generated fractured and healed adhesive by FTIR, SEM and fluorescence microscopy that appears that amine may be released from the scratched nano-containers and healing happens at the damaged region.

Recuperating competence of the compounds was confirmed via the affect analysis. The virgin example shows no alters in shape, and there is the nonappearance of split line that can tolerate a weight up to 200 g. While the mended example was once more affect stacked and undergoes repairing within the same circumstances, the same results were indeed shown at third impact-healing tests. The performed study obviously shows the epoxy tar stuff to reparation constantly.

9.3.7 Other Epoxy Polymers

Celikbag et al. [30] introduced an innovative self-curing bio-based epoxy tar utilizing bio-oil. Bio-oil has been created via aqueous loblolly pine liquefaction and had utilized it as a bio-polyol for BOBER union (bio-oil-based epoxy gum). It was calculated that along with the whole hydroxyl quantity of bio-oil modifying the yield and epoxy equivalent weight of BOBER, the configuration of hydroxyl groups controlled by bio-oil components (aliphatic, phenolic and acidic OH) also plays a crucial role for the computation of the ideal quantity of catalyst in the development of BOBER. Self-curing wonders of BOBER has been explored by differential filtering calorimetry examination, although Fourier transfer infrared spectroscopy (FTIR) demonstrated that etherification response was the foremost noticeable response amid the self-curing process.

References

1 Khajouei, M., Najafi, M., and Jafari, S.A. (2019). Development of ultrafiltration membrane via in-situ grafting of nano-GO/PSF with anti-biofouling properties. *Chem. Eng. Res. Des.* 142: 34–43.

2 Siró, I. and Plackett, D. (2010). Microfibrillated cellulose and new nanocomposite materials: a review. *Cellulose* 17 (3): 459–494.

3 Nakagaito, A.N. and Yano, H. (2005). Novel high-strength biocomposites based on microfibrillated cellulose having nano-order-unit web-like network structure. *Appl. Phys. A* 80 (1): 155–159.

4 Zarrintaj, P., Saeb, M.R., Ramakrishna, S., and Mozafari, M. (2018). Biomaterials selection for neuroprosthetics. *Curr. Opini. Biomed. Eng.* 6: 99–109.

5 Fattahi, P., Yang, G., Kim, G., and Abidian, M.R. (2014). A review of organic and inorganic biomaterials for neural interfaces. *Adv. Mater.* 26 (12): 1846–1885.

6 Donoghue, J.P. (2008). Bridging the brain to the world: a perspective on neural interface systems. *Neuron* 60 (3): 511–521.

7 Jonoobi, M., Aitomäki, Y., Mathew, A.P., and Oksman, K. (2014). Thermoplastic polymer impregnation of cellulose nanofibre networks: morphology, mechanical and optical properties. *Composites Part A* 58: 30–35.

8 Manouchehri, S., Bagheri, B., Rad, S.H. et al. (2019). Electroactive bio-epoxy incorporated chitosan-oligoaniline as an advanced hydrogel coating for neural interfaces. *Prog. Org. Coat.* 131: 389–396.

9 Dannenberg, H. and Harp, W. (1956). Determination of cure and analysis of cured epoxy resins. *Anal. Chem.* 28 (1): 86–90.

10 Mijovic, J. and Andjelic, S. (1995). A study of reaction kinetics by near-infrared spectroscopy. 1. Comprehensive analysis of a model epoxy/amine system. *Macromolecules* 28 (8): 2787–2796.

11 Poisson, N., Lachenal, G., and Sautereau, H. (1996). Near-and mid-infrared spectroscopy studies of an epoxy reactive system. *Vib. Spectrosc.* 12 (2): 237–247.

12 Xu, L. and Schlup, J.R. (1998). Etherification versus amine addition during epoxy resin/amine cure: an in situ study using near-infrared spectroscopy. *J. Appl. Polym. Sci.* 67 (5): 895–901.

13 Paz-Abuin, S., Lopez-Quintela, A., Varela, M. et al. (1997). Method for determination of the ratio of rate constants, secondary to primary amine, in epoxy-amine systems. *Polymer* 38 (12): 3117–3120.

14 Musto, P., Mascia, L., Ragosta, G. et al. (2000). The transport of water in a tetrafunctional epoxy resin by near-infrared Fourier transform spectroscopy. *Polymer* 41 (2): 565–574.

15 Xu, C. and Webb, W.W. (1996). Measurement of two-photon excitation cross sections of molecular fluorophores with data from 690 to 1050 nm. *JOSA B.* 13 (3): 481–491.

16 Lipton M. Purdue CHM (2014). 26505: Organic chemistry for chemistry majors (1st Semester). https://chem.libretexts.org/Courses/Purdue/Purdue%3A_Chem_26505%3A_Organic_Chemistry_I_(Lipton)/Chapter_5._

Spectroscopy/5.3_Nuclear_Magnetic_Resonance_(NMR)_Spectroscopy (accessed 14 August 2020).

17 Saadatkhah, N., Garcia, A.C., Ackermann, S. et al. (2020). Experimental methods in chemical engineering: thermogravimetric Analysis – TGA. *Can. J. Chem. Eng.* 98 (1): 34–43.

18 Qi, M., Xu, Y.-J., Rao, W.-H. et al. (2018). Epoxidized soybean oil cured with tannic acid for fully bio-based epoxy resin. *RSC Adv.* 8 (47): 26948–26958.

19 Wang, S., Ma, S., Xu, C. et al. (2017). Vanillin-derived high-performance flame retardant epoxy resins: facile synthesis and properties. *Macromolecules* 50 (5): 1892–1901.

20 Deng, L., Shen, M., Yu, J. et al. (2012). Preparation, characterization, and flame retardancy of novel rosin-based siloxane epoxy resins. *Ind. Eng. Chem. Res.* 51 (24): 8178–8184.

21 Chatel, G. and Rogers, R. (2014). ACS sustain. Oxidation of lignin model compounds using single-electron-transfer catalysts. *ACS Sustainable Chem. Eng.* 2: 322–339.

22 Zhao, S. and Abu-Omar, M.M. (2015). Biobased epoxy nanocomposites derived from lignin-based monomers. *Biomacromolecules* 16 (7): 2025–2031.

23 Nikafshar, S., Zabihi, O., Hamidi, S. et al. (2017). A renewable bio-based epoxy resin with improved mechanical performance that can compete with DGEBA. *RSC Adv.* 7 (14): 8694–8701.

24 Guzman, D., Ramis, X., Fernandez-Francos, X. et al. (2018). Preparation of new biobased coatings from a triglycidyl eugenol derivative through thiol-epoxy click reaction. *Prog. Org. Coat.* 114: 259–267.

25 Shen, X., Liu, X., Wang, J. et al. (2017). Synthesis of an epoxy monomer from bio-based 2,5-furandimethanol and its toughening via Diels–Alder reaction. *Ind. Eng. Chem. Res.* 56 (30): 8508–8516.

26 Werpy, T. and Petersen, G. (2004). *Top value added chemicals from biomass: volume I--results of screening for potential candidates from sugars and synthesis gas.* Golden, CO: National Renewable Energy Lab.

27 Tian, Q., Yuan, Y.C., Rong, M.Z., and Zhang, M.Q. (2009). A thermally remendable epoxy resin. *J. Mater. Chem.* 19 (9): 1289–1296.

28 Billiet, S., Van Camp, W., Hillewaere, X.K. et al. (2012). Development of optimized autonomous self-healing systems for epoxy materials based on maleimide chemistry. *Polymer* 53 (12): 2320–2326.

29 Khoee, S. and Kachoei, Z. (2015). Design and development of novel reactive amine nanocontainers for a self-healing epoxy adhesive: self-repairing investigation using the lap shear test. *RSC Adv.* 5 (27): 21023–21032.

30 Celikbag, Y., Meadows, S., Barde, M. et al. (2017). Synthesis and characterization of bio-oil-based self-curing epoxy resin. *Ind. Eng. Chem. Res.* 56 (33): 9389–9400.

31 Weyer, L.G. and Lo, S.C. (2006). Spectra–structure correlations in the near-infrared. In: *Handbook of Vibrational Spectroscopy.*

10

Flame Retardancy of Bioepoxy Polymers, Their Blends, and Composites

Young-O Kim and Yong Chae Jung

Korea Institute of Science and Technology (KIST), Institute of Advanced Composite Materials, 92, Chudong-ro, Bongdong-eup, Wanju-gun, Jeonbuk 55324, Republic of Korea

10.1 Introduction

At present, over 90% of the world's epoxy resin materials are made of bisphenol-A-type diglycidyl ether (DGEBA) [1–3]. It is well established that the epoxy resins of bisphenol A type offer extraordinary physical properties including remarkable mechanical properties, chemical resistance, and form stability [1–5]. In general, epoxy resins cross-link into a rigid three-dimensional network structure by chemical reaction between monomers and hardeners [6]. These characteristics are employed in a wide range of applications such as coatings, adhesives, solar cells, automotive, and aerospace [7–9].

In recent years, problems associated with high combustibility, environmental issues, global warming, and oil depletion from the use of polymer materials as well as composite materials from epoxy have been observed. In addition, it has been recently demonstrated that the toxic effects of combustion severely affect humans as well as other living organisms [10–13]. Accordingly, significant research is being conducted on the development of renewable bio-based polymer materials from plants [14–17].

Technologies for developing bio-based polymer materials derived from plants or recycled materials can reduce the consumption of petrochemical products by petroleum-based engineering plastic producers [2, 14, 15, 17]. These technologies can also effectively reduce various environmental pollutants generated during production. Therefore, research on bio-based epoxy is both an important development direction in the field of polymer materials as well as an important means for saving energy, reducing pollutant emissions, and developing a low-carbon economy. In order to effectively solve these problems, in recent years, research has been performed on bio-based epoxy (i.e. from vegetable oil [18–20], lignin [21–24], rosin acid [25], tannin [26–28], and cardanol [29–33]) rather than on traditional petroleum-based epoxy. Significant attention has also been focused on the discovery of bio-based epoxy resins from renewable resources such as itaconic acid [34].

Bio-Based Epoxy Polymers, Blends, and Composites: Synthesis, Properties, Characterization, and Applications, First Edition. Edited by Jyotishkumar Parameswaranpillai, Sanjay Mavinkere Rangappa, Suchart Siengchin, and Seno Jose.

Because of the rapid advances in the synthesis of bioepoxy resins, these materials have been identified as potential candidates for replacing the commercial petroleum-based epoxy [14–34]. However, bio-based polymers are established to exhibit inferior mechanical and thermal properties owing to their long aliphatic chains and low cross-linking density [35]. In addition, the low curing behavior of resins results in the fatal degradation of bio-based epoxy resins.

Meanwhile, epoxy resins are vulnerable to heat. An increase in the temperature adversely affects the mechanical properties, and the combustion process generates smoke and hazardous substances. These are obstacles for the commercialization of composite materials. This problem highlights the need to improve the flame retardancy index of bioepoxy resins as well as petroleum-based epoxy resins [36–39]. Bioepoxy resins are also highly flammable. This problem must be resolved before these materials can be used in transportation, construction, and electronics. An effective strategy to improve the flame retardancy index of a resin is to add a flame retardant or to introduce a functional group capable of inducing flame retardancy to the matrix main chain or end group. However, the adverse effects of uniform dispersion or movement to the surface within the matrix still remain unresolved.

Bromine-based flame retardants are the most widely used flame retardants among the halogen-based flame retardants, accounting for 25% of the flame-retardant (FP) market in 2014 [40, 41]. These include Tetrabromobisphenol A (TBBPA), hexabromocyclododecane (HBCD), penta-BDE (bromodiphenyl ether), octa-BDE, and deca-BDE. However, halogen-based flame retardants are hazardous to humans as they cause cancers (octa-BDE and penta-BDE) or damage human lungs by generating toxic substances such as furan and dioxins or corrosive gases during combustion. They can also cause mechanical and electrical failures and severe problems such as accumulation in soil and air, which can destroy the ecosystem. In the early 2000s, international environmental regulations such as the Stockholm Convention banned the production and use of certain brominated flame retardants worldwide owing to the environmental debate. Accordingly, the development of eco-friendly bio-based epoxy resins and bio-based flame retardants with remarkable flame-retardant properties in the future can be considered to be an effective approach to solving the above problems [42–44].

Against this backdrop, we report on the present research into flame-retardant bio-based epoxy resins, in this chapter. Specifically, we describe the flame retardancy evaluation method, recent market trend, and flame retardancy of epoxy based on lignin and tannic acid (TA). Finally, the recent research trend of the flame retardancy of bioepoxy using nano- or micro-sized organic–inorganic additives is described.

10.2 Methods for Analyzing Flame-Retardant Properties

A flame-retardant material burns when it comes in contact with a flame. However, when the flame is removed, it prevents or suppresses combustion, i.e. the material is not burned. It is a property that lowers or halts the propagation capability of fire. In general, flame-retardant materials have a limit of 25 or more. Combustion of plastics proceeds in the order of pyrolysis, and the combustion process can be classified into micro, macro, and mass, respectively. The macro, micro, and mass scales are with regard to the molecular behavior in plastics, the material behavior, and the behavior in real systems such as space or structure, respectively. Generally, flame-retardant combustion can be divided into five stages:

(1) *Stage 1*: It is heating. The specific heat, thermal conductivity, latent heat, and phase transition (melting, sublimation, and evaporation) are the determining factors. Specific heat is the amount of heat required to increase the temperature of unit weight of material. The higher the specific heat, the more gradually the temperature increases. Furthermore, the higher the thermal conductivity is for a specified temperature variation and the material thickness, the more straightforward is the heat transfer.
(2) *Stage 2*: It is the decomposition process. Here, combustible gas, noncombustible gas, liquid, solid, and smoke are generated. The main determinants are the initial decomposition temperature, light heat, and the decomposition behavior of polymer resins.
(3) *Stage 3*: It is an ignition process, where combustible gases are inflamed in the presence of oxygen or an oxidant. The flash ignition temperature, self-ignition temperature, and limit oxygen concentration are the determining factors.
(4) *Stages 4 and 5*: They are the processes of combustion and propagation. In the combustion stage, the combustion heat or the heat of emission emerges, and in the progress stage, the combustion is apparently diffused.

The flame-retardant test methods involve the measurement of the oxygen index, combustion rate, combustion time, and smoke generation. Countries use either their own standards or the universally accepted UL (Underwriters Laboratories) method. UL is an independent, not-for-profit organization that sets or tests standards for safety. This section introduces the most commonly used vertical test, 5V test, and oxygen index. The specifications of other countries are briefly described in Table 10.1.

Table 10.1 The oxygen index of representative plastics.

Plastic	LOI	Plastic	LOI
Polyoxymethylene (ACETAL)	14.9	polyethylene terephthalate (PET)	20.6
Poly methyl methacrylate (PMMA)	17.3	Poly(phenylene oxide) (PPO)/HIPS	24.3
Polyethylene (PE)	17.4	NYLON	24.3
Polypropylene (PP)	17.4	Polycarbonate (PC)	24.9
Polystyrene (PS)	17.8	Poly(vinyl chloride) (PVC)	45.0
High impact polystyrene (HIPS)	18.2	Amorphous silica (AS)	18.8
Styrene-acrylonitrile (SAN)	19.1		

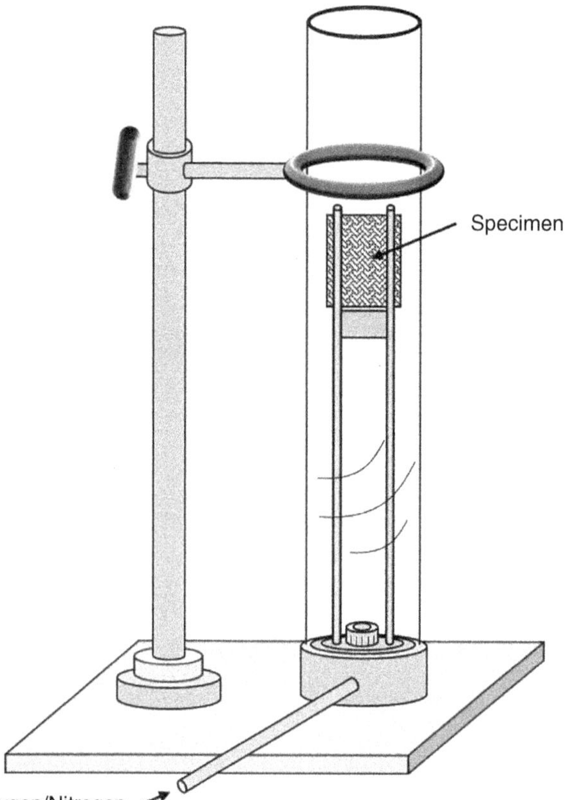

Figure 10.1 LOI test method.

10.2.1 LOI (Limiting Oxygen Index)

The limiting oxygen index (LOI) is a representative measurement method for evaluating the combustibility of a material as shown in Figure 10.1. It is the minimum oxygen concentration (of a nitrogen–oxygen mixture) required that supports the combustion of the material. In general, the higher the ignition resistance of a material is, the higher is the oxygen index [45].

The temperature of the gas mixture affects the oxygen index. The standard atmosphere at sea level has 21% oxygen. The LOI evaluation is performed after 24 hours of pretreatment in a constant temperature/humidity chamber at 23 ± 2 °C, 50 ± 5% RH, with specimens of 10 mm width and 120 mm length (according to ISO 4589, ASTM D 2863, and NFT 51-071) [46]. In the measuring method, the specimen is first fixed perpendicularly to the measuring equipment. Then, a certain amount of high-purity nitrogen and oxygen is introduced into the lower portion. At this time, the combustion source is ignited at the tip of the sample to assess the combustion behavior, and the likelihood that the combustion state can be maintained even after the heat source is removed. The limit oxygen index is defined as follows:

$$\text{LOI} = [O_2]/[(O_2) + (N_2)] \times 100 \tag{10.1}$$

where O_2 is the oxygen flow rate (l/min) and N_2 is the nitrogen flow rate (l/min). The limit oxygen index is defined as the minimum oxygen concentration of the oxygen–nitrogen mixture air required for a sample to ignite without burning for 90 seconds.

Consider the example of PET in Table 10.1. An LOI of 20.6% implies that if the oxygen concentration in the air is reduced below 20.6%, the combustion cannot be continued after the heat source is removed. In general, a material with an LOI of 20.9 or less burns well in the air; a material with an LOI between 20.9 and 27 burns gradually in air. If the LOI is 27 or more, the material can be considered as a self-extinguishing test piece that is difficult to burn in air.

The marginal oxygen index represents the burning characteristics of plastics. However, it differs slightly from the UL-94 flame retardant. This is discussed subsequently. The flame retardancy of various plastics exhibits a tendency to increase the LOI as the amount of flame retardant used increases. However, UL-94 flame retardancy is different in that the flame retardancy is realized only when a certain amount of flame retardant is present.

10.2.2 UL-94

Developed by the Underwriter's Laboratory (UL), a nonprofit organization founded by the American fire insurer in 1894, it is one of the most commonly used methods for evaluating the flame retardancy. The UL-94 test method is divided into the horizontal combustion test (HB) applied to noncombustible resins, vertical combustion test applied to flame-retardant resins, and flat combustion test. These are described in Figure 10.2 and are rated by combustion time, flame drop, and afterglowing extinguishing time.

10.2.2.1 Horizontal Testing (UL-94 HB)

UL-94 HB measurement provides the HB rating by measuring the flame propagation velocity for verifying the flame propagation pattern in the horizontal direction after providing the heat source to the specimen. The evaluation method comprises measuring the combustion time after the flame is applied to a 1 in. (25 mm) section after applying the flame for 30 seconds or until the mark is

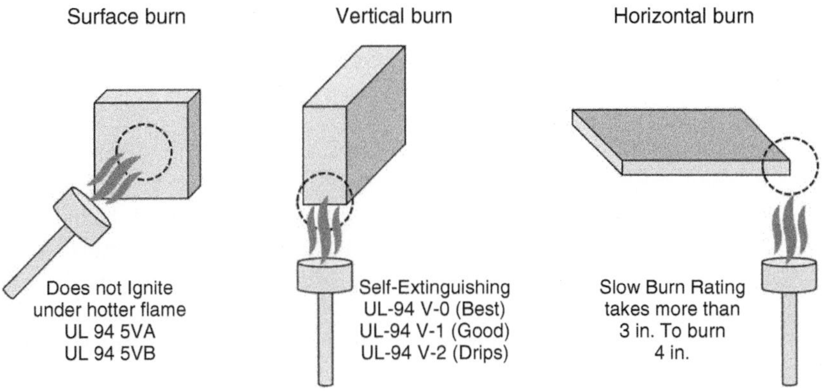

Figure 10.2 UL-94 method.

Table 10.2 UL-94 rating.

	V-0	V-1	V-2
Individual after-flame time, $t1$ or $t2$	≤10 s	≤30 s	≤30 s
Total after-flame time for any condition set, $t1 + t2$ for these five specimens	≤50 s	≤250 s	≤250 s
After-flame plus afterglow time for each individual specimen after the second flame application, $t2 + t3$	≤30 s	≤60 s	≤60 s
Burning up to the holding clamp (125 mm under line)	No		
Cotton ignition	No		Yes

reached. Moreover, it measures the combustion speed within the range up to 4 in. (100 mm). The HB rating is satisfied when specimens of thickness larger than 3.0 mm exhibit a burn rate of 40 mm/min or less and those of thickness less than 3.0 mm exhibit a burn rate below 75 mm/min.

10.2.2.2 Vertical Testing (UL-94 V)

It is a method of evaluating the combustion pattern of the product and the degree of flame propagation to the surroundings, upon the application of a flame to the test piece in the vertical direction. Install the specimen vertically and apply a flame under it for 10 seconds to measure the time until the fire on the specimen and the additional ignition caused by the falling drops are extinguished. If combustion stops within 30 seconds of applying the first flame, apply the flame again for another 10 seconds and measure the second combustion time. Grades are assigned based on this evaluation as illustrated in Table 10.2.

10.2.3 Cone Calorimeter

Cone calorimeter is a bench-scale test widely used for evaluating a material's fire hazard or flame retardancy. It measures the heat release rate (HRR) and the smoke and gas production as evaluation indices to quantitatively express the fire risk and hazard of new materials. In addition, because a combustion process initiated by thermal stimulation (generally called external heat flux) and spark during cone calorimeter analysis represents a typical fire-developing scenario, it is also effective for establishing the fire model occurring in materials by using a small specimen and reduced time and cost.

10.2.3.1 Configuration

A cone calorimeter for horizontal testing is generally used for analyzing the flame-retardant properties of samples and its structure is shown in Figure 10.3. The required sample is of size 100 × 100 mm and thickness between 2 and 50 mm. The edge of the sample is covered by a steel-framed holder, reducing the exposed area to 94 × 94 mm. The holder containing the sample is mounted on a load cell and is exposed to a conical heater, which controls the radiant heat flux. An electric spark is used to ignite the center of the specimens. The flames

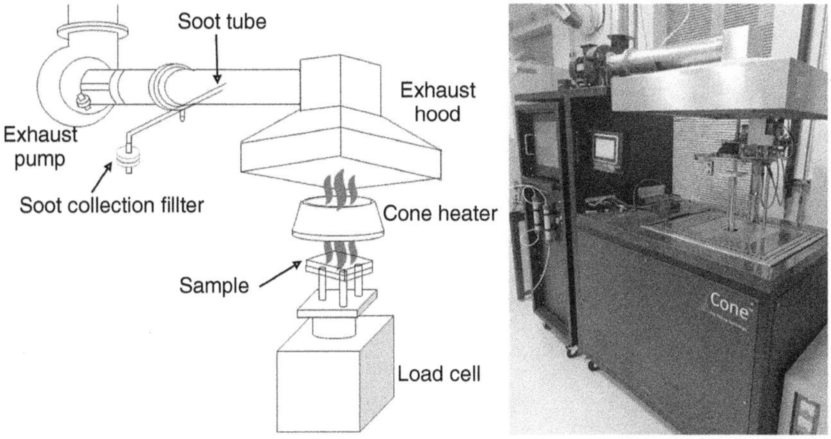

Figure 10.3 Cone calorimeter configuration.

and combustion products pass through the circular opening at the center of the heater and the exhaust hood. The oxygen depleted during the burning process of the sample is the basis for calculating the heat release. It is analyzed using a paramagnetic analyzer. In addition, CO and CO_2 are analyzed by a nondispersive infrared analyzer. Provided the effluent flow through the exhaust is effectively controlled, the heat release will be proportional to the oxygen depletion. Procedures for conducting cone calorimeter tests are described in ASTM E 1354 and ISO 5660.

10.2.3.2 Controlling Factors: Heat Flux, Thickness, and Distance Between Sample Surface and Cone Heater

Heat Flux Heat flux is the flow of energy per unit area per unit time. Therefore, different external heat fluxes can alter the fire behavior of samples. An increase in the external heat flux increases the time for attaining the ignition temperature, the expansion rate of decomposition, and fuel production rate. Ultimately, an increase in the external heat flux can increase the rate of release of heat from the burning samples. In general, as shown in Figure 10.4, with increasing external heat flux, the peak of the heat release rate (PHRR) appears at earlier time and increases linearly [47]. Meanwhile, the total heat release (THR), which is obtained by integration over time, is constant.

Thickness The thickness of the specimen on the cone calorimeter influences the flammability. Therefore, the thickness details must be discussed for analyzing the results of the cone calorimeter. The thickness-dependent flame retarding response of the specimen is illustrated in Figure 10.5 [48]. Thermally thin poly(methyl methacrylate) (PMMA) as a noncharring polymer exhibits a sharp HRR peak and short ignition time. As the thickness increases, the PHRR decreases, and the flaming time increases. In addition, an increase in the sample thickness increases the time to PHRR, owing to the increase in the time when the pyrolysis zone approaches the glass wool supporting the sample. This can

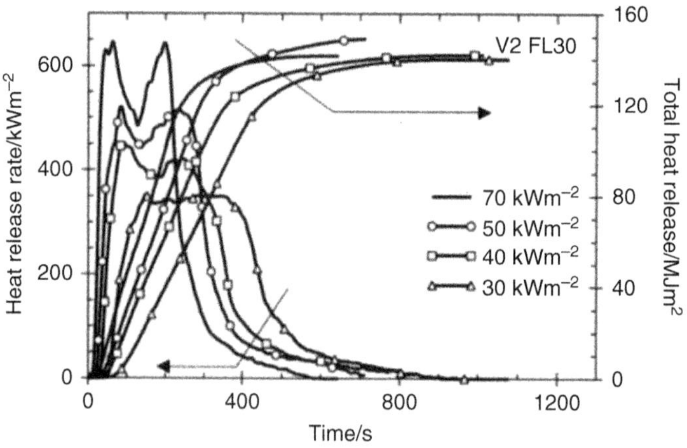

Figure 10.4 Heat flux-dependent HRR and THR for glass fiber-reinforced PA 66. Source: Reproduced from Schartel et al. [47]. © 2003, Elsevier.

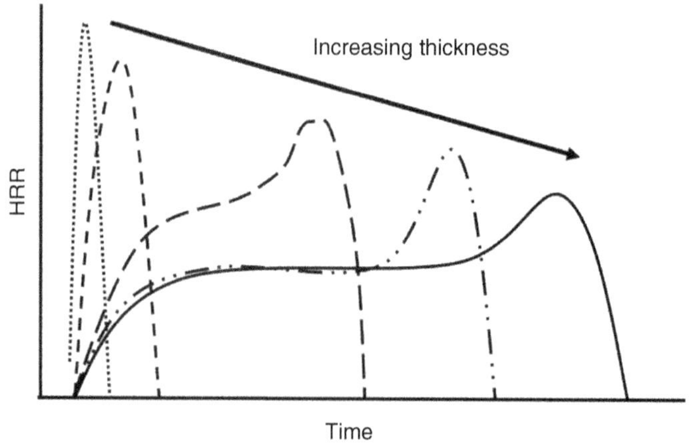

Figure 10.5 Sample thickness-dependent HRR for PMMA [48].

reduce the conductive heat flux. In contrast, for the char-forming material, the time to PHRR is constant, which is ascribed to the formation of PHRR by the ignition-initiated char formation. In general, a comparison of cone calorimeter results between new materials has been conducted with identical thickness of the samples. As shown in Figure 10.6, the characteristic response of the HRR vs. time can be determined by the type of the sample and thickness. This aids us in analyzing the flame retardancies of materials.

Distance Between Sample Surface and Cone Heater During the cone calorimeter analysis, a constant heat flux is applied to the sample such that its surface is gradually heated to temperatures above the decomposition temperature. The distance below the cone heater can affect the heat flux on the sample surface, i.e. the effective heat flux (Figure 10.7) [49]. The commercialized cone calorimeter

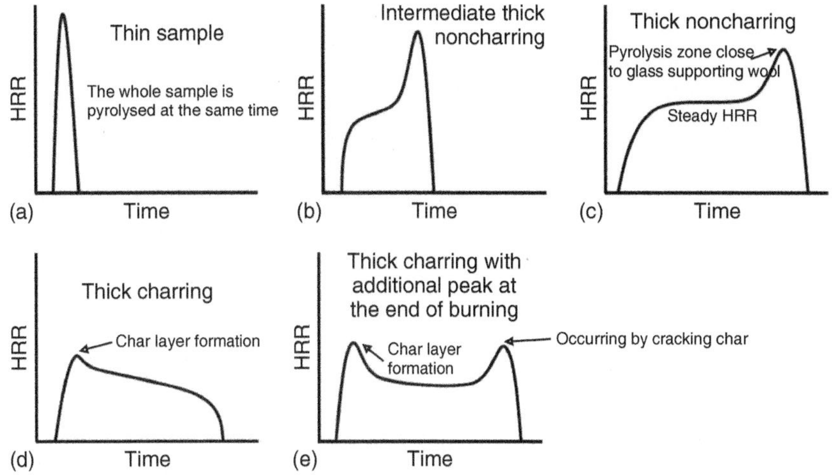

Figure 10.6 HRR curves for different types of samples. (a) Thin samples, (b) Intermediate thickness non-charring sample, (c) Thick non-charring sample, (d) Thick charring sample, and (e) Thick charring and non-charring sample.

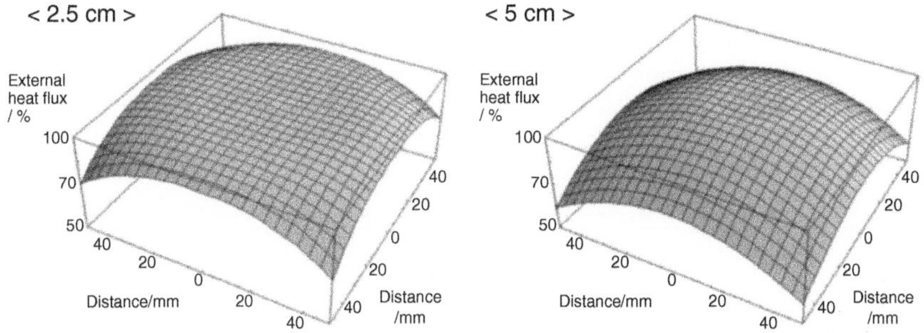

Figure 10.7 Effect of distance between a sample surface and the cone heater to external heat flux. Source: Reproduced from Schartel et al. [49]. © 2005, Elsevier.

is designed and optimized for a homogeneous heat flux on the whole sample surface. However, the distance between the sample surface and cone heater must be selected according to the purpose.

The distance between the sample surface and cone heater can also change based on the type of the sample. This change originates from the deformation of the sample during the combustion process. In thermoplastic materials, melting is generally induced before ignition, resulting in an increase in the distance. In most intumescent materials, whose flame-retarding mechanism arises from the expansion of the thermal insulating layer, the distance between the sample surface and cone heater decreases. Both cases can change the HRR pattern, leading to inaccurate data generation. Therefore, to improve the reproducibility and obtain accurate data, a protecting grid is used to reduce the deformation of the intumescent samples. In addition, if the combustion product is a liquid or a melt, which can cause material loss, an aluminum foil is used.

10.2.4 Microscale Combustion Calorimeter

The microscale combustion calorimeter (MCC) (pyrolysis combustion flow calorimetry (PCFC) was developed for the rapid and simple development of fire-resistant polymers by the U.S. Federal Aviation Administration (FAA). It is a miniature version of a cone calorimeter and requires only a milligram-size specimen. In addition, many factors considered for an accurate interpretation of the results can be excluded in this instrument. However, the obtained information is limited to that related to the HRR, such as the peak of the HRR, THR, and heat release capacity (HRC). The results obtained from the MCC are the HRR per mass unit of fuel volatiles (W/g) and specific HRR (Q) rather than the HRR per unit specimen area (W/m^2). The HRC (J/g K) can be calculated by dividing the maximum Q by the heating rate (β, K/s).

Figure 10.8 shows the processes that a specimen undergoes in an MCC for gaining the flame-retardant properties. The specimen is subjected to an anaerobic (Method A) or aerobic (Method B) environment in a pyrolysis chamber. The sample is heated using linearly increasing temperature with a constant rate between 0.2 and 2 °C/s. This results in gaseous pyrolysis products and char formation. The gaseous pyrolysis products are completely oxidized at 900 °C, which is the temperature of the combustor. The gas stream exiting the combustor contains nitrogen, combustion products, and unreacted oxygen. This combustion gas stream cools and enters a Teflon tube tightly packed with anhydrous calcium sulfate (Drierite™) to remove moisture and acid gases from the sample stream, which can dilute the oxygen concentration measured by the oxygen analyzers.

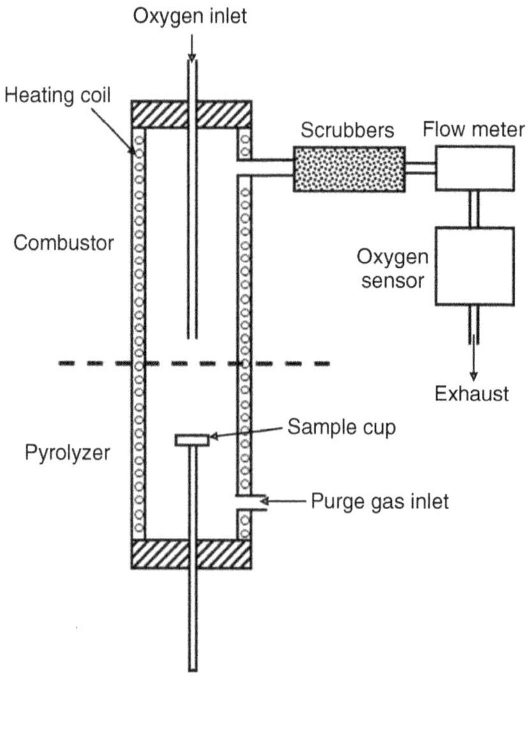

Figure 10.8 MCC configuration.

The transient HRR is calculated from the measured flow rate and oxygen concentration after correcting for flow dispersion. The HRR can be calculated from the depletion rate of oxygen required for completely oxidizing the pyrolysis gases using a polarographic or a zirconia oxygen analyzer. The heat of combustion of the volatile component and the net calorific value of the volatiles and solid residue during MCC are determined by Methods A and B, respectively.

10.3 Halogen-Free Flame-Retardant Market

The flame-retardant market was US$8752.9 million and 2361.7 kilotons in 2015 [40, 41]. The market size is estimated at $9385.7 million and 2489.5 kilotons in 2016 and is likely to attain $12 813.1 million and 3331.5 kilotons at a compound annual growth rate (CAGR) of 6.4% by 2021. In particular, in 2015, ATH (aluminum trihydrate) dominated the flame-retardant market, accounting for 35.6% of the total market. This is owing to its low cost, high efficiency, and minimal adverse impact on the environment compared to halogenated flame retardants. Among nonhalogenated flame retardants, phosphorus flame retardants are likely to have a maximum CAGR of 7.12% (volume) and 7.9% (value) between 2016 and 2021. Because the use of flame retardants for halogen compounds has been gradually reduced and banned since the 2000 Stockholm Convention, the development and use of flame retardants for nonhalogen compounds is likely to experience high demand in North America and the developed countries. Flame retardants are being and will be used in addition to various types of plastics and composites. Therefore, the demands for the use and development of flame retardants are increasing in developed and developing countries owing to stringent regulations to protect environmental safety and health.

Globally, China, Japan, and India in the Asia–Pacific region form a representative market (49.8% of the world market) that consumes flame retardants. Among these, China (which accounts for 38.2% of the total market in 2015) has a large ripple effect. The use of flame-retardant chemicals is increasing in the Asia–Pacific region owing to the presence of major developing countries such as China, India, and Korea. In all the countries in the Asia–Pacific region, regulations on fire safety and the demand for flame retardants are increasing.

10.4 Bioepoxy Polymers with Flame-Retardant Properties

Diglycidylether bisphenol A (DGEBA) constitutes 90% of the epoxy market. However, their inherently low flame retardancy induces the use of many flame retardants that are generally toxic halogen-containing materials. Therefore, to increase the intrinsic flame retardancy and decrease toxicity, greener halogen-free epoxy monomers have been considered, such as phosphorus- or aromatic structure-containing epoxy monomers. Phosphorus-containing epoxy does not release toxic gases during the combustion process. Moreover, the phosphorus can spark the formation of the char layer as a gas and heat barrier. Aromatic structure-rich epoxy polymers are similar to the phosphorus-containing epoxy polymers with regard to the mechanism of improving the flame retardancy. With regard to char layer formation, an important information for designing epoxy polymers with flame retardancy is the char-forming tendency (CFT) [50]. It represents the degree of contribution to the amount of char residue depending on the molecular structures of polymers. According to Table 10.3, the aromatic structures exhibit a higher yield of combustion residue than the aliphatic structure. Moreover, they exert a higher flame-retardant effect in the solid phase.

There are abundant plant-based aromatic resources such as lignocellulosic and sugar biomass, furan, and tannins, with various functional groups. These have been chemically modified for producing bioepoxy polymers to be used as an epoxy monomer or a hardener. Hydroxyl, ethylene, and carboxylic acid groups are generally used to functionalize epoxy groups to bio-resources. In addition, phosphorus atoms are introduced into bio-resources for enhancing the flame-retardant properties of bioepoxy polymers by forming a char layer. In this section, we address the recently reported bio-resource-derived epoxy polymers without addition of flame retardant or other additives.

10.4.1 Lignocellulosic Biomass-Derived Epoxy Polymers

From lignocellulose, lignin, cellulose, and hemicellulose can be extracted. Lignin as the second most abundant natural organic material constitutes aromatic monomers. Monoaromatic compounds such as eugenol and vanillin having ethylene and aldehyde groups, respectively, in the *para*-position of 2-methoxylphenol (the so-called guaiacol) can be functionalized to a bioepoxy monomer form. Cellulose and hemicellulose are hydrolyzed and dehydrated, which can produce furan derivatives and can be modified for producing epoxy resins.

10.4.1.1 Eugenol

Eugenol-derived epoxy monomers for bioepoxy matrix with high flame retardancy can be conveniently obtained by oxidizing the double bond of an ethylene part using H_2O_2/acetic acid or *meta*-chloroperoxybenzoic acid (mCPBA). Wang and co-workers [51] synthesized a eugenol-based epoxy monomer with full aromatic ester backbone (TPEU-EP [eugenol-based epoxy resin]) (Figure 10.9). They had it react with 3,3′-diamonodiphenylsulfone (33DDS) as a curing agent to produce a bio-based epoxy resin. Owing to its remarkable intumescent and high char yielding (31.7 wt%) capabilities, it exhibits a UL-94 V1 rating and has 1.14 times higher LOI than DGEBA/33DDS (26.8 vs. 23.5). The cone calorimetry data revealed that the PHRR, THR, total smoke release (TSR), and total smoke production (TSP) of TPEU-EP/33DDS decreased to 20%, 28%, 62%, and 61%, respectively, of those of DGEBA/33DDS. Resultantly, TPEU-EP/33DDS exhibits higher flame retardancy with low smoke production during combustion.

Table 10.3 CFT factors with variation of polymer molecular structures.

Structure	CFT	Structure	CFT
Aliphatic		*Heterocyclic*	
–CHOH–	1/3	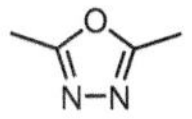	1
Others	0		
Aromatic		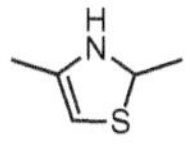	3 × 1/2
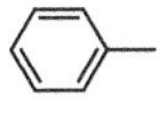	1		
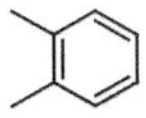	2	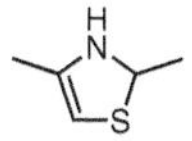	3 × 1/2
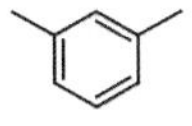	3	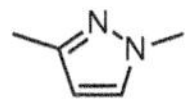	3 × 1/2
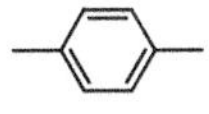	4	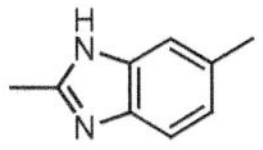	7
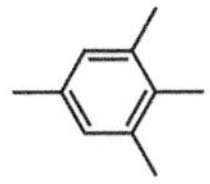	6	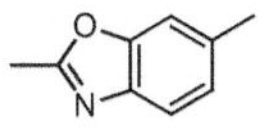	7
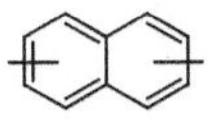	6	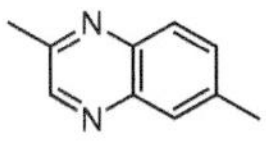	9
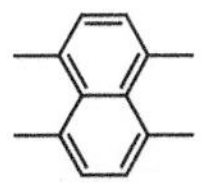	10	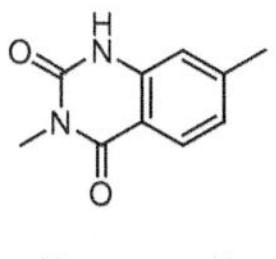	11
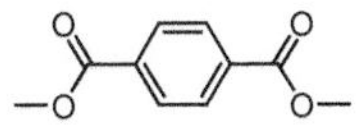	1 1/4	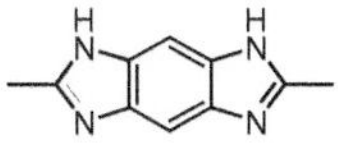	10
Aromatic side chain		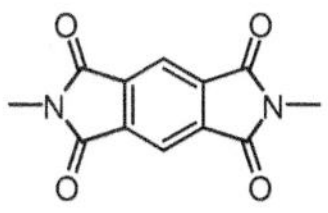	12
$>CH_2$, $>CH–CH_2–$	–1	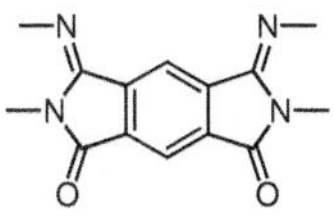	10
$–CH_3$	–1 1/2		
$>C(CH_3)_2$	–3		
$–CH(CH_3)_2$	–1	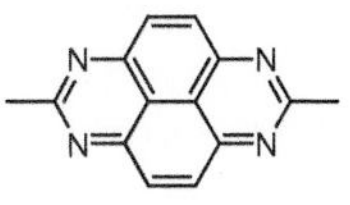	15

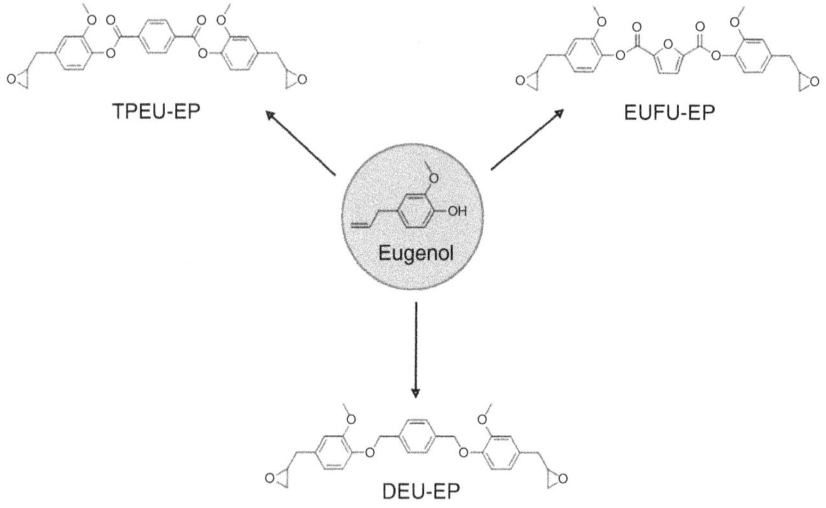

Figure 10.9 Eugenol-derived epoxy monomers.

The same group of authors developed a bio-based flame-retardant epoxy resin (DEU-EP) with a marginally different structure (Figure 10.9). It was synthesized from the start where two α,α′-dichloro-*p*-xylene molecules reacted with eugenol [52]. The net biomass content of the eugenol-based epoxy matrix from DEU-EP and 4,4′-diaminodiphenyl methane (DDM; curing agent) was 70.2%. It has a higher char yield (38 wt%) than the previously reported TPEU-EP/33DDS (31.7 wt%). Therefore, DEU-EP/DDM exhibits an improved inhibition of the flame propagation (self-extinguished at 10 seconds) compared to TPEU-EP/33DDS (24 seconds). From the microscale combustion calorimeter (MCC) data, the stark decreases (55% and 39%, respectively) in the PHRR and THR of DEU-EP/DDM compared to those of DGEBA/DDM are registered.

Gu and co-workers [53] used both 2,5-furfuran dicarboxylic acid (FDCA) and eugenol to increase the biomass content of a bio-based flame-retardant epoxy matrix. In this group, methyl hexahydrophthalic anhydride (MHHPA) as a curing agent is reacted with eugenol and furan-based epoxy resin (EUFU-EP). This produced EUFU-EP/MHHPA epoxy polymer with 93.3% of biomass content and 10.9 wt% of char yield. With regard to this material, only MCC is used as an instrument for the flame retardancy test. The PHRR and THR of DGEBA/MHHPA thermoset are 359.7 W/g and 24.7 kJ/g, respectively. However, when the epoxy resin was changed from DGEBA to EUFU-EP, the PHRR and THR decreased to 291.3 W/g and 20.0 kJ/g, respectively.

10.4.1.2 Vanillin

Zhu and co-workers [36] used Schiff-base structure formation and phosphorus–hydrogen addition to synthesize two vanillin-based epoxy monomers with flame retardancy (EP1 and EP2) (Figure 10.10). Each monomer was reacted with a DDM hardener, and their flame retardancy was compared to that of a DGEBA/DDM polymer. The char yields of EP1/DDM and EP2/DDM obtained

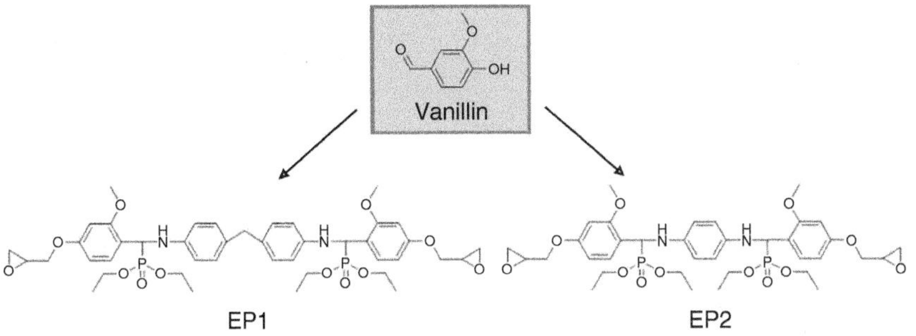

Figure 10.10 Vanillin-derived epoxy monomers.

from thermogravimetric analysis (TGA) under a nitrogen atmosphere are 53% and 58%, respectively. These values are significantly higher than those of nonphosphorous bio-based epoxy polymers. The phosphorus and nitrogen retention in their char residues is also high (EP1/DDM: 38.5%, EP2/DDM: 53.5%), resulting in remarkable flame retardancy with UL-94 V0 rating and LOI of 31.4% and 32.8%, respectively. Meanwhile, DGEBA/DDM has an LOI of 24.6% and no UL-94 rating.

10.4.2 Furan

We have introduced the EUFU-EP in Section 10.4.1.1. In the following context, only furan-based epoxy resins are described. In general, furan-based epoxy resins are synthesized from hydroxyethyl furfural (HMF) as starting materials (Figure 10.11). Guo and co-workers [54] produced 2,5-bis [(2-oxiranylmethoxy)methyl]furan (BOF) and difuranic diepoxide monomer (OmbFdE). They also compared the flame retardancies of epoxy polymers made

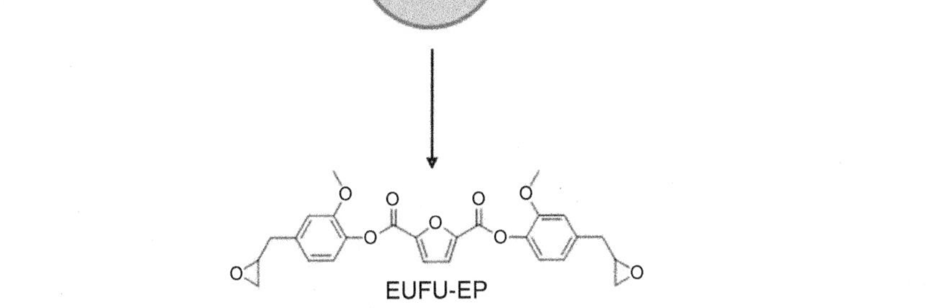

Figure 10.11 Furan-derived epoxy monomers.

from each furan-based monomer and the aliphatic diamine hardener (TEGA). From the TGA data obtained in an N_2 atmosphere, the char yields of BOF/TEGA and OmbFdE/TEGA are 19% and 19.8%, respectively. In addition, the MCC results represent the PHRR and HRR of BOF/TEGA and OmbFdE/TEGA. It is noteworthy that the OmbFdE/TEGA exhibited a higher PHRR value (174.7 [1st]/85.6 [2nd] vs. 71.2 [1st]/140.5 [2nd] W/g) and a lower THR value (13.7 vs. 15.6 kJ/g) than BOF/TEGA. This could be due to the degradation of the more numerous ether linkages in the matrix of OmbFdE accounted for the first PHRR. For the second PHRR, the contribution of more stable furan motifs should have caused the second peak heat release. Because the effect of the furan moieties is superior to that of the ether linkages, the THR of the OmbFdE/TEGA is lower than that of BOF/TEGA. However, because commercial epoxy resin such as DGEBA was not used as a control in this study, it is infeasible to objectively evaluate the flame retardancy of the furan-based epoxy monomers.

10.4.3 Tannins

Although tannins as polyphenolic compounds extracted from plants are considered as bio-based epoxy monomers with flame retardancy, there has not been any report on it. However, Jung and co-workers [55] used hydrolyzable tannin or tannic acid as a multifunctional epoxy hardener with flame retardancy. Using DGEBA as an epoxy resin, the flame retardancies of TA-based epoxy polymers were evaluated depending on the content of TA. A Jeffamine-type hardener (D230) was selected as a control. The char yield of the sample with a 1.2 TA molar ratio (TD 1.2; the maximum TA molar ratio in this study) is 21.9%. This is exceptionally higher than that of the control (4.1%). Based on the LOI and MCC instruments, TD1.2 exhibits 1.46 times higher LOI than the control, and 24.8% and 33.8% of decrease in the PHRR and THR as compared with DGEBA/D230. The flame suppression of the TD thermoset may be ascribed to the quenching effect of the phenoxy radicals formed during combustion and the formation of a char layer, which functions as a barrier obstructing heat flux and gas transport.

10.5 Use of Fillers for Improving Flame-Retardant Properties of Bioepoxy Polymers

Organic/inorganic nanocomposites are prepared by uniformly dispersing nano-sized organic and inorganic particles in a polymer matrix. The mechanical and thermal properties, impact resistance, chemical resistance, heat resistance, and flame retardancy can be improved over conventional single polymer materials. Nanomaterials (particles) have a small particle size and are well dispersed in the polymer matrix. Moreover, the specific surface area is large, so that the addition of even a small amount is likely to yield remarkable properties.

Conventionally, to increase the flame retardancy of a polymer and thus improve its physical properties, a flame retardant is incorporated as a filler or additive. However, as described above, the use of halogens and phosphorus-based

Figure 10.12 Conceptual diagram of (a) single-walled carbon nanotube (SWCNT), (b) multiwalled carbon nanotube (MWCNT) [58], and (c) graphene layer. Source: Reproduced from Geim and Novoselov [59]. © 2007, Springer Nature.

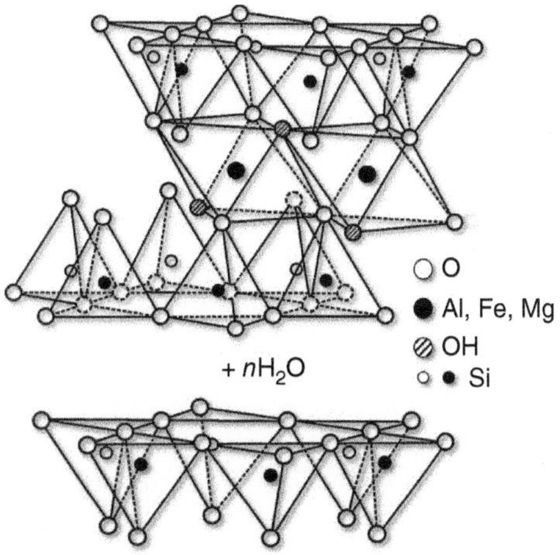

Figure 10.13 Structure of 2 : 1 layered silicate such as montmorillonite clay [56, 57].

substances gradually decreased owing to the emission of toxic gases in the fire and the unsafe lighting. Recently, studies on improving flame retardancy using various organic fillers have been reported. Clay [56, 57], carbon nanotubes [58], graphene [59], and expanded graphite are used as representative nanoadditives to improve flame retardancy (Figures 10.12 and 10.13).

Bioepoxy has also been reported to improve the flame retardancy index by using the aforementioned additives, and a few examples are provided. Lijun Qian's research group [60] coated phosphor–nitrogen flame retardant on carbon nanotubes (PDAP-CNT) and used it as a filler in epoxy (Figure 10.14). In summary, PDAP-CNT exhibited remarkable flame-retardant performance in flame-retardant EP composites (FR-EP). Moreover, the incorporation of 5.0 wt% PDAP-CNT improved the LOI of EP from 26.0% to 31.8%. Moreover, UL-94 exhibited a V-0 rating. This improvement can be considered as owing to the thermal stability of the carbon nanotubes and the synergistic effect of the phosphorus–nitrogen-containing coating layer coated on the surface. In addition, because of the flame retardant coated on the nanofiller, a continuous and compact char layer is formed in the condensation step. This suppresses the second- and third-order heat transfer.

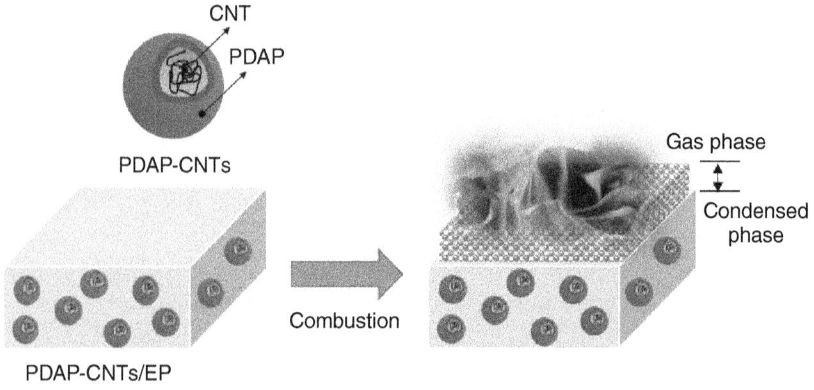

Figure 10.14 Carbon nanotubes coated with phosphorus–nitrogen flame retardant and its application in epoxy thermosets. Source: Reproduced from Fei et al. [60]. © 2019, Taylor & Francis.

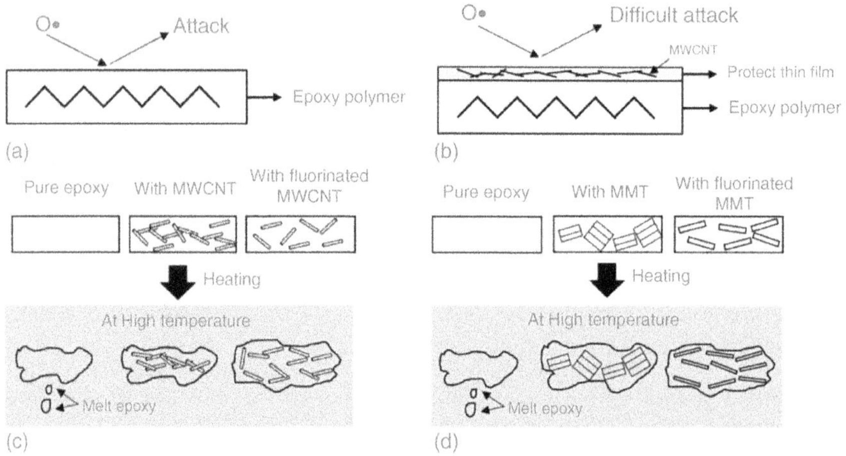

Figure 10.15 Postulated mechanism of flame-retardant additives. al. [61, 62].

Lee and co-workers [61, 62] reported the results of adding montmorillonite (MMT) and multiwalled carbon nanotubes as flame retardants to epoxy resins to improve the oxidation resistance and flame retardancy of the resins. MMT used as a flame-retardant additive functioned as an energy storage medium to prevent heat transfer in epoxy composites. Multiwalled carbon nanotubes were observed to be a medium for effectively reducing the decomposition rate of epoxy and increasing the yield of charcoal. Moreover, the combined use of these two additives increased the thermal activity energy of the epoxy resin, which contributed significantly to the improvement in the antioxidant and flame retardancy index. In addition, it has been verified that the addition of epoxy resin through the fluorination treatment to the existing MMT increased the heat activation energy to more than two times that of the value in the previous research results. The mechanism is described in Figure 10.15 [61, 62].

Meanwhile, there are a few examples of utilization of the structural properties of additives. As the use of wood-based building materials increases, flame-retardant treatment technology gains importance. This intumescent system is a nonhalogen flame-retardant system that combines the technology of polymer foaming and carbonization layer formation [63].

Intumescent–nanoclay (INC) composites [64–66] are highly advanced flame-retardant materials that combine foam, carbonization, and clay insulation technology in an integrated system. The use of INC composites has expanded in Europe since the late 2000s. INC composite materials exhibit a mechanism wherein a carbonization layer is formed simultaneously with initial foaming. A system is also implemented simultaneously to form an isolation layer that blocks external energy during rearrangement of the stacking of nanoclay that is spread on the surface. Therefore, the combustion process exhibits a delay and a combustion blocking effect simultaneously. Intumescent system is a technology that is directly applied to fire protection paints. It has a structure that maximizes performance by integrating with nanotechnology.

B. Szolnoki and co-workers [67] has developed a new flame-retardant carbon-fiber-reinforced composite using glucofuranoside-based flame-retardant bioepoxy resin trifunctional epoxy monomer (GFTE) and aromatic amine as a curing agent (Figure 10.16). The flame-retardant epoxy carbon-fiber-reinforced composite material was prepared by combining a liquid resorcinol bis(diphenyl phosphate) (RDP) and a solid ammonium polyphosphate (APP) to prepare

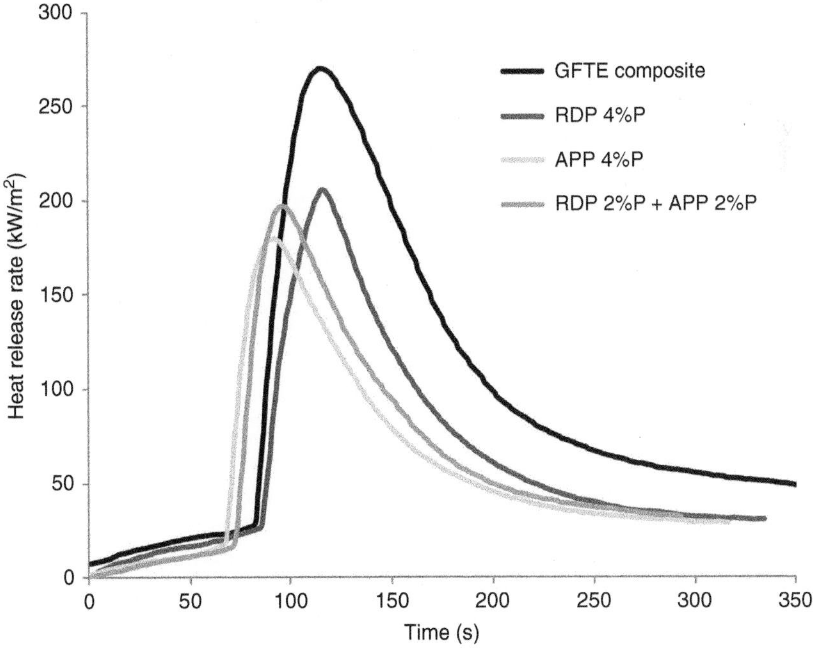

Figure 10.16 Heat release rate of reference and flame-retarded GFTE composite. Source: Reproduced from [67]. © 2017, Elsevier.

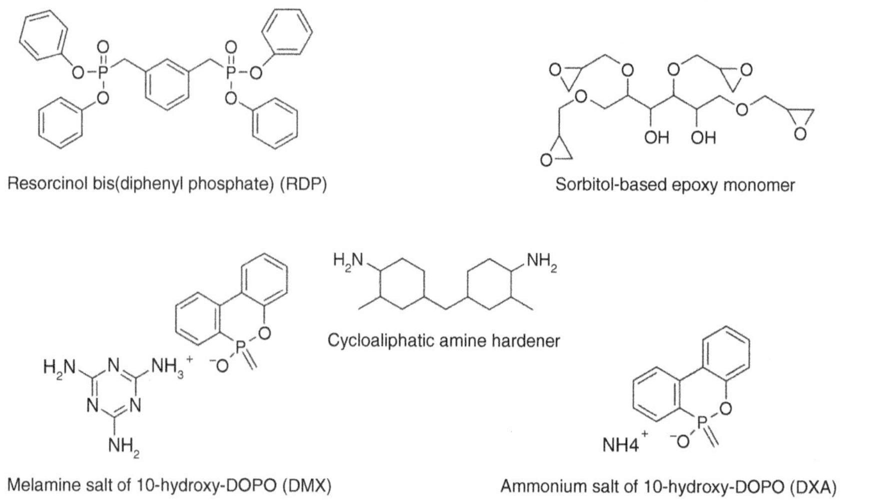

Figure 10.17 Chemical structure of the epoxy resin components and applied flame retardants.

a sample containing approximately 4% phosphorus. This combination could effectively compensate for the plasticization effect with the P-containing and improved thermal mechanical properties. UL-94 has a V-0 rating. In addition, when manufacturing a carbon-fiber-reinforced composite material using a conventional bioepoxy resin, the physical properties exhibited a tendency to weaken owing to the attenuated mechanical properties and decrease in plasticization temperature. However, this study can be said to be the result of solving the existing problems effectively.

György Maro and co-workers [68] prepared a bioepoxy using sorbitol monomer. Moreover, APP and RDP were added, resulting in the synthesis of flame-retardant bioepoxy. The synthesized bioepoxy exhibited a phosphorus content of approximately 3%. In particular, red phosphorus (RP) exhibit UL-94's V-0 grade, whereas melamine salt exhibits HB grade because of the lower phosphorus content in the sample. As the flame retardant was added to the matrix, the LOI of the sample increased significantly. However, the results revealed that the effects of different types of flame retardants varied (Figure 10.17).

10.6 Conclusion

Countries worldwide are deeply concerned about the severe pollution of the global environment. Under these circumstances, the substitution of halogen flame-retardant additives and petroleum-based epoxy materials, which have been causing severe environmental issues and human hazards, is a problem that must be solved to ensure a healthy future. In this context, the researchers' efforts to develop or replace eco-friendly epoxy-based or plant-based materials are highly commendable. This chapter describes recent research trends on bio-based epoxy development and bio-based retardants. In particular, bio-based flame retardants were reported to improve the material properties and synergistic effects. Moreover, the flame-retardant mechanism for improving the flame retardancy index was discussed. At present, the newly developed materials are ineffective for replacing petroleum-based epoxy, which is being widely used now. However, the efforts of several researchers are likely to contribute to good results soon.

Acknowledgment

This work was supported by the KIST Institutional Program.

References

1 Hajime, K., Akihiro, M., Kiichi, H. et al. (1998). Epoxy resin cured by bisphenol a based benzoxazine. *J. Appl. Polym. Sci.* 68 (12): 1903–1910.

2 Elaheh, R.R., Henri, V., Agustin, R.A. et al. (2019). Bio-epoxy resins with inherent flame retardancy. *Prog. Org. Coat.* 135: 608–612.

3 Levchik, S., Piotrowskia, A., Weil, E., and Yao, Q. (2005). New developments in flame retardancy of epoxy resins. *Polym. Degrad. Stab.* 88 (1): 57–62.

4 Fregert, S. and Trulsson, L. (1978). Simple methods for demonstration of epoxy resins of bisphenol A type. *Contact Dermatitis* 4 (2): 69–72.

5 Ochi, M., Tsuyuno, N., Sakaga, K. et al. (1995). Effect of network structure on thermal and mechanical properties of biphenol-type epoxy resins cured with phenols. *J. Appl. Polym. Sci.* 56 (9): 1161–1167.

6 Anthony, R.P. and Richard, J.F. (1990). Evolution of residual stresses in three-dimensionally constrained epoxy resins. *Polymer* 31 (10): 1932–1936.

7 Fan-Long, J., Xiang, L., and Soo-Jin, P. (2015). Synthesis and application of epoxy resins: a review. *J. Ind. Eng. Chem.* 29: 1–11.

8 Chun-Shan, W. and Jeng-Yueh, S. (1999). Phosphorus-containing epoxy resin for an electronic application. *J. Appl. Polym. Sci.* 73 (3): 353–361.

9 Tsung-Han, H. and Chun-Shan, W. (2001). Modification of epoxy resin with siloxane containing phenol aralkyl epoxy resin for electronic encapsulation application. *Eur. Polym. J.* 37 (2): 267–274.

10 Clayton, M. (2018). *Epoxy Resins: Chemistry and Technology*. CRC Press.

11 Azar, N.G., Heidari, M., Bahrami, Z.S., and Shokri, F. (2000). In vitro cytotoxicity of a new epoxy resin root canal sealer. *J. Endodontic.* 26 (8): 462–465.

12 Gabriele, L., Jürgen, H., Georg, R. et al. (1999). Genotoxicity and cytotoxicity of the epoxy resin-based root canal sealer AH plus. *J. Endodontic.* 25 (2): 109–113.

13 Sushanta, K.S., Smita, M., and Sanjay, K.N. (2015). Synthesis and characterization of bio-based epoxy blends from renewable resource based epoxidized soybean oil as reactive diluent. *Chin. J. Polym. Sci.* 33 (1): 137–152.

14 Elyse, A.B., Santosh, K.Y., Giuseppe, R.P., and Joseph, F.S. III, (2016). Recent advances in bio-based epoxy resins and bio-based epoxy curing agents. *J. Appl. Polym. Sci.* 133: 45.

15 Saeid, N., Omid, Z., Susan, H. et al. (2017). A renewable bio-based epoxy resin with improved mechanical performance that can compete with DGEBA. *RSC Adv.* 7 (14): 8694–8701.
16 Jiang, Z., Chandrashekhara, K., Virgil, F., and Shubhender, K. (2004). Curing and mechanical characterization of a soy-based epoxy resin system. *J. Appl. Polym. Sci.* 91 (6): 3513–3518.
17 Jean-marie, R., Deléglisea, M., Marie, F.L., and Patricia, K. (2010). Thermosetting (bio) materials derived from renewable resources: a critical review. *Prog. Polym. Sci.* 35 (4): 487–509.
18 Shida, M., Ping, W., Zhiguo, S., and Songping, Z. (2014). Vegetable-oil-based polymers as future polymeric biomaterials. *Acta Biomater.* 10 (4): 1692–1704.
19 Soo-Jin, P., Fan-Long, J., and Jae-Rock, L. (2004). Synthesis and thermal properties of epoxidized vegetable oil. *Macromol. Rapid. Commun.* 25 (6): 724–727.
20 Sylwia, D., Adrien, C., Dariusz, B. et al. (2015). Formulation of bio-based epoxy foams from epoxidized cardanol and vegetable oil amine. *Eur. J. Lipid. Sci. Technol.* 117 (11): 1893–1902.
21 Fatemeh, F., Yongsheng, Z., Zhongshun, Y. et al. (2016). Curing kinetics and mechanical properties of bio-based epoxy composites comprising lignin-based epoxy resins. *Eur. Polym. J.* 82: 153–165.
22 Fatemeh, F., Zhongshun, Y., and Mark, A. (2014). Synthesis of lignin-based epoxy resins: optimization of reaction parameters using response surface methodology. *RSC Adv.* 4 (60): 31745–31753.
23 Chahinez, A., Hélène, N., Maxence, F. et al. (2013). Multi-functionalization of gallic acid. Synthesis of a novel bio-based epoxy resin. *Eur. Polym. J.* 49 (6): 1185–1195.
24 Fatemeh, F., Zhongshun, Y., Mark, A., and Chunbao (Charles)X. (2015). Sustainable lignin-based epoxy resins cured with aromatic and aliphatic amine curing agents: curing kinetics and thermal properties. *Thermochim. Acta.* 618: 48–55.
25 Ayman, M.A., Mansour, R., Mahmoud, I.A., and Ashraf, M.S. (2004). Epoxy resins from rosin acids: synthesis and characterization. *Polym. Adv. Technol.* 15 (9): 514–522.
26 Rémi, A., Sylvain, C., Ghislain, D. et al. (2013). Biobased thermosetting epoxy: present and future. *Chem. Rev.* 114 (2): 1082–1115.
27 Wasiuddin, A.K., Zhiyuan, W., Mohammad, T.A. et al. (1988). Inhibition of the skin tumorigenicity of (±)-7β, 8α-dihydroxy-9α, 10α-epoxy-7, 8, 9, 10-tetrahydrobenzo [a] pyrene by tannic acid, green tea polyphenols and quercetin in Sencar mice. *Cancer Lett.* 42 (1-2): 7–12.
28 Mitsuhiro, S. and Nakai, K. (2010). Preparation and properties of biocomposites composed of bio-based epoxy resin, tannic acid, and microfibrillated cellulose. *J. Polym. Sci. Part B: Polym. Phys.* 48 (4): 425–433.
29 Fanny, J., Emilie, D., Amédée, R. et al. (2014). New biobased epoxy materials from cardanol. *Eur. J. Lipid. Sci. Technol.* 116 (1): 63–73.
30 Pietro, C., Daniele, D.'A., Luigia, L. et al. (2009). Cardanol-based novolac resins as curing agents of epoxy resins. *J. Appl. Polym. Sci.* 114 (6): 3585–3591.

31 Aggarwal, L.K., Thapliyal, P.C., and Sukhdeo, K. (2007). Anticorrosive properties of the epoxy–cardanol resin based paints. *Prog. Org. Coat.* 59 (1): 76–80.
32 Xin, W., Ehsan, N.K., and De-Yi, W. (2015). Renewable cardanol-based surfactant modified layered double hydroxide as a flame retardant for epoxy resin. *ACS Sustain. Chem. Eng.* 3 (12): 3281–3290.
33 Emilie, D., Nelly, D., Bernard, B., and Sylvain, C. (2015). New cardanol/sucrose epoxy blends for biobased coatings. *Prog. Org. Coat.* 83: 47–54.
34 Songqi, M., Xiaoqing, L., Yanhua, J. et al. (2013). Bio-based epoxy resin from itaconic acid and its thermosets cured with anhydride and comonomers. *Green Chem.* 15 (1): 245–254.
35 Henri, V., Maryam, J., Marianne, C. et al. (2018). Short-lasting fire in partially and completely cured epoxy coatings containing expandable graphite and halloysite nanotube additives. *Prog. Org. Coat.* 123: 160–167.
36 Sheng, W., Songqi, M., Chenxiang, X. et al. (2017). Vanillin-derived high-performance flame retardant epoxy resins: facile synthesis and properties. *Macromolecules* 50 (5): 1892–1901.
37 Raphaël, M., Claire, N., Maxence, F. et al. (2015). From a bio-based phosphorus-containing epoxy monomer to fully bio-based flame-retardant thermosets. *RSC Adv.* 5 (87): 70856–70867.
38 Xin, W., Shun, Z., Wen-Wen, G. et al. (2017). Renewable cardanol-based phosphate as a flame retardant toughening agent for epoxy resins. *ACS Sustainable Chem. Eng.* 5 (4): 3409–3416.
39 Naheed, S., Mohammad, J., and Othman, Y.A. (2016). Recent advances in epoxy resin, natural fiber-reinforced epoxy composites and their applications. *J. Reinf. Plast. Compos.* 35 (6): 447–470.
40 Halogen-free flame retardants market, global forecasts to 2021, markets and markets. www.marketsandmarkets.com
41 Flame retardants market, global forecasts to 2021, markets and markets. www.marketsandmarkets.com
42 Kumar, K., Sukhila, K., Smita, M., and Sanjay, K.N. (2018). Synthesis and characterization of petroleum and biobased epoxy resins: a review. *Polym. Int.* 67 (7): 815–839.
43 Ramon, E., Carmen, S., and Pedro, M. (2018). A review of recent research on bio-based epoxy systems for engineering applications and potentialities in the aviation sector. *Aerospace* 5 (4): 110.
44 Christopher, E.H. (2019). Recent advances in bio-based flame retardant additives for synthetic polymeric materials. *Polymers* 11 (2): 224.
45 Norman, G.M., Buckley, C.P., Clive, B.B., and Bucknall, C.B. (1997). *Principles of Polymer Engineering*. USA: Oxford University Press, New York.
46 Maya, J.J. (2019). Green composite for automotive applications, chapter 2. In: *Flammability Performance of Biocomposites* (ed. J.J. Maya), 43–58. Woodhead Publishing.
47 Schartel, B., Braun, U., Schwarz, U., and Reinemann, S. (2003). Fire retardancy of polypropylene/flax blends. *Polymer* 44 (20): 6241–6250.
48 Babrauskas, V. (2002). Heat release rates. In: *The SFPE Handbook of Fire Protection Engineering*, 3e), (Chapter 3.1) (eds. D.N. PJ, D. Drysdale, C.L. Beyler, et al.), 3–8. National Fire Protection Association, Inc.

49 Schartel, B., Bartholmai, M., and Uta, K. (2005). Some comments on the use of cone calorimeter data. *Polym. Degrad. Stab.* 88 (3): 540–547.
50 (1975). Some basic aspects of flame resistance of polymeric materials. *Polymer* 16 (8): 615–620.
51 Jintao, W., Bin, G., Cheng, L. et al. (2015). A novel biobased epoxy resin with high mechanical stiffness and low flammability: synthesis, characterization and properties. *J. Mater. Chem. A* 3 (43): 21907–21921.
52 Jintao, W., Bin, G., Cheng, L. et al. (2016). A sustainable, eugenol-derived epoxy resin with high biobased content, modulus, hardness and low flammability: synthesis, curing kinetics and structure–property relationship. *Chem. Eng. J.* 284: 1080–1093.
53 Jia-Tao, M., Li, Y., Qingbao, G. et al. (2017). Biobased heat resistant epoxy resin with extremely high biomass content from 2, 5-furandicarboxylic acid and eugenol. *ACS Sustainable Chem. Eng.* 5 (8): 7003–7011.
54 Jingjing, M., Yushun, Z., Guiqin, Z. et al. (2019). Sustainable bio-based furan epoxy resin with flame retardancy. *Polym. Chem.* 10 (19): 2370–2375.
55 Young-O, K., Jaehyun, C., Hyeonuk, Y. et al. (2019). Flame retardant epoxy derived from tannic acid as biobased hardener. *ACS Sustainable Chem. Eng.* 7 (4): 3858–3865.
56 Feng, Z., Yong, W., Shaoxiang, L., and Jun, Z. (2013). Influence of thermophysical properties on burning behavior of intumescent fire-retardant materials. *J. Therm. Anal. Calorim.* 113 (2): 803–810.
57 Anderson, J.R., Charles, E., Donald, E.K., and William, P.M. (1988). Thermal conductivity of intumescent chars. *J. Fire. Sci.* 6 (6): 390–410.
58 Peter, J.F.H. (2009). *Carbon Nanotube Science: Synthesis, Properties and Application.* USA: Cambridge University Press, New York.
59 Geim, A.K. and Novoselov, K.S. (2007). The rise of graphene. *Nat. Mater.* 6: 183–191.
60 Fei, X., Congcong, Z., Chao, G. et al. (2019). Carbon nanotubes coated with phosphorus-nitrogen flame retardant and its application in epoxy thermosets. *Polymer-Plast. Technol. Mater.* 58 (17): 1889–1899.
61 Lee, S.K. and Bai, B.C. (2010). Flame retardant epoxy complex produced by addition of montmorillonite and carbon nanotube. *J. Ind. Eng. Chem.* 16 (6): 891–895.
62 Im, J.S., Lee, S.K., In, S.J., and Lee, Y.-S. (2010). Improved flame retardant properties of epoxy resin by fluorinated MMT/MWCNT additives. *J. Anal. Appl. Pyrolysis.* 89 (2): 225–232.
63 Haiyun, M., Lifang, T., Zhongbin, X. et al. (2007). A novel intumescent flame retardant: synthesis and application in ABS copolymer. *Polym. Degrad. Stab.* 92 (4): 720–726.
64 Zhen-yu, W., En-hou, H., and Wei, K. (2007). Fire-resistant effect of nanoclay on intumescent nanocomposite coatings. *J. Appl. Polym. Sci.* 103 (3): 1681–1689.
65 Hao, W., Mourad, K., and Joseph, H.K. (2014). Flame retardant polyamide 6/nanoclay/intumescent nanocomposite fibers through electrospinning. *Text. Res. J.* 84 (10): 1106–1118.

66 Doğan, M. and Erdal, B. (2011). Synergistic effect of boron containing substances on flame retardancy and thermal stability of clay containing intumescent polypropylene nanoclay composites. *Polym. Adv. Technol.* 22 (12): 1628–1632.
67 Toldy, A., Péter, N., Zs, R., and Beáta, S. (2017). Flame retardancy of glucofuranoside based bioepoxy and carbon fibre reinforced composites made thereof. *Polym. Degrad. Stab.* 142: 62–68.
68 Beáta, S., Andrea, T., and György, M. (2019). Effect of phosphorus flame retardants on the flammability of sugar-based bioepoxy resin. *Phosphor. Sulfur Silicon Related Element.* 194 (4-6): 309–312.

11

Water Sorption and Solvent Sorption of Bio-epoxy Polymers, Their Blends, and Composites

Amirthalingam V. Kiruthika

Seethalakshmi Achi College for Women, Physics Department, Pallathur, Sivagangai Dt, Tamilnadu 630 107, India

11.1 Introduction

In recent years, environmental consciousness, economic concerns, and governmental legislations throughout the world have driven the researchers to develop biodegradable, renewable, lightweight, sustainable, and recyclable polymers referred as bio-based epoxy polymers. Bio-epoxy is defined as a polymer made from a wide variety of renewable resources, and it can be used as an adhesive coating and lamination material in a wide range of industrial applications.

The global market potential of epoxy resin is enhanced by the increasing demand in paint industry, wind energy construction, composites, automotives, and aerospace. The most commercially available epoxy polymers were bisphenol-A or BPA and diglycidyl ether of bisphenol A (DGEBA). For the production of polymeric products, BPA could be used as a raw material, especially in the production of DGEBA, because of its deleterious effects on human health and environment and is also shown to be toxic to living organisms and to function as endocrine disruptors; however, BPA is widely used in the United States and in European countries, but in France and Canada, the usage of BPA in food-related materials such as children's bottle, cups, etc., has been banned.

In recent years, it is of great interest that researchers and industrialists are mainly focused on sustainable, renewable, and environmentally friendly materials to develop safer epoxy polymers. It includes plant oils (soybean, clove, sunflower, and linseed), polysaccharides (tamarind seed gum, plant extracts, etc.), cardonals (from cashew nut shell liquid (CSNL)), rosins, lignin, terpenes, ferulic acid, itaconic acid (IA), gallic acid (GA) (oak bark, sumac, and tea leaves), vanillic acid, natural rubber (NR), etc., and the development of bio-epoxy polymers from renewable resources has gained much interest in academic and industrial research to enable the production of polymers with increased renewable content and reduced carbon foot print.

Bio-Based Epoxy Polymers, Blends, and Composites: Synthesis, Properties, Characterization, and Applications, First Edition. Edited by Jyotishkumar Parameswaranpillai, Sanjay Mavinkere Rangappa, Suchart Siengchin, and Seno Jose.

The advantages of bio-epoxy resins are as follows:

(a) Superior fatigue strength
(b) Remarkable resistance to corrosion
(c) Low shrinkage on cure
(d) Achieve higher mechanical properties such as tensile strength, compressive strength, and flexural modulus
(e) Good electrical insulation properties and their durability in tough conditions or exposure
(f) The performance of interfacial bonding with types of substrates is good
(g) High degree of resistance to physical abuse
(h) Good stability
(i) Excellent stability in many solvents and alkali solutions

Water absorption is an important parameter of structural materials that may be subjected to rain and humid environments. The diffusion coefficient and water uptake behavior are used to determine the suitability of the material for applications that involve prolonged exposure to water. Three important mechanisms of water absorption are observed in polymer composite materials: (i) diffusion behavior of water molecules that occur inside the voids between the polymeric chains. It was further subdivided into three methods: Fickian diffusion model, anomalous or non-Fickian model, and intermediate stages between Fickian and non-Fickian. (ii) capillary transport into gaps and flows of water molecules at the interfacial bonding between the fillers and the resins, and (iii) transport of water molecules through the transmission of microcracks in the matrix, arising due to fiber swelling, which is highly stringent in the case of plant fiber composites. In this chapter, the water contact angle (WCA), diffusion coefficient, and maximum percentage of weight gain of bio-epoxy, their blends as well as composites were briefly reviewed.

11.2 Bio-epoxy Resins

Bio-epoxy resin is one of the most promising options in the polymer industry because of its low eco-toxicity, and it can be easily derived from various kinds of bio-based materials (soybean, lignin, hempseed, cardonal, rosin, gallic acid, etc.). Here, Section 11.2.1 highlights the water and solvent uptake behavior of bio-epoxy resins, their blends, and composites.

11.2.1 Soybean Oil (SO)-Based Epoxy Resins

Traditionally, epoxidized soybean oil (ESO) has received greater attention in scientific sectors and is widely used as a green plasticizer in the plastic industry. It is a triglyceride made up of oleic, linoleic, and linolenic acid methyl ester as well as saturated fatty acid compositions.

Bio-based ESO composites were fabricated with the introduction of *n*-octadecyl isocyanate-modified microcrystalline cellulose (MCC) as the reinforcing fillers. The surface wettability of 18C-g-MCC from 0% to 25 wt% was found by static contact angle measurement. From the analysis, the results indicated that the hydrophobicity of MCC was enhanced with the addition of 18C-g-MCC. Also, the contact angle of a neat ESO matrix was greater than that of both ESO/MCC and ESO/18C-g-MCC composites. The water absorption property of the ESO composites was determined by the difference between the initial weight samples and the water-immersed samples. The neat ESO polymer exhibited a water repelling characteristic that indicates the amount of water absorption (WA) to be less than 0.2% after 10 days of exposure. After attaining a saturation point (from 0 to 6 days), the composites absorb more water and later the amount of WA increases slowly in the composites. This is attributed to the hydrophilic nature of the cellulose used, which leads a larger content of WA with the influence of 18C-g-MCC. Moreover, mechanical properties such as tensile strength, Young's modulus, impact strength, flexural strength, and glass transition temperature of composites increase gradually with the increasing content of 18C-g-MCC [1].

Under controlled reaction conditions, waterborne epoxy (WBE) was synthesized through the modification of ESO by using diethanolamine [2]. At room temperature, the swelling characteristics of both ESO and WBE specimens were studied to determine the effective cross-linking density (Q), average molecular weight (M_c), and the swelling coefficient (S) values. The presence of a large number of OH groups in ESO samples indicated a higher swelling coefficient of 1.60 and WBE samples showed 1.39. Compared to WBE (1.39), ESO samples show higher swelling coefficient (1.60), which may be due to the presence of a large number of OH groups in ESO. Furthermore, M_c is greater for ESO (783.72 g/mol) and lower for WBE (613.03 g/mol) samples. It was revealed that the Q value (0.816×10^{-3} mol/cm^3) is greater for WBE, suggesting that WBE samples have linkages between the carboxylic end of the curing agent and oxirane, ester, and OH groups, but in the case of ESO (Q value is $0.635\,816 \times 10^{-3}$ mol/cm^3), it is cross-linked with carboxyl end group and oxirane. Both ESO and WBE followed the Fickian transport. As compared to WBE, ESO samples possessed relatively greater WA properties. A high diffusion coefficient (D is 0.11) was also observed in ESO, indicating that a larger number of water molecules diffused into the ESO film and hence low cross-linking than WBE (D is 0.06).

Munoz and García-Manrique [3] used flax and bio-epoxy for the preparation of composite materials and studied the influence of WA on the mechanical properties of prepared samples. The following equation is used to determine the WA, and it possesses Fickian behavior at room temperature:

$$c(t) = c_s - c_s \frac{8}{\pi^2} \sum_{k=1}^{20} \frac{1}{(2k-1)^2} \exp\left[\frac{-(2k-1)^2 D\pi^2}{d^2} t\right]$$

The maximum percentage weight gain and "D" values for the sample with six layers of flax (fiber vol. 40%) were 6.23% and 1.63×10^{-6} mm^2/s for tensile samples and 6.56% and 1.47×10^{-6} mm^2/s for flexural samples, respectively. In the same way, the specimen with eight layers of flax (fiber vol. 55%) has 8.71% (weight gain) and 2.32 (D value) for tensile and 9.76% and 1.85 for flexural samples, respectively. Compared to the dried samples, the tensile strength for all water-immersed specimens was higher because of the more adhesive nature between the fiber and the matrix used. Contrarily, the amount of WA increases, and the flexural properties decrease. As a hydrophilic nature of the fibers tends the tensile modulus was found to decrease substantially. Similarly after water absorption, flexural modulus decreases in higher fiber content specimens. Hence, the authors concluded that WA does not negatively affect the flax fiber bio-epoxy composite's mechanical properties.

11.2.2 Cardanol-Based Epoxy

Cardanol is an agro by-product of cashew nut industry and a source of phenolic lipids. It mainly consists of four main phenolic ingredients, namely, cardanol, cardol, anacardic acid, and 6-methyl cardol. A biologically based phenol with a meta-substituted saturated/unsaturated aliphatic chain derived from natural cashew nut shell liquid (CNSL).

A series of epoxy cured by cardanol/phenol novolac (CN) with different compositions were regulated at 3 : 1(3C-PN), 1 : 1(C-PN), or 1 : 3(C-3PN) and synthesized by Zhuoyu Liu et al. [4]. The prepared specimens were subjected to WA tests by using gravimetric measurement as a simulation of normal room temperature (at 25 °C) and at high temperature (at 85 °C). Initially, microcracks were observed on phenolic novalac (PN)/epoxy and phenolic novalac acetate (PNA)/epoxy samples at 85 °C for 80 days, so the experiments were terminated. Cardanol-based epoxies such as CN/Epoxy, 3C-PN/Epoxy, C-PN/Epoxy, and C-3PN/Epoxy remained intact because of hydrothermal influence, which was induced degradation and occurred at later stage.

Cardanol-based epoxies were more water repellent, and thus, it makes less water sensitive. The effect of WA on thermomechanical properties of cured samples depends primarily on temperature. Diffused water acts as a plasticizer at low temperature (25 °C), thereby reducing T_g of cured samples. At 85 °C, water molecules improved strong dipole–dipole interaction, thus increasing the T_g of some resins. This form of healing agent that contains cardanol content has greater potential for application in electronic packages.

Particle board were fabricated [5] by using CNSL as a matrix material with the influence of (i) resin content, (ii) P : F molar ratio (ratio of total phenol to total formaldehyde), and (iii) CNSL : P ratio (CNSL and phenol used for condensation with hexamethylenetetramine). The properties of tensile strength (parallel and perpendicular to the surface), water absorption, and compressibility of the prepared particle board were analyzed. The effect of resin content varied from 5% to 25% with a P : F ratio of 1 : 2.9 and exhibited a gradual reduction in WA. As the composition of formaldehyde content changes from 1 : 1.1, 1 : 1.7, 1 : 2.1, and 1 : 2.9, WA does not vary greatly. Because of the water-repellant potential of CNSL, a slight reduction in WA is noted with the increase in resin content beyond 50%. An increment in tensile strength is noticed for the effect of resin and formaldehyde content and decreases dramatically as the amount of CNSL rises by up to 50%. Similarly, compressibility decreases with the increase of formaldehyde content and increases with the increase of CNSL content.

11.2.3 Lignin-Based Epoxy

Lignin is a promising renewable substitute of bisphenol-A in the synthesis of epoxy resin available in the cell walls of various plants and the second frequent natural polymer on earth after cellulose. It is an amorphous and high molecular weight polymer composed of aromatic, hydroxyl, methoxyl, carbonyl, and carboxyl groups. Lignin is hydrophobic that is insoluble in H_2O and soluble in some alkali solutions and acids. The morphology, structure, and properties are highly dependable on the lignin content present in the fiber.

The effect of water absorption (using ISO 62:2008) on lignin-based epoxy resin (LER) prepared from phenolated lignin (PL) and epichlorohydrin (ECH) in the presence of sodium hydroxide(NaOH) was investigated [6]. In order to incur LER in aqueous NaOH, first eucalyptus acetic acid lignin (AAL) was reacted to PL with phenol and then obtained PL reacted with ECH. The prepared samples were immersed in water at 30 °C for about 24 hours. It was found that increasing the percentage (0, 10, 20, 30, 40, and 50) of LER increased the percentage of water uptake (0.30, 1.07, 1.39, 2.14, 2.24, and 2.33) of epoxy blends. The increasing content of hydroxyl groups present in polymers can enhance the sorption of polar penetratants. In this aspect, when the LER content increased, epoxy blends became highly porous. As a result, the specific surface area decreased, and the water uptake increased.

Natural fiber-reinforced lignin phenol formaldehyde (LPF) resin in which 30 wt% was substituted by methylolated lignosulfonate was prepared and characterized [7] the WA studies. From the experimental results, randomly arranged fiber resin interaction or higher void content leads to a higher WA of 35% for the flax fiber composites. Simultaneously, the flax/wood and flax/kenaf composites are 25–27%, which is tolerable for 30 wt% binder composition. Contrarily, the thickness swelling rate of flax/wood composites has 26% because water molecules penetrated into fiber cell walls. The packing density will be reduced, leading to high swelling ratios. Flax and kenaf fiber slowly lose their packaging density relative to wood, which ensures that flax/kenaf has the lowest swelling values of 13%.

Oil palm empty fruit bunch (OPEFB) fiber-reinforced lignin composites with varying resin contents of 15%, 20%, 25%, and 30% were prepared using a compression molding technique and analyzed the WA tests. In this study, lignin resin derived from soda-anthraquinone black liquor of OPEFB was used as the curing agent in the epoxy resin matrix. The water uptake test was performed at ambient temperature (25 ± 3 °C) using ASTM D570. The inclusion of lignin content reduced the WA of the composites. In the case of epoxy +15% lignin, the WA is higher, whereas it was lower for epoxy +25% lignin. Sufficient cross-linking between the fiber and the matrix was hydrophobic and will prevent accessibility of water. Naturally, oil palm fibers are hydrophilic with the presence of hydroxyl, acetal, and ether linkages in the cellulosic structure of fibers. The increase in WA results an improvement in the dimensional properties of the composites. As a result of swelling of fibers, a crack can form in the matrix. This could help to penetrate more water molecules into the composites during prolonged exposure [8].

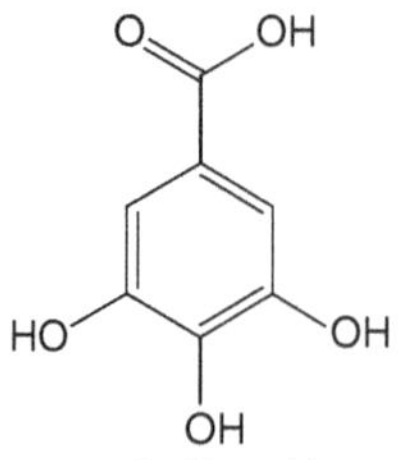

Figure 11.1 Structure of gallic acid. Source: Reena et al. [9].

11.2.4 Gallic Acid ($C_7H_6O_5$)-Based Epoxy

It is a bio-based trihydroxybenzoic acid compound with low molecular weight widely present in grapes, gallnuts, oak bark, vegetables, and in plant tissues. The IUPAC name of gallic acid (GA) is 3,4,5-trihydroxybenzoic acid and can be easily soluble in H_2O and in some organic solvents such as alcohol, ether, dimethyl formamide (DMF), and ethyl acetate. The structure of gallic acid is shown in Figure 11.1 [9].

Water barrier properties of zein–oleic acid (0–4%) composite films with the influence of gallic acid was prepared and investigated [10]. The results showed that a considerable reduction in solubility with the incorporation of oleic acid in the zein films and further reduction is noted with the increased oleic acid concentration. Solubility was decreased from 10.2% for films treated with gallic acid to 6.3% for GA treated with zein–oleic acid films. After gallic acid treatment, the decrement in water vapor permeability was noticed from 0.60 $gmmm^2/(h\,kPa)$ (for the control film) to 0.41 $gmmm^2/(h\,kPa)$ (for the GA treated films), which is attributed to the GA-induced cross-link formation that results in a more compact film structure that limits water movement. Gallic acid treatment was considerably affected by the moisture content ($p < 0.05$). In gallic acid treated films, low moisture content of 14.5% was recorded compared to 18.0% recorded in control films. The effect of GA-treated samples showed a significant increase (45%) in tensile strength when compared to control films, whereas elongation at break was not significantly affected, which means marginally raised to 4.2% for GA treated from 3.7% control films.

11.2.5 Itaconic Acid ($C_5H_6O_4$)-Based Epoxy

IA or methylene succinic acid is an unsaturated C5 carboxylic acid with a molecular weight of 130.1 g/mol. It is a colorless solid soluble in H_2O, ethanol, methanol, and acetone. There are two main routes for the production of biobased IA: the direct fermentation into the acid or the production of citric acid, followed by its conversion into IA by pyrolysis. IA has a density of 1.573 g/ml at 25 °C, a melting point of 165–168 °C, and a flash point of 268 °C [11], and IA could be used as building blocks for many materials, such as resins, paints, plastics, and synthetic fibers. The mass swelling rate and diffusion coefficient (D) of poly(N-isopropyl acrylamide/itaconic acid [NIPAAm/IA]) containing 0–3 mol% irradiated at 48 kGy were studied by Betul Tasdelen et al. [12]. The hydrogels of poly(NIPAAm) and poly(NIPAAm/IA) were synthesized in order to emphasize the influence on their swelling kinetics and solute absorption mechanisms of the ionization degree of polymeric networks. The swelling mass percentage was calculated by using the formula $k_s t^{0.5}$, where k_s denotes the swelling rate. The mass swelling capacity of NIPAAm/IA gels was raised (from 1260 to 11 280) with an increase of IA (from 0 to 3 mol%) content because more specific acid groups were integrated into the polymer network and with higher swelling ratios of gels. The result indicated that at 25 °C, swelling exponents are in the range of 0.49–0.62, followed by a non-Fickian behavior. With an increase in the content of IA, the diffusion coefficients (D) for copolymeric gels increase, so that the water molecule easily infiltrates the hydrogels for gels containing higher IA content.

11.2.6 Natural Rubber (NR)-Based Epoxy

NR is also known as latex, a polymer of *cis*-1,4-polyisoprene, the natural commodity resource, because of its excellent physical properties. The rubber formed from *cis*-polyisoprene is called NR and that formed from trans-polymerization is called synthetic rubber (Gutta Percha). It is a hydrophobic polymer, very low in polarity, and its structure has unsaturated double bonds that can be degraded from heat, UV, O_2, and ozone. A green polymeric composite based on NR reinforced with natural fibers (flax and sawdust) was designed by using benzoyl peroxide as a vulcanizing agent [13]. The percentage of WA of this composite depends on the amount of fiber taken and the composition of fiber used. The hydrophobicity of fiber and the highest interbonding between these fibers and NR enhance the highest WA. It was noticed that before incubation with *Aspergillus niger*, the NR sample has a WA of 1.76%, while for NR/flax composites, it was 3.82% and 6.8% (for 10 and 20 phr) and for NR/sawdust composites, it was 6.09% and 8.08% (for 10 and 20 phr), respectively. Then, after incubation with *A. niger*, the NR sample has a WA of 169.3%, while 79.6% and 44.74% (for 10 and 20 phr) for NR/flax composites and 20.53% and 4.58% (for 10 and 20 phr) for NR/sawdust composites, respectively. Hence, from this experiment, it was seen that the incubated samples have increased the values of WA compared to unincubated samples. Microbial degradation with *A. niger* might lead to (i) substantial decrement in molecular mass and (ii) the production of various functional polar groups –OH, $-O^-$, –COOH, $-COO^-$, etc.

Biocomposites reinforced with hemp fiber (length of 3 mm) and natural rubber have been processed [14] by electron beam radiation (with 75, 150, 300, and 600 kGy) in accordance with SR EN ISO 20344/2004. The results predicted an increase in gel content (G = 97.20%), volume fractions (ν_{2m} = 0.2128), and cross-linking density ($\nu = 1.6544 \times 10^{-4}$ mol/cm^3) of composites with an increase of electron beam dose and amount of hemp fiber content. A similar trend could be observed in the WA studies (Figure 11.2a–d).

It was obtained that the WA characteristic was increased for NR/hemp composites with an increment in hemp fiber content, and it was decreased while electron beam dose absorption. Irradiation dose might change the soluble property of hemp fiber. Activation of samples by low-dose irradiation could be

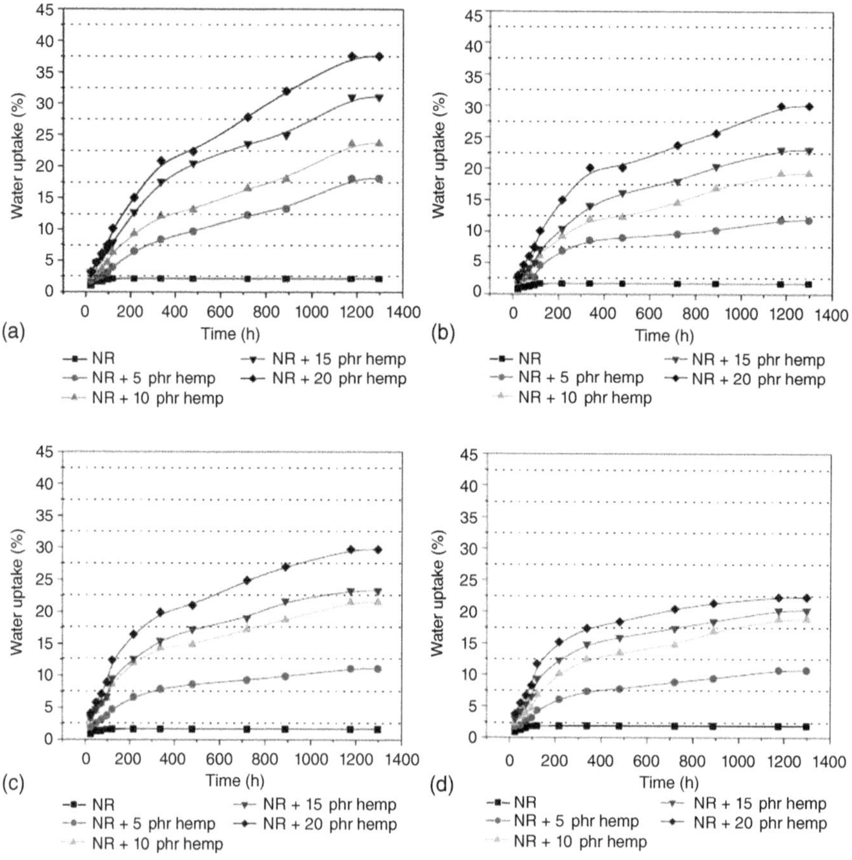

Figure 11.2 Water uptake of polymeric composites at an absorbed dose of (a) 75 kGy, (b) 150 kGy, (c) 300 kGy, and (d) 600 kGy. Source: (a–d) Maria-Daniela et al. [14].

achieved in terms of increased accessibility for the solvent and weak hydrogen bond networks, which promote into better solubility, whereas for higher irradiation, this effect was suppressed by intra- and intermolecular cross-linking.

Using a Schwabenthan laboratory two-roll mill with a speed of 13.1 and 11.6 rpm, sugarcane bagasse cellulose (CB) and natural rubber (RB) composites were prepared. The ingredients were added in the following order: elastomers, activators, sulfur, accelerators, inhibitors, and at last fibers. Samples of known weight were immersed in xylene and toluene as the solvents. The swelling coefficient (Q) is computed by the equation

$$Q = \frac{m_s}{m_1} \times \frac{1}{d}$$

where m_s, m_1, and d were the weight of solvent sorbed at equilibrium swelling, mass of sample before swelling, and density of solvent used, respectively. The RB and RB-SB (sugarcane bagasse) composites have greater values of Q, whereas RB-CA (acid-treated cellulose) and RB-CB have lesser values of Q in toluene and xylene. The transport mechanism of solvent diffusion suggested that irregular particles are responsible for solvent absorption. Compared to neat natural rubber, the composites show a decrease in swelling rate, which was attributed to the virtue of its path and the lowest transport area in the composites [15].

11.2.7 Rosin-Based Epoxy

Rosin is a major component of pine resin abundant in nature, which is composed of 90% diterpene base acids ($C_{20}H_{30}O_2$). The remaining 10% is a mixture of esters, alcohols, aldehydes, and hydrocarbons. Isomeric abietic and primaric types are the two acids composed in resin. Thermoplastic polymers synthesized from rosin can be categorized into two: (i) main-chain polymers (formed by the process of step-growth polymerization) and (ii) side-chain polymers (by the process of radical polymerization). The latter process can be converted into vinyl, acrylic, or allyl ester groups.

The influence of different contents of maleopimaric acid polyester polyols (MAPP), which was derived from MPA (maleopimaric acid), on rosin-based waterborne polyurethane (RWPU) was investigated. In this study, it was noticed that water uptake decreased from 78.6% to 14.2% for RWPU. The rigid structure of RWPU composite has the advantage of decreasing WA; hence, the inclusion of MAPP reduced WA. This is due to the reason that rosin acids have a hydrophobic skeleton in combination with hydrophilic carboxyl groups. RWPU with MAPP attained the above-said properties than WPU and thus constitute more obstacles to migration of water, resulting in much better water resistance. In addition, the authors concluded that rosin is a potential candidate to increase the use of natural and renewable resources [16].

11.2.8 Furan-Based Epoxy

The agro-industrial by-product, also a heterocyclic organic compound, has an outstanding heat resistance and a heavy volatile liquid. Owing to the aromatic and environmentally friendly character, furan derivatives are potential candidates to prepare the epoxy/amine thermosets. Furan is insoluble in H_2O and soluble in some organic solvents such as alcohol, ether, and acetone. Furfuryl alcohol and furfuraldehyde are prepared by the synthesis of furan, and its derivatives are from saccharide renewable resources (especially from corn cobs). Dehydration and condensation reactions are the two methods to polymerize the furan resins. Moreover, compared to other common matrix materials, these furan resins provided an effective improvement in the carbon storage potential (in terms of 1.3 Kg-CO_2-9) for the preparation of biodegradable composites.

The comparative study of composites prepared from (i) furan resin/glass fiber-reinforced polymer (bio-sourced resin) and (ii) epoxy/glass fiber-reinforced polymer (petro-sourced resin) was investigated [17]. Using ASTM D570, the WA tests were performed by immersing specimens (bio and petro-resin composites) in deionized water (at 60 °C). The diffusion coefficient was calculated using $D = \left(\frac{\pi\lambda^2}{4t}\right)\left(\frac{M_t}{M_{max}}\right)^2$, where "$\lambda$" denotes the thickness, "t" denotes the time, "M_t" denotes the mass of water absorbed, and "M_{max}" denotes the mass gain at equilibrium. The bioresin composite showed the Fickian behavior in which the mass transportation is fully controlled by diffusion of water molecules, whereas petro-sourced resin showed non-Fickian behavior. The water molecules that pave the way to facilitate the diffusion movement of chain segments have disrupted the polymer network. At saturation point, the absorption level of bio-sourced resin was higher when compared to petro-sourced resin because the former one is hydrophilic in nature.

Thus, the results also clearly indicate that moisture absorption at saturation and the rate of absorption at each time for petro-sourced resin was lower than bio-sourced resin because the latter one is less hydrophobic and encompasses greater amount of porosity. Thus, from these experimental results, it was proved that Furan resin/glass fiber-reinforced polymer causes an increase in WA and a significant reduction in the alkaline solution degradation.

11.2.9 Hempseed Oil-Based Epoxy

Hempseed oil (HO) includes a variety of beneficial unsaturated triglycerides with nutritional benefits, and it can act as a curing agent for epoxies. HO contains 30–35% oil by its weight and is a rich source of digestible protein (20–25%). A good amount of other esterified fatty acids (EFAs), α-linolenic acid (18,3*n*–3), and omega 6, γ-linolenic acid (18,4*n*–6) were found in this oil. In recent years, the use of HO is more because of its nutritional benefits and not only that it consists of essential fatty acids and higher iodine value.

WA tests were monitored for the biocomposites prepared from jute fiber and HO (as a matrix). The saturation moisture level (M_m) and diffusion coefficient (D) were analyzed for the epoxidized hemp oil-based bioresins and biocomposites. Sample dimensions of 76.2 × 25.4 × 5 mm were taken for both neat resin and jute fiber biocomposite. The absorption properties of biocomposites followed a linear Fickian behavior and were found to increase with increasing bioresin loading. The "D" and M_m values of epoxidized hemp oil resin samples with a weight ratio of 40–60 were $1.99 \times 10^{-6}(\text{mm}^2/\text{s})$ and 1.47%, respectively. Similarly, for jute fiber composites, it was $5.09 \times 10^{-6}(\text{mm}^2/\text{s})$ and 24.13%. Compared to neat resins, the highest WA was noticed for the composites because of the fact that the fiber provided an additional water transport mechanism throughout the material. Thus, fibers dramatically increase the "D" as well as the "M_m" of the composites as they absorb water as a result of their hydrophilic nature provided by the hydroxyl groups in the cellulose portion of the fibers and swelling, allowing water to be transported through microcracks in the matrix resulting from the swelling of fibers [18].

A hardener and sometimes a cross-linking accelerator are used to convert an epoxidized hempseed oil (EHSO) into a curable thermoset polymer. Novel biocomposites made of an EHSO reinforced with natural fibers were fabricated (Andrea [19]) by using an anhydride hardener methyl tetrahydrophthalic anhydride (MTHPA). The moisture sensitivity of EHSO was measured for the optimized resin formulation (stoichiometric ratio of $R = 0.84$ and 2.8 wt% of 2-ethylimidazole). Hence, the resin showed a 57.5 wt% bio-based carbon fraction. At water immersing temperatures 25 °C and 60 °C, the resin exhibits a water

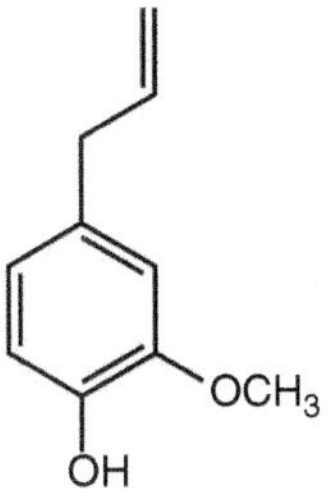

Figure 11.3 Structure of eugenol. Source: Nascimento et al. [20].

uptake of 1.4 and 3.1 wt%, respectively, in the cured state. The incorporated water caused the thermoset to geriatric effect mechanically and/or chemically, and thus, the mechanical performance of the thermoset was deteriorated. Moreover, the optimized material portrayed an innovative and sustainable substitute for fully petrochemical-based matrix materials.

11.2.10 Eugenol ($C_{10}H_{12}O_2$)-Based Epoxy

A naturally occurring phenol 2-methoxy-4-(prop-2-enyl)phenol is an "essential oil" that can be obtained from several plants including cinnamon, clove, and bay leaves. Because of its relatively cheaper cost and abundant material availability, this eugenol is widely used in medicine, food industry, and dentistry. The chemical structure of eugenol is shown in Figure 11.3 [20].

Eugenol-modified polysiloxanes P_1 and P_2 were combined at a low loading percent of 2.5 wt% with commercial epoxy resins as anticorrosion additives and studied the properties of the epoxy coating surfaces by WCA (θ) measurements. From the results, it was clear that surface wettability depends on the surface roughness and the chemical composition of the surface. The WCA is 60° for an unmodified epoxy coatings, which is an indication of hydrophilicity, whereas the value of WCA is raised to 78° and 90° for the polysiloxanes P_1 and P_2 blended in the epoxy coatings at a low loading percent of 2.5 wt%. It is also evident that 2.5 wt% of P_1 and P_2 is adequate to cause a significant increase in hydrophobicity. In the presence of epoxy resin P_2, the nature of the epoxy-coating surface is changed from hydrophilic ($\theta < 90°$) to hydrophobic ($\theta > 90°$). The resulting coating resulted in a new class of effective anticorrosion substances with potential marine corrosion control and protection applications [21].

11.3 Conclusion

This chapter briefly reviewed the works related to water sorption and solvent sorption of bio-epoxy polymers, their blends, and composites. Bio-epoxy, an important polymer, derived from renewable resources, have environmental impact throughout its end of life. Also, the WA properties of various bio-epoxies, such as eugenol, isosorbide, rosin, natural rubber, furan, itaconic, gallic acid, CSNL, lignin, hemp seed oil, and soybean oil, were discussed. From this concise review, it was noted that higher water uptake characteristics were observed while using the natural fibers (as a reinforcing agent) because of the higher cellulosic content present in the fibrous materials. Bio-epoxies, an important challenger for the development of the future agro-based products because of its economically viable, commercially available, low cost, renewable, and environmentally friendly characteristics alternative to the use of petrochemical polymers. Thus, the effect of various copolymers on bio-epoxy on water and solvent uptake is of great interest for practical applications in functional coatings where the materials sustained in humid environments.

References

1 Zhang, H., Guo, Y., Yao, J., and He, M. (2016). Epoxidised soybean oil polymer composites reinforced with modified microcrystalline cellulose. *J. Exp. Nanosci.* 11 (15): 1213–1226.

2 Pradhan, S., Priyanka, P., Smita, M., and Nayak, S.K. (2017). Synthesis and characterization of waterborne epoxy derived from epoxidized soybean oil and bioderived C-36 dicarboxylic acid. *J. Coat. Technol. Res.* 14 (4): 915–926.

3 Munoz, E. and García-Manrique, J.A. (2015). Water absorption behaviour and its effect on the mechanical properties of flax fibre reinforced bioepoxy composites. *Int. J. Polym. Sci.* 2015: 390275.

4 Liu, Z., Jizhen, H., and Yingfeng, Y. (2017). Water absorption behavior and thermal-mechanical properties of epoxy resins cured with cardanol-based novolac resins and their esterified ramifications. *Mater. Today Commun.* 10: 80–94.

5 Mary Lubi, C. and Eby Thomas, T. (2007). Particleboard from cashew nut shell liquid. *Polym. Polym. Compos.* 15 (1): 75–82.

6 Feng, P. and Chen, F. (2012). Preparation and characterization of acetic acid lignin-based epoxy blends. *BioResources* 7 (3): 2860–2870.

7 Arunjunai Raj, M., Wuzella, G., Aust, N. et al. (2012). Processing and characterization of natural fibre reinforced composites using lignin phenolic binder. *Polym. Polym. Compos.* 21 (4): 199–206.

8 Abdul Khalil, H.P.S., Marliana, M.M., and Alshammari, T. (2011). Material properties of epoxy-reinforced biocomposites with lignin from empty fruit bunch as curing agent. *BioResources* 6 (4): 5206–5223.

9 Reena, K., Pankaj, S., and Lal, B. (2013). Natural pigments from plants used as sensitizers for TiO_2 based dye-sensitized solar cells. *J. Energy* 2013: 654953.

10 Masamba, K., Li, Y., Joseph, H. et al. (2016). Effect of gallic acid on mechanical and water barrier properties of zein-oleic acid composite films. *J. Food Sci. Technol.* 53 (5): 2227–2235.

11 Pomogailo, A.D., Kestelman, V.N., and Dzhardimalieva, G.I. (2010). Monomeric and polymeric carboxylic acids. In: *Macromolecular Metal Carboxylates and Their Nanocomposites* (eds. R. Hull, C. Jagadish, R.M. Osgood, et al.), 7–25. Berlin Heidelberg: Springer.

12 Tasdelen, B., Kayaman-Apohan, N., Guven, O., and Baysal, B.M. (2004). Swelling and diffusion studies of poly(*N*-isopropylacrylamide/itaconic acid)

copolymeric hydrogels in water and aqueous solutions of drugs. *J. Appl. Polym. Sci.* 91: 911–915.
13 Maria-Daniela, S., Elena, M., Gabriela, C., and Corina, C. (2017). Development and characterization of polymer eco-composites based on natural rubber reinforced with natural fibers. *Materials* 2017 (10): 787. https://doi.org/10.3390/ma10070787.
14 Maria-Daniela, S., Elena, M., Gabriela, C., and Dumitrascu, M. (2014). New green polymeric composites based on hemp and natural rubber processed by electron beam irradiation. *Sci. World J.* 2014: 684047.
15 Sibiya, N.N., Mochane, M.J., Motaung, T.E. et al. (2018). Morphology and properties of sugarcane bagasse cellulose – natural rubber composites. *Wood Res.* 63 (5): 821–832.
16 Xu, X., Shang, S., Song, Z. et al. (2011). Preparation and characterization of rosin-based waterborne polyurethane from maleopimaric acid polyester polyol. *BioResources* 6 (3): 2460–2470.
17 Mohammadreza, F., Elkoun, S., Fam, A., and Robert, M. (2015). Degradation characteristics of new bio-resin based-fiber-reinforced polymers for external rehabilitation of structures. *J. Compos. Mater.* 50 (9): 1227–1239.
18 Manthey, N.W., Cardona, F., Francucci, G., and Aravinthan, T. (2013). Thermo-mechanical properties of epoxidized hemp oil-based bioresins and biocomposites. *J. Reinf. Plast. Compos.* 32 (19): 1444–1456.
19 Anusic, A., Resch-Fauster, K., Arunjunai Raj, M., and Gunter, W. (2019). Anhydride cured bio-based epoxy resin: effect of moisture on thermal and mechanical properties. *Macromol. Mater. Eng.* 304: 1–9.
20 Joice Nascimento, B., da Carlos, S., Silva, R.O. et al. (2018). An overview on the anti-inflammatory potential and antioxidant profile of eugenol. *Oxid. Med. Cell. Longevity* 2018: 3957262.
21 Chen, G., Feng, J., Qiu, W., and Zhao, Y. (2017). Eugenol-modified polysiloxanes as effective anticorrosion additives for epoxy resin coatings. *RSC Adv.* 7: 55967–55976.

12

Biobased Epoxy: Applications in Mendable and Reprocessable Thermosets, Pressure-Sensitive Adhesives and Thermosetting Foams

Roxana A. Ruseckaite, Pablo M. Stefani, and Facundo I. Altuna

Universidad Nacional de Mar del Plata – CONICET, Instituto de Investigaciones en Ciencia y Tecnología de Materiales (INTEMA), Av. Colón 10850, Mar del Plata 7600, Argentina

12.1 Introduction

The substitution of synthetic (i.e. produced from petroleum-based platforms) polymers by their biobased counterparts in our everyday life is a gradual process that does not depend merely on technical aspects, namely the properties of the biobased polymer and its performance in a given application. It is, instead, a much more complex matter that also involves economic, environmental, and societal considerations, among many others. Mario Pagliaro proposed two main driving forces that could lead to a paradigmatic change in the chemical industry in the near future: "1) the global demand for better, lighter, more durable, healthier, and greener functional products by the industry's largest customers and 2) the uptake of decentralized production based on clean chemical technology." [1]. The first set of factors includes, among other aspects, a customer-driven change to more environmentally friendly products that could stimulate biobased polymers to reach the market and finally the users.

Significant efforts have been devoted to study how biobased raw materials can be used to supply today's market needs, resulting in a vast and growing body of literature [2, 3]. We understand that the first and main goal of research and development on biobased polymeric materials, and biobased epoxies in this particular case, is to obtain materials with the same properties (or as similar as possible) to those of the petroleum-based ones. Another challenge of biobased materials is cost-competitiveness because at present, most of the biobased materials are not economical compared to petrochemical derived equivalents. However, sometimes, the use of biobased raw materials leads to improvements that should also be considered, besides the obvious environmental benefits associated with its usage.

This contribution is focused on the technical features of some of the more recent advances on applications of biobased epoxies, regardless of the economic or social motivations for biobased materials to substitute the synthetic ones. These subjects, as stated above, are also necessary conditions but largely exceed the scope of this chapter.

Bio-Based Epoxy Polymers, Blends, and Composites: Synthesis, Properties, Characterization, and Applications, First Edition. Edited by Jyotishkumar Parameswaranpillai, Sanjay Mavinkere Rangappa, Suchart Siengchin, and Seno Jose.

This chapter covers three appealing applications that promote the use of biobased epoxy resins and precursors. Self-healing thermosetting polymers and composite matrices are the first applications described. The use of dynamic and reversible bonds is a relatively new research field. Such materials have not reached the market yet, but they are being developed at the same time when the use of biobased raw materials for synthesizing polymers is gaining momentum. Pressure-sensitive adhesives (PSAs) are the second chosen topic. The growing PSA market, together with the ever-increasing restrictions on the use of fossil-derived precursors, gives a very good opportunity for biobased epoxy PSAs, stimulating the research efforts in this area. Finally, the most relevant applications of epoxy foams, both syntactic and those obtained through a foaming process, are described. The technical potential of biobased systems to replace at least partially the nonrenewable feedstocks is discussed. Epoxy systems derived from natural polyphenols or vegetable oils are highlighted, and some examples are used to illustrate the advantages of such formulations.

12.2 Mendable and Reprocessable Biobased Epoxy Polymers

Mendable (often also called self-healing) cross-linked polymers and composites based on thermosetting polymer matrices have been an intensively growing subject of research in the past decades. They emerged from mimicking nature, where self-healing ability is a common feature of most living systems, and are aimed to extend the service life of polymeric materials, thus reducing the costs, the consumption of raw materials, and the amount of generated waste.

Wool [4] and Williams and coworkers [5] carried out seminal studies on the self-healing ability of some polymers and studied the reversibility of cracks in viscoelastic polymers and the crack healing in thermoplastic blends, respectively. Since then, a good amount of work has been devoted to the study of crack healing mechanisms [6, 7]. However, it was not until several years later that thermosetting polymers purposely designed to display self-healing ability were reported. An extrinsic polymer composite was developed by Dry et al., who adapted their method for healing concrete matrices [8, 9]. The authors proposed the utilization of hollow fibers to deliver a repairing agent (i.e. a reactive monomer) to the crack. The repairing agent then polymerizes into the crack, binding the crack faces together and restoring the mechanical strength of the material [9, 10]. In a modification of this approach, the group of White and coworkers [11–16] used microcapsules (and later also hollow microfibers) filled with reactive monomers that are released when a growing crack provokes the braking of the microcapsules (or microfibers). Figure 12.1 depicts both methodologies.

White and coworkers [17] and Du Prez and coworkers [18] proposed a classification for self-healing polymers with two broad categories: extrinsic and intrinsic healable systems. Both are shown schematically in Figure 12.2.

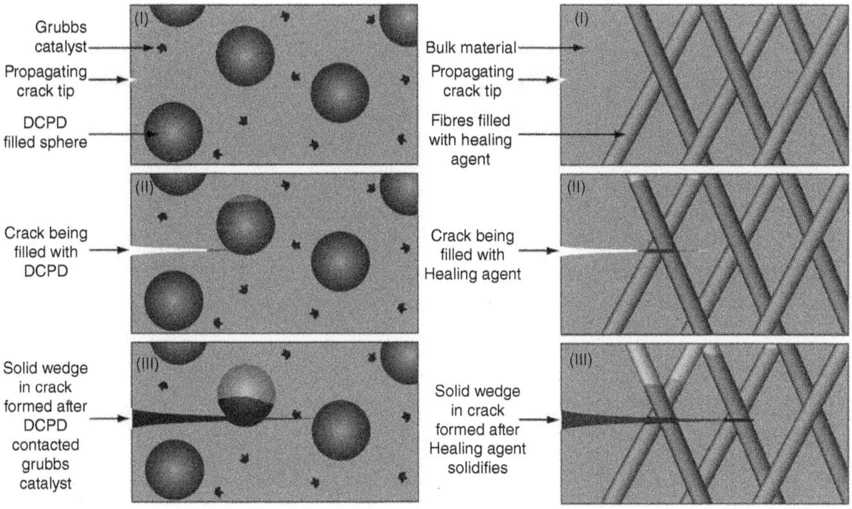

Figure 12.1 Schematic representation of the most common extrinsic self-healing systems. Source: Reproduced with permission from Wu et al. [6]. © 2008, Elsevier.

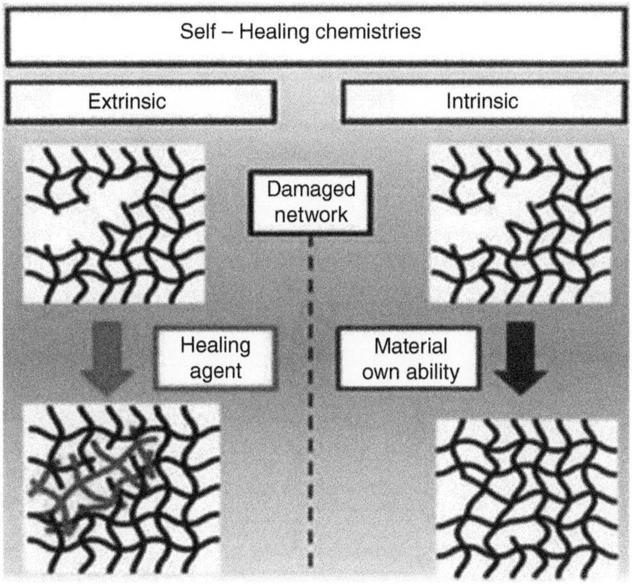

Figure 12.2 Difference between extrinsic and intrinsic healing polymeric network systems. Source: Reprinted from Billiet et al. [18]. © 2013, John Wiley & Sons.

Extrinsic healable systems comprise mainly polymeric matrices with embedded microcapsules or vascular networks containing the external repairing agent as those described above. These materials have been studied for a longer time than their intrinsic counterparts, and a number of chemistries have been evaluated, including cross-linked epoxy resins [6, 19]. The most important advantage of extrinsic healable systems is that they do not require the healing to be induced, but it is instead triggered automatically when the material is damaged.

Intrinsic systems are based either on combinations of thermosetting and thermoplastic polymers or on polymeric networks with reversible or dynamic bonds that allow the material to modify its topology at a molecular scale and achieve its healing [20, 21]. Its research and development is relatively quiet recent, and despite the need for an external stimulus – such as heat or light – it has several advantages. The most important are that the properties of the healing material are the same as those of the original one (especially for networks with reversible or dynamic bonds, where essentially the same material is obtained after reparation) and that multiple reparations are possible. Moreover, intrinsic healable systems display reprocessing and recycling capability, which was not possible for thermosetting polymers some time ago [22, 23].

12.2.1 Extrinsic Self-healing Biobased Epoxies

The use of capsules or hollow fibers to contain a repairing agent for epoxy polymers could be readily extended to biogenic systems because it does not rely on the nature of the polymeric matrix or on the monomers used to obtain it. However, there are only a few examples in the literature of such self-healable composites with some of the raw materials having a renewable origin. Moreover, they are mainly based on synthetic epoxy resins. Xiao et al. used a microencapsulated epoxy monomer dispersed in a matrix based on diglycidyl ether of bisphenol A (DGEBA) cross-linked with triethylenetetramine (TETA). Instead of dissolving the $(C_2H_5)_2O{\cdot}BF_3$ catalyst in the monomer mixture, it was absorbed in sisal fibers, in order to prevent its deactivation during the synthesis, and very good healing efficiencies were obtained [24]. In an attempt to use biogenic components, the performance of alginate microcapsules to encapsulate the epoxy monomer was assessed [25]. They proved to have some advantages in comparison with their synthetic counterparts: beside having a renewable origin and being free of toxic VOCs (e.g. formaldehyde), they allow multiple reparations because of its multicore structure [25]. Vegetable oils and derived alkyd resins were also evaluated as healing agents for anticorrosive films [26–28]. These materials can autonomously self-repair and efficiently protect the surface from corrosion (Figure 12.3), but in some cases, the addition of the microcapsules was detrimental to other properties of the materials, such as its adhesive strength [27, 28].

The examples listed in this section can be considered as important precedents in the way to the production of autonomous self-healing epoxy materials based on biogenic resources. The – at least partial – replacement of the synthetic resins for biobased monomers could give place to numerous possibilities, from the modulation of some specific characteristics of the unreacted system that could ease its processing to the design of materials with specific final properties. Although it is very likely that in the near future biobased epoxy monomers will be used to develop extrinsic self-healing polymers, this research field remains largely unexplored.

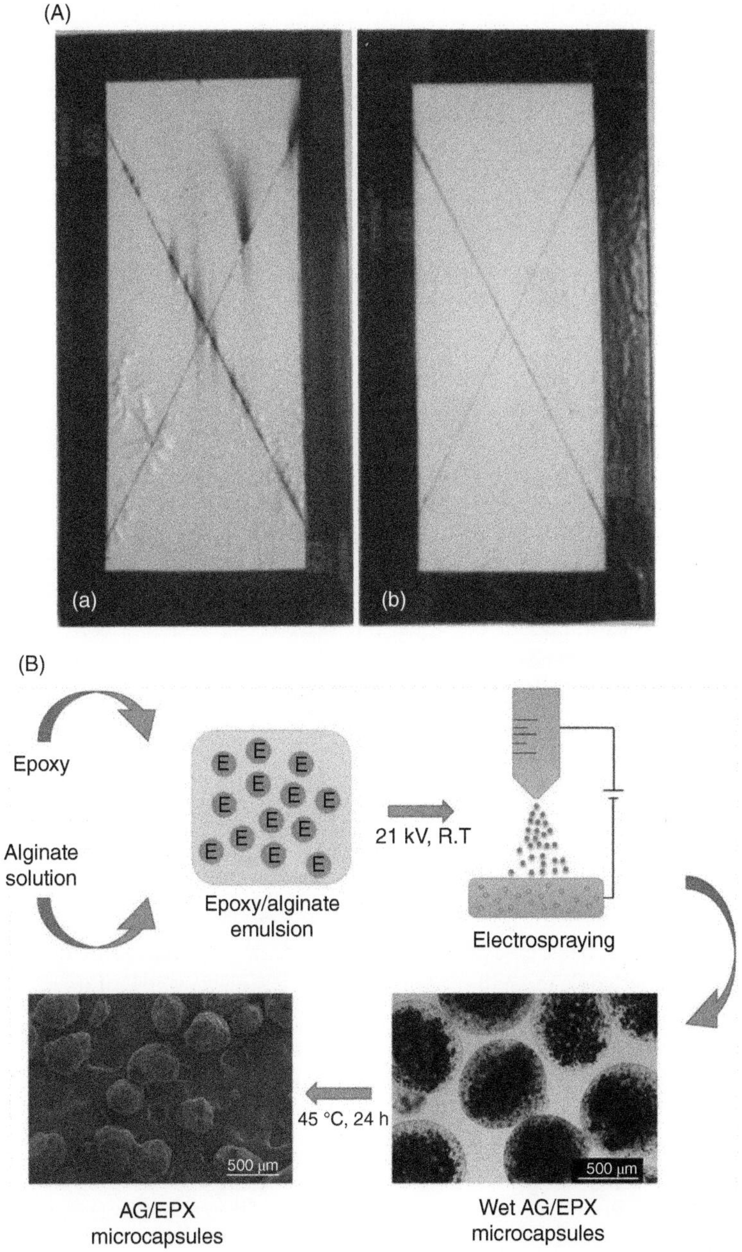

Figure 12.3 (A) Coating surface after immersion for 500 hours in 3.5% NaCl solution: (a) control epoxy coating and (b) self-healing epoxy coating. Source: Reproduced from Thanawala et al. [26] with permission from MDPI. (B) Scheme showing the process to obtain epoxy/alginate multicore microcapsules that can be used in self-healing systems. Source: Reproduced with permission from Hia et al. [25].

12.2.2 Intrinsic Self-healing Biobased Epoxies

Development on intrinsic self-healing polymers was first centered on polymers with reversible bonds, such as Diels–Alder (DA) adducts or supramolecular interactions. [29–35]. The healing mechanism of these materials was based on a "depolymerization–polymerization" process, as depicted in Figure 12.4. The bonds acting as cross-links are thermodynamically favored only at low temperatures, but upon heating – up to a temperature that depends on the chemical nature of the polymer – the equilibrium is shifted and the network breaks down into smaller species. Intrinsic self-healing thermosetting polymers experienced a remarkable boost following the groundbreaking work of L. Leibler and his team [37–39]. They used a mixture of poly-carboxylic acids with biological origin (Pripol 1040, produced by Croda) to cross-link DGEBA, producing β-hydroxyester groups linking the monomeric precursors. With the aid of a Zn^{+2} salt, triazabicyclodecene (TBD), or triphenylphosphine (PPh_3) as a catalyst [38], exchange reactions between the β-hydroxyester groups took place at an appreciable rate, allowing the cross-links to shuffle and thus changing the topology of the network. The term "vitrimers" was coined to name these materials, whose main advantage over those with reversible bonds is that their

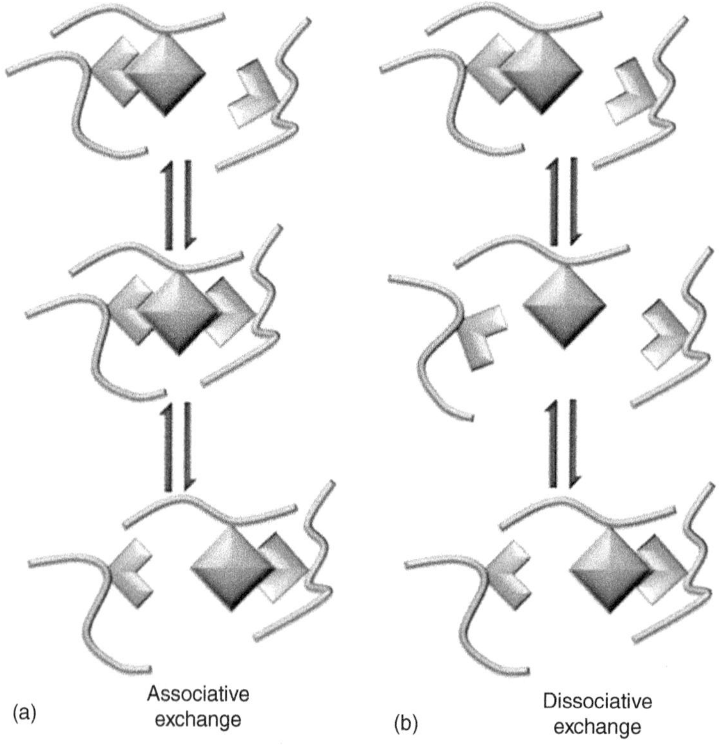

Figure 12.4 Rearrangement of cross-links through reversible covalent bonds (a) and dynamic covalent bonds (b). Source: Reproduced from Chakma and Konkolewicz [36]. © 2019, John Wiley & Sons.

cross-linking density remains constant (or at least nearly constant) during the process, preserving their integrity [36, 40].

There are some notable examples of biobased epoxy polymeric networks with reversible cross-links, but the use of synthetic resins such as DGEBA persists in almost all of them, and only partially biobased systems have been studied so far. Karami et al. studied an epoxy resin (DGEBA) modified with furfuryl alcohol (FA) and used bismaleimide (BMI) as a reversible cross-linking agent [41]. This polymer showed lower mechanical properties than an amine-cured DGEBA, but they displayed a better thermal stability. The remarkable feature is that the adducts formed between FA moieties and BMI underwent the retro-DA reaction at around 140 °C, making it possible to weld two pieces of this material. After one hour at 140 °C and 48 hours at 65 °C, a good healing of a scratch was observed, and two separate pieces were welded together (Figure 12.5). Previously, Pin and coworkers had demonstrated that FA could be successfully used with epoxidized linseed oil (ELO) to obtain fully biobased polymeric networks that could be used as a platform for biobased epoxy thermosets with self-healing ability [42]. Other epoxy precursors with DA moieties have also been synthesized from biogenic raw materials, paving the way for the synthesis of self-healing thermosets with different properties [43, 44].

Xu et al. [45] successfully prepared a self-healing supramolecular rubber using epoxidized natural rubber (ENR) and carboxymethyl chitosan (CMCS) and showed that 12 hours of room temperature healing was enough to achieve healing efficiencies of about 90%. It was demonstrated that the addition of CMCS

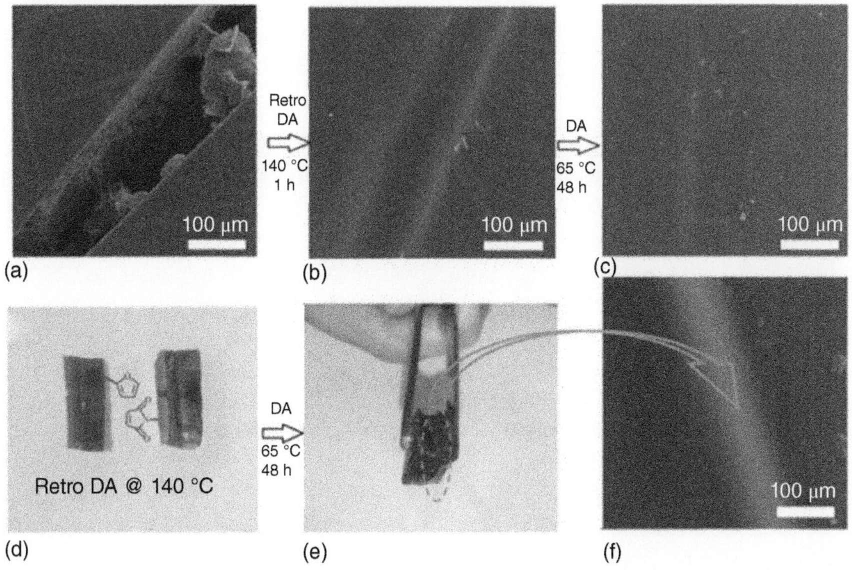

Figure 12.5 (a–c) Scratch healing process on a DGEBA-FA-BMI cross-linked polymer observed through scanning electron microscopy (SEM). (d, e) Optical images, and (f) SEM image showing the welding of two separate pieces of the same material. Source: (a–f) Reproduced from Karami et al. [41]. © 2018, John Wiley & Sons.

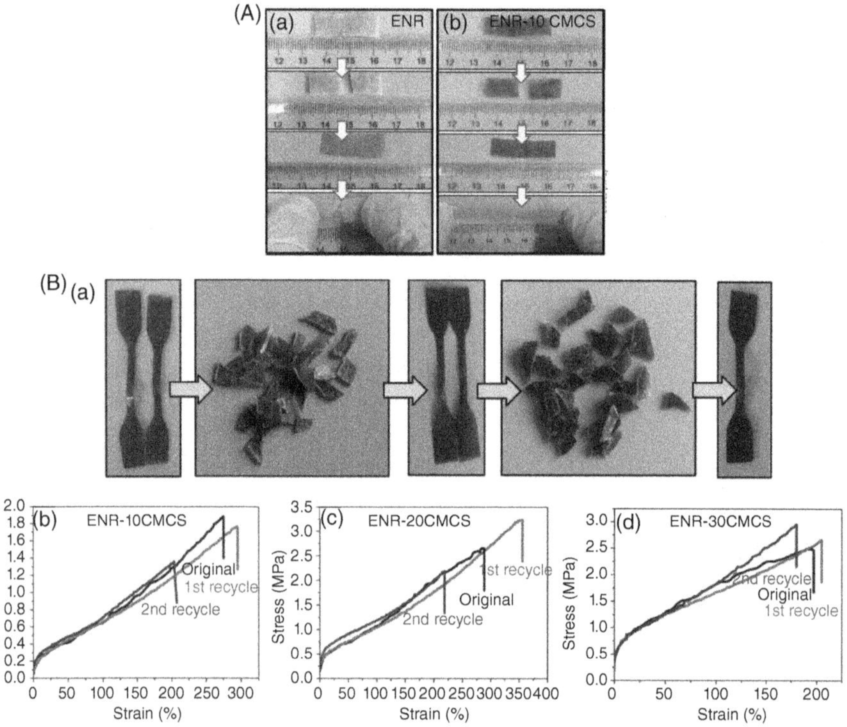

Figure 12.6 (A): (a) Neat ENR and (b) ENR-CMCS self-healing qualitative tests. (B): Recycling tests of ENR-CMCS composites with different contents of CMCS. (a) photography of the grinding-recycling process, and stress-strain curves for virgin and recycled samples of ENR supramolecular network with (b) 10, (c) 20 and (d) 30 wt% CMCS. Source: (A,B) Reprinted with permission from Xu et al. [45]. © 2019, American Chemical Society.

generates hydrogen bond interactions that are responsible for the self-healing behavior (Figure 12.6A). Moreover, the recycling ability of the materials was evaluated as well (Figure 12.6B), but the results showed a reduction in the mechanical properties after three cycles, probably because of oxidation reactions of CMCS. A biobased epoxy network with both covalent and supramolecular reversible cross-links was developed by Cao et al. [46]. They cross-linked an ENR reinforced with cellulose nanocrystals that provided large amounts of –OH groups, giving place to strong supramolecular interactions. Maleic anhydride grafted onto the cellulose nanocrystals was the covalent cross-linker. The reversible physical interactions were intended here to be used as a sacrificing element to absorb energy and improve the mechanical performance, increasing the strength, toughness, and deformation. The authors admit that the physical cross-links could also allow the materials to self-heal, but no tests were performed to prove this hypothesis.

Most biobased epoxy intrinsic self-healing thermosets, however, owe their self-healing ability to transesterification reactions. Altuna et al. prepared a completely biobased covalent network by cross-linking an epoxidized soybean oil (ESO) with an aqueous solution of citric acid (CA) [47]. The obtained networks

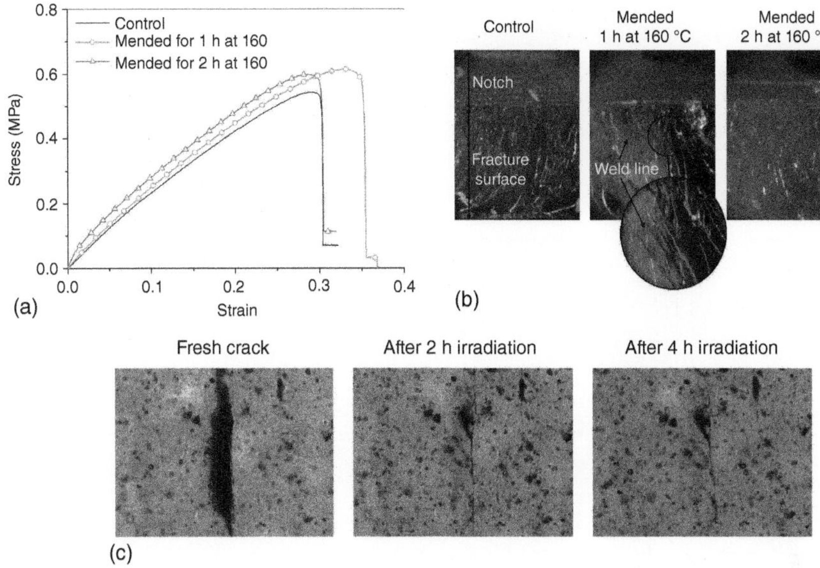

Figure 12.7 (a) Stress–strain curves for modified lap-shear tests of control and mended samples. (b) Fracture surfaces of the samples after the modified lap-shear tests. Source: (a) and (b) Reproduced from Altuna et al. [47], with permission from The Royal Society of Chemistry. (c) TOM images (100×) of a crack in an ESO-CA-Au NP nanocomposite before and after healing with laser irradiation. Source: Dr. Facundo I. Altuna.

showed a glass transition temperature (T_g) value around 5 °C and a very good healing efficiency at 160 °C, even in the absence of any specific catalyst, thanks to transesterification reactions. Figure 12.7a,b shows the healing experiments and photos of the fracture surface showing the bonding between the welded pieces. A further improvement for these materials was the introduction of small amounts (around 0.02–0.08 wt%) of gold nanoparticles (NPs) that allowed to remotely activate the healing through irradiation with a green laser [48]. Almost complete healing was observed after two hours of irradiation, as shown by transmission optical microscopy shown in Figure 12.7c; the evolution with further irradiation was negligible.

A similar concept with different precursors was adopted by Hao et al. [49, 50]. The coating that they synthesized was based on modified lignin with grafted –COOH groups and an epoxy-terminated polyethylene glycol (PEG) with a Zn^{+2} catalyst to aid the transesterification reactions. The reparation at 200 °C strongly depended on the initial –COOH/epoxy ratio. The authors explained that a better performance was measured for systems with higher concentrations of hydroxyl and carboxyl groups. Furthermore, the addition of ethylene glycol was needed when the reparation was tested with the material used as coating for tin plates. The crucial role of the –OH groups in the transesterification reactions had been previously highlighted by Leibler and coworkers [39]. Another vitrimer was developed by researchers of the same group through the cross-linking of an eugenol-derived diepoxy monomer with succinic anhydride in the presence of Zn^{+2} [51]. Again, the –OH concentration, determined by the

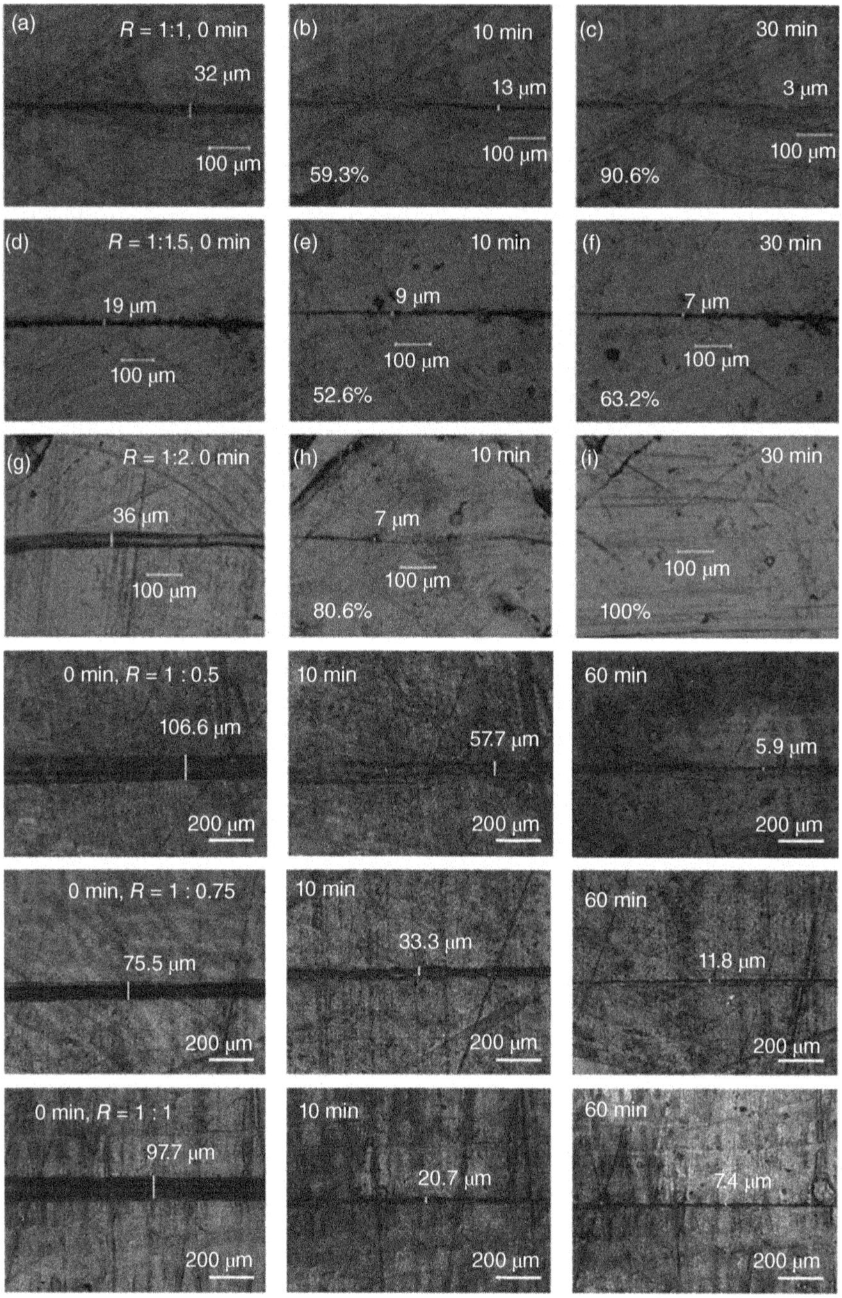

Figure 12.8 Rows 1–3: Reparation of a crack in lignin-PEG epoxy thermosets with different stoichiometric ratios at 200 °C. (a)–(c) R=1; (d)–(f) R=1.5; (g)–(i) R=2. Source: Reproduced from Hao et al. [49]. © 2019, John Wiley & Sons. Rows 4–6: Reparation of a crack in eugenol-based epoxy thermosets with different stoichiometric ratios at 190 °C. Source: Reproduced from Liu et al. [51]. © 2017, American Chemical Society.

initial stoichiometry, resulted in a critical parameter affecting the reparability of the material, which was evaluated qualitatively through transmission optical microscopy (TOM) observation. Figure 12.8 shows some examples of the healing produced in the lignin–PEG and eugenol-based epoxies.

The exchange reaction between disulfides can also be harnessed to obtain self-healing polymer networks. Ma and coworkers used a diamine with disulfide groups to cross-link an isosorbide-based diepoxy [52]. The resulting network showed a rapid stress relaxation, very good self-healing ability, and an acceptable recyclability. At temperatures as low as 100 °C, these materials can mend cracks and be reprocessed through compression molding (Figure 12.9A), although the mechanical properties suffer an important deterioration with successive recycling. Cheng et al. proposed the use of both disulfide bonds and reversible supramolecular interactions (H bonds) to achieve a better combination of mendability and mechanical properties. They produced a network by cross-linking an ENR with sulfur, 2,2′-dithiodibenzoic acid (DTSA), and 4,4′-dithiodianiline (DTDA) [53]. The disulfide groups of the DTSA and DTDA have different dissociation energies than the SS bonds derived from the sulfur vulcanization of the ENR double bonds and additionally gave place to –OH groups that contribute to form supramolecular interactions. The synergy of both self-healing mechanisms was assessed by mechanical tests of samples cured with different amounts of S, DTSA, and DTDA. The formulation showing the best compromise between mechanical properties and self-healing efficiency was further assessed by healing at 120 °C for different times (Figure 12.9B). The change in the value of the mechanical modulus shows that the material still evolves after curing for 20 minutes at 155 °C and for 48 hours at 120 °C. The remarkable fact is that healing efficiencies as high as 98% were achieved.

12.3 Pressure-Sensitive Adhesives (PSAs) From Biobased Epoxy Building Blocks

PSAs are a special family of adhesives that adhere to a substrate without any chemical reaction or phase change upon a light external pressure (pressure hand) [54–60] and has enough cohesive strength and elasticity to be cleanly removed from a substrate surface. PSAs are viscoelastic soft materials with T_g much lower that the usage temperature, an elastic modulus below certain threshold called the Dahlquist criterion [61], and a large value of the viscous component of the elastic modulus to dissipate energy upon debonding [56].

The tackiness, peel strength, shear resistance, and aging determine the adhesion performance [59]. Tack is the ability of an adhesive to form a bond of measurable strength to another material under conditions of low contact pressure and short contact time [58]. Peel strength is the force required to separate the adhesive from a rigid standardized panel by peeling the flexible carrier in a controlled manner at either 90° or 180° [54, 58]. Shear strength is tested under static load applied to a known surface area of the adhesive-coated product until failure is produced (Figure 12.10) [58]. Adhesive properties can be tuned in a wide range by selecting

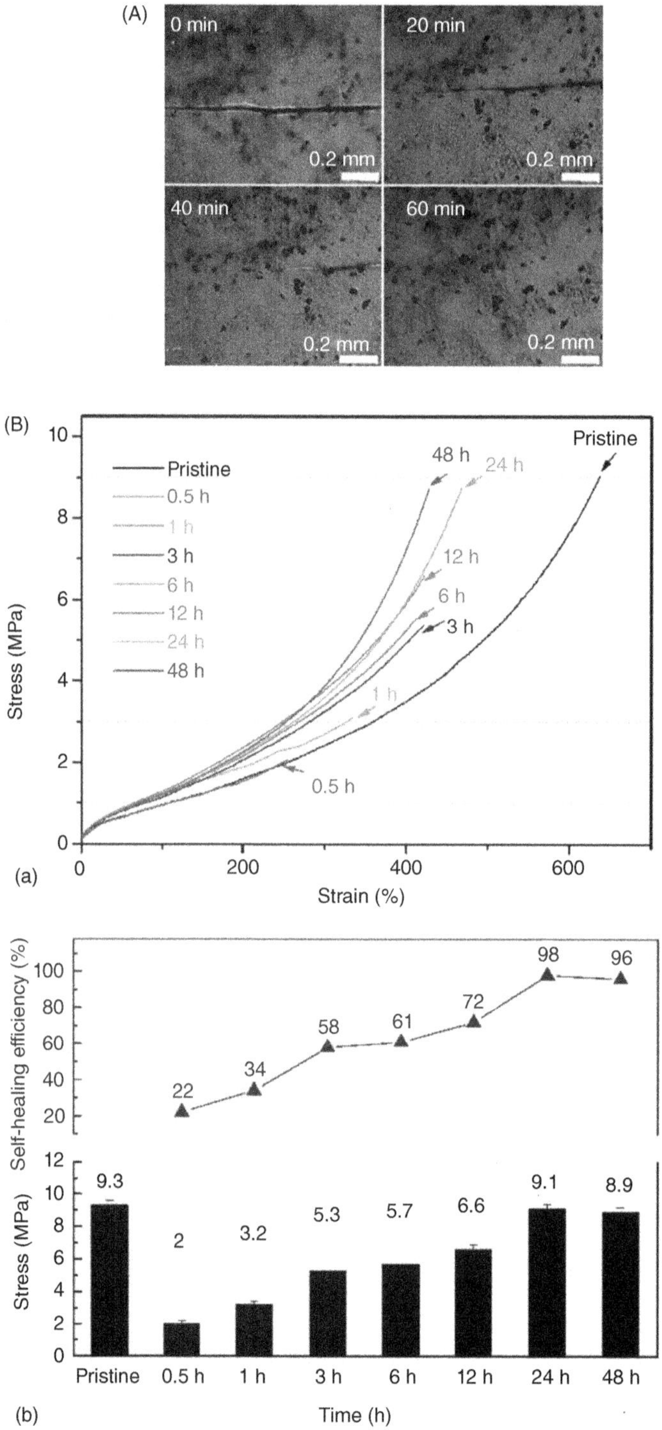
(A)
0 min
20 min
40 min
60 min
0.2 mm
(B)
Pristine
0.5 h
1 h
3 h
6 h
12 h
24 h
48 h
Stress (MPa)
Strain (%)
(a)
Self-healing efficiency (%)
22
34
58
61
72
98
96
9.3
2
3.2
5.3
5.7
6.6
9.1
8.9
Time (h)
(b)

Figure 12.9 (A) Pictures depicting a crack mending in an isosorbide-based epoxy with disulfide dynamic bonds at 100 °C. Source: Reproduced from Ma et al. [52]. © 2017, John Wiley & Sons. (B) (a) Stress–strain curves of samples with different times of healing at 120 °C and (b) evolution of the maximum stress and the healing efficiency with the healing time. Source: Reproduced from Cheng et al. [53]. © 2019, American Chemical Society.

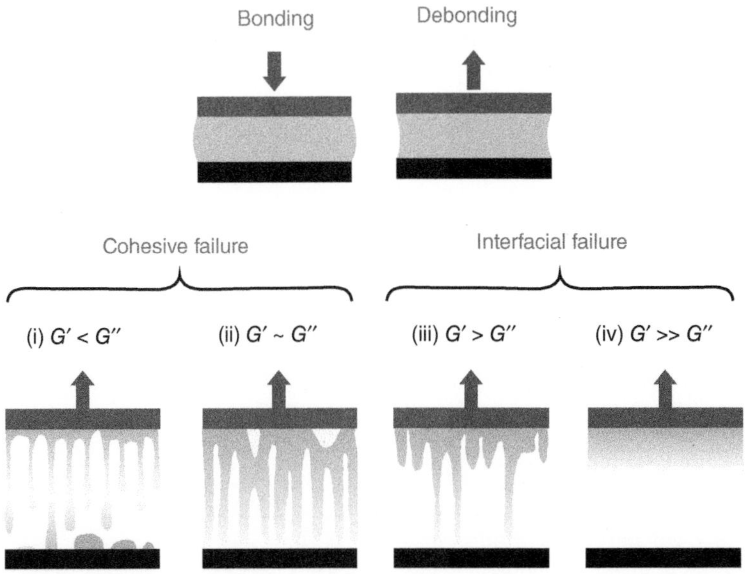

Figure 12.10 Schematic representation of the phases of bonding under light pressure and debonding at small strain, (i) viscous fibrils, (ii) cohesive fibrils, (iii) interfacial fibrils, and (iv) crack propagation. Source: Reproduced from Vendamme et al. [58]. © 2014, John Wiley & Sons.

suitable comonomers affecting T_g or the surface energy and most importantly by adjusting the molecular weight as well as the degree of cross-linking and branching [54] to fulfill the requirement of the intended application, including packaging, labels, sticky notes, and plastic wraps.

The bonding, adhesion strength, and failure mode of a PSA can be correlated with its rheological profile [58], and oscillatory frequency sweeps are well suited for such determinations. The lower frequencies of the PSA window (c. 0.01 Hz) characterize the bond formation, whereas higher frequencies are related to the debonding behavior.

The global PSA market size was valued at US$ 11.11 billion in 2018 and is expected to register a compound annual growth rate (CAGR) of 4.3% from 2019 to 2025 [62]. Currently, the market of PSAs is dominated by petrochemical-based platforms, such as acrylic, natural rubbers, block copolymers, styrene–butadiene–styrene (SBS) block copolymers, butadiene rubber (SBR), and polysiloxanes [58, 60]. However, modern PSAs are subject to some requirements that must meet materials in a fast and safe way according to the technical application, and at the same time, they must also have to be environmentally sound and commercially viable [58].

The use of renewable resources in the manufacture of PSAs competitive with those derived from petrochemical platforms represents an appealing opportunity to expand their scope to further applications with the additional advantage of being sustainable [58–60]. For practical uses, PSAs need a carrier. Accordingly, the design of biobased PSAs stimulates the search for carrier tapes derived from renewable resources, such as cellulose-based films and bioplastics such as poly(lactic acid) or polyhydroxyalkanotes [60].

Plant oils, fatty acids, and fatty acid esters offer many attractive advantages apart from their renewability, including their worldwide availability and relatively low prices [59, 60]. Most importantly, they have intrinsic low glass transition temperatures and need minor modification reactions to obtain a variety of polymer precursors for many different applications. The increasing number on patent applications [63–68] on PSAs from vegetable oils and fatty acids (VOs and FAs, respectively) provides evidence of the potential of these renewable feedstocks as PSA precursors. Different synthesis procedures for the functionalization of triglycerides and fatty acids have been studied, including epoxidation (chemical and enzymatic), esterification, or acrylation [59, 60].

In the recent past, attempts have been made to obtain PSAs with good adhesive properties by reacting epoxidized vegetable oils (EVOs) or fatty acid esters with petroleum-based acrylates or acrylic acid, followed by free-radical polymerization. Wool and his groups focused their work on the acrylation of methyl oleate (MO) to obtain high molecular weight latexes by emulsion polymerization [60, 69–71]. MO was epoxidized and then treated with acrylic acid to produce an acrylated monomer (AMO) that was homopolymerized and copolymerized with 2-ethylhexyl acrylate (2-EHA)-co-methyl methacrylate (MMA) and pure MMA. Copolymers of AMO-MMA [60] showed tack values against polyethylene comparable to Acronal® A220 (www.basf.de/dispersions), which is known for its high transparency, excellent water resistance, and outstanding adhesion to polyolefin substrates. These materials were also biocompatible, as claimed by the authors [71]. In another synthetic pathway disclosed in a patent application [63], EVOs reacted with acrylic copolymers via photocatalyzed cationic polymerization to afford tacky resins with a suitable high molecular weight between cross-links. Maaßen et al. [72] polymerized AMO and acrylate methyl erucate (AME) in bulk, miniemulsion, and solution to create high molecular weight materials intended for PSAs. Polymers were all bulk homopolymers with very low T_g values at −60 °C. Several of the formulated polymers fulfilled the criterion of Dahlquist, but the storage modulus of the obtained products was lower than that of the commercial synthetic control, Acrona V212. Registered tack values were lower than the control but were in a reasonable range for PSA applications. Carrier foils coated with the adhesives were peeled off by cohesive failure, leaving the residue on a glass substrate. Authors proposed to improve the adhesive properties by introducing long-chain branching or suitable comonomers. Nevertheless, the use of petrochemical-based acrylic acid and acrylic comonomers such as MMA remains a drawback.

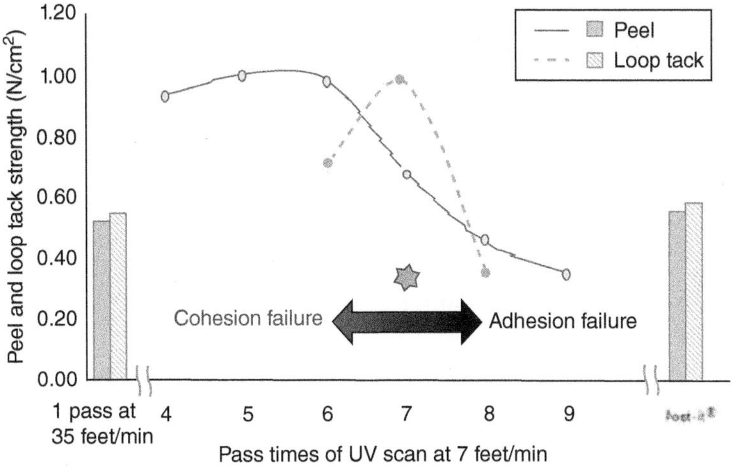

Figure 12.11 Peel and loop tack strength with cohesion/adhesion balance. Source: Reproduced from Ahn et al. [73]. © 2013, Taylor & Francis.

Sun et al. [66, 73–78] focused on the design of PSAs with high biobased content. Anh et al. [73] synthesized solvent-free PSAs with high biobased content (97–100%) by UV free radical polymerization of acrylated ESO (AESO). Balanced cohesion and adhesion strength were obtained by controlling the amount of UV irradiation (Figure 12.11). AESO PSA with seven UV scans (UV radiation dose: 1505–1617 mJ/cm^2) had the most balanced cohesion/adhesion strength (Figure 12.11). The tape on polyester carrier had relatively low peel and tack strength of around 1 N/cm and 0.98 N/cm^2, respectively, which is comparable with general purpose PSA tapes and stronger peel and tack strength than low-tack reusable PSA tape such as Post-it (3M, St. Paul, MN) with a strength of around 0.5 N/c m^2. Besides, these PSA tapes recorded excellent shear strength (+30 000 minutes) on a glass plate. Results were consistent with the viscoelastic behavior that defined AESO polymer as high-shear PSA.

The same research group [74] proposed the synthesis of fast curing copolymer from ESO and dihydroxy soybean oil (DSO) using H_3PO_4 as an environmentally friendly harder. ESO–DSO copolymers exhibited thermal stability ($T_g \sim 34\,°C$, $T_m \sim 250\,°C$, and degradation at $T > 388\,°C$) in the same range than commercial flexible PSAs based on polyacrylate or polyethylene. Peel strength of samples dried at 110 °C on a glass substrate was 2.18 N/cm after 30 seconds drying and reusability was 10 times, comparable to commercial adhesive tape and removable notes. The same system was additivated with rosin ester as a tackifier [75]. PSA's structure consisted of ether-cross-linked triglycerides functionalized with diols and rosin esters and covalent incorporation of the rosin ester as a copolymer in the ESO/DSO polymeric matrix. The PSA with an optimized ratio 1 : 1 : 0.7 of ESO : DSO rosin (by weight) displayed the best set of properties and steady tendency with an appropriate balance between elastic modulus (cohesion) and viscous modulus (viscosity/adhesion) at a frequency sweep from 0.01 to 100.01 rad/s. Peel strength (on aluminum foil) was 1.7 times higher than the commercial adhesive tape without cohesion failure, while the shear strength was of about three weeks, performing better than Magic tape (10 000 minutes) and Post-it (one minutes). All PSAs were thermally stable with T_g around −25 °C and a long rubbery region from −10 to 150 °C, being larger than the values reported for commercial PSAs (100–165 °C). ESO and DSO with different molecular weights and hydroxyl equivalent values were copolymerized by UV curing procedure in the presence of rosin ester [76]. These resins containing ESO and DSO can involve homopolymerization of epoxy and hydroxyl groups in DSO and copolymerization between epoxy from ESO and OH from DSO, both resulting in ester linkages. T_g varied in a narrow range (−14 to −16 °C) being still appropriate for usage at room temperature. ESO alone displayed poor peel strength (0.22 N/in.). The maximum peeling strength was 1.91 N/in. for ESO–DSO with an optimal OH/epoxy molar ratio of 0.55. Following the same idea, Camelina sativa oil (CO) was epoxidized under controlled conditions and polymerized via UV cationic polymerization with ESO and dihydroxylated CO [77]. Samples based on epoxidized castor oil (ECO) and dihydroxy CO had a peel strength of 2.1 N/in., but combining ESO with dihydroxyl Camelina oil, the value increased about three times, in line with viscoelastic properties.

ESO and polysaccharide-derived lactic acid were also combined to afford PSAs with almost 100% biobased carbon [78]. ESO was copolymerized with lactic acid oligomers (OLAs) containing carboxyl and hydroxyl groups able to open the oxirane ring in ESO through UV polymerization (Figure 12.12). Several formulations attained T_g values meeting the requirement for PSAs applications, depending on the components molar ratio (i.e. T_g = −15 °C for a formulation with molar ratio OLA3/ESO = 2.5).

Most of the ESO/OLA copolymers exhibited good tack, peel, and shear resistance, depending on the chain length of OLAs and their proportion. For an ESO/OLA3 molar ratio of 2.5, the peel strength was 1.6 N/cm, tack 1.7 N/cm, and shear resistance larger than 30 000 minutes. Increasing the content of OLA enhanced peel and tack but decreased the shear resistance by one order, and also cohesive failure appeared, indicating that some residual adhesives remained on the substrate (stainless steel panel), which is not desirable.

Kinchag Li and his group formulated a variety of eco-friendly and commercially viable PSAs based on vegetable oils, epoxidated fatty acids and esters, and EVOs cured by a variety of biobased hardeners or UV radiation [67, 68, 79–82]. ESO was selectively hydrolyzed, resulting in a mixture of epoxidized fatty acids

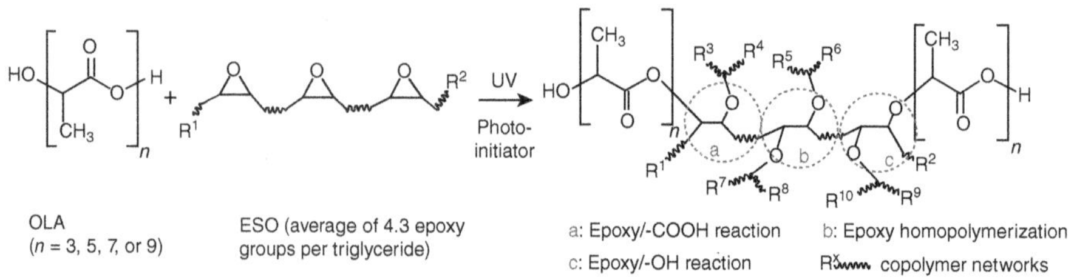

Figure 12.12 Synthesis pathway of ESO/OLA copolymers [78].

(EFA) containing both epoxy and carboxylic acid groups and further polymerized with a variety of diacids (adipic, sebacic, and itaconic) and anhydrides (succinic, phthalic, and terephthalic) to afford hydroxyl-functionalized polymers [66, 79]. Storage modulus of cured polymers were highly dependent on the frequency, suggesting that PSAs had low cross-linking density. The value of the dynamic modulus was 2.7×10^4 at room temperature, satisfying the Dahlquist criterion [61]. The presence of OH and unreacted COOH groups present in the resin could also contribute to the good adhesion and tack through noncovalent interactions. ESO was polymerized with sebacic (SA), adipic acid (AA), a dimeric acid (DA), and a difunctional polymeric carboxylic acid (PA) to produce hydroxyl-functionalized polymers with tailored properties by selecting the carboxylic acid [67, 80]. PSAs were tacky, soft, and flexible mostly because of the long hydrocarbon chains of the precursors. For a molar ratio ESO/DA system and for a molar ratio COOH : epoxy > 1, the peel strength (measured by peeling the PSA from stainless steel slabs) was significantly increased from 1.6 N/cm up to 2.9 N/cm. The presence of unreacted COOH groups (as evidenced by Fourier transform inferred spectroscopy [FTIR]) improved the wetting onto the substrate and enhanced the adhesion through hydrogen bonding. The chain length and molar mass of the difunctional acids had a great impact on the performance of PSAs. Increasing the molar mass decreased the cross-linking density (for a similar COOH/epoxy molar ratio), which in turn decreased the peeling strength (i.e. from 2.1 N/cm for ESO-DA to 1.6 N/cm for ESO-SA). The peel strength of the produced PSAs comprising values from 1.4 to 5 N/cm was comparable to the values of commercially available office tapes (3.5–8.8 N/cm) and Post-it notes, demonstrating the potential commercial viability of these biobased PSAs.

Based on the same concept, Ciannamea and Ruseckaite [82] proposed the synthesis of completely biobased PSAs from ESO and sebacic acid with tunable viscoelastic properties. ESO-SA-based PSAs were produced by a solvent-free one-step reaction in the absence of any catalyst. Curing conditions and pot-life were determined for formulations with stoichiometric molar acid/epoxy ratio. Curing conditions of stoichiometric mixtures were fixed from calorimetric and rheological experiments; by curing at 170 °C, the gel point (t_{gel}) was attained at 55.9 minutes. All PSAs met Dahlquist's criterion (Figure 12.13). Specimens cured at shorter times, such as 45 and 55 minutes, acted as viscous liquids ($G'' > G'$) in the entire frequency range, being very tacky at touching with the formation of numerous fibrils when debonding, in line with the low degree of cross-linking (0–10%). Increasing curing time, i.e. 65 minutes (t_{gel} + 10 minutes), gave rise to sticky and cohesive gels. Fibrils were formed when debonding but to a significantly lesser extent than for 55 minutes cured samples. Samples cured for 65 minutes exhibited a frequency-dependent behavior: elastic behavior ($G' > G''$) at low frequencies, a crossover ($G' = G''$) at approximately 1 Hz, and viscous behavior at higher frequencies. The relatively high viscous component at 1 Hz suggested that PSA can dissipate enough energy through deformation. Finally, longer curing times resulted in gels with higher cohesion but less tackiness. Results probed that viscoelastic profile and adhesion properties can be tailored by controlling curing parameters.

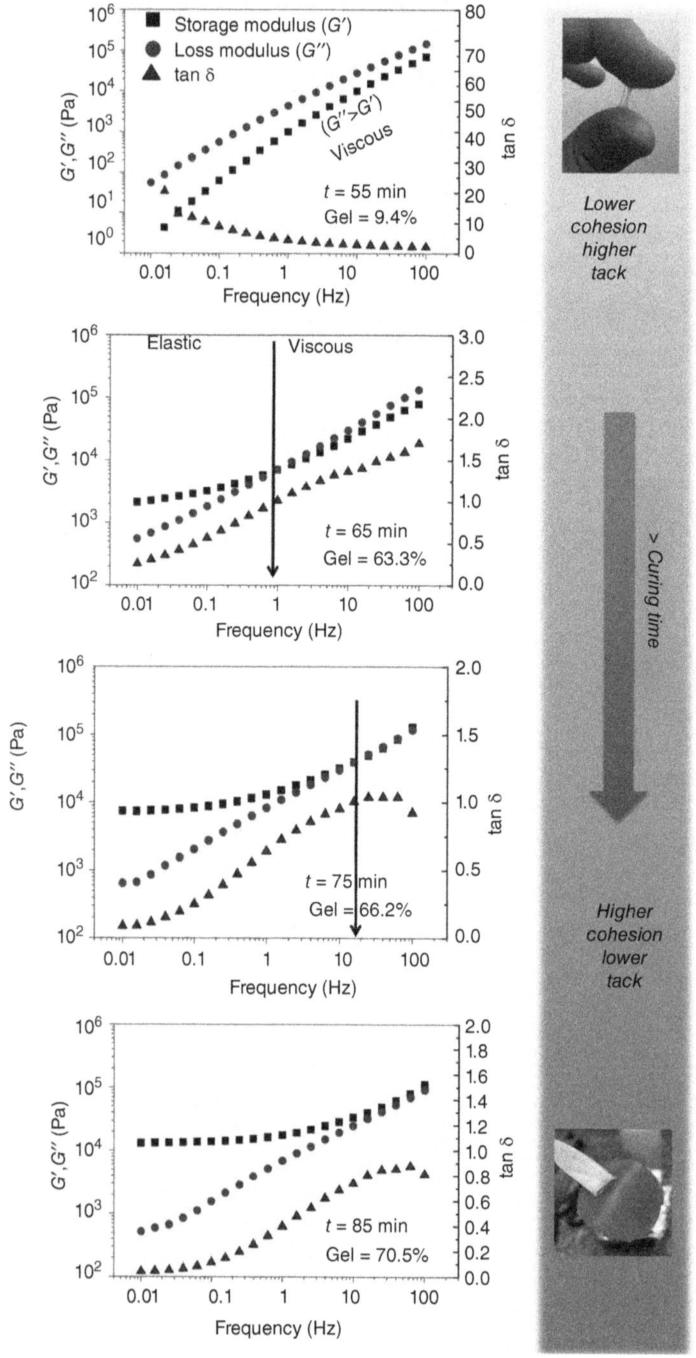

Figure 12.13 Rheological behavior of ESO-SA samples (epoxy : COOH 1 : 1). Source: Adapted from Ciannamea and Ruseckaite [82]. © 2018, John Wiley & Sons.

Vendamme and Eevers [83] suggested a "sweet solution for sticky problems" by synthesizing PSAs derived from EVOs with tailored viscoelastic and adhesion properties through the incorporation of 1,4:3,6-dianhydro-D-glucitol (isosorbide IS), a commercially available diol derived from starch. A series of carboxylic acid-terminated-polyesters were prepared by bulk polycondensation of dimerized fatty acids with several diols including IS. The obtained polymers were then cured commercially available EVOs (c. ESO and ELO) with different oxirane functionalities to afford a range of viscoelastic elastomers. The renewable cured materials combined the intrinsic flexibility of lipids with the polarity of sugars and demonstrate interesting performances. Authors probed that formulation can be used to adjust the gel content of the PSAs, which in turn provide a way to tune the cohesive strength of gels and the peel force. The inclusion of IS in PSA formulation impacted on T_g, rheological behavior, and adhesive profile in a nontrivial manner. The study showed that well-established design principles of petrochemical-based PSAs, such as adjusting cohesion by adding a small fraction of high-T_g monomer to a low-T_g monomers, can be applied in the formulation of biobased adhesives. A similar approach was formulated by Torron et al. [84] who examined the impact of adding rigid segments such as sugar derivatives into fatty acid-based network. Controlled macromolecular structure was achieved through the combination of different sugars and different fatty acids through lipase-catalyzed condensation, resulting in monomeric structures with varying hydroxyl and epoxy functionalities. These monomers were reacted with a sustainable dicarboxylic acid at different epoxy/acid ratios to control the macromolecular structure and consequently mechanical and viscoelastic properties. The two selected sugars used for condensation with the fatty acid methyl esters, i.e. sorbitol and erythritol, were compared with an aliphatic analog molecule, i.e. 1,4-butanediol. The strong frequency dependence of G' with increasing OH content in the sugar, i.e. sorbitol > erythritol > 1,4-butanediol, evidenced the beneficial effect of using sugars on the viscoelastic properties of the materials (Table 12.1) [83].

With the key goal being to improve the flame retardance of biobased PSAs, Wang et al. [85] created a flame-retardant adhesive by incorporating phosphorous-based compounds to PSA formulations derived from ESO and

Table 12.1 Detailed data from microcombustion calorimetry of the PSA.

Sample	DDP (molar)	DOPO-HQ-HE (molar)	HRR (W/g)	PHRR (J/gK)	THR (kj/g)	Peak Temp (°C)
E30-P1	—		316	317	17.8	407.6
E30-P2	10		304	304	17.3	402.6
E30-P3	20		291	291	16.3	399.5
E30-P4		10	252	253	17.8	408.8
E30-P5		20	197	198	16.9	409.7
E30-P6		30	175	162	16.5	415.1

Source: Reproduced from Wang et al. [85]. © 2017, American Chemical Society.

carboxylic acid-terminated polyesters. Authors used 9,10-dihydro-10-[2,3-di(hydroxycarbonyl)propyl]-10-phosphaphenantrene-10-oxide (DDP) and 2-(6-oxido-6*H*-dibenzc<c,e><1,2>oxaphosphorin-6-yl)-1,4-hydroxyethoxyphenylene (DOPO-HQ-HE) to synthesize a variety of PSAs exhibiting viscoelastic and adhesive properties compatible with high-performance PSAs. The incorporation of rigid DOPO pendant groups increased T_g, cohesive strength, and thermal stability as evidenced by the increment in the onset and maximum decomposition temperature of the flame-retardant PSAs. The two DOPO-derived PSAs showed different modes of action, but both resulted in comparable flame-retardant performance. The heat release rate (HRR) and total heat release (THR) as determined by microscale combustion calorimetry showed that both flame retardants induced a small decrease in both PHRR and THR, and such a drop was proportional to the content of the flame-retardant monomer. The LOI and UL94 tests confirmed the increased flame resistance of de-biobased PSAs.

12.4 Biobased Epoxy Foams

12.4.1 Syntactic Foams from Biobased Epoxy Resins

Syntactic epoxy foams are light materials that display exceptional properties such as low shrinkage, low moisture absorption, and good mechanical behavior [86, 87]. These materials have been extensively used since 1970s on applications requiring the properties of low density and high damage tolerance such as the fabrication of naval vessels, military vehicles, aircraft, buildings, and offshore structures because of their unique characteristic of closed pores and their characteristic wide range of mechanical properties and mechanical isotropy [88, 89]. In addition to a low initial viscosity, the potential polymer matrix must meet certain requirements to obtain syntactic foams (SFs) to achieve a good dispersion and wettability of the microballoons, controllable gelation times, and low shrinkage during curing [88]. Epoxy resins, such as those based on DGEBA, satisfy these conditions and are commonly used in the production of SFs combined with commercial curing agent such as anhydrides or amines (aromatic and cycloaliphatic) [86, 89–94]. Indeed, most of the available literature on epoxy SFs is devoted to synthetic formulation, and little information on biobased epoxy matrices is available [89]. Accordingly, a brief description on SF from petroleum-based raw materials is provided.

SF can be prepared using a variety of polymeric matrices; however, final density, mechanical, and water absorption properties can be finely tuned just by varying the size, wall thickness, and volumetric fraction of the microspheres without changing the matrix [89, 91–93, 95]. Gupta and Woldesenbet reported SF based on synthetic DGEBA with improved mechanical properties by increasing the wall thickness of the filler, reaching values even higher than those of the neat matrix do [96]. This last is an advantage compared to conventional thermosetting foams that always exhibit lower properties than the neat matrix. The occlusion of air bubbles during the mixing stage might induce increased porosity in the matrix,

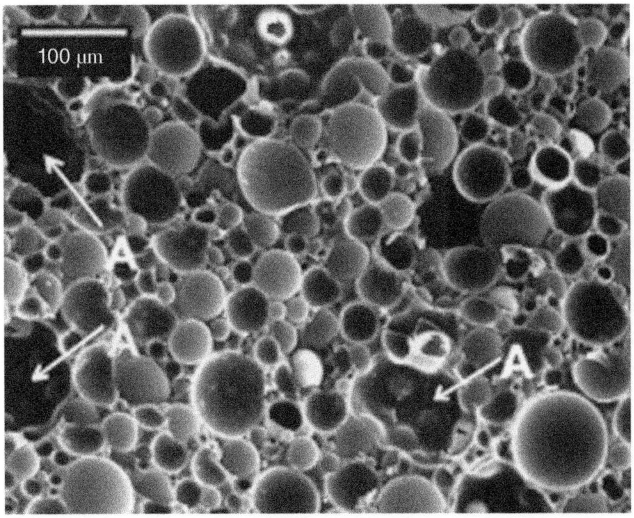

Figure 12.14 Structure of syntactic foam. Entrapped voids are marked as "A." Source: Reproduced from Gupta and Woldesenbet [96]. © 2003, Elsevier.

which, in turn, have a negative impact on the mechanical properties and water absorption (Figure 12.14) [96].

Some works reported an increase in porosity with the microballoon content reaching values as high as 10% [97]. In this sense, the initial viscosity of the resin has a dominant role because prepolymers with a high viscosity make the mixing operation with the microballoons more difficult. This leads to an undesirable increment in the volume of air entrapped in the matrix [98]. Altuna et al. designed a mixing procedure under reduced pressure to avoid or at least reduce the occlusion of air and keep the porosity to a minimum (lower than 3%) [98].

Syntactic epoxy foams have been traditionally used for marine applications. Its buoyant behavior, very low moisture absorption, thermal insulation capacity, and high hydrostatic compressive strength provided significant advantage over conventional epoxy foams [89]. BMTI-Alseamar company designed SFs suitable for deep water applications (−1000 to −8000 m), particularly to build rudders and flaps of submarines [99]. Engineered Syntactic Systems company reported the use of SF in remotely operated vehicles (ROVs) and human-operated vehicles (HOVs) [100]. The Alvin HOV was used for exploration of Titanic and several other missions, using SF buoyancy aids. No information about raw materials or processing conditions is available. Recently, the team led by Nikhil Gupta produced new SF filaments using a commercial 3D printer. The 3D printable SFs are made from a mixture of microscopic hollow glass or ceramics embedded in epoxy or polyethylene resin. This material provides good buoyancy and strength and can be used in submarines [101].

Another interesting application of syntactic epoxy foams is due to their particular acoustic properties that can be tailored to yield interesting underwater acoustic characteristics. These materials are used in transducer isolators, anechoic test tank liners, and subsea damping structures. Each application is unique and is dependent on the transparency, absorption, reflection, and refraction of the material. Because the acoustic impedance of SF is very close to that of water, acoustic waves are transmitted through the material without loss at the SF–water boundary. Trelleborg Emerson & Cuming reported the use of their SFs in US nuclear submarines because of their buoyancy and adequate acoustic profile [102].

SFs are also applied in anechoic chambers for room designed to absorb the reflections produced by acoustic or electromagnetic waves on any of the surfaces that comprise it. The British Aerospace Dynamics Group has manufactured radomes for broadband microwave transmission from SF using glass microspheres and epoxy matrix [103]. SFs are transparent to microwave radiation, which is useful in these applications because it protects sensitive equipment without distorting signals and can help military aircraft for stealth purposes.

As above stated, most formulations of SF are based on DGEBA that is obtained by reacting epichlorohydrin (ECH) with bisphenol A (BPA) in the presence of an alkaline medium [104]. In recent years, the ECH can be industrially produced from biobased glycerol. The conversion of glycerol to ECH is economically attractive and can be used in the production of DGEBA with about 25% of biobased content [105].

Several formulations of biobased epoxy resins with the potential to replace synthetic counterparts have been reported in the literature. Unfortunately, not all biobased systems meet the technical requirements for being used in SF, including low water absorption, high strength, low viscosity, and hydrophobicity. Epoxidized cardanol is a commercial product produced and distributed by Cardolite company. This bioresin is obtained from Cashew nut shell liquid (CNSL). The CNSL global market was US$ 250.2 million in 2018, indicating that is an attractive economic source of natural phenols [106]. Cardanol combines aromatic and aliphatic structures, and it seems to be a promising candidate for the substitution of petroleum-based phenol derivate. Diepoxy cardanol is produced by phenolation of cardanol, followed by the reaction with resulting diphenol with ECH (biobased). The chemical structure of epoxidized cardanol is shown in Figure 12.15.

Diepoxidized cardanol resulted as versatile as DGEBA in terms of chemical compounds that can act as curing agents, thanks to the high reactivity of the terminal oxirane rings toward nucleophilic and electrophilic agents such as amines or anhydrides [107]. Kaur and Jayakumari SF foams dispersing glass cenospheres

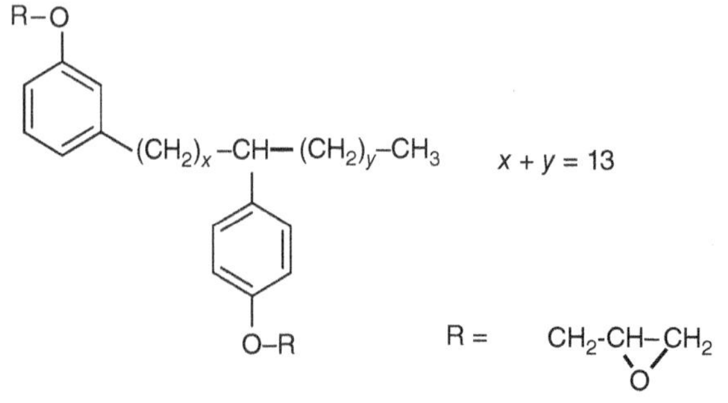

Figure 12.15 Structure of diepoxy cardanol. Source: Reproduced from Auvergne et al. [3]. © 2013, American Chemical Society.

(10–40 wt%) in a matrix based in epoxidized cardanol–TETA system. The curing reaction was carried out at different steps from room temperature to 100 °C, which is a low curing temperature for epoxy systems [97]. The hydrothermal studies revealed an extremely lower water absorption (distilled and sea water) under condition of high and room temperatures (<0.9 and 3.5 wt%, respectively), which would be associated with high hydrophobicity of cardanol. Compression tests conducted on the wet samples revealed a negligible drop in compressive strength (<2%) regardless of the cenosphere content. Authors concluded that epoxidized cardanol is an excellent candidate to replace oil-derived epoxy systems used in marine applications, unless it is necessary to address the study of this polymer system with microballoons of greater wall thickness in order to increase the hydrostatic resistance of SFs.

Other polyphenols have also been used as precursors of natural epoxy systems such as gallic acid, catechin, lignin, etc. [3]. In general, these epoxy monomers are obtained by phenolation and subsequent reaction with epichlorhydrin (EPC). No reports have been found on the use of these systems in SFs, despite their potential as a modifier of DEGBA-based systems.

Functionalized vegetable oils (FVOs) are considered as other attractive candidates to replace at least a proportion of the traditional epoxy resins in thermosetting formulations. Epoxidized oil (EVO) reacts more rapidly with anhydrides than amine hardeners because of hindered position of oxirane groups in the molecule [108, 109]. This fact explains why methyl-tetrahydrophthalic anhydride (MTHPA) and tetrahydrophthalic anhydride (THPA) among other anhydrides are more used as hardeners for EVOs. Altuna et al. manufactured SF dispersing glass microballoons (~50% v/v) in blends of DGEBA with an increasing amount of ESO (0–100%) and with MTHPA as a hardener and 1-methylimidazole (1-MI) as an initiator [98]. The mixing process was upgraded by using reduced pressure to attain low levels of porosity (less than 3%). Figure 12.16 shows that scanning electron microscopy (SEM) provides evidence of the absence of entrapped air bubbles, indicating that the mixing process under reduced pressure was efficient.

Foams containing up to 60 wt% of ESO exhibited comparable mechanical properties than that from neat DGEBA, with a small decrease in T_g values (Figure 12.17). These results indicate the potential utility of these foams with a high biogenic content for further exploration as marine applications.

12.4.2 Thermosetting Epoxy Foams

It is well known that thermosetting rigid epoxy foams have high mechanical strength, adequate thermal and chemical stability, and very good adhesion to a wide variety of substrates [110]. There are a number of parameters that must be considered to achieve stable foaming and finally to obtain a consolidated thermosetting epoxy foam. One of the most important constraints is the processing temperature because the competition between foaming and curing rate of the system depends directly on it. High curing temperature accelerates the reaction, and thus, the system reaches gelation before cremation (when foaming process starts). On the other hand, if the processing temperature is excessively low, the

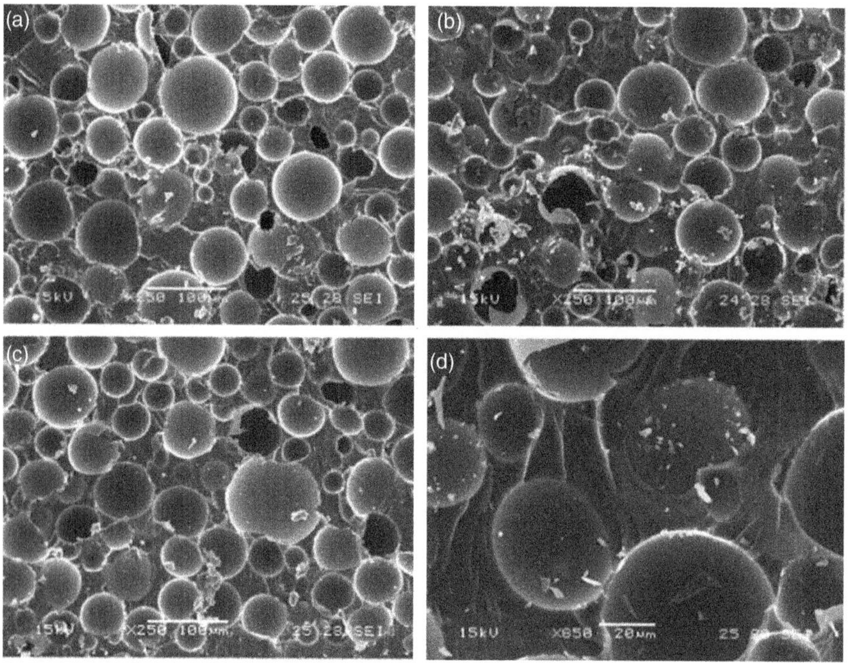

Figure 12.16 SEM image syntactic foams with different ESO content: (a) 0% (250×), (b) 100% (250×), (c) 40% (250×), and (d) 40% (850×). Source: (a–d) Reproduced from Altuna et al. [98]. © 2010, Elsevier.

curing reaction proceeds too slowly, and the mixture does not achieve adequate viscosity, so it cannot retain the gas released by the foaming agent, and the foam initially formed collapses [111, 112].

The initial viscosity of the mixture as well as the surface tension is key parameters when attempting foaming. The inclusion of surfactants in the formulation allows the system to attain a stable foaming even at relatively low viscosities by stabilizing the gas–liquid interfaces formed during foaming of the reactive mixture [111]. Finally, the amount of foaming agent (FA) in the formulation is central because it defines the final density and structure of the material [112–114].

Foaming of epoxy systems can be achieved by incorporating a FA (physical or chemical) into the reactive mixture [86, 111, 112, 115]. In the case of physical FAs, as the exothermic hardening reaction progresses, its boiling point (cremation time) is reached [116]. The evaporation produces bubbles that are trapped in the viscous reactive mass, inducing the expansion [117, 118]. In recent years, chemical FAs, such as azo compounds, alkali metal bicarbonates, and CO_2 have gained ground as green alternatives to physical foaming agents based on chlorofluorocarbon [119] because of the deterioration caused by the latter in the ozone layer. Li et al. proposed a batch foaming process at ~120 °C with supercritical CO_2 to obtain epoxy foam based on DGEBA [120]. CO_2 is stable, nontoxic, low cost, and has an easily attainable low critical temperature and moderate critical pressure. Bubbles can also be generated using a chemical agent that releases a gas by thermal decomposition or by reaction with any of the formulation reagents [110–

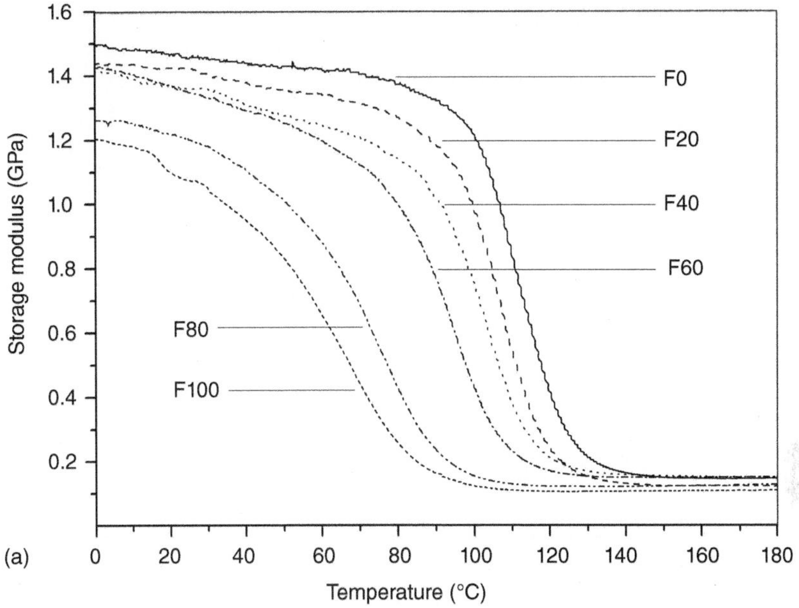

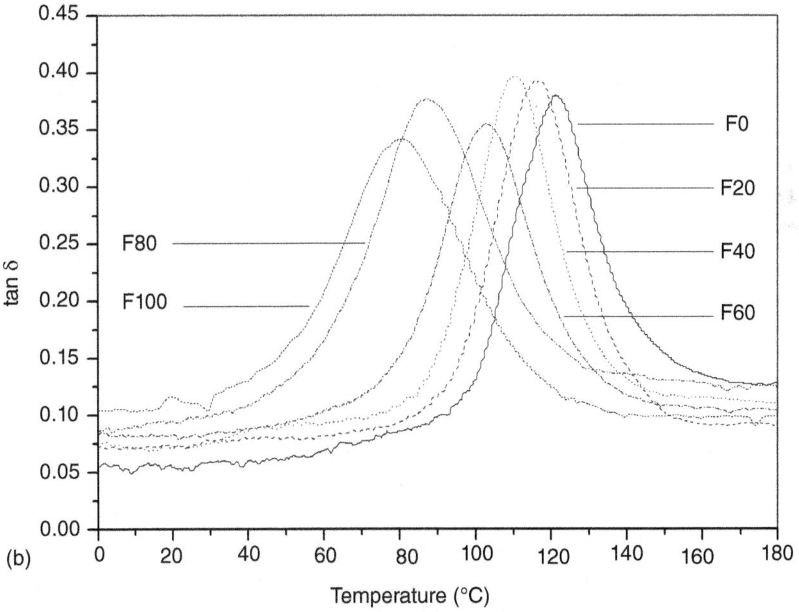

Figure 12.17 (a) Storage modulus and (b) tan δ curves of syntactic foams with different ESO content. Source: (a,b) Reproduced from Altuna et al. [98]. © 2010, Elsevier.

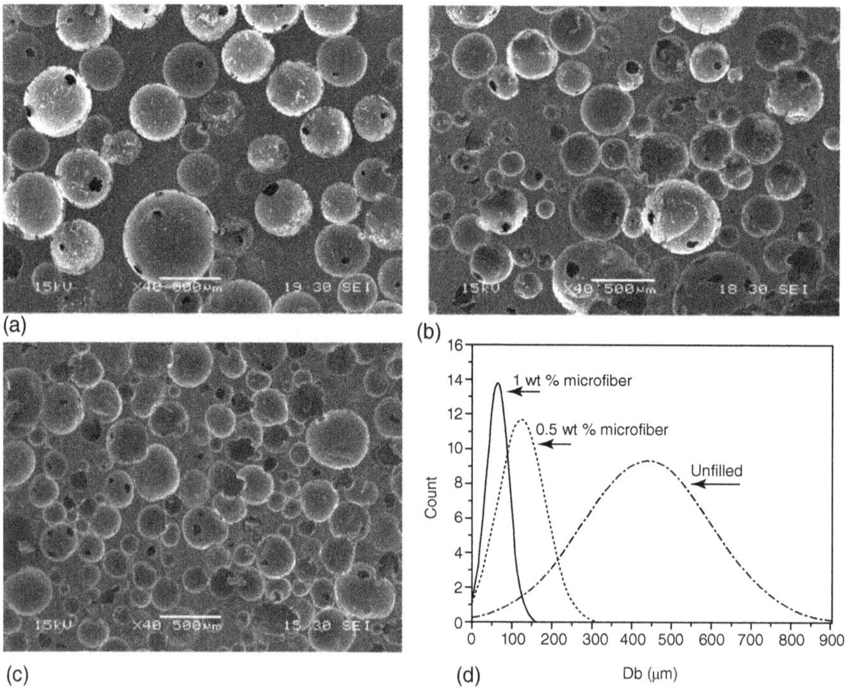

Figure 12.18 SEM micrographs of the pure and reinforced epoxy foams with different amounts of microfibers: (a) unfilled; (b) 0.5 wt% of microfiber, and (c) 1 wt% of microfiber. (d) Bubbles diameter distribution for each sample. Source: Reproduced from Stefani et al. [114]. © 2008, John Wiley & Sons.

114, 121]. Stefani et al. obtained epoxy foams based on DGEBA and a mixture of amines as a curing agent using siloxane as a chemical blowing agent releasing H_2 through the reaction between siloxane and amine hardener [112–114]. Curing was performed at relatively low temperature (~40 °C), reaching T_g values ~85 °C for densities between 150 and 500 Kg/m^3. They reach good compression resistance, but this property can be improved up to 50% by incorporating 1 wt% of microcellulose or ~20 wt% filler derived from rice husk ash [113, 114]. The microcellulose and rice husk ash had a nucleating effect in the foaming process and dramatically reduced the average size of bubbles improving mechanical performance. Typical structure of an epoxy foam and the effect of incorporating microfibers on the bubble size distribution is depicted in Figure 12.18.

Alonso et al. used a similar epoxy-amine reactive system blown with H_2 to obtain epoxy foam reinforced with glass and aramid fibers. This modification allowed higher mechanical and thermal stability [121].

Thermosetting epoxy foams have found many applications in the automotive industry. For example, Back et al. developed epoxy foam using DGEBA, dicyandiamide as a curing agent, and a modified sodium bicarbonate as a chemical foaming agent [110]. This system is placed between two automotive panels and, after foaming, fill the gap between them (Figure 12.19). In this case, epoxy foam provides both reinforcement of panels and the development of a watertight structure

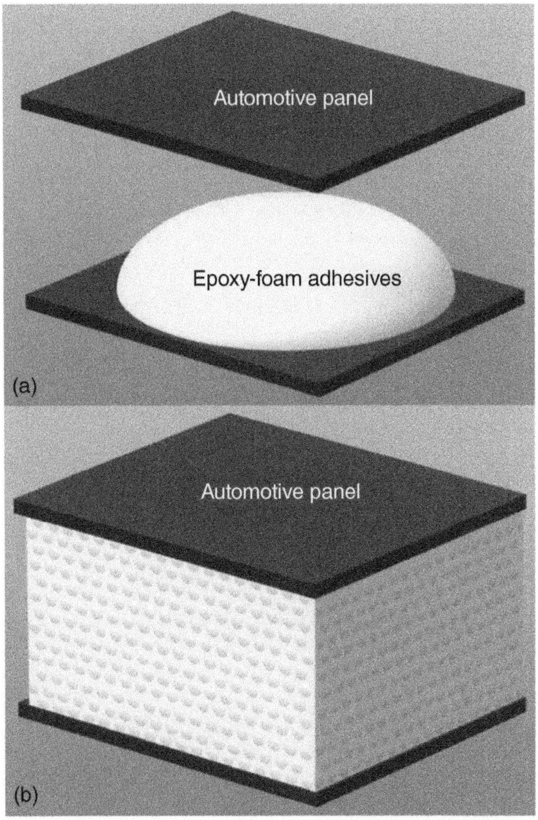

Figure 12.19 Epoxy–foam adhesives between automotive panels (a) before and (b) after foaming. Source: (a,b) Reproduced from Back et al. [110]. © 2018, Elsevier.

[110]. Hence, epoxy foam must have good mechanical strength for reinforcement of the panel, high T_g, and capacity to generate a watertight structure. It is important to find the optimal composition that would provide a suitable mechanical strength for automotive applications.

The efforts to replace raw materials derived from fossil resources by others of renewable origin have been partially extended to the production of thermosetting epoxy foams. Yadav et al. synthesized thermosetting epoxy foams using decafluoropentane as a foaming agent using epoxidized soybean oil esters (EAS) as a modifier of DGEBA [117]. The amount of biobased materials varied from 12.5 to 25 wt%, and the foaming was performed in a closed mold. Bonnaillie and Wool produced foams with high biobased content and interesting mechanical properties from epoxidized and acrylated soybean oil [118]. Foaming was achieved by insufflating CO_2, followed by vacuum application that can be considered a drawback for a potential industrial application. Dogan and Kusefoglu used malonic acid (MA) as a simultaneous foaming and cross-linking agent for ESO [122]. The decarboxylation of malonic acid monoester produced CO_2, which acted as a physical foaming agent. Although the mechanical properties of these foams turned out to be poor, they have the advantage of being obtained by a simple method. Among the aforementioned studies, only a few of them were successful in producing foams with mechanical properties comparable to

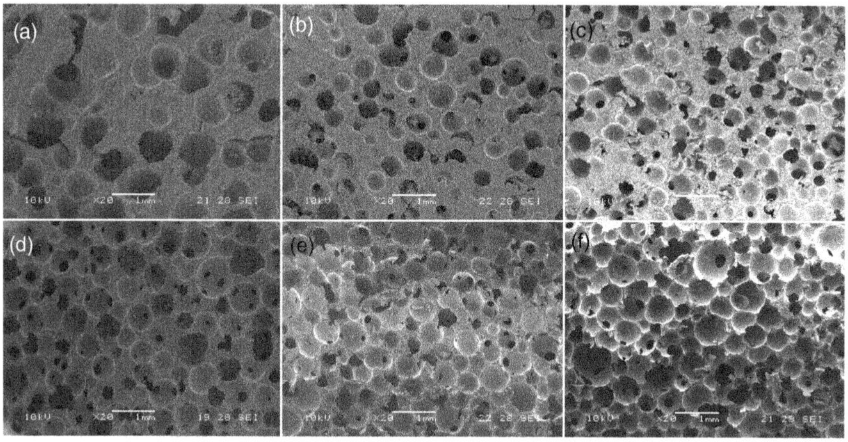

Figure 12.20 Horizontal plane SEM micrographs of foams with different foaming agent content (FA): (a) 1%FA; (b) 1.5% FA; (c) 2.5% FA; (d) 3.5% FA; (e) 5% FA; and (f) 7% FA. Source: Reproduced from Altuna et al. [111]. © 2015, American Chemical Society.

those of commercially available systems and a reproducible foaming process. In this sense, Altuna et al. reported the develop epoxy foams based on ESO with a high content of biogenic component (not less than 55 wt%) [111]. Sodium bicarbonate was selected as a low cost, nontoxic, and safe thermally latent foaming agent, and a low toxicity anhydride (MTHPA) was chosen as a curing agent (Figure 12.20). ESO foams reach mechanical properties comparable to the synthetic one with compressive modulus between 168 and 28 MPa for foams with density between 501 and 193 kg/m^3, respectively, while the T_g value of the foams (measured as tanδ peak temperature) was ~70 °C. From mechanical and thermal properties measured, it is possible to affirm that these biobased foams can be used for the manufacture of lightweight structure with potential application in automotive and furniture industry. Based on the previous work by Altuna et al. [111], Khundamri et al. developed flexible biobased epoxy foams from a mixture of ESO and epoxidized mangosteen tannin (EMT) using MTHPA as a hardener and azodicarbonamide as a foaming agent. In this work, EMT was obtained by glycidylation of mangosteen tannin extract with EPC [123]. In this work, flexible foams were achieved by using a low amount of MTHPA (lower than stoichiometric) and did not make use of a catalyst.

Recently, Esmaeili et al. afforded epoxidized tannic acid foam (ETA foam), for being used as a thermally insulating material as well as a wastewater remediation product [124]. Herein, water was used as both a green solvent and a blowing agent. ETA was obtained by glycidylation of tannic acid with ECH. These foams showed an open porosity (Figure 12.21), and it was proved for wastewater remediation, being competitive respect to synthetic foams.

Negrell et al. developed a novel epoxy system based on epoxidized algal oil (EAO) extracted from *Schizochytrium* microalgae that were applied in biobased foams from triepoxide phloroglucinol and ELO and ESO as comonomers [125]. Priamine was used as a hardener and the foaming agent was

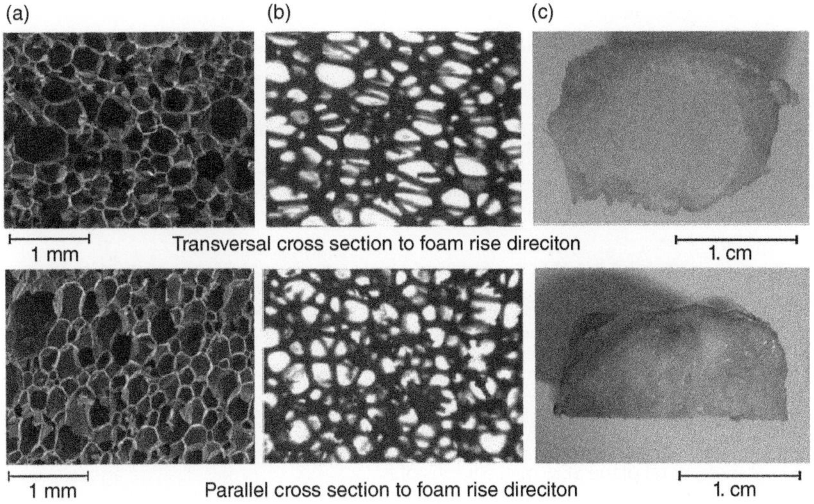

Figure 12.21 Photo (a), light microscopy ((b) magnification: 50×), and SEM (c) images of the ETA foam. Source: Reproduced from Esmaeili et al. [124]. © 2018, Elsevier.

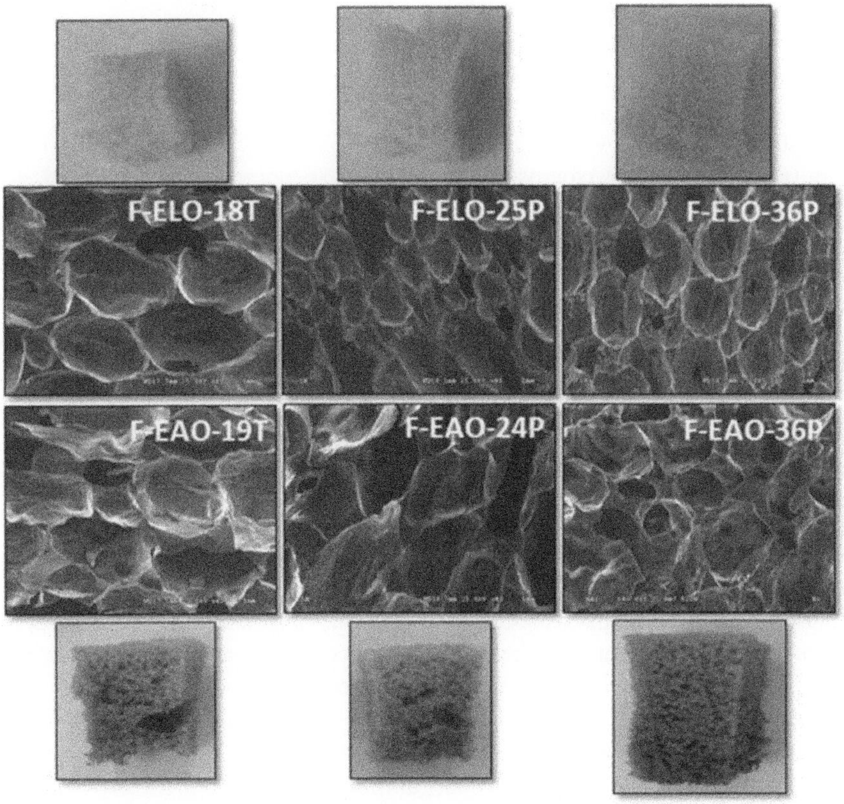

Figure 12.22 Foams based on EAO and ELO and SEM pictures. Source: Reproduced from Negrell et al. [125]. © 2017, John Wiley & Sons.

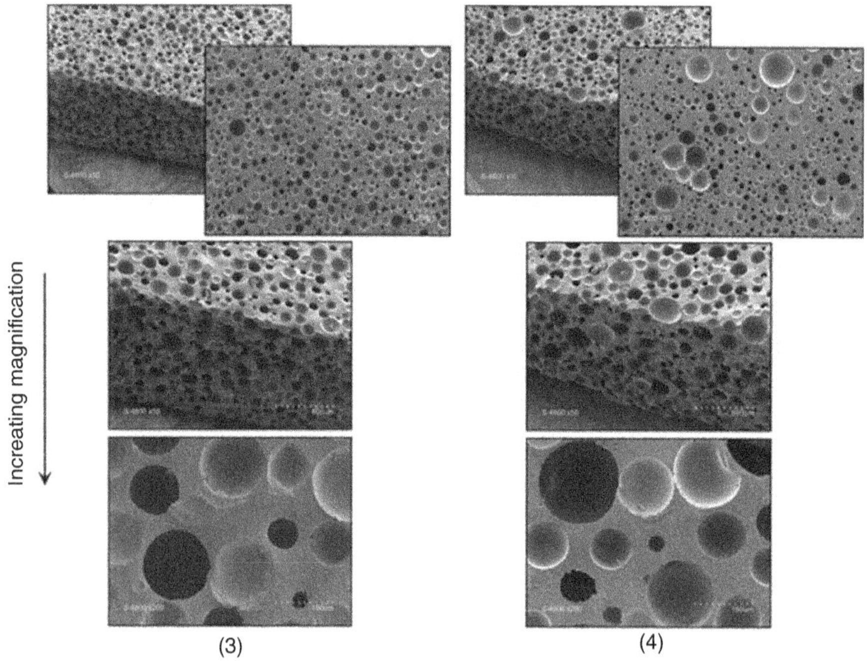

Figure 12.23 Epoxy foams based on epoxidized cardanol. Source: Reproduced from Dworakowska et al. [126]. © 2015, John Wiley & Sons.

polymethylhydrogenosiloxane (PMS). Density values of ~200 kg/m^3 and T_gs between −10 and 47 °C were obtained for different formulations. As shown in Figure 12.22, an open cell structure was observed for this system. Dworakowska et al. developed biobased epoxy foams based on epoxidized cardanol using poly-(methylhydrosiloxane) (PMHS) as a foaming agent and priamine as a hardener [126]. The resulting foams exhibited a homogeneous cell distribution, high thermal stability, and densities between 193 and 507 kg/m^3 when 0.05 (sample 3) and 0.06 (sample 4) molar equivalents of PMHS were used (Figure 12.23). The maximum T_g attained was 20 °C, which implies that the foams would remain in a rubbery state at room temperature and beyond, rendering them suitable for most structural applications.

It is clear from the reported works that it is possible to expand the field of application of biobased epoxy foams to high-performance materials. There are two promising emerging systems: (i) bioresins with terminal epoxy groups such as epoxidized cardanol, epoxidized polyphenols, etc., that can be combined with multifunctional amines (including biobased) and siloxane derivatives as a foaming agent and (ii) bioresins from epoxidized vegetables. Because of the hindered position of oxirane groups, EVOs are more suitable to react with anhydride-based curing agents. The foaming agents also need to be environmentally benign; sodium bicarbonate or azocarbamide are the most convenient because their decomposition temperature allows an adequate balance between the curing reaction and the foaming process.

References

1 Pagliaro, M. (2019). An industry in transition: the chemical industry and the megatrends driving its forthcoming transformation. *Angew. Chem. Int. Ed.* 58 (33): 11154–11159.
2 Nakajima, H., Dijkstra, P., and Loos, K. (2017). The recent developments in biobased polymers toward general and engineering applications: polymers that are upgraded from biodegradable polymers, analogous to petroleum-derived polymers, and newly developed. *Polymers* 9 (12): 523.
3 Auvergne, R., Caillol, S., David, G. et al. (2013). Biobased thermosetting epoxy: present and future. *Chem. Rev.* 114 (2): 1082–1115.
4 Wool, R.P. (1978). Material response and reversible cracks in viscoelastic polymers. *Polym. Eng. Sci.* 18 (14): 1057–1061.
5 Jud, K., Kausch, H.H., and Williams, J.G. (1981). Fracture mechanics studies of crack healing and welding of polymers. *J. Mater. Sci.* 16 (1): 204–210.
6 Wu, D.Y., Meure, S., and Solomon, D. (2008). Self-healing polymeric materials: a review of recent developments. *Prog. Polym. Sci.* 33 (5): 479–522.
7 Wool, R.P. (2008). Self-healing materials: a review. *Soft Matter* 4 (3): 400.
8 Dry, C. (1994). Matrix cracking repair and filling using active and passive modes for smart timed release of chemicals from fibers into cement matrices. *Smart Mater. Struct.* 3 (2): 118–123.
9 Dry, C.M. and Sottos, N.R. (1993). Passive smart self-repair in polymer matrix composite materials. *Proc. SPIE Int. Soc. Opt. Eng.* 1916: 438–444.
10 Dry, C.M. and McMillan, W. (1996). Crack and damage assessment in concrete and polymer matrices using liquids released internally from hollow optical fibers. *Proc. SPIE Int. Soc. Opt. Eng.* 2718: 448–451.
11 Patel, A.J., Sottos, N.R., Wetzel, E.D., and White, S.R. (2010). Autonomic healing of low-velocity impact damage in fiber-reinforced composites. *Compos. Part Appl. Sci. Manuf.* 41 (3): 360–368.
12 Kessler, M.R. and White, S.R. (2001). Self-activated healing of delamination damage in woven composites. *Compos. Part A Appl. Sci. Manuf.* 32 (5): 683–699.
13 Brown, E.N., Sottos, N.R., and White, S.R. (2002). Fracture testing of a self-healing polymer composite. *Exp. Mech.* 42 (4): 372–379.
14 Kessler, M.R., Sottos, N.R., and White, S.R. (2003). Self-healing structural composite materials. *Compos. Part Appl. Sci. Manuf.* 34 (8): 743–753.
15 Toohey, K.S., Sottos, N.R., Lewis, J.A. et al. (2007). Self-healing materials with microvascular networks. *Nat. Mater.* 6 (8): 581–585.
16 Hansen, C.J., Wu, W., Toohey, K.S. et al. (2009). Self-healing materials with interpenetrating microvascular networks. *Adv. Mater.* 21 (41): 4143–4147.
17 Blaiszik, B.J., Kramer, S.L.B., Olugebefola, S.C. et al. (2010). Self-healing polymers and composites. *Annu. Rev. Mater. Res.* 40 (1): 179–211.
18 Billiet, S., Hillewaere, X.K.D., Teixeira, R.F.A., and Du Prez, F.E. (2013). Chemistry of crosslinking processes for self-healing polymers. *Macromol. Rapid Commun.* 34 (4): 290–309.

19 Keller, M.W. (2010). Self-healing epoxy composites. In: *Epoxy Polymers* (eds. J.-P. Pascault and R.J.J. Williams), 325–344. Weinheim: Wiley-VCH.

20 Roy, N., Bruchmann, B., and Lehn, J.-M. (2015). DYNAMERS: dynamic polymers as self-healing materials. *Chem. Soc. Rev.* 44 (11): 3786–3807.

21 Luo, X., Ou, R., Eberly, D.E. et al. (2009). A thermoplastic/thermoset blend exhibiting thermal mending and reversible adhesion. *ACS Appl. Mater. Interfaces* 1 (3): 612–620.

22 Shi, Q., Yu, K., Kuang, X. et al. (2017). Recyclable 3D printing of vitrimer epoxy. *Mater. Horiz.* 4 (4): 598–607.

23 Taynton, P., Yu, K., Shoemaker, R.K. et al. (2014). Heat- or water-driven malleability in a highly recyclable covalent network polymer. *Adv. Mater.* 26 (23): 3938–3942.

24 Xiao, D.S., Yuan, Y.C., Rong, M.Z., and Zhang, M.Q. (2009). A facile strategy for preparing self-healing polymer composites by incorporation of cationic catalyst-loaded vegetable fibers. *Adv. Funct. Mater.* 19 (14): 2289–2296.

25 Hia, I.L., Pasbakhsh, P., Chan, E.-S., and Chai, S.-P. (2016). Electrosprayed multi-core alginate microcapsules as novel self-healing containers. *Sci. Rep.* 6 (1): 34674.

26 Thanawala, K., Mutneja, N., Khanna, A., and Raman, R. (2014). Development of self-healing coatings based on linseed oil as autonomous repairing agent for corrosion resistance. *Materials* 7 (11): 7324–7338.

27 Ataei, S., Khorasani, S.N., Torkaman, R. et al. (2018). Self-healing performance of an epoxy coating containing microencapsulated alkyd resin based on coconut oil. *Prog. Org. Coat.* 120: 160–166.

28 Kurt Çömlekçi, G. and Ulutan, S. (2019). Acquired self-healing ability of an epoxy coating through microcapsules having linseed oil and its alkyd. *Prog. Org. Coat.* 129: 292–299.

29 Lehn, J.-M. (2005). Dynamers: dynamic molecular and supramolecular polymers. *Prog. Polym. Sci.* 30 (8–9): 814–831.

30 Bosman, A.W., Sijbesma, R.P., and Meijer, E.W. (2004). Supramolecular polymers at work. *Mater. Today* 7 (4): 34–39.

31 Wojtecki, R.J., Meador, M.A., and Rowan, S.J. (2011). Using the dynamic bond to access macroscopically responsive structurally dynamic polymers. *Nat. Mater.* 10 (1): 14–27.

32 Murphy, E.B., Bolanos, E., Schaffner-Hamann, C. et al. (2008). Synthesis and characterization of a single-component thermally remendable polymer network: staudinger and stille revisited. *Macromolecules* 41 (14): 5203–5209.

33 Cordier, P., Tournilhac, F., Soulié-Ziakovic, C., and Leibler, L. (2008). Self-healing and thermoreversible rubber from supramolecular assembly. *Nature* 451 (7181): 977–980.

34 Peterson, A.M., Jensen, R.E., and Palmese, G.R. (2010). Room-temperature healing of a thermosetting polymer using the Diels–Alder reaction. *ACS Appl. Mater. Interfaces* 2 (4): 1141–1149.

35 Herbst, F., Döhler, D., Michael, P., and Binder, W.H. (2013). Self-healing polymers via supramolecular forces. *Macromol. Rapid Commun.* 34 (3): 203–220.

36 Chakma, P. and Konkolewicz, D. (2019). Dynamic covalent bonds in polymeric materials. *Angew. Chem. Int. Ed.* 58 (29): 9682–9695.

37 Montarnal, D., Capelot, M., Tournilhac, F., and Leibler, L. (2011). Silica-like malleable materials from permanent organic networks. *Science* 334 (6058): 965–968.

38 Capelot, M., Unterlass, M.M., Tournilhac, F., and Leibler, L. (2012). Catalytic control of the vitrimer glass transition. *ACS Macro Lett.*: 789–792.

39 Capelot, M., Montarnal, D., Tournilhac, F., and Leibler, L. (2012). Metal-catalyzed transesterification for healing and assembling of thermosets. *J. Am. Chem. Soc.* 134 (18): 7664–7667.

40 Bowman, C.N. and Kloxin, C.J. (2012). Covalent adaptable networks: reversible bond structures incorporated in polymer networks. *Angew. Chem. Int. Ed.* 51 (18): 4272–4274.

41 Karami, Z., Zohuriaan-Mehr, M.J., and Rostami, A. (2018). Biobased Diels-Alder engineered network from furfuryl alcohol and epoxy resin: preparation and mechano-physical characteristics. *ChemistrySelect* 3 (1): 40–46.

42 Pin, J.-M., Guigo, N., Vincent, L. et al. (2015). Copolymerization as a strategy to combine epoxidized linseed oil and furfuryl alcohol: the design of a fully bio-based thermoset. *ChemSusChem* 8 (24): 4149–4161.

43 Shen, X., Liu, X., Wang, J. et al. (2017). Synthesis of an epoxy monomer from bio-based 2,5-furandimethanol and its toughening via Diels–Alder reaction. *Ind. Eng. Chem. Res.* 56 (30): 8508–8516.

44 Chen, Y., Xi, Z., and Zhao, L. (2016). New bio-based polymeric thermosets synthesized by ring-opening polymerization of epoxidized soybean oil with a green curing agent. *Eur. Polym. J.* 84: 435–447.

45 Xu, C., Nie, J., Wu, W. et al. (2019). Self-healable, recyclable, and strengthened epoxidized natural rubber/carboxymethyl chitosan biobased composites with hydrogen bonding supramolecular hybrid networks. *ACS Sustainable Chem. Eng.* 7 (18): 15778–15789.

46 Cao, L., Huang, J., and Chen, Y. (2018). Dual cross-linked epoxidized natural rubber reinforced by tunicate cellulose nanocrystals with improved strength and extensibility. *ACS Sustainable Chem. Eng.* 6 (11): 14802–14811.

47 Altuna, F.I., Pettarin, V., and Williams, R.J.J. (2013). Self-healable polymer networks based on the cross-linking of epoxidised soybean oil by an aqueous citric acid solution. *Green Chem.* 15 (12): 3360–3366.

48 Altuna, F.I., Antonacci, J., Arenas, G.F. et al. (2016). Photothermal triggering of self-healing processes applied to the reparation of bio-based polymer networks. *Mater. Res. Express* 3 (4): 045003.

49 Hao, C., Liu, T., Zhang, S. et al. (2019). A high-lignin-content, removable, and glycol-assisted repairable coating based on dynamic covalent bonds. *ChemSusChem* 12 (5): 1049–1058.

50 Zhang, S., Liu, T., Hao, C. et al. (2018). Preparation of a lignin-based vitrimer material and its potential use for recoverable adhesives. *Green Chem.* 20 (13): 2995–3000.

51 Liu, T., Hao, C., Wang, L. et al. (2017). Eugenol-derived biobased epoxy: shape memory, repairing, and recyclability. *Macromolecules* 50 (21): 8588–8597.

52 Ma, Z., Wang, Y., Zhu, J. et al. (2017). Bio-based epoxy vitrimers: reprocessibility, controllable shape memory, and degradability. *J. Polym. Sci. Part A Polym. Chem.* 55 (10): 1790–1799.

53 Cheng, B., Lu, X., Zhou, J. et al. (2019). Dual cross-linked self-healing and recyclable epoxidized natural rubber based on multiple reversible effects. *ACS Sustainable Chem. Eng.* 7 (4): 4443–4455.

54 Chu, S.G. (1989). *Handbook of Pressure Sensitive Adhesive Technology*. New York: Springer Science+Business Media.

55 Cohen, E., Binshtok, O., Dotan, A., and Dodiuk, H. (2013). Prospective materials for biodegradable and/or biobased pressure-sensitive adhesives: a review. *J. Adhes. Sci. Technol.* 27 (18–19): 1998–2013.

56 Creton, C. (2003). Pressure-sensitive adhesives: an introductory course. *MRS Bull.* 28 (6): 434–439.

57 Johnston, J. (2018). The physical testing of pressure sensitive adhesive systems. In: *Handbook of Adhesive Technology* (eds. A. Pizzi and K.L. Mittal). Boca Raton, FL: CRC Press.

58 Vendamme, R., Schüwer, N., and Eevers, W. (2014). Recent synthetic approaches and emerging bio-inspired strategies for the development of sustainable pressure-sensitive adhesives derived from renewable building blocks. *J. Appl. Polym. Sci.* 131 (17): 40669.

59 Wool, R.P. (2005). Pressure-sensitive adhesives, elastomers and coatings from plant oils. In: *Bio-Based Polymers and Composites* (eds. R.P. Wool and X.S. Sun). Amsterdam; Boston, MA: Elsevier Academic Press.

60 Wool, R.P. (2013). Pressure-sensitive adhesives, elastomers, and coatings from plant oil. In: *Handbook of Biopolymers and Biodegradable Plastics: Properties, Processing and Applications* (ed. S. Ebnesajjad). Oxford ; Waltham, MA: Elsevier/William Andrew.

61 Dahlquist, C.A. (1989). *Treatise on Adhesion and Adhesives*, vol. 6 (ed. R.L. Patrick), 219. New York: Dekker.

62 Adhesives & Sealants Industry (ASI) magazine. (2018). *Bio-Based Adhesives Market Continues to Grow*. ASI – Adhes. Sealants Ind. https://www.adhesivesmag.com/ (accessed 2 August 2019)

63 Koch, C.A. (2008). Pressure-sensitive adhesives made from renewable epoxidized triglyceride and epoxidized fatty ester. WO 2008144703, filed 2008.

64 Koch, C.A. and Pathak, S. (2017). Pressure-sensitive adhesives based on renewable resources, UV curing and related methods. US 9453151 B2, filed 2017.

65 Wool, R.P. and Bunker, S. (2003). Pressure-sensitive adhesives from plant oils. US 6646033 B2, filed 2003.

66 Sun, X.S., Ahn, B.K., and Wang, D. (2013). Pressure-sensitive adhesives, coatings, and films from plant oils. US2013/0330549 A1, filed 2013.

67 Li, K. and Li, A. (2018). Pressure-sensitive adhesives based on fatty acids. US10030182 B2, filed 2018.

68 Li, K. and Li, A. (2017). Vegetable oil-based pressure-sensitive adhesives. US 9556368 B2, filed 2017.

69 Bunker, S.P. and Wool, R.P. (2002). Synthesis and characterization of monomers and polymers for adhesives from methyl oleate. *J. Polym. Sci. Part Polym. Chem.* 40 (4): 451–458.

70 Bunker, S., Staller, C., Willenbacher, N., and Wool, R. (2003). Miniemulsion polymerization of acrylated methyl oleate for pressure sensitive adhesives. *Int. J. Adhes. Adhes.* 23 (1): 29–38.

71 Klapperich, C.M., Noack, C.L., Kaufman, J.D. et al. (2009). A novel biocompatible adhesive incorporating plant-derived monomers. *J. Biomed. Mater. Res. A* 91A (2): 378–384.

72 Maaßen, W., Oelmann, S., Peter, D. et al. (2015). Novel insights into pressure-sensitive adhesives based on plant oils. *Macromol. Chem. Phys.* 216 (15): 1609–1618.

73 Ahn, B.K., Sung, J., Rahmani, N. et al. (2013). UV-curable, high-shear pressure-sensitive adhesives derived from acrylated epoxidized soybean oil. *J. Adhes.* 89 (4): 323–338.

74 Ahn, B.-J.K., Kraft, S., Wang, D., and Sun, X.S. (2011). Thermally stable, transparent, pressure-sensitive adhesives from epoxidized and dihydroxyl soybean oil. *Biomacromolecules* 12 (5): 1839–1843.

75 Ahn, B.K., Sung, J., Kim, N. et al. (2013). UV-curable pressure-sensitive adhesives derived from functionalized soybean oils and rosin ester: UV-curable pressure-sensitive adhesives. *Polym. Int.* 62 (9): 1293–1301.

76 Li, Y. and Sun, X.S. (2014). Di-hydroxylated soybean oil polyols with varied hydroxyl values and their influence on UV-curable pressure-sensitive adhesives. *J. Am. Oil Chem. Soc.* 91 (8): 1425–1432.

77 Kim, N., Li, Y., and Sun, X.S. (2015). Epoxidation of Camelina sativa oil and peel adhesion properties. *Ind. Crops Prod.* 64: 1–8.

78 Li, Y., Wang, D., and Sun, X.S. (2015). Copolymers from epoxidized soybean oil and lactic acid oligomers for pressure-sensitive adhesives. *RSC Adv.* 5 (35): 27256–27265.

79 Li, A. and Li, K. (2014). Pressure-sensitive adhesives based on soybean fatty acids. *RSC Adv.* 4 (41): 21521.

80 Li, A. and Li, K. (2014). Pressure-sensitive adhesives based on epoxidized soybean oil and dicarboxylic acids. *ACS Sustainable Chem. Eng.* 2 (8): 2090–2096.

81 Wu, Y., Li, A., and Li, K. (2014). Development and evaluation of pressure sensitive adhesives from a fatty ester. *J. Appl. Polym. Sci.* 131 (23): 41143.

82 Ciannamea, E.M. and Ruseckaite, R.A. (2018). Pressure sensitive adhesives based on epoxidized soybean oil: correlation between curing conditions and rheological properties. *J. Am. Oil Chem. Soc.* 95 (4): 525–532.

83 Vendamme, R. and Eevers, W. (2013). Sweet solution for sticky problems: chemoreological design of self-adhesive gel materials derived from lipid biofeedstocks and adhesion tailoring via incorporation of isosorbide. *Macromolecules* 46 (9): 3395–3405.

84 Torron, S., Hult, D., Pettersson, T., and Johansson, M. (2017). Tailoring soft polymer networks based on sugars and fatty acids toward pressure sensitive adhesive applications. *ACS Sustainable Chem. Eng.* 5 (3): 2632–2638.
85 Wang, X.-L., Chen, L., Wu, J.-N. et al. (2017). Flame-retardant pressure-sensitive adhesives derived from epoxidized soybean oil and phosphorus-containing dicarboxylic acids. *ACS Sustainable Chem. Eng.* 5 (4): 3353–3361.
86 Shutov, F.A., Henrici-Olivé, G., and Olivé, S. (1985). *Integral/Structural Polymer Foams: Technology, Properties, and Applications.* Berlin ; New York: Springer-Verlag.
87 Gupta, N., Kishore, Woldesenbet, E., and Sankaran, S. (2001). Studies on compressive failure features in syntactic foam material. *J. Mater. Sci.* 36 (18): 4485–4491.
88 Bibin, J. and Reghunadhan, N.C.P. (2010). *Update on Syntactic Foams.* Shawbury: Smithers Rapra.
89 Gupta, N., Zeltmann, S.E., Shunmugasamy, V.C., and Pinisetty, D. (2014). Applications of polymer matrix syntactic foams. *JOM* 66 (2): 245–254.
90 Gupta, N., Woldesenbet, E., and Mensah, P. (2004). Compression properties of syntactic foams: effect of cenosphere radius ratio and specimen aspect ratio. *Compos. Part Appl. Sci. Manuf.* 35 (1): 103–111.
91 Kishore, S. and R., and Sankaran, S. (2005). Short-beam three-point bend tests in syntactic foams. Part II: Effect of microballoons content on shear strength. *J. Appl. Polym. Sci.* 98 (2): 680–686.
92 Gupta, N. and Ricci, W. (2006). Comparison of compressive properties of layered syntactic foams having gradient in microballoon volume fraction and wall thickness. *Mater. Sci. Eng., A* 427 (1–2): 331–342.
93 Porfiri, M. and Gupta, N. (2009). Effect of volume fraction and wall thickness on the elastic properties of hollow particle filled composites. *Compos. Part B Eng.* 40 (2): 166–173.
94 Dando, K.R., Cross, W.M., Robinson, M.J., and Salem, D.R. (2019). Characterization of mixture epoxy syntactic foams highly loaded with thermoplastic and glass microballoons. *J. Compos. Mater.* 53 (13): 1737–1749.
95 Gupta, N., Ye, R., and Porfiri, M. (2010). Comparison of tensile and compressive characteristics of vinyl ester/glass microballoon syntactic foams. *Compos. Part B Eng.* 41 (3): 236–245.
96 Gupta, N. and Woldesenbet, E. (2003). Hygrothermal studies on syntactic foams and compressive strength determination. *Compos. Struct.* 61 (4): 311–320.
97 Kaur, M. and Jayakumari, L. (2017). Novel bio-based epoxidized cardanol/cenosphere syntactic foams: thermal, hygrothermal and compression characterization. *High Perform. Polym.* 29 (7): 785–796.
98 Altuna, F.I., Espósito, L., Ruseckaite, R.A., and Stefani, P.M. (2010). Syntactic foams from copolymers based on epoxidized soybean oil. *Compos. Part Appl. Sci. Manuf.* 41 (9): 1238–1244.
99 Naval Technology. BMTI. Composite materials and parts for submarines. *Nav. Technol.* https://www.naval-technology.com/contractors/advanced_materials/bmti2 (accessed 15 July 2020).

100 Engineered Syntactic Systems. Esyntactic.com (accessed 15 July 2020).
101 Singh, A.K., Saltonstall, B., Patil, B. et al. (2018). Additive manufacturing of syntactic foams: Part 2: Specimen printing and mechanical property characterization. *JOM* 70 (3): 310–314.
102 Trelleborg. Marine and Infrastructure. Trelleborg Marine and Infrastructure. https://www.trelleborg.com/en/marine-and-infrastructure (accessed 15 July 2020).
103 BAE Systems. BAE Systems. https://www.baesystems.com/en/home (accessed 15 July 2020).
104 Grouchard, F. (1993). Process for the manufacture of epichlorohydrin. EP 561441, filed 1993.
105 Tuck, C.O., Perez, E., Horvath, I.T. et al. (2012). Valorization of biomass: deriving more value from waste. *Science* 337 (6095): 695–699.
106 Cashew Nut Shell Liquid (CNSL) Market To Reach USD 450.4 Million By 2026. Rep. Data. https://www.globenewswire.com/news-release/2019/04/17/1805582/0/en/Cashew-Nut-Shell-Liquid-CNSL-Market-To-Reach-USD-450-4-Million-By-2026-Reports-And-Data.html (accessed 15 July 2020).
107 Pascault, J.-P. and Williams, R.J.J. (eds.) (2010). *Epoxy Polymers: New Materials and Innovations*. Weinheim: Wiley-VCH.
108 Altuna, F.I., Espósito, L.H., Ruseckaite, R.A., and Stefani, P.M. (2011). Thermal and mechanical properties of anhydride-cured epoxy resins with different contents of biobased epoxidized soybean oil. *J. Appl. Polym. Sci.* 120 (2): 789–798.
109 Altuna, F.I., Pettarin, V., Martin, L. et al. (2014). Copolymers based on epoxidized soy bean oil and diglycidyl ether of bisphenol a: relation between morphology and fracture behavior. *Polym. Eng. Sci.* 54 (3): 569–578.
110 Back, J.-H., Hwang, J.-U., Lee, Y.-H. et al. (2018). Morphological study and mechanical property of epoxy-foam adhesives based on epoxy composites for automotive applications. *Int. J. Adhes. Adhes.* 87: 124–129.
111 Altuna, F.I., Ruseckaite, R.A., and Stefani, P.M. (2015). Biobased thermosetting epoxy foams: mechanical and thermal characterization. *ACS Sustainable Chem. Eng.* 3 (7): 1406–1411.
112 Stefani, P.M., Tejeira Barchi, A., Sabugal, J., and Vazquez, A. (2003). Characterization of epoxy foams. *J. Appl. Polym. Sci.* 90 (11): 2992–2996.
113 Stefani, P.M., Cyras, V., Tejeira Barchi, A., and Vazquez, A. (2006). Mechanical properties and thermal stability of rice husk ash filled epoxy foams. *J. Appl. Polym. Sci.* 99 (6): 2957–2965.
114 Stefani, P.M., Perez, C.J., Alvarez, V.A., and Vazquez, A. (2008). Microcellulose fibers-filled epoxy foams. *J. Appl. Polym. Sci.* 109 (2): 1009–1013.
115 Berlin, A.A., Shutov, F.A., Zhitinkina, A.K., and Odinak, A. (1982). *Foam Based on Reactive Oligomers*. Westport, CT: Technomic Pub.
116 Mondy, L.A., Rao, R.R., Moffat, H. et al. (2010). Structural epoxy foams. In: *Epoxy Polymers: New Materials and Innovations* (eds. J.-P. Pascault and R.J.J. Williams), 303–324. Weinheim: Wiley-VCH.
117 Yadav, R., Shabeer, A., Sundararaman, S. et al. (2006). Development and characterization of soy-based epoxy foams. *Proceeding SAMPE Conf*erence, pp. 1–10.

118 Bonnaillie, L.M. and Wool, R.P. (2007). Thermosetting foam with a high bio-based content from acrylated epoxidized soybean oil and carbon dioxide. *J. Appl. Polym. Sci.* 105 (3): 1042–1052.
119 Eaves, D. (2004). *Handbook of Polymer Foams*. Shawbury: Rapra Technology.
120 Li, J., Zhang, G., Fan, X. et al. (2017). Preparation and mechanical properties of thermosetting epoxy foams based on epoxy/2-ethyl-4-methylimidazol system with different curing agent contents. *J. Cell. Plast.* 53 (6): 663–681.
121 Alonso, M.V., Auad, M.L., and Nutt, S. (2006). Short-fiber-reinforced epoxy foams. *Compos. Part Appl. Sci. Manuf.* 37 (11): 1952–1960.
122 Doğan, E. and Küsefoğlu, S. (2008). Synthesis and in situ foaming of biodegradable malonic acid ESO polymers. *J. Appl. Polym. Sci.* 110 (2): 1129–1135.
123 Khundamri, N., Aouf, C., Fulcrand, H. et al. (2019). Bio-based flexible epoxy foam synthesized from epoxidized soybean oil and epoxidized mangosteen tannin. *Ind. Crops Prod.* 128: 556–565.
124 Esmaeili, N., Salimi, A., Zohuriaan-Mehr, M.J. et al. (2018). Bio-based thermosetting epoxy foam: tannic acid valorization toward dye-decontaminating and thermo-protecting applications. *J. Hazard. Mater.* 357: 30–39.
125 Negrell, C., Cornille, A., de Andrade Nascimento, P. et al. (2017). New bio-based epoxy materials and foams from microalgal oil: agal oil epoxy materials and foams. *Eur. J. Lipid Sci. Technol.* 119 (4): 1600214.
126 Dworakowska, S., Cornille, A., Bogdał, D. et al. (2015). Formulation of bio-based epoxy foams from epoxidized cardanol and vegetable oil amine: cardanol-based epoxy foams. *Eur. J. Lipid Sci. Technol.* 117 (11): 1893–1902.

Index

Bio-Based Epoxy Polymers, Blends, and Composites: Synthesis, Properties, Characterization, and Applications, First Edition. Edited by Jyotishkumar Parameswaranpillai, Sanjay Mavinkere Rangappa, Suchart Siengchin, and Seno Jose.

d

e

n